Sets and Transfinite Numbers

Under the Editorship
of Carl B. Allendoerfer

Sets and Transfinite Numbers

Martin M. Zuckerman
Associate Professor of Mathematics
City College of the City University
of New York

Macmillan Publishing Co., Inc.
NEW YORK
Collier Macmillan Publishers
LONDON

Printed in the United States of America

Macmillan Publishing Co., Inc.
866 Third Avenue, New York, New York 10022

Collier-Macmillan Canada, Ltd.

Library of Congress Cataloging in Publication Data

Zuckerman, Martin M.
Sets and transfinite numbers.
Bibliography:
1. Set theory. 2. Numbers, Transfinite.
I. Title.
QA248.Z8 511'.3 72-12455
ISBN 0-02-432110-9

Printing: 1 2 3 4 5 6 7 8 Year: 4 5 6 7 8 9 0

To Musia

Preface

Sets and Transfinite Numbers is a semiformal development of (essentially) Zermelo–Fraenkel set theory. Logical prerequisites are minimal; these are given in Sections 1.2 and 1.3. Even these sections can be postponed until Chapter 5, if desired. (See footnote 3, p. 191.) There are no other mathematical prerequisites for understanding the material, except that the student is expected to reason abstractly. In many instances parenthetical remarks are included that relate the abstract situation to topics the student is likely to have encountered already. Occasionally, concepts are discussed before they are properly introduced; we indicate this usage by means of quotation marks.

The book has been written with the beginning student in mind. A key feature is the presence of a large number of solved problems, which should greatly facilitate understanding the more difficult concepts.

The text has been class-tested at City College several times; selected topics from the first three chapters have also served as the basis of N.S.F-sponsored High School Institute courses at the City University.

The basic axioms of (our version of) Zermelo set theory are presented in Chapter 1. They are followed in Chapter 2 by a discussion of various types of relations: functions, equivalence relations, partial and linear orderings, and well-orderings. In Chapter 3, real and complex numbers are constructed within the framework of set theory. This material, which does not fit into set theory proper, can be skipped; a brief introduction in Section 3.1 prepares the reader for considerations of denumerability (of the rationals) and nondenumerability (of the reals) in Chapter 4. Ordinal numbers and their arithmetic are the subject of Chapter 5; order types and cardinal numbers and their arithmetics are discussed in Chapter 6. Cardinal numbers are introduced in terms of the notion of rank; we first discuss cardinal arithmetic without, then with, the Axiom of Choice. Various equivalences of the Axiom of Choice are presented in Chapters 4, 5, and 6. The reader is briefly introduced to the concepts of ordinal cofinality, the Generalized Continuum Hypothesis, and inaccessible cardinals. Some recent independence and consistency results are outlined.

There is a general indebtedness to standard works on set theory, mentioned in the bibliography, particularly to those of Kuratowski and Mostowski, Rubin, Rubin and Rubin, Sierpiński (1958), and Smullyan (1965). I would also like to express my appreciation to the following colleagues and students for their suggestions: A. Agard, R. Cowen, M. Ecker, S. Ellberger, M. Ellis, E. Feig, M. Fitting, S. Franco, R. Grewe, J. D. Halpern, N. Hanges, W. Hans, B. Horowitz, K. Hrbacek, R. Kopperman, B. Lee, G. Loullis, M. Matzka, E. Mendelson, R. Rosenkrantz, W. Sanchez, F. Sansone, and L. Sonn. Special thanks are due to Anil Nerode, Kenneth Danhof, and John Olmsted for their many suggestions and their proficient editing of the manuscript. I am grateful to Everett Smethurst and Elaine Wetterau of the Macmillan Publishing Co., Inc., for their diligent attention to publishing and production. Finally I greatly appreciate the excellent typing of the manuscript by William Rose.

M. M. Z.

Contents

2
Relations

3
Real and Complex Numbers as Sets

4
Denumerability and Choice

5
Ordinal Numbers

6
Order Types and Cardinal Numbers

Usage and Organization of the Book

We outline several courses for which the text can be used; these suggestions can be modified to suit individual needs, as indicated below.

(A) A one-quarter course in axiomatic set theory and the construction of real and complex numbers:

Chapters 1, 2, and 3

One or more of the following topics can be omitted without interrupting the continuity:

Logic: Sections 1.2 and 1.3

Integers as equivalence classes: Section 3.5

Complex numbers: Sections 3.11 and 3.12

(B) A one-semester course in axiomatic set theory and the construction of real and complex numbers:

Chapters 1, 2, and 3

(with the above qualifications)

Chapter 4: Sections 4.1–4.6

Chapter 5: Sections 5.1 and 5.2

As time permits, the following can be added:

Sections 5.3 and 5.4

and one or more among

1. Sections 4.7, 4.8, and 5.13
2. Sections 5.5 and 5.6
3. Sections 6.1 and either 6.2 or 6.4

(C) A one-semester course in axiomatic set theory:

Chapters 1 and 2
Chapter 3: Sections 3.1–3.3
Chapter 4
Chapter 5: Sections 5.1–5.6
Chapter 6: Sections 6.1 and 6.4

As time permits, use the accompanying guide to sectional interdependence to choose topics from the latter halves of Chapters 5 and 6.

(D) A full-year course in axiomatic set theory and transfinite arithmetic. Ideally, the entire book should be covered. However, the following sections can be skipped without interrupting the continuity:

Sections 3.4–3.12
Section 5.13
Sections 6.2 and 6.3

As these course outlines indicate, the book contains more material than can be covered in most one-semester courses. The sections not covered are suitable for use in independent readings courses.

Obviously, the simplest method of using the book is to take sections in the order written. Nevertheless, there is considerable leeway in the choice and order of topics.

CHAPTERS 1 AND 2: These chapters are prerequisite for the rest of the text.

Sections 1.2–1.3: This material on logic can be postponed until prior to Section 5.3. Students familiar with elementary logic can omit these sections altogether, except for the definition of "formula of set theory", at the beginning of Section 1.2.

Sections 1.4–2.4: These sections can be taken (in order) directly after Section 1.1.

Sections 2.5–2.8: Section 2.8 can precede Sections 2.5–2.7.

CHAPTER 3: Sections 3.4–3.12 can be postponed until prior to Section 6.2, or even omitted altogether. Many readers will have covered this material in other courses. These people should read the introductory section (3.1) before Section 4.5. However, they should read Sections 3.2–3.3 before Chapter 4. Those who read Chapter 3 can omit Section 3.5; also, if familiar with elementary properties of complex numbers and their arithmetic, they should omit Section 3.11.

CHAPTER 4: Sections 4.2, 4.3, 4.7, and 4.8, on the Axiom of Choice, can be postponed. They can be covered together with Section 5.13 just before Section 6.9. (There are some inessential references to this material in Sections 5.3 and 6.1.) Section 4.4 can be taken directly after Section 4.1 or after Section 4.5.

Section 4.5: Theorem 4.3.2 is the only information needed from Sections 4.2–4.3. A superficial acquaintance with Sections 3.4, 3.6, and 3.7 is desirable.

Section 4.6: Sections 4.4–4.5 are prerequisite. Some acquaintance with Sections 3.8–3.12 is desirable.

Sections 4.7–4.8: Sections 4.5–4.6 are not needed.

Sections 5.1–5.2: Section 4.1 and some familiarity with real and complex numbers are prerequisite.

Section 5.3: Sections 1.2–1.3 and 5.2 are prerequisite. Also, Section 4.2 and the statement of the Well-ordering Theorem (Section 4.8) are referred to in the concluding paragraphs.

Section 5.4: Sections 4.4 and 5.3 are prerequisite.

Sections 5.5–5.6: Section 5.4 is prerequisite.

Section 5.7: Section 5.6 is prerequisite.

Section 5.8: Section 5.6 is prerequisite, but not Section 5.7.

Sections 5.9–5.10: These sections concern ordinal addition and multiplication with respect to infinitely many summands (respectively, factors). This material can be omitted from courses that do not intend to cover generalized addition of order types—the last part of Section 6.3—and that end before Section 6.8. Sections 5.7–5.8 are prerequisite for Section 5.9; Sections 4.7 and 5.9 are prerequisite for Section 5.10.

Section 5.11: Section 5.8 is prerequisite.

Section 5.12: Section 5.11 is prerequisite. Also, if Section 5.9 has been omitted, only the text material through Example 5.12.2 should be read; only the first four exercises apply.

Section 5.13: This section can be read directly after Section 5.3.

Section 6.1:	This can be read directly after Section 5.4.
Sections 6.2–6.3:	Except for the discussion of Souslin's Hypothesis in Section 6.12, material from Section 6.2 is utilized only in the beginning of Section 6.4. Section 6.3 can be omitted. Sections 4.5 and 6.1 are prerequisite for Section 6.2. There is also considerable reference to Sections 3.4–3.12. Section 6.2 is prerequisite for Section 6.3. Sections 5.1–5.6 are needed prior to the discussion of generalized addition of order types. For the remainder of Section 6.3, Sections 5.7–5.9 are required.
Section 6.4:	Sections 4.6 and 6.1 are prerequisite. If Section 6.2 has been omitted, then in order to read Example 6.4.1 and the paragraph preceding it, first read the beginning of Section 6.2, through Example 6.2.1.
Section 6.5:	Sections 5.5–5.7 and 6.4 are prerequisite.
Section 6.6:	Sections 5.5, 5.6, 5.8, and 6.5 are prerequisite.
Section 6.7:	Sections 4.8, 5.12, and 6.5 are prerequisite.
Sections 6.8–6.9:	Sections 6.6–6.7 are prerequisite.
Sections 6.10–6.11:	Sections 5.10 and 6.9 are prerequisite.
Section 6.12:	Sections 6.10–6.11 are prerequisite. The discussion of the Souslin Hypothesis depends upon Section 6.2.

It is hoped that the diversity of worked-out examples and exercises will facilitate mastery of the material. However, many of the illustrative examples and exercises are routine and should be bypassed when the underlying concepts are clearly understood.

The examples are set off from the text by indentation and can easily be skipped. Examples that are used or referred to are marked with a dagger (†).

Certain exercises, or parts thereof, are also marked with a dagger (†); such exercises express basic properties that are important for later sections. We exclude from this group the verification of theorems and corollaries (or parts of these). Other exercises, though not necessarily of fundamental significance, are useful for a few subsequent sections; we mention where they apply. The exercises indicated in either of these ways are not necessarily the most interesting; however, their statements should be noted for later use. The more challenging exercises are marked with a (*); there are considerable differences among these exercises in the level of difficulty.

Sets and Transfinite Numbers

1

The Basic Axioms of Set Theory

1.1 Elements and Subsets

The fundamental mathematical entity with which we are concerned is known as a *set*. In general mathematical usage a set is regarded simply as a "*collection of objects*". We speak of "the set of complex roots of unity" or of "the set of points in a closed, bounded interval at which a continuous (real-valued) function attains a relative maximum". We will presently discuss the shortcomings in such a "naïve" approach to set theory.

Here, in contrast, a set is to be regarded as an *undefined notion*. Sets are basic to our treatment in the sense that all our subsequent notions will be defined in terms of sets together with an *undefined "relationship" among sets*, denoted by "$\in$". We will use the term "set" exclusively in this technical sense; however, the student may often find it heuristically useful to retain the intuitive conception of a set as a "collection of (mathematical) objects". In fact, many of the sets to be discussed will correspond to various collections that the reader will have encountered in previous mathematical studies.

(An exception to this rule concerning the usage of "set" will be made, occasionally, in examples. There, it will be convenient to illustrate notions about sets by considering collections of people, collections of points on a line, etc. Thus, we might refer to the "set of Englishmen" when, strictly speaking, we mean the "collection of Englishmen".)

Our approach will be "*axiomatic*" in the sense that *a collection of objects will be considered to be a set if and only if one of our axioms guarantees this.* Thus, some of our axioms postulate that specific collections—such as the one consisting of all natural numbers—are to be considered as sets. Other axioms guarantee that various related collections must also be sets. For example, having established that the numbers "two" and "three" are sets, the "pairing" axiom will enable us to speak of the set consisting of the numbers "two" and "three".

EXAMPLE 1.1.1. The following are examples of collections, frequently encountered in mathematics, whose existence (as sets) we will consider in Chapter 3:

(1) $\mathscr{Z}$, the set of integers; $\mathscr{Z}_+$, the set of positive integers; $\mathscr{Z}_-$, the set of negative integers.
(2) $\mathscr{Q}$, the set of rational numbers; $\mathscr{R}$, the set of real numbers.
(3) The set of nonsingular 2×2 matrices whose entries are rational.

For the remainder of this section we will let a, A, B, and C denote sets.

We symbolize the intuitive notion of *being an element* (*member*) *of* by "$\in$". More precisely, we regard $\in$ as an **undefined relationship between two sets.** Thus, "$a \in A$" means (intuitively) that a **is an element of** A; we also say that a **belongs to** A, or simply that (**the element**) a **is in** (**is contained in**) A. If a **does not belong to** A, we write "$a \notin A$". Furthermore, *for any two sets a and A, exactly one of the relationships*

$$(1)\quad a \in A, \qquad (2)\quad a \notin A$$

is understood to hold. Finally, *we will only employ the notation "$x \in y$" and "$x \notin y$" when both x and y are sets.*

We define the equality of two sets, A and B, as follows: A **equals** B (equivalently, A **and** B **are identical, the same**) means that for every set C, $A \in C$ if and only if $B \in C$. In this case we write "$A = B$". Thus, equality of sets is defined in terms of the undefined relationship, $\in$. If A and B are not equal, we say that they are **distinct** (**different**); we write "$A \neq B$".

If A and B are sets such that every element of A is an element of B, we say that A **is a subset of** B or that A **is included in** B; we symbolize this by

"$A \subseteq B$". Clearly, *for every set* A, $A \subseteq A$; also, $A \subseteq B$ *and* $B \subseteq C$ *together imply that* $A \subseteq C$.

Note the distinction between saying "A is contained in B." and saying "A is included in B." The former is symbolized by "$A \in B$" and means that A is an element of B. The latter is symbolized by "$A \subseteq B$" and means that every element of A is an element of B.

Our first axiom specifies the following additional property of equality of sets:

Axiom of Extensionality: *The sets A and B are equal if and only if both A is a subset of B and B is a subset of A.* ($A = B$ *if and only if both* $A \subseteq B$ *and* $B \subseteq A$.)

In other words, $A = B$ *if and only if either of the following holds*:

(1.1.1) *For every set* X, $A \in X$ *if and only if* $B \in X$.
(1.1.2) *For every set* Y, $Y \in A$ *if and only if* $Y \in B$.

Moreover, (1.1.1) *holds if and only if* (1.1.2) *holds*. Thus, *a set is completely determined by its elements as well as by the sets of which it is an element.*

A **proper subset of** B is a subset A of B that is distinct from B; in this case we also say that A **is properly included in** B, and we write "$A \subset B$".

We employ braces to stand for the phrase "**the set whose elements are** —". Thus, $\{a, b\}$ designates the set whose elements are a and b and only these. The order in which the elements are given plays no role in such descriptions of sets—nor does repetition of an element (within the braces). We also use the notation $\{x : s(x)\}$ to denote **the set of** x **for which** $s(x)$ **is true, where** $s(x)$ **is a statement pertaining to** (the set) x. These notations are used only when the "sethood" of the specified collection has been guaranteed by the axioms already introduced. For instance, assuming that 0, 1, 2, and $\{0, 1\}$ have already been shown to be sets and that the expressions "natural number" and "$<$" have been properly defined, then

$$\{x : x \text{ is a natural number and } x < 2\}$$

would also denote $\{0, 1\}$. (Note that we regard 0 as a natural number.) Also (assuming we know of the existence of the indicated sets),

$$\{0, 1, 2, \ldots, 9\}$$

designates the set consisting of the first 10 natural numbers, and

$$\{0, 1, 2, \ldots\}$$

designates the set consisting of the natural numbers 0, 1, 2, etc. (i.e., the set of all natural numbers).

The notation "$A, B \in C$" will abbreviate "$A \in C$ and $B \in C$". The notation "$A = B = C$" will sometimes be used for "$A = B$ and $B = C$". Similarly, "$A \in B \in C$" will stand for "$A \in B$ and $B \in C$"; "$A \in B = C$" will stand for "$A \in B$ and $B = C$". Other abbreviations of this sort will be used. Finally, "$A \nsubseteq B$" will mean that (A and B are sets and) A is not a subset of B. A similar remark holds for "$A \not\subset B$".

EXAMPLE 1.1.2. Suppose that a, b, c, d, e, f, and g are distinct sets and that we have the sets $A = \{a, c, e\}$, $B = \{a, b, c, d, e\}$, $C = \{a, f, g\}$, and $D = \{x : x \text{ appears in the word "cabbage"}\}$. Then $A = \{c, a, e\} = \{a, c, c, e, a\}$. Moreover, $A \subset B$, $A \subset D$. In each case the inclusion is proper because no two of these sets contain the same elements. The set C is neither included in, nor does it include, any of the other sets. Observe that $D = \{a, b, c, e, g\}$. Furthermore, $d \in B$, but $d \notin D$; thus, $B \nsubseteq D$.

†EXAMPLE 1.1.3.[1] Although it would seem a somewhat unusual circumstance for a set, X, to be an element of itself, we have not yet prohibited this possibility; we have only required that exactly one of the statements $X \in X$, $X \notin X$, be true. Say that a set X is **ordinary** if $X \notin X$. Let $\mathscr{A}$ be the collection of all ordinary sets, and assume that $\mathscr{A}$ is a set. Is $\mathscr{A}$ ordinary?

This is the famous **Russell Paradox** (after Bertrand Russell, 1872–1969, the English philosopher–mathematician). If $\mathscr{A} \in \mathscr{A}$, then $\mathscr{A}$ is not ordinary, and by the definition of $\mathscr{A}$, it follows that $\mathscr{A} \notin \mathscr{A}$. Contradiction! If $\mathscr{A} \notin \mathscr{A}$, then $\mathscr{A}$ is ordinary and, hence, belongs to $\mathscr{A}$. Again we have a contradiction.

This paradox results from assuming that $\mathscr{A}$ is a set; thus, not every describable collection of elements can be considered to be a set.

The discovery of the Russell Paradox and other anomalies of set theory, at about the turn of the century, had a profound effect on the subject. Georg Cantor had introduced the notion of a set in the 1870s to distinguish between different kinds of infinite numbers and to extend the ordinary arithmetic operations to these. To Cantor, a set was simply a collection of definite, distinct objects to be conceived of as a whole. The paradoxes of Cantorian set theory necessitated the refinement of this naïve concept of a set.

Precise rules were sought that would determine which objects were to be regarded as sets. In this way, it was (and is) hoped, contradictions could be avoided in set theory. Moreover, because different people do not necessarily have the same intuition, the axiomatic approach has the advantage that the properties of sets are simply the logical consequences of our axioms.

[1] A dagger (†) before an example indicates that the example will be used or referred to. Unmarked examples may be omitted.

Set theory aims at unifying mathematics; it is possible to express most of known mathematics in terms of sets (and, of course, an underlying language). Thus, it is especially important that we clarify what we mean by a set. In an axiomatic system we begin with a few primitive (i.e., undefined) concepts and a list of rules for qualifying objects as sets. The axioms must be sufficiently powerful to express advanced mathematics, and yet not so permissive that a statement, *s*, together with its negation, not *s*, can be derived.

The first "successful" axiomatic treatment was that of Zermelo (1904–1908), as refined by Fraenkel and Skolem (1921–1929). This will be, essentially, the approach that we will be following. In Zermelo set theory the only objects studied are sets. The Russell Paradox is avoided by eliminating certain collections that are "too large". Thus, a collection is admitted only if an axiom either specifically proclaims its existence or relates it to a known set, possibly by means of a statement of set theory. These matters will be amplified in subsequent discussion. In Section 1.7, after we have introduced our basic axioms, we will indicate how (our version of) Zermelo set theory conforms with our intuition about sets. When the Axiom of Substitution (Section 5.3) is added, we speak of Zermelo–Fraenkel set theory.

In the theory of types, which was the treatment in Russell and Whitehead's *Principia Mathematica* (1910–1913), sets are introduced in a hierarchical structure. $X \in Y$ can hold only when Y is one level higher than X. In particular, a set cannot be an element of itself. Furthermore, we cannot consider a "set of all sets"; we can only speak of a set of all sets that are on the same level. Hence, there is no such set of all ordinary sets. We can consider the set A_α of all ordinary sets of level α. Clearly, $A_\alpha \notin A_\alpha$; however, the paradox is avoided, for A_α is of level $\alpha + 1$.

A third major axiomatic development is that of class-set theory—due to von Neumann (1925–1929), Bernays (1937–1954), and Gödel (1940). Here, certain "large" collections, called "proper classes", are considered as well as sets (improper classes). $X \in Y$ cannot hold if X is a proper class. The collection $\mathscr{A}$ consisting of all ordinary sets is a proper class, and thus $\mathscr{A} \in \mathscr{A}$ cannot hold.

Exercises[2]

For some exercises in Chapter 1, there is insufficient information given.

1.1.1. Let a, b, c, d, e, f, and g be distinct sets and suppose that $A = \{a, c, e\}$, $B = \{a, e\}$, $C = \{b, d, f\}$, and $D = \{a, b, c, d, e\}$. Which of the following statements are true?

(a) $c \in B$.
(b) $c \in C$.
(c) $b \notin A$.
(d) $B \subseteq A$.
(e) $B \subset A$.
(f) $A \subset C$.
(g) $A \subset D$.
(h) $A \in D$.

(Used in Example 1.8.1.)

[2] See the remarks concerning the exercises in the introductory material, Usage and Organization of the Book.

1.1.2. Suppose that A is the set of all Londoners, B is the set of all Parisian women, C is the set of all Europeans, D is the set of all Englishmen, E is the set of all English men, and F is the set of all people of the world. Determine all proper inclusions among these sets.

†1.1.3. Let A, B, and C be any sets. Prove (a) and (b):

(a) If $A \subset B$ and $B \subseteq C$, then $A \subset C$.

(b) If $A \subseteq B$ and $B \subset C$, then $A \subset C$.

(c) Are either (a) or (b) true if "$\nsubseteq$" replaces each occurrence of "$\subseteq$" and "$\not\subset$" replaces each occurrence of "$\subset$"?

(d) Which of the following are true?

(1) If $A \nsubseteq B$, then $A \not\subset B$.

(2) If $A \not\subset B$, then $A \nsubseteq B$.

(3) If $A \nsubseteq B$, then $A \neq B$.

†1.1.4. Prove that equality of sets has the following properties: Let A, B, and C be arbitrary sets.

(1) $A = A$ (**reflexivity** of equality).

(2) If $A = B$, then $B = A$ (**symmetry** of equality).

(3) If $A = B$ and $B = C$, then $A = C$ (**transitivity** of equality).

†1.1.5. Which of the properties mentioned in the preceding problem apply to

(a) $\subset$? **(b)** $\subseteq$? **(c)** $\nsubseteq$?

(d) $\not\subset$? **(e)** $\in$?

[In other words, replace each instance of "$=$" by "$\subset$" and indicate whether each of the modified statements (1), (2), and (3) of Exercise 1.1.4 is true. Then replace each instance of "$=$" by "$\subseteq$", etc.]

★1.1.6. Suppose that A and a are distinct sets and that $A = \{A, a\}$. Which of the following are true?

(a) $A \in A$. **(b)** $A \subset A$.

(c) $A \subseteq A$. **(d)** $\{A\} \in A$.

(e) $\{A\} \subset A$. **(f)** $\{A\} = A$.

1.2 Language of Set Theory[3]

In every mathematical discipline, even in so basic a study as set theory, we must utilize certain rules of logic. In this section we briefly analyze our underlying logical assumptions as well as our English usage.

[3] The main use in the first four chapters of Sections 1.2 and 1.3 is in defining the Axiom of Subsets (Section 1.4). Readers who prefer not to interrupt the development of set theory can omit these sections at this time. Such readers can temporarily read "property" in place of "formula". More specifically, "formula that has x as its one and only free variable" should be replaced by "property pertaining to x". Sentences of set theory should be regarded as ordinary English sentences pertaining to sets. Logical terminology is seldom utilized; reference should be made, when necessary, to Sections 1.2 and 1.3. For formal treatments of the language of set theory see (Cohen) and (Takeuti and Zaring).

We first distinguish between constants and variables. A **constant** is a symbol that designates a specific object. In particular, a **set constant** is a symbol that denotes a specific set. A **variable** is a symbol that designates any one of a given collection of objects; when these objects are all sets, we refer to the variable as a **set variable**.

†EXAMPLE 1.2.1. In Section 1.6 we will indicate how natural numbers as well as certain collections of natural numbers can be regarded as sets. Suppose that we have in mind the set consisting of the natural numbers 1, 3, and 5. We may denote this set by the letter "A". Here, A stands for a specific set; A is therefore a set constant.

However, when we write an expression such as

$$\text{Let } x \in A,$$

x refers to each of the sets that are elements of A, rather than to any particular set. Here x is a set variable; we also say that x **varies over** A.

[Strictly speaking, when we say "The set A ...", we really mean "The set denoted by (the letter) A ...". Similarly, when we use an expression such as "$A = B$", we really are indicating that the sets denoted by A and by B are identical.]

Let a, A, etc., each be either a set constant or else a set variable. Let x, y, etc., each be a set variable. An expression that has the form

$$a \in A$$

is called an **atomic formula**. By a **formula (of set theory)** we mean each of the following expressions:

(1) An atomic formula.
(2) An expression that can be constructed from atomic formulas by repeated application (a "finite number" of times) of the following **connectives** [(1)–(5)] and **quantifiers** [(6) and (7)]:
   (a) not,
   (b) or,
   (c) and,
   (d) if ..., then,
   (e) if and only if,
   (f) for some (set) x, for some (set) y, etc.,
   (g) for every (set) x, for every (set) y, etc.

(3) An expression that has been defined to mean the same as one of type (1) or (2).
(4) Only those expressions given by (1), (2), or (3).

In expressions 2(f) and (g) we generally omit the word "set", because the only objects referred to in this way will be sets.

†EXAMPLE 1.2.2. Each of the following is a formula of set theory:

(1) $$x \in A \text{ or for some } y, y \in A.$$

Here, the atomic formulas are "$x \in A$", "$y \in A$". The formula with which we are concerned is constructed by applying the quantifier "for some y" to "$y \in A$" and then combining "$x \in A$" with "for some y, $y \in A$" by means of the connective "or".

(2) $$x \in A \text{ or for some } y, y \text{ is not in } A.$$

Here, "y is not in A" means "Not '$y \in A$' " (which we would write as "$y \notin A$"). The remainder of the construction is similar to that of (1).

(3) $$x = y.$$

This has been defined to mean "For every set A, $x \in A$ if and only if $y \in A$".

(4) For technical reasons we have allowed

$$\text{For some } x, a \in A$$

to be a formula. (It will turn out that the application of the quantifier in such cases will not affect the "meaning" of the formula; see p. 21.)

In English usage it is sometimes necessary to modify the wording of these connectives and quantifiers. We now list some alternative expressions that are in common mathematical usage. Here, s_1 and s_2 denote formulas of set theory.

(a) Not s_1:
- (a)$_{\text{i}}$ It is not the case that s_1.
- (a)$_{\text{ii}}$ s_1 does not hold.
- (a)$_{\text{iii}}$ (If the formula s_1 has a unique verb, negating this verb amounts to forming "not s_1".)

(b) s_1 *or* s_2:
- (b)$_{\text{i}}$ Either s_1 or s_2.
- (b)$_{\text{ii}}$ s_1 or s_2 or both.
- (b)$_{\text{iii}}$ s_1 and/or s_2.

(c) s_1 *and* s_2:
  (c)$_{\text{i}}$ Both s_1 and s_2.
  (c)$_{\text{ii}}$ s_1 as well as s_2.
  (c)$_{\text{iii}}$ Not only s_1, but also s_2.
(d) *If* s_1, *then* s_2:
  (d)$_{\text{i}}$ s_1 implies s_2.
  (d)$_{\text{ii}}$ In case s_1, then s_2.
  (d)$_{\text{iii}}$ s_1 is (a) sufficient (condition) for s_2.
  (d)$_{\text{iv}}$ s_2 is (a) necessary (condition) for s_1.
  (d)$_{\text{v}}$ s_1 only if s_2.
  (d)$_{\text{vi}}$ s_2 whenever s_1.
(e) s_1 *if and only if* s_2:
  (e)$_{\text{i}}$ s_1 iff s_2.
  (e)$_{\text{ii}}$ s_1 is (a) necessary and sufficient (condition) for s_2.
(f) *For some* x:
  (f)$_{\text{i}}$ There exists an x such that ....
  (f)$_{\text{ii}}$ There is an x such that ....
  (f)$_{\text{iii}}$ There is some x such that ....
(g) *For every* x:
  (g)$_{\text{i}}$ For each x.
  (g)$_{\text{ii}}$ For all x.
  (g)$_{\text{iii}}$ For any x.

For example, we will regard "s_1 implies s_2." as meaning the same thing as "If s_1, then s_2".

Note that we have indicated the "*inclusive*" sense of "or"; i.e., "s_1 or s_2" allows any of the possibilities

$$s_1, s_2, \text{both } s_1 \text{ and } s_2.$$

In the English language there is also an "*exclusive*" sense of "or" that prohibits the third possibility.

†EXAMPLE 1.2.3. Consider the use of the word "or" in the following sentences:

(1) In order to qualify, you must speak Spanish or French.

Presumably, you qualify if you speak both languages. "Or" is used here in the inclusive sense.

(2) In contrast, when ordering a complete dinner in a Chinese restaurant, you may read in the menu:

Choose one dish from column A or one from column B.

Here the restaurant, undoubtedly, *intends* that you select only one item. "Or" is used here in the exclusive sense.

(3) Columbus was born in Spain or in Italy.

Here it is not *possible* for both of these countries to be the birthplace. This is a slightly different usage of "or" than in (2); there it was possible to select two dishes, although that was contrary to the obvious intention. The usage of "or" in (3) is not regarded as the exclusive sense.

For the most part the intended usage will be clear from the context. Occasionally, we will write "*or* (*excl.*)" to clarify the intention; "s_1 or (excl.) s_2" means "s_1 or s_2, and not both s_1 and s_2".

Strictly speaking, "or (excl.)" is also a connective, although we did not include it in our list on page 7.

The usual convention is to use the word "if" in lieu of "if and only if" in definitions; we will sometimes, but not always, adhere to this. Thus, in Example 1.1.3, we defined a set X to be ordinary if $X \notin X$; clearly, the intent is that X is ordinary if and only if $X \notin X$. On the other hand, in defining the rules governing the truth values of quantified formulas (pp. 21–22), we utilize "if and only if" for the purpose of clarity.

The order in which we apply these connectives and quantifiers, in general, affects the meaning. In questionable cases we will indicate the intended order by means of punctuation marks. Thus,

$$x \in A \text{ or } x \in B, \text{ and } x \in C$$

indicates that "or" is applied first, in contrast with the order of application of the connectives in

$$x \in A, \text{ or } x \in B \text{ and } x \in C.$$

The following **Venn diagram** illustrates the distinction between these formulas:

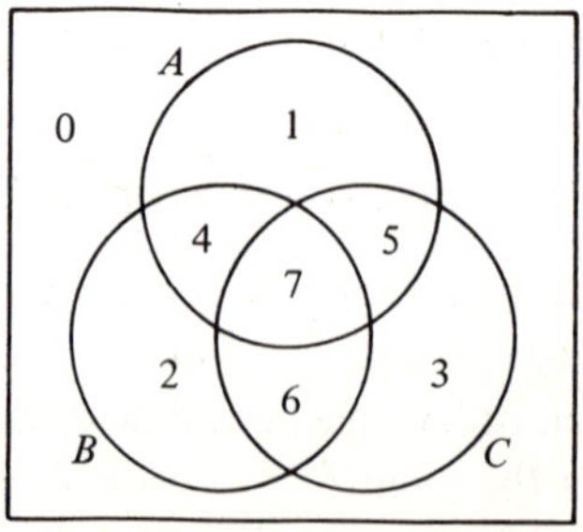

Fig. 1.2.1

Here, "$x \in A$ or $x \in B$" corresponds to regions 1, 2, 4, 5, 6, and 7, and "$x \in C$" corresponds to regions 3, 5, 6, and 7; hence "$x \in A$ or $x \in B$, and $x \in C$" corresponds to regions 5, 6, and 7.

Also, "$x \in A$" corresponds to regions 1, 4, 5, and 7 and "$x \in B$ and $x \in C$" corresponds to regions 6 and 7; hence, "$x \in A$, or $x \in B$ and $x \in C$" corresponds to regions 1, 4, 5, 6, and 7.

On the other hand,

$$x \in A \text{ or } x \in B, \text{ or } x \in C$$

has the same meaning as

$$x \in A, \text{ or } x \in B \text{ or } x \in C.$$

Each of these asserts that x is an element of at least one of the set A, B, and C; in a Venn diagram each corresponds to regions 1, 2, 3, 4, 5, 6, and 7. In such a case we agree to omit the commas and to write each of these as

$$x \in A \text{ or } x \in B \text{ or } x \in C.$$

We will sometimes, but not always, denote a formula in which the set variable x occurs by "$s(x)$". Also, there may be occurrences of set variables other than x in the formula $s(x)$.

Let x be a set variable and let $s(x)$ denote a formula in which x occurs. Then x is said to be **free in** $s(x)$ if x occurs at least once without being introduced by a quantifier involving x, i.e., if x occurs at least once without being introduced by a phrase such as "for some x", "for every x", etc. Such an occurrence of x in $s(x)$ is called a **free occurrence of** x. If each time x occurs in $s(x)$ it is introduced by a quantifier, we say that x **is not free in** $s(x)$. If s is a formula in which x does not occur, we also say that x **is not free in** s.

EXAMPLE 1.2.4. Let x and y be set variables and let A be a set constant.

(1) The variable x is free in

$$x \in A.$$

(2) Furthermore, x is free in

$$x \in A \text{ and for every } x,\ x \in B;$$

here, the first, but not the second, occurrence of x is free.

(3) We note that x is free in

$$\text{For some } x,\ x \in A; \text{ or } x \in B.$$

Here, the semicolon is used to indicate that "or $x \in B$" is not to be considered within the scope of the quantifier "For some x".

(4) On the other hand, y is free, but x is not free, in

For every x, $x \in y$.

(5) Neither x nor y is free in

For every x there is some y for which $x \in y$.

Let the set variable x occur in the formula $s(x)$, and let c be a set constant. Let "$s\begin{pmatrix} x \\ c \end{pmatrix}$" denote the formula that is obtained from $s(x)$ by replacing each free occurrence of x by "c"; each occurrence of x that is not free in $s(x)$ is left alone. Thus, if x occurs in $s(x)$ but is not free, then "$s\begin{pmatrix} x \\ c \end{pmatrix}$" denotes the same formula as $s(x)$ denotes. We will write "$s_1 \equiv s_2$" to indicate that s_1 and s_2 denote the same formula. It will be expedient to define "$s_3\begin{pmatrix} x \\ c \end{pmatrix}$" to be "$s_3$" in case x does not occur in the formula s_3. For any formula s_4, we will write

$$\text{“}s_4\begin{pmatrix} x_1 & x_2 & \cdots & x_N \\ c_1 & c_2 & \cdots & c_N \end{pmatrix}\text{”} \quad \text{for} \quad \text{“}\left(\cdots\left(s_4\begin{pmatrix} x_1 \\ c_1 \end{pmatrix}\right)\begin{pmatrix} x_2 \\ c_2 \end{pmatrix}\cdots\right)\begin{pmatrix} x_N \\ c_N \end{pmatrix}\text{”},$$

where $x_1, x_2, \ldots, x_N$ are distinct set variables and $c_1, c_2, \ldots, c_N$ are (not necessarily distinct) set constants.

EXAMPLE 1.2.5. Let x and y be set variables and let c_1, c_2, A, B, and C be set constants. Consider the following formulas:

s_1: $x \in A$.

s_2: For every x, $x \in A$.

s_3: $x \notin A$ implies there is some x such that $x \in B$.

s_4: $x \in A$ or $y \in B$.

s_5: There exists an x such that for all y, $x \in A$ implies either $x \in B$ or $y \in C$.

Then

(1)
$$s_1\begin{pmatrix} x \\ c_1 \end{pmatrix} \equiv c_1 \in A; s_1\begin{pmatrix} y \\ c_1 \end{pmatrix} \equiv s_1;$$

$$s_1\begin{pmatrix} x\ y \\ c_1 c_2 \end{pmatrix} \equiv s_1\begin{pmatrix} x\ y \\ c_1 c_1 \end{pmatrix} \equiv s_1\begin{pmatrix} x \\ c_1 \end{pmatrix}.$$

(2)
$$s_2\begin{pmatrix} x \\ c_1 \end{pmatrix} \equiv s_2.$$

(3) $$s_3\begin{pmatrix} x \\ c_1 \end{pmatrix} \equiv \text{“}c_1 \notin A \text{ implies there is an } x \text{ such that } x \in B.\text{”}$$

(4) $$s_4\begin{pmatrix} x \\ c_1 \end{pmatrix} \equiv \text{“}c_1 \in A \text{ or } y \in B\text{”};$$

$$s_4\begin{pmatrix} y \\ c_2 \end{pmatrix} \equiv \text{“}x \in A \text{ or } c_2 \in B\text{”};$$

$$s_4\begin{pmatrix} x\ y \\ c_1 c_2 \end{pmatrix} \equiv \text{“}c_1 \in A \text{ or } c_2 \in B\text{”};$$

$$s_4\begin{pmatrix} x\ y \\ c_1 c_1 \end{pmatrix} \equiv \text{“}c_1 \in A \text{ or } c_1 \in B\text{”}.$$

(5) $$s_5\begin{pmatrix} x \\ c_1 \end{pmatrix} \equiv s_5\begin{pmatrix} y \\ c_2 \end{pmatrix} \equiv s_5\begin{pmatrix} x\ y \\ c_1 c_2 \end{pmatrix} \equiv s_5.$$

We note that this substitution process obeys the following rules: *Let x be a set variable; let s_1 and s_2 denote arbitrary formulas, possibly having occurrences of x; and let c be a set constant. Then*

(a) $$\text{“}Not\ s_1\text{”}\begin{pmatrix} x \\ c \end{pmatrix} \equiv Not\ \text{“}s_1\begin{pmatrix} x \\ c \end{pmatrix}\text{”}.$$

(b) $$\text{“}s_1\ or\ s_2\text{”}\begin{pmatrix} x \\ c \end{pmatrix} \equiv \text{“}s_1\begin{pmatrix} x \\ c \end{pmatrix}\ or\ s_2\begin{pmatrix} x \\ c \end{pmatrix}\text{”}.$$

(b′) $$\text{“}s_1\ or\ (excl.)\ s_2\text{”}\begin{pmatrix} x \\ c \end{pmatrix} \equiv \text{“}s_1\begin{pmatrix} x \\ c \end{pmatrix}\ or\ (excl.)\ s_2\begin{pmatrix} x \\ c \end{pmatrix}\text{”}.$$

(c) $$\text{“}s_1\ and\ s_2\text{”}\begin{pmatrix} x \\ c \end{pmatrix} \equiv \text{“}s_1\begin{pmatrix} x \\ c \end{pmatrix}\ and\ s_2\begin{pmatrix} x \\ c \end{pmatrix}\text{”}.$$

(d) $$\text{“}If\ s_1,\ then\ s_2\text{”}\begin{pmatrix} x \\ c \end{pmatrix} \equiv \text{“}If\ s_1\begin{pmatrix} x \\ c \end{pmatrix},\ then\ s_2\begin{pmatrix} x \\ c \end{pmatrix}\text{”}.$$

(e) $$\text{“}s_1\ if\ and\ only\ if\ s_2\text{”}\begin{pmatrix} x \\ c \end{pmatrix} \equiv \text{“}s_1\begin{pmatrix} x \\ c \end{pmatrix}\ if\ and\ only\ if\ s_2\begin{pmatrix} x \\ c \end{pmatrix}\text{”}.$$

Thus, *applying connectives and substituting constants for variables can be done interchangeably*. In contrast to this,

(f)$_i$ $$\text{“}For\ some\ x,\ s_1\text{”}\begin{pmatrix} x \\ c \end{pmatrix} \equiv For\ some\ x,\ s_1$$

—whether or not x is free in s_1 (because x is not free in “For some x, s_1”).

$(\text{f})_{\text{ii}}$ *If x is not free in s_1, then*

$$\textit{For some } x, \text{“}s_1\binom{x}{c}\text{”} \equiv \textit{For some } x, s_1 \equiv s_1.$$

$(\text{f})_{\text{iii}}$ *If x is free in s, then*

$$\textit{For some } x, \text{“}s_1\binom{x}{c}\text{”} \not\equiv \textit{For some } x, s_1.$$

In fact, here

$(\text{g})_{\text{i}}$
$$\textit{For some } x, \text{“}s_1\binom{x}{c}\text{”} \equiv s_1\binom{x}{c}.$$

$$\text{“}\textit{For every } x, s_1\text{”}\binom{x}{c} \equiv \textit{For every } x, s_1$$

—whether or not x is free in s_1.

$(\text{g})_{\text{ii}}$ *If x is not free in s_1, then*

$$\textit{For every } x, \text{“}s_1\binom{x}{c}\text{”} \equiv \textit{For every } x, s_1 \equiv s_1.$$

$(\text{g})_{\text{iii}}$ *If x is free in s_1, then*

$$\textit{For every } x, \text{“}s_1\binom{x}{c}\text{”} \not\equiv \textit{For every } x, s_1.$$

Here,

$$\textit{For every } x, \text{“}s_1\binom{x}{c}\text{”} \equiv s_1\binom{x}{c}.$$

Thus, *when x is free in a formula, the order of applying a quantifier* (*involving x*) *and of substituting for x—is essential.*

EXAMPLE 1.2.6. Let x be a variable and c a constant. Let s_1 denote “$x \in A$”; let s_2 denote “$B \in x$”; let s_3 denote “For some $x, x \in A$”.

(1) $\text{“}s_1 \text{ and } s_2\text{”}\binom{x}{c} \equiv \text{“}c \in A \text{ and } B \in c\text{”} \equiv \text{“}s_1\binom{x}{c} \text{ and } s_2\binom{x}{c}\text{”}.$

(2)
$$s_3 \equiv \text{For some } x, s_1.$$

Therefore,

$$\text{“For some } x, s_1\text{”}\binom{x}{c} \equiv s_3\binom{x}{c} \equiv s_3$$

(because x is not free in s_3).

$$s_1\binom{x}{c} \equiv \text{“}c \in A\text{”}.$$

Therefore,

$$\text{“For some } x, s_1\binom{x}{c}\text{”} \equiv s_1\binom{x}{c}.$$

Exercises

Let x, y, and z be set variables and let A, B, C, and c be set constants.

†1.2.1. Show how each of the following formulas can be constructed from atomic formulas by repeated application of connectives and quantifiers. (Expressions previously analyzed, such as "$x = y$", may be used.)

(a) A is a subset of B.

(b) A is a proper subset of B.

(c) A and B are distinct subsets of C.

1.2.2. Suppose that we already have the set of natural numbers $\{0, 1, 2, \ldots\}$ (and that the usual arithmetic and ordering have been defined). Consider the following formulas with one free variable, x:

$$s_1 : x^2 > 0;$$

$$s_2 : x + 3 \geq 5;$$

$$s_3 : 1 < x < 2.$$

(a) Express "Not s_1" using the usual mathematical symbolism.

(b) Do the same for "Not s_2".

(c) Express s_3 in the form "p and q".

(d) Determine an expression of the form "not p, or not q" that is equivalent to "not s_3".

(Used in Exercises 1.3.4–1.3.9.)

1.2.3. Express each of the following formulas in several other ways.

(a) $x \in A$ implies $y \in B$.

(b) $x \in A$ if and only if $y \in B$.

(c) For every x, if $x \in A$, then $x \in B$.

1.2.4. **(a)** Let s be "$x \in A$ and $y \in B$". Determine

$$s\binom{x}{c}, \quad s\binom{y}{c}.$$

(b) Let s' be "For some x, $x = y$; and for all x, $x \in A$". Determine

$$s'\begin{pmatrix} x \\ c \end{pmatrix}, \quad s'\begin{pmatrix} y \\ c \end{pmatrix}.$$

(c) Let s'' be "For every x there is a y such that if $x \in A$ and $y \in B$, then either $z \in C$ or $z = x$." Determine

$$s''\begin{pmatrix} x \\ c \end{pmatrix}, \quad s''\begin{pmatrix} y \\ c \end{pmatrix}, \quad \text{and} \quad s''\begin{pmatrix} z \\ c \end{pmatrix}.$$

1.2.5. To which regions of a Venn diagram (Figure 1.2.1) does each of the following formulas correspond?

(a) If $x \in A$, then either $x \in B$ or $x \in C$.

(b) $x \notin A$ and if $x \in B$, then $x \notin C$.

(c) $x \in A$ if and only if $x \notin B$, and $x \in C$.

(d) $x \in A$ if and only if both $x \notin B$ and $x \in C$.

1.2.6. Determine two formulas that are constructed—by application of connectives—from the atomic formulas "$x \in A$", "$x \in B$", and "$x \in C$" and that correspond to the indicated regions of a Venn diagram (Figure 1.2.1).

(a) 0.

(b) 4, 5, 6.

(c) 0, 1.

(d) 1, 3, 4, 7.

1.3 Truth

Before discussing the "truth" of sentences, we must consider some preliminary notions.

Let A be a set constant and let x vary over A. Let $s(x)$ be the formula, "$x \subseteq A$". For a certain element a of A it may be true that every element of a is an element of A; for such a set a we wish to regard $s\begin{pmatrix} x \\ a \end{pmatrix}$ as true. For another element b of A, there may exist a set c such that $c \in b$, but $c \notin A$. We wish to regard $s\begin{pmatrix} x \\ b \end{pmatrix}$ as false. Thus, the original formula, "$x \subseteq A$", may have a different meaning under two different substitutions. The occurrence of a free variable in a formula may leave room for various possibilities. With this in mind *we assign "truth values" only to those formulas without free variables.*

Let $s(x)$ be a formula in which the variable x has a free occurrence. There are two ways in which $s(x)$ can be modified so that x will no longer be free in the new formulas.

(1) We can substitute the same set constant for each occurrence of x; for example, if c is a set constant, we obtain the formula $s\begin{pmatrix} x \\ c \end{pmatrix}$.

(2) We can quantify over x (by means of a quantifier involving x); for example, we might obtain "For some x, $s(x)$" or "For every x, $s(x)$".

Clearly, in a "finite number" of modifications of these types, we can eliminate all free occurrences of variables from a given formula.

EXAMPLE 1.3.1. Let w, x, y, and z be free variables and let a, b, c, and d be set constants. Let s be the formula

$$x \in a \text{ and } w \in z\text{; or for some } y,\ y \in z.$$

Each of the following is a modification of s in which there are no free occurrences of variables:

(1) $s\begin{pmatrix} w & x & z \\ a & b & d \end{pmatrix}$.

(2) $s\begin{pmatrix} w & x & y & z \\ a & b & c & d \end{pmatrix}$.

(Note that $s\begin{pmatrix} w & x & y & z \\ a & b & c & d \end{pmatrix} \equiv s\begin{pmatrix} w & x & z \\ a & b & d \end{pmatrix}$. Why?)

(3) For some x there is a w such that for all z, s.

(4) For some x, $s\begin{pmatrix} w & z \\ a & a \end{pmatrix}$.

A formula is called a **sentence** if it has no free variables; in particular, an **atomic sentence** is a sentence that is also an atomic formula.

In ordinary usage some sentences would be regarded as neither true nor false. Our sole concern is with **2-valued logic**; here, *we assume that every sentence asserts something that is either true or (excl.) false*. Thus, we assign truth values—"truth" (t) or (excl.) "falsehood" (f)—to all sentences—however bizarre this may appear. Moreover, the connectives and quantifiers are to operate with regard to truth as they do in ordinary mathematical usage. We first present the **rules governing** the **connectives**: Let s_1 and s_2 denote sentences.

(1) *Exactly one of the sentences*

$$s_1$$

$$not\ s_1$$

is to be true.

(2) *The sentence*

(1.3.1) s_1 *or* s_2

is to be true in case at least one of the sentences

(1.3.2) s_1

(1.3.3) s_2

is true; otherwise (1.3.1) *is to be false.*

(3) *The sentence*

(1.3.4) s_1 *or (excl.)* s_2

is to be true in case exactly one of the sentences (1.3.2), (1.3.3) *is true; otherwise* (1.3.4) *is to be false.*

(4) *The sentence*

(1.3.5) s_1 *and* s_2

is to be true in case both of the sentences (1.3.2), (1.3.3) *are true; otherwise* (1.3.5) *is to be false.*

(5) *The sentence*

If s_1, *then* s_2

is to be true except in the case where (1.3.2) *is true and* (1.3.3) *is false.*

(6) *The sentence*

(1.3.6) s_1 *if and only if* s_2

is to be true in case (1.3.2) *and* (1.3.3) *have the same truth values; otherwise* (1.3.6) *is to be false.*

In ordinary usage, we sometimes do not consider "If s_1, then s_2" as either true or false when s_1 is false. Here, we wish to consider every sentence as either true or (excl.) false. One reason for preferring truth to falsehood in case s_1 should be false is that sentences of the forms

If p and q, then q

If p, then p or q

—which are intuitively plausible—would then be true regardless of the truth of p and q. For example, according to rule (5), the only way the first sentence could be false is for "p and q" to be true but q to be false. This would violate rule (4). In order to prove the truth of a sentence "If s_1, then s_2", it is only necessary to consider the case in which s_1 is true.

A **truth table** is a convenient way of expressing the truth-functional rules governing these connectives.

There are four (2^2) possible pairs of truth values for two sentences—s_1 and s_2:

s_1 true, s_2 true;

s_1 true, s_2 false;

s_1 false, s_2 true;

s_1 false, s_2 false.

A truth table indicates the corresponding truth values of sentences compounded from s_1, s_2 by means of connectives.

We first consider a truth table for the following basic sentences:

Not s_1.

s_1 or s_2.

s_1 or (excl.) s_2.

s_1 and s_2.

If s_1, then s_2.

s_1 if and only if s_2 (s_1 iff s_2).

The possibilities for each of these six sentences are listed in a separate column. The four rows of the truth table represent the various possibilities for s_1, s_2.

s_1	s_2	Not s_1	s_1 or s_2	s_1 or (excl.) s_2	s_1 and s_2	If s_1, then s_2	s_1 iff s_2
t	t	f	t	f	t	t	t
t	f	f	t	t	f	f	f
f	t	t	t	t	f	t	f
f	f	t	f	f	f	t	t

Thus, an "f" appears in the "s_1 or s_2" column only in the case in which the sentences s_1, s_2 are each false. This corresponds to rule (2) on page 18. "Not s_1" involves only one of the original sentences; here, the truth values are repeated and we ignore the value of s_2.

We illustrate how truth tables are utilized in ascertaining the truth values of more complicated sentences.

†EXAMPLE 1.3.2
(1) If not s_2, then not s_1.
(2) Not s_2—if and only if—either s_1 or s_2.
(3) It is not the case that if s_1, then s_2.

(i)	(ii)	(iii)	(iv)	(v)	(vi)	(vii)	(viii)	(ix)	(x)
s_1	s_2	Not s_1	Not s_2	If not s_2, then not s_1	Both s_1 and not s_2	Either s_1 or s_2	Not s_2 iff either s_1 or s_2	If s_1, then s_2	It is not the case that if s_1, then s_2.
t	t	f	f	t	f	t	f	t	f
t	f	f	t	f	t	t	t	f	t
f	t	t	f	t	f	t	f	t	f
f	f	t	t	t	f	f	f	t	f

(a) The truth values of sentence (1) are in column (v); the only f value occurs in the second row, where there is a "t" in column (iv) and an "f" in column (iii).
(b) The truth values of sentence (2) are in column (viii). In order for a "t" to appear, the corresponding entries in columns (iv) and (vii) must agree.
(c) The truth values of sentence (3) are in column (x). Observe that the corresponding entries in columns (viii) and (x) agree. Such sentences are said to be **truth-functionally equivalent**. Every time a "t" appears in column (vi), the corresponding entry in column (vii) is also a "t". In such a case we say that sentence (vi) **logically implies** sentence (vii). Note that truth-functionally equivalent sentences always logically imply one another. Find another sentence among the above that is logically implied by, but that is not truth-functionally equivalent to, sentence (vi).
(d) Sentence (1), "If not s_2, then not s_1" is known as **the contrapositive of** "If s_1, then s_2". Note that these are truth-functionally equivalent.
(e) **The converse of** "If s_1, then s_2" is the sentence, "If s_2, then s_1". Show that this is not truth-functionally equivalent to "If s_1, then s_2".

†EXAMPLE 1.3.3. In constructing truth tables for more complicated sentences, it is often convenient to add intermediate auxiliary columns. Consider

$$s_1\text{, or } s_1 \text{ implies } s_2\text{; if and only if } s_2.$$

s_1	s_2	s_1 implies s_2	s_1, or s_1 implies s_2	s_1, or s_1 implies s_2; iff s_2
t	t	t	t	t
t	f	f	t	f
f	t	t	t	t
f	f	t	t	f

A sentence, such as "s_1, or s_1 implies s_2", that is true in all possible cases is called a **tautology**.

EXAMPLE 1.3.4. A truth table for a sentence compounded (by means of connectives) from three sentences—s_1, s_2, s_3—will have eight (2^3) rows. Consider the sentence

$$s_1 \text{ or } s_2\text{; and } s_2 \text{ if and only if not } s_3.$$

s_1	s_2	s_3	s_1 or s_2	Not s_3	s_2 iff not s_3	s_1 or s_2; and s_2 iff not s_3
t	t	t	t	f	f	f
t	t	f	t	t	t	t
t	f	t	t	f	t	t
t	f	f	t	t	f	f
f	t	t	t	f	f	f
f	t	f	t	t	t	t
f	f	t	f	f	t	f
f	f	f	f	t	f	f

We now consider the **rules for quantification.** Let s_1 again denote a sentence and let $s_2(x)$ denote a formula with (exactly) one free variable, x.

$(7)_a$ *The sentence*

$$For\ some\ x,\ s_1$$

is to be true if and only if s_1 is true. (Note that x is not free in s_1; thus, quantifying over x does not affect s_1.)

$(7)_b$ *The sentence*

$$For\ some\ x,\ s_2(x)$$

is to be true if and only if $s_2\begin{pmatrix}x\\c\end{pmatrix}$ *is true for at least one (constant) set c.*

$(8)_a$ *The sentence*

$$\textit{For every } x, s_1$$

is to be true if and only if s_1 *is true.*

$(8)_b$ *The sentence*

$$\textit{For every } x, s_2(x)$$

is to be true if and only if $s_2\binom{x}{c}$ *is true for every set c.*

EXAMPLE 1.3.5. Suppose that we have the set $A = \{a, b, c, d\}$ and that B is a set, distinct from a, b, c, and d. Then

(1)
$$\textit{For some } x, x \in A$$

is true because

$$\text{“}x \in A\text{”}\binom{x}{a}$$

is true.

(2)
$$\textit{For every } x, x \in A$$

is false because

$$\text{“}x \in A\text{”}\binom{x}{B}$$

is false.

(3)
$$\text{For some } x, a \in A$$

and

$$\text{For every } x, a \in A$$

are both true because

$$a \in A$$

is true.

Exercises

Let x and y be set variables and let A, B, C, and a be set constants.

1.3.1. Assume that “$a \in A$” is true but that “$a \in B$” is false. Which of the following sentences are true?

(a) $a \in A$ implies $a \in B$.
(b) $a \in B$ implies $a \in A$.
(c) $a \in A$ and $a \notin B$.
(d) $a \notin A$ iff both $a \in A$ and $a \in B$.
(e) $B \subseteq A$.
(f) For some y, $y \in A$ and $y \notin B$.

1.3.2. Suppose that $A \subset B$, $B \subset C$. Which of the following sentences are true?
(a) There is some x such that both $x \in B$ and $x \notin A$.
(b) For all x, $x \in A$ implies $x \in C$.
(c) For all x, $x \in A$ iff $x \in C$.
(d) If $A = B$, then $B = C$.
(e) For every x, $A \in x$ iff $B \in x$.

1.3.3. Assume that $A \subset B$, $B \subset C$. Which of the following sentences are true?
(a) $A \subset B$ or $B \subset C$, and $C \subset B$.
(b) $A \subset B$, or $B \subset C$ and $C \subset B$.
(c) $A \subset B$ and $B \subset C$, or $C \subset B$.
(d) $A \subset B$, and $B \subset C$ or $C \subset B$.

1.3.4. **(a)** Use a truth table to generalize Exercise 1.2.2(d) by showing that for any two sentences s_1 and s_2,

Not—s_1 and s_2

is (truth-functionally) equivalent to

Not s_1, or not s_2.

(b) Now show, by means of a truth table, that

Not—s_1 or s_2

is equivalent to

Not s_1, and not s_2.

1.3.5. Let s_1 be as in Exercise 1.2.2. Which of the following sentences are true?
(a) For some x, x is a natural number and s_1.
(b) For every x, if x is a natural number, then s_1.
(c) For some x, x is a natural number and not s_1.
(d) For every x, if x is a natural number, then not s_1.
(e) It is not the case that for some x, x is a natural number and s_1.
(f) It is not the case that for every x, if x is a natural number, then s_1.
(g) $s_1\begin{pmatrix}x\\0\end{pmatrix}$. **(h)** $s_1\begin{pmatrix}x\\1\end{pmatrix}$.

1.3.6. Let s_2 be as in Exercise 1.2.2, and let (a)–(h) be as in Exercise 1.3.5 but with "s_2" in lieu of "s_1". Which of these are true?

1.3.7. Let s_3 be as in Exercise 1.2.2, and let (a)–(h) be as in Exercise 1.3.5 but with "s_3" in lieu of "s_1". Which of these are true?

1.3.8. **(A)** Answer Exercise 1.3.5(a)–(f) with "nonzero natural number" replacing "natural number".

(B) Answer Exercise 1.3.6(a)–(f) with this modification.

(C) Answer Exercise 1.3.7(a)–(f) with this modification.

1.3.9. Assume that we have the set of real numbers together with its arithmetic and ordering.

(A) Answer Exercise 1.3.5(a)–(f) with "real number" in lieu of "natural number".

(B) Answer Exercise 1.3.6(a)–(f) subject to this modification.

(C) Answer Exercise 1.3.7(a)–(f) subject to this modification.

1.3.10. Construct truth tables for each:

(a) s_1 if and only if not s_2—implies s_1.

(b) s_1 if and only if—not s_2 implies s_1.

(c) s_1 and s_2—if and only if—s_1 implies not s_3.

(d) s_1 or s_2—implies—both not s_3 and s_4.

1.3.11. Let s_1 be "$x \in A$", s_2 be "$x \in B$", and s_3 be "$x \in C$". To which rows of a truth table do the regions $0, 1, \ldots, 7$ of a Venn diagram (Figure 1.2.1) correspond?

***1.3.12.**

Up to and including this statement there are an odd number of incorrect statements in this book.

Is the preceding statement true?

1.4 Specification of Subsets

Let A and B be sets. We know what is meant by saying that B is a subset of A. Suppose that A alone is a given set. We now formulate a principle by means of which we guarantee the existence of certain subsets B of A. In other words, this principle will guarantee that various subcollections of A will be sets.

Axiom of Subsets (Aussonderungsaxiom):[4] *Let A be any set and let s(x) be any formula that has x as its one (and only) free variable. Then there exists a set, S, consisting of all elements a in A for which*

$$s\begin{pmatrix} x \\ a \end{pmatrix}$$

is true.

Note that the formulas $s(x)$ may have occurrences of constants (other than S); each such constant refers to a collection whose existence (as a set) has been previously guaranteed.

[4] For a formulation of the Axiom of Subsets in a formal treatment see the Axiom of Separation in (Cohen, p. 55); see also (Levy).

We modify our previous notation by writing "S" as "$\{a \in A : s\binom{x}{a}\}$".

†EXAMPLE 1.4.1. Suppose that A and B are sets. By the Axiom of Subsets, the following are also sets.
(1) $\{a \in A : a \notin B\}$; here, $s(x)$ is "$x \notin B$".
(2) $\{C \in A : C \subseteq B\}$.
(3) $\{C \in A : C \subseteq A\}$.

The following is not as yet guaranteed to be a set: The collection of x such that $x \in A$ or $x \in B$. This collection might not be included in either A or in B; hence, we need a new principle to guarantee its existence as a set. We consider this in the next section.

Note that we have not yet guaranteed the existence of a single set. The Axiom of Subsets merely asserts that once we have a set A, certain other sets—subcollections of A—exist.

†EXAMPLE 1.4.2. Suppose that A and B are sets and $B \in A$. According to the Axiom of Subsets, each of the following is a set:
(1) $\{B\} = \{a \in A : a = B\}$.
(2) $\{a \in A : a \neq B\}$.
Note that with this information we cannot as yet establish the sethood of any of the following:
(3) The collection whose sole element is A.
(4) The collection whose sole element is $\{B\}$.
(5) The collection consisting of the two elements A and B.
(6) The collection consisting of all subsets of A.

Exercises

1.4.1. Suppose that we already have the sets a, b, c, and d and suppose that $A = \{a, b, c, d\}$. Use the Axiom of Subsets to show that each of the following is also a set.
(a) $\{a\}$. **(b)** $\{a, b\}$. **(c)** $\{a, b, c\}$.

1.4.2. Let A be a set and let $s(x)$ be the formula "$x \notin x$". Let

$$S = \left\{a \in A : s\binom{x}{a}\right\}.$$

Show that
(a) $S \subseteq A$. **(b)** $A \notin S$.
(c) If $A \in A$, then $S \subset A$. ★**(d)** $S \notin A$.

★,†**1.4.3.** Prove that there is no set of all sets (**Cantor's Paradox**).

1.5 Axioms of Null Set, Pairing, and Union

Axiom of the Null Set: *There exists a set, $\varnothing$, that contains no element.*

The set $\varnothing$ is called the **empty set (null set)**. *The empty set is unique*; i.e., if A is a set for which $a \in A$ never holds, then $A = \varnothing$ (Exercise 1.5.1).

Intuitively, we can think of the empty set as any collection described by a property that does not pertain to any object. For example, we might think of it as the aggregate of all people over thirty who are teenagers—or as the collection of all real roots of the equation "$x^2 + 1 = 0$".

EXAMPLE 1.5.1. Let X be any set. Then by the Axiom of Subsets, each of the following equals the empty set.

(1) $\{a \in X : a \notin X\}$.

(2) $\{a \in X : a \neq a\}$.

(3) $\{A \in X : A \subset X \text{ and } X \subset A\}$.

Thus far we have guaranteed the existence of only one set—the empty set. We now introduce another axiom that will yield many new sets.

Axiom of Pairing: *Let a and b be sets, distinct or not. There exists a set whose elements are a, b, and only these.*

We denote this set by "$\{a, b\}$"; it is called an **(unordered) pair**. In particular, if $a = b$, then $\{a, b\} = \{a, a\} = \{a\}$; we call $\{a\}$ a **unit set**.

Let $a = b = \varnothing$. By means of the Axiom of Pairing we obtain the unit set $\{\varnothing\}$. Now let $a = \varnothing$ and $b = \{\varnothing\}$; we obtain $\{\varnothing, \{\varnothing\}\}$. Similarly, the Axiom of Pairing yields such sets as $\{\{\varnothing\}\}$, $\{\{\{\varnothing\}\}\}$, $\{\varnothing, \{\{\varnothing\}\}\}$, $\{\{\varnothing\}, \{\{\varnothing\}\}\}, \ldots$. In fact, the Pairing Axiom now guarantees "infinitely" many different sets $-\varnothing$, $\{\varnothing\}, \{\{\varnothing\}\}, \{\{\{\varnothing\}\}\}, \ldots$. However, we still have not guaranteed the existence of a set A such that for distinct sets x, y, z—$x \in A$, $y \in A$, and $z \in A$ (i.e., as yet, we know of no set A that contains "more than two" distinct elements).

†EXAMPLE 1.5.2. Assume that A, B, and C are distinct sets. By the Axiom of Pairing, each of the following is a set:

(1) $\{A, B\}$.

(2) $\{A, C\}$.

(3) $\{B, C\}$.

Any two of the sets (1), (2), and (3) are distinct. For example, $B \in \{A, B\}$, but $B \notin \{A, C\}$; $A \in \{A, B\}$, but $A \notin \{B, C\}$. Thus, (1) is different from (2) and from (3).

We could also form such sets as the following:

(4) $\{A, \{B, C\}\}$.

(5) $\{A, \{A, B\}\}$.

(6) $\{\{A, B\}, \{A, C\}\}$.

Clearly, for all sets a, b, and A—$a \in A$ and $b \in A$, iff $\{a, b\} \subseteq A$.

It is conceivable that (on the basis of our previous axioms) all nonempty sets are of the form $\{a, b\}$, where a and b are sets, not necessarily distinct. Moreover, although the Axiom of Pairing guarantees the existence of a set that *contains* the sets a and b, we do not know of a set that *includes* both a and b. We now remedy these deficiencies.

Axiom of Union: *Let A be any set. Let $\cup A$ be the collection of b for which there is some a in A such that $b \in a$. Then $\cup A$ is a set.*

Thus, $b \in \cup A$ *iff there is an a in A such that $b \in a \in A$.* The set $\cup A$ is called **union** A.

In case $X = \{A, B\}$ we write "$A \cup B$" in lieu of $\cup X$, and we speak of "the union of A and B"; thus $\{a, b\} \cup \{a, c\} = \{a, b, c\}$. Clearly, $A \cup B = A$ iff $B \subseteq A$.

The set $\cup A$ can be thought of as representing the collection of objects each of which is an element of an element of A. In particular, $A \cup B$ stands for the collection of objects that are in A or in B.

By the Axioms of the Null Set and Pairing we have the sets $A = \{\varnothing, \{\varnothing\}\}$, $B = \{\{\{\varnothing\}\}\}$, and $C = \{\{\{\{\varnothing\}\}\}\}$. The Axiom of Union yields sets

$$D = A \cup B = \{\varnothing, \{\varnothing\}, \{\{\varnothing\}\}\}$$

and

$$E = C \cup D = \{\varnothing, \{\varnothing\}, \{\{\varnothing\}\}, \{\{\{\varnothing\}\}\}\}.$$

Intuitively speaking, D is a "3-element" set, E is a "4-element" set; it is clear that we can obtain larger and larger sets by repetition of these processes. Observe that such sets will still be "finite". We will make these notions more precise in subsequent sections.

Example 1.5.3. Let $A = \{B, C, D, \varnothing\}$, where

$$B = \{a, b, c\}, C = \{a, c, d\}, D = \{\varnothing, a\},$$

Then $\cup A = \{a, b, c, d, \varnothing\}$.

Exercises

†**1.5.1.** Prove the uniqueness of $\varnothing$.

1.5.2. Let A be an arbitrary set.

(a) Prove that $A \subset \varnothing$ is false.

(b) Suppose that $A \subseteq \varnothing$. What can be said about A?

†**(c)** Prove that $\varnothing \subseteq A$.

1.5.3. Prove that (in the presence of the Axiom of Subsets) the Axiom of the Null Set is equivalent to the sentence, "There exists a set." in the sense that the two sentences are either simultaneously true or false. (This is referred to in Section 1.6; the terminology is used in other exercises.)

1.5.4. Let a and b be sets and suppose that

$$A = \{a, \{a, b\}, b, \{b\}, \{a, \{b\}\}\}.$$

Which of the following are true?

(a) $b \in A$. **(b)** $\{b\} \in A$.

(c) $\{b\} \subseteq A$. **(d)** $\{a, b\} \in A$.

(e) $\{a, b\} \subseteq A$. **(f)** $\{a, \{b\}\} \in A$.

(g) $\{a, \{b\}\} \subseteq A$.

1.5.5. Consider Example 1.5.2. Are each of the sets (1), (2), and (3) distinct from the sets A, B, and C?

1.5.6. Determine sets A, B, and C such that $\varnothing \in A$, $A \in B$, $A \subset C$, and $B \subset C$.

★**1.5.7.** Prove that (in the presence of the Axioms of Extensionality and Subsets) the Axiom of Pairing is equivalent to the sentence, "For any sets a and b, there exists a set A such that $a \in A$ and $b \in A$." (In the second form A might contain other elements.) (Referred to in Section 1.6.)

1.5.8. Formulate an equivalent of the Axiom of Union.

1.5.9. Let a, b, c, d, and e be distinct sets and let $A = \{a, b\}$, $B = \{a, c, d\}$, $C = \{a, b, e\}$, and $D = \{a, \{a\}, b\}$. Determine the following sets:

(a) $A \cup B$. **(b)** $A \cup C$.

(c) $\cup\{A, B, C\}$. **(d)** $B \cup D$.

(e) $\cup D$. **(f)** $\cup\{D\}$.

1.5.10. Determine the following sets:

(a) $\cup\varnothing$. **(b)** $\cup\{\varnothing\}$.

(c) $\cup\{\varnothing, \{\varnothing\}\}$. **(d)** $\cup\{\{\varnothing\}\}$.

(e) $\{\varnothing, \{\varnothing\}\} \cup \{\{\varnothing, \{\varnothing\}\}\}$.

1.5.11. Let A and B be sets.

†**(a)** Prove that if $A \in B$, then $A \subseteq \cup B$.

(b) Is the converse true—i.e., if $A \subseteq \cup B$, does $A \in B$ follow?

★**1.5.12.** Assume the Axioms of Extensionality and Subsets. Which sets necessarily exist on the basis of the following additional axioms?

(a) Null set.

(b) Null set and Union.

(c) Union and Pairing.
(d) s_1, where $s_1 \equiv$ "There exists a set."
(e) s_1 and Pairing.

★1.5.13. Prove that in the presence of the Axioms of Subsets and Pairing, the Axiom of Extensionality is equivalent to the sentence, "$A = B$ if both A is a subset of B and B is a subset of A." (Here, we have substituted "if" for "if and only if" in the original axiom.)

1.6 Natural Numbers

We now define the natural numbers, 0, 1, 2, . . . , as sets.

If A is any set, we define A^+, **the successor of** A, to be the set $A \cup \{A\}$. We define

$$\text{"0" to be "}\varnothing\text{";}$$

$$\text{"1" to be "}0^+\text{";}$$

$$\text{"2" to be "}1^+\text{";}$$

$$\text{"3" to be "}2^+\text{"; etc.}$$

Thus,

$$1 = 0^+ = 0 \cup \{0\} = \varnothing \cup \{0\} = \{0\},$$

$$2 = 1^+ = 1 \cup \{1\} = \{0, 1\},$$

$$3 = 2^+ = 2 \cup \{2\} = \{0, 1, 2\}, \text{ etc.}$$

Our definition of natural numbers will be such that these will be identified with the sets obtained from 0 by iteration of this process.

A set X is said to be **inductive** if both of the following are true:

(1) $0 \in X$.
(2) For any set A, if $A \in X$, then $A^+ \in X$.

We do not as yet know, on the basis of the axioms thus far introduced, that any such inductive set exists.

†EXAMPLE 1.6.1

(1) 0 is not inductive; in fact, $0 \notin 0$.
(2) 1 is not inductive, although $0 \in 1$. However, $0^+ = 1 \notin 1$.
(3) Similarly, 2 is not inductive. Here, $0 \in 2$, $0^+ = 1 \in 2$, but $1^+ = 2 \notin 2$.

A set Y is called a **natural number** if Y is an element of every inductive set. Regardless of whether inductive sets exist, each of the sets 0, 1, 2, 3, as well as every other "iterated successor of 0", is a natural number. (Why?) We will subsequently prove as a *theorem* that the natural numbers obey the familiar Peano Postulates.

Axiom of Infinity: *There is a set, to be denoted by "ω", that consists of all the natural numbers.*

Instead of the Axiom of Infinity, we could have considered the following:

Weak Axiom of Infinity: *There exists a set that contains all the natural numbers.*

Clearly, *in the presence of the Axiom of Subsets, the Axiom of Infinity and the Weak Axiom of Infinity are equivalent.* (Recall the comparable situation for other axioms that we considered in Exercises 1.5.3 and 1.5.7.)

At this point we might ask whether ω really consists of the numbers 0, 1, 2, ... (and only these). Although we will eventually indicate that this is indeed the case, there are several difficulties in attempting to answer this so quickly. We have already remarked that each "iterated successor of 0" is a natural number. We might be tempted to use the Axiom of Subsets to obtain $W = \{x \in \omega : x$ is an "iterated successor of 0"$\}$, and then to use the fact that ω is inductive to show that $W = \omega$. The trouble with this approach is in formulating the notion "is an iterated successor of 0" as a sentence of set theory. [Remember that each constant occurring in such a sentence must correspond to a collection whose existence (as a set) has already been established.] The second problem is in showing that if a set A is not an "iterated successor of 0", then $A \notin \omega$. We will return to these matters at the end of this section and in Section 2.7.

Theorem 1.6.1

(1) *The set ω is inductive.*

(2) *If A is an inductive set, then $\omega \subseteq A$; thus, ω is the "smallest" inductive set.*

PROOF: (1) By definition, 0 belongs to every inductive set; i.e., 0 is a natural number. Thus, $0 \in \omega$.

Suppose that $A \in \omega$; then A is a natural number. Thus, given any inductive set X, we have $A \in X$. By the definition of "inductive", we must have $A^+ \in X$, as well. Because X is an arbitrary inductive set, A^+ is a natural number. Hence, $A^+ \in \omega$. It follows that ω is inductive.

(2) Let A be an arbitrary inductive set and let $X \in \omega$. Then X, as a natural number, must be an element of A. It follows that $w \subseteq A$.

Corollary. *If A is an inductive subset of ω, then $A = \omega$.*

A set A is called **transitive** if whenever $x \in A$, then $x \subseteq A$ (i.e., if every element of A is also a subset of A).

†EXAMPLE 1.6.2

(1) The natural number 0 is transitive because $x \in 0$ never holds.
(2) 1 is transitive because $x \in 1$ is true iff $x = 0 = \varnothing$, and $\varnothing$ is a subset of every set.
(3) $A = \{0, 1, 3\}$ is not transitive because $2 \in 3 \in A$, but $2 \notin A$.
(4) Let $B = \{0, 1, \{1\}\}$. Then B is transitive because $0 \subseteq B$, $1 = \{0\} \subseteq B$, and $\{1\} \subseteq B$. We will shortly establish that B is not a natural number.

Theorem 1.6.2
(1) *Every natural number is transitive.*
(2) *ω is transitive.*

PROOF: (1) Suppose that A is the collection of transitive natural numbers. Then, by the Axiom of Subsets, A is a subset of ω; in fact, because a natural number is just an element of ω, $A = \{X \in \omega : \text{for all } y, \text{if } y \in X, \text{then } y \subseteq X\}$. We utilize the corollary of Theorem 1.6.1 to show that $A = \omega$. We know that 0 is transitive; i.e., $0 \in A$. Suppose that $B \in A$. Then by the transitivity of B, it follows that if $b \in B$, then $b \subseteq B$. Also, $B^+ = B \cup \{B\}$. Thus, if $c \in B^+$, either $c \in B$ or $c = B$. In either case, $c \subseteq B \subseteq B^+$; hence, B^+ is transitive. We have shown that $B \in A$ implies $B^+ \in A$; therefore, A is an inductive subset of ω. It follows that $A = \omega$; i.e., every natural number is transitive.

(2) Here, let $A = \{X \in \omega : X \subseteq \omega\}$. Clearly $0 \in A$. Suppose that $B \in A$. Then $B^+ = B \cup \{B\}$; $B \subseteq \omega$ because $B \in A$, and $\{B\} \subseteq \omega$ because $B \in \omega$. Thus, B^+, as the union of two subsets of ω, must be a subset of ω. As above, $A = \omega$. Consequently, each element of ω is a subset of ω, and ω is transitive.

Note that in (2), as in (1), the sethood of A follows from the Axiom of Subsets. We will often omit explicit mention of such applications of the Axiom of Subsets in defining sets in this manner.

Recall that a set A is called "ordinary" provided $A \notin A$.

In mathematics we generally consider only ordinary collections of objects. Many systems of set theory include an axiom that specifically prohibits "non-ordinary" sets. Indeed, it is difficult to conceive, intuitively, of a collection that is not ordinary. In the next example we consider a set that is not ordinary—but in order to assure that this is the case, we must make a rather bizarre assumption about the set.

EXAMPLE 1.6.3. Let A be a set and suppose that $A = \{A\}$. Then A is not ordinary because $A \in \{A\} = A$. However, A is transitive. For if $X \in A$, then $X \in \{A\}$; hence $X = A$. Therefore, $X \subseteq A$.

Theorem 1.6.3

(1) *Every natural number is ordinary.*
(2) *ω is ordinary.*

Corollary 1. *The set ω is not a natural number.*

Corollary 2. *For any natural numbers N and M, either*

$$N \notin M \text{ or } M \notin N.$$

(The proofs of Theorem 1.6.3 and its corollaries will be left as an exercise.)

Theorem 1.6.4. The Peano Postulates

(1) $0 \in \omega$.
(2) *For every set N—if $N \in \omega$, then $N^+ \in \omega$.*
(3) *For no N in ω does $N^+ = 0$.*
(4) *For all sets N and M—$N \in \omega$, $M \in \omega$, $N^+ = M^+$ together imply $N = M$.*
(5) **(The Principle of Induction)** *If X is a set that contains* 0 *and that contains A^+ whenever it contains the set A, then $\omega \subseteq X$.*

Observe that parts (1), (2), and (5) follow immediately from the definitions of this section together with Theorem 1.6.1; part (3) follows from the fact that $N \in N^+$, but $N \notin 0$.

PROOF OF (4): Assume that $M \in \omega$, $N \in \omega$, $M^+ = N^+$, and $M \neq N$. Then $M \in M^+ = N^+ = N \cup \{N\}$; because $M \neq N$, it follows that $M \in N$. Similarly, $N \in M^+$, and thus $N \in M$. But by Corollary 2 of Theorem 1.6.3, $M \notin N$ or $N \notin M$. (Contradiction.)

In applications of the Principle of Induction, the assumption that $A \in X$, which is used to prove that $A^+ \in X$, is known as the **inductive hypothesis**.

Theorem 1.6.5. *If N is a natural number, then either $N = 0$ or (excl.) there is a (unique) natural number $M(\neq N)$ for which $N = M^+$.*

The natural number M, here, is called the **predecessor of** N.

PROOF: Let $A = \{0\} \cup \{N \in \omega\colon \text{ for some } M \text{ in } \omega,\ N = M^+\}$. Then $0 \in A$ because $0 \in \{0\}$. For each M in ω, $M^+ \in A$ because M^+ is the successor of M.

Thus, if $M \in A$, then $M^+ \in A$ because "if s_1, then s_2" is true except when s_1 is true and s_2 is false. By induction, $A = \omega$; hence, every nonzero natural number is the successor of some natural number.

For each N in ω, $N \neq N^+$ because $N \in N^+$ and $N \notin N$.

If $N = M^+ = R^+$, $M, R \in \omega$, then $M = R$, by Theorem 1.6.4(4).

†EXAMPLE 1.6.4. Consider the set $B = \{0, 1, \{1\}\}$ of Example 1.6.2(4). Surely $B \neq \varnothing$; thus, according to Theorem 1.6.5, if B were a natural number, then $B = A^+$ for some natural number A. Consequently, $A \in B$ because $B = A \cup \{A\}$. However, $B \neq 0^+(= 1)$, $B \neq 1^+(= 2)$, and $B \neq \{1\}^+ = \{1, \{1\}\}$. Thus, B is (a transitive set that is) not a natural number. (We reconsider this in Example 2.2.3.)

Theorem 1.6.5 *aids us in our quest to show that our definition of "natural number" conforms with our intuition. If $N \neq 0$, N must be M^+ for M in ω. The natural number M, in turn, is either* 0 *or is L^+ for L in ω; in the latter case $N = (L^+)^+$, and $L \in M \in N$. If we continue in this manner, we see that either N conforms to our intuitive notion of being an "iterated successor of* 0*" or else there exists an "infinite descending $\in$-chain"*

$$\ldots \in L \in M \in N.$$

In Section 2.7 *we will rule out the second possibility.*

We will continue to study the natural numbers in Chapters 2 and 3.

Note, finally, that our axioms guarantee the existence of "infinitely many" iterated successors of ω:

$$\omega^+, (\omega^+)^+, ((\omega^+)^+)^+, \ldots.$$

But we cannot as yet deduce that the above collection is a set. (See Section 5.3.) Nor can we, at this time, guarantee the existence of an inductive set other than ω.

Exercises

†1.6.1. Suppose that A is a set. Prove that A^+ is also a set.

1.6.2. Of which of the following sets is 5 an element?

(a) 4. **(b)** 5. **(c)** 6.
(d) $\{5\}$. **(e)** $\{6\}$. **(f)** $\cup\{5\}$.
(g) $\cup\{6\}$. **(h)** $\cup 5$. **(i)** $\cup 6$.
(j) $\cup 7$.

1.6.3. Determine all natural numbers N satisfying the following:

(a) $2 \in N \in 4$. **(b)** $5 \subseteq N \in 10$.

(c) $5 \subseteq N \subseteq 10$. **(d)** $N \in (N^+)^+$.

(e) $N \in \omega^+$.

★1.6.4. **(a)** Prove that the sentence, "There is an inductive set." is equivalent to the Axiom of Infinity.

(b) Prove that the sentence, "$\{1\}$ is not a natural number." is equivalent to the Axiom of Infinity.

†1.6.5. Determine **(a)** $\cup\omega$; **(b)** $\cup\omega^+$.

1.6.6. Prove the corollary of Theorem 1.6.1; i.e., prove that if A is an inductive subset of ω, then $A = \omega$.

1.6.7. Prove Theorem 1.6.3 and its corollaries; i.e., prove the following:

(a) Every natural number is ordinary.

(b) ω is not a natural number.

(c) ω is ordinary.

(d) For any natural numbers N and M, $N \notin M$ or $M \notin N$.

1.6.8. **(a)** Express 4 using only the symbols 0, { }.

(b) Determine $\cup 4$.

(c) Without using the Axiom of Infinity prove that 4 is a natural number.

(d) Show directly that 4 is transitive.

(e) Show directly that 4 is ordinary.

†1.6.9. **(a)** Prove that 0 is an element of every nonzero natural number.

(b) Prove that for every natural number N and for every set K, if $K \in N$, then either $K^+ \in N$ or (excl.) $K^+ = N$.

1.6.10. Prove that the collection consisting of the natural numbers

$$1, 2, 4, 8, 16, 32, \text{ and } 64$$

is a set by utilizing the Axioms of **(a)** Pairing and Union and **(b)** Infinity and Subsets. (Referred to in Section 1.7.)

†1.6.11. **(a)** Prove that a set A is transitive iff $A = \cup(A^+)$.

(b) Prove that if M is the predecessor of the natural number N, then $M = \cup N$.

†1.6.12. **(a)** Prove that if A is a transitive set, then so is A^+.

(b) Prove that if B is a set of transitive sets, then $\cup B$ is transitive.

1.6.13. For an arbitrary set A, let "$A^\star$" denote $\{A\}$. Then

$$0, 0^\star, (0^\star)^\star, \ldots$$

are each sets. On the basis of the axioms thus far introduced we cannot establish the sethood of the collection of all sets obtained from 0 in this way. (In Section 5.3 we will introduce an axiom that will yield this sethood.) By modifying the procedure in this section, introduce an axiom that would guarantee this sethood. (Of course, we will not assume this axiom in what follows.)

1.7 Axiom of Power Set

Axiom of the Power Set: *Let A be a set. Then there exists a set, $\mathscr{P}(A)$, whose elements are precisely the subsets of A.*

Thus, *a set B is in $\mathscr{P}(A)$ iff $B \subseteq A$.* The set $\mathscr{P}(A)$ is known as the **power set of** A. *The sets A and $\varnothing$ are always among the elements of $\mathscr{P}(A)$. Moreover, $a \in A$ iff $\{a\} \in \mathscr{P}(A)$.*

†EXAMPLE 1.7.1. Suppose that $A = \{a, b, c\}$. Then

$$\mathscr{P}(A) = \{\varnothing, \{a\}, \{b\}, \{c\}, \{a, b\}, \{a, c\}, \{b, c\}, A\}.$$

This is true regardless of whether or not a, b, and c are distinct.

In the remainder of the section, we will take the liberty of discussing certain notions (indicated by quotation marks) before they have been properly introduced.

EXAMPLE 1.7.2. Consider the subcollection of 6 consisting of the natural numbers 1, 2, 4, and 5. In order to establish that this collection is a set S, we consider either the formula

$$s_1(x) : x = 1 \text{ or } x = 2 \text{ or } x = 4 \text{ or } x = 5$$

or, alternatively, the formula

$$s_2(x) : x \neq 0 \text{ and } x \neq 3.$$

Thus, $S = \{x \in 6 : s_1(x)\} = \{x \in 6 : s_2(x)\}$. (We could also have utilized the Axioms of Pairing and Union, as in Exercise 1.6.10, to establish the sethood of S.) Therefore, $S \in \mathscr{P}(6)$. Similarly, we find that such collections as

$$\{1, 3, 4\}, \{0, 2, 4, 5\}, \{1, 2, 3, 4, 5\}$$

are subsets of 6. "How many" elements are in $\mathscr{P}(6)$? in $\mathscr{P}(N)$, $N \in \omega$?

If A is a "finite" set, the preceding example suggests the possibility of establishing the sethood of $\mathscr{P}(A)$ without utilizing the Axiom of the Power Set. The difficulty with this suggestion is that we will utilize the Power Set Axiom to define "finite". (Precise definitions of "finite" and "infinite" will be given in Section 2.4, where we will reconsider this matter.) The next example indicates a further need for the new axiom.

EXAMPLE 1.7.3. Among the elements of $\mathscr{P}(\omega)$ are the distinct sets

$$\{0\}, \{1\}, \{2\}, \ldots;$$
$$\{0, 1\}, \{0, 2\}, \{0, 3\}, \ldots;$$
$$\omega \sim 0, \omega \sim 1, \omega \sim 2, \ldots.$$

Thus, clearly, $\mathscr{P}(\omega)$ contains "infinitely many" elements. We can determine "finite" subsets of ω much the same as we did subsets of 6 in the preceding example. We have indicated some "infinite" subsets of ω, of whose existence we already know. The existence of other subsets of ω—such as that of "the prime numbers"—will follow readily, once we have established the usual arithmetic processes. However, there will still remain the question of specifying *all* subsets of ω. This is an extremely difficult question. Recall that in order to utilize the Axiom of Subsets, we must specify a formula. We will discuss this question further in Section 4.1 and in Chapter 6.

Now that we have introduced the Power Set Axiom, we are in a better position to understand the "generation" of sets in accordance with our intuition. Let

$$\mathbf{R}_0 = 0;$$
$$\mathbf{R}_1 = \mathscr{P}(\mathbf{R}_0) = \mathscr{P}\{0\} = 1;$$
$$\mathbf{R}_2 = \mathscr{P}(\mathbf{R}_1) = \mathscr{P}(1) = \{0, 1\} = 2;$$
$$\mathbf{R}_3 = \mathscr{P}(\mathbf{R}_2) = \mathscr{P}(2) = \{0, 1, \{1\}, 2\} = 3 \cup \{1\}.$$

In general,

$$\mathbf{R}_{N^+} = \mathscr{P}(\mathbf{R}_N).$$

For each nonzero N, $\mathbf{R}_N$ is a "2^{N-1}-element" set. We could also apply this process beginning with ω. Thus, let

$$\mathbf{S}_0 = \omega;$$
$$\mathbf{S}_1 = \mathscr{P}(\omega);$$
$$\mathbf{S}_2 = \mathscr{P}(S_1) = \mathscr{P}(\mathscr{P}(\omega)).$$

In general,

$$\mathbf{S}_{N^+} = \mathscr{P}(\mathbf{S}_N) = \mathscr{P}(\ldots \mathscr{P}(\mathscr{P}(\omega)) \ldots).$$

If we were to think of "sets" as being elements of either an $\mathbf{R}_N$ or of an $\mathbf{S}_N$, we would have sufficient sets at our disposal to do large chunks of mathematics. (See Exercise 3.12.10.) We are not yet in a position to extend this method, nor even to "take the union" of all the $\mathbf{R}_N$ and $\mathbf{S}_N$. In Section 5.3 we will introduce an axiom that will permit us to "collect" all these sets. In Section 6.1 we will extend the hierarchy of $\mathbf{R}_N$'s by utilizing "ordinal numbers"—a generalization of the natural numbers. There we will introduce an axiom that asserts that every set is an element of an $\mathbf{R}_\alpha$ for some ordinal number α. This axiom guarantees that the only sets are those that arise essentially as we have indicated.

Exercises

1.7.1. Let $A = \{1, 2, 3, 4\}$. Determine $\mathscr{P}(A)$.

1.7.2. Let $A = \{1, \{1, 2\}, 2, \{\{1, 2\}, \{1\}\}\}$. Determine $\mathscr{P}(A)$.

1.7.3. Suppose that $A = \{a, b\}$. Determine $\mathscr{P}(\mathscr{P}(A))$.

1.7.4. Prove that for every set X, $2 \in \mathscr{P}(\mathscr{P}(\mathscr{P}(X)))$.

1.7.5. Let $A = \{\{3, 4, 5\}\}$. Determine $\mathscr{P}(A)$, $\mathscr{P}(\cup A)$.

†1.7.6. Let X be an arbitrary set. Which of the following are necessarily true?

(a) $\{X\} \in \mathscr{P}(X)$. **(b)** $X \subseteq \mathscr{P}(X)$.

(c) $\{X\} \subseteq \mathscr{P}(X)$. **(d)** $X \notin \mathscr{P}(X)$.

†1.7.7. **(a)** Prove that $A \subseteq B$ iff $\mathscr{P}(A) \subseteq \mathscr{P}(B)$.

(b) Prove that $A \subset B$ iff $\mathscr{P}(A) \subset \mathscr{P}(B)$.

†1.7.8. Prove that every set A has properties (a) and (b).

(a) $\cup(\mathscr{P}(A)) = A$.

(b) $A \subseteq \mathscr{P}(\cup A)$.

(c) Determine sets A and B for which $A = \mathscr{P}(\cup A)$ but $B \subset \mathscr{P}(\cup B)$.

†1.7.9. Let A and B be arbitrary sets.

(a) Prove that $\mathscr{P}(A) \cup \mathscr{P}(B) \subseteq \mathscr{P}(A \cup B)$.

(b) Determine a necessary and sufficient condition for equality in (a).

(c) Prove (b).

1.7.10. For which natural numbers N is $\mathscr{P}(N)$

(a) inductive?

(b) a natural number?

(c) transitive?

(d) ordinary?

†1.7.11. Let A be a set.

(a) Prove that A is transitive iff $A \subseteq \mathscr{P}(A)$.

(b) Does the transitivity of A imply the transitivity of $\mathscr{P}(A)$?

1.7.12. Without utilizing the Axiom of the Power Set, show that the collection denoted by "$\mathscr{P}(2)$" is a set.

1.7.13. **(a)** Show that the collection **E** consisting of all "even numbers",

$$0, 2, 4, \ldots,$$

is a set.

(b) Show that the collection **O** consisting of all "odd numbers",

$$1, 3, 5, \ldots,$$

is a set.

1.8 Algebra of Sets

Let A be any nonempty set. Define $\cap A$ to be $\{b : \text{for every } a \text{ in } A, b \in a\}$. The set $\cap A$ is called **intersection** A. We note that it is unnecessary to introduce a new axiom in order to guarantee that $\cap A$ is a set; this is assured by the Axiom of Subsets.

EXAMPLE 1.8.1. As in Exercise 1.1.1, suppose that a, b, c, d, e, f, and g are distinct sets and $A = \{a, c, e\}$, $B = \{a, e\}$, $C = \{b, d, \varnothing\}$, and $D = \{a, b, c, d, e\}$. Then $\cap\{A, B, C, D\} = \varnothing$ and $\cap\{A, B, D\} = \cap\{A, B\} = \cap\{B\} = B$. We cannot determine $\cap B$ without further information about the sets a and e.

Note that *if* $\varnothing \in A$, *then* $\cap A = \varnothing$.

In case $A = \{B, C\}$, we write "$B \cap C$" in lieu of "$\cap A$"; we speak of "**the intersection of** A **and** B". Thus, if $a \neq d$, then $\{a, b, c\} \cap \{b, c, d\} = \{b, c\}$. Clearly, $B \cap C = B$ *iff* $B \subseteq C$.

EXAMPLE 1.8.2. Let $A = \{0, 1, 5\}$, $B = \{0, 1, \{1\}\}$, and $C = \{\{1\}, \{5\}\}$.

(1) $A \cap B = \{0, 1\} = 2$.
(2) $A \cap 3 = 2$.
(3) $A \cap 6 = A$.
(4) $B \cap 6 = 2$.
(5) $A \cap C = \varnothing$.
(6) $B \cap C = \{\{1\}\}$.

The sets A **and** B **are** said to be **disjoint** if $A \cap B = \varnothing$; thus A and C of the preceding example are disjoint. Also, X is said to be **a set of pairwise disjoint sets** if for every $A, B(\neq A)$ in X, $A \cap B = \varnothing$. For example, the set $X = \{\{1\}, \{2\}, \{3\}\}$ is a set of pairwise disjoint sets.

When we confine our attention to the elements and subsets of a particular set, we call the set in question a **universe** and denote it by "$\mathcal{U}$". Thus, in discussing natural numbers and sets of these, $\mathcal{U}$ would be ω. The set $\mathcal{U}$ is *not* to be thought of as the set of all sets. Indeed, no such set exists, by Cantor's Paradox (Exercise 1.4.3); $\mathcal{U}$ is simply the set of all sets under consideration. When our universe is fixed, it is not necessary to write "$x \in \mathcal{U}$" in descriptions of sets; thus, when $\mathcal{U} = \omega$, "$\{X : 1 \in X\}$" will mean "$\{2, 3, 4, \ldots\}$".

No matter which universe $\mathcal{U}$ is considered, we always have $\varnothing \subseteq \mathcal{U}$ (but not necessarily $\varnothing \in \mathcal{U}$).

If A is a subset of a universe $\mathcal{U}$ we denote by "$\tilde{A}$" the set of elements of $\mathcal{U}$ that are not elements of A. Thus, $\tilde{A} = \{x : x \notin A\}$; we call $\tilde{A}$ the **complement of** A. If B is also a subset of $\mathcal{U}$, we let "$B \sim A$" stand for "$B \cap \tilde{A}$"; we call $B \sim A$ the **relative complement of** A **in** B. The sets *A and $B \sim A$ are always disjoint. Moreover,* $\tilde{A} = \mathcal{U} \dot{\sim} A$ *and* $\tilde{\tilde{A}} = A$.

EXAMPLE 1.8.3. Let a, b, c, d, and e be distinct sets and let $A = \{a, b, c, d\}$, $B = \{a, c\}$, and $C = \{a, b, e\}$.

(1) $A \sim B = \{b, d\}$.
(2) $B \sim A = \varnothing$.
(3) $A \sim C = \{c, d\}$.
(4) $C \sim A = \{e\}$.
(5) $B \sim C = \{c\}$.
(6) $C \sim B = \{b, e\}$.

We conclude Chapter 1 by considering the following laws governing union, intersection, and complementation.

Theorem 1.8.1. *Let P, Q, and R be subsets of a universe $\mathcal{U}$.*

(1) **(Commutative laws)**
   (a) $P \cup Q = Q \cup P$.
   (b) $P \cap Q = Q \cap P$.
(2) **(Associative laws)**
   (a) $(P \cup Q) \cup R = P \cup (Q \cup R)$.
   (b) $(P \cap Q) \cap R = P \cap (Q \cap R)$.
(3) **(Distributive laws)**
   (a) $P \cup (Q \cap R) = (P \cup Q) \cap (P \cup R)$.
   (b) $P \cap (Q \cup R) = (P \cap Q) \cup (P \cap R)$.
   (c) $P \cup (\cap Q) = \cap\{P \cup q : q \in Q\}$.
   (d) $P \cap (\cup Q) = \cup\{P \cap q : q \in Q\}$.

(4) **(De Morgan's Laws)**

(a) $\widetilde{P \cup Q} = \tilde{P} \cap \tilde{Q}$.

(b) $\widetilde{P \cap Q} = \tilde{P} \cup \tilde{Q}$.

(5) **(Laws for Relative Complements)**

(a) $(P \sim Q) \sim R = P \sim (Q \cup R)$.

(b) $P \sim (Q \sim R) = (P \sim Q) \cup (P \cap R)$.

(c) $(P \sim Q) \cup R = (P \cup R) \sim (Q \sim R)$.

(d) $(P \sim Q) \cap R = (P \cap R) \sim (Q \cap R)$.

Rules (2a) and (2b) generalize to any "finite number" of sets and to any combinations of parentheses. We therefore agree to omit parentheses, and we write "$P \cup Q \cup R$" for both "$(P \cup Q) \cup R$" and "$P \cup (Q \cup R)$". Similarly, we write "$P \cap Q \cap R \cap S$" for "$(P \cap (Q \cap R)) \cap S$", for "$(P \cap Q) \cap (R \cap S)$", etc.

PROOF: We prove parts (1a), (2b), (3a), (4a), and (5b); other parts are left as an exercise.

(1a) $a \in P \cup Q$ iff $a \in P$ or $a \in Q$, iff $a \in Q$ or $a \in P$, iff $a \in Q \cup P$.

(2b) $a \in (P \cap Q) \cap R$ iff $a \in P \cap Q$ and $a \in R$; iff $a \in P$ as well as $a \in Q$, and $a \in R$; iff $a \in P$, as well as $a \in Q$ and $a \in R$; iff $a \in P$ as well as $a \in Q \cap R$; iff $a \in P \cap (Q \cap R)$.

(3a) We note that for sentences s_1, s_2, and s_3, the truth of "s_1 or s_2—and—s_1 or s_3" assures that both "s_1 or s_2" and "s_1 or s_3" are true. This is the case iff either s_1 is true or else both s_2 and s_3 are true. Thus, $a \in P \cup (Q \cap R)$ iff $a \in P$ or $a \in Q \cap R$, iff $a \in P$—or both $a \in Q$ and $a \in R$, iff $a \in P$ or $a \in Q$—and $a \in P$ or $a \in R$, iff $a \in P \cup Q$ and $a \in P \cup R$, iff $a \in (P \cup Q) \cap (P \cup R)$.

(4a) $a \in \widetilde{P \cup Q}$ iff $a \notin P \cup Q$, iff $a \notin P$ and $a \notin Q$, iff $a \in \tilde{P}$ and $a \in \tilde{Q}$, iff $a \in \tilde{P} \cap \tilde{Q}$.

(5b) $P \sim (Q \sim R) = P \cap (\widetilde{Q \sim R}) = P \cap (\widetilde{Q \cap \tilde{R}}) = P \cap (\tilde{Q} \cup \tilde{\tilde{R}}) =$ $P \cap (\tilde{Q} \cup R) = (P \cap \tilde{Q}) \cup (P \cap R) = (P \sim Q) \cup (P \cap R)$. [We have utilized (3b), whose proof is similar to (3a).]

Exercises

1.8.1. What happens if we try to use the same definition of $\cap A$ when $A = \varnothing$?

1.8.2. Let

$$A = 9; \qquad B = \{0, 3, 5, 8, 10\};$$

$$C = \{2, 4, 8, 9, 10\}; \qquad D = \{0, 2, 3, 7\}.$$

Determine the sets:

(a) $B \cap C$. **(b)** $A \cap B \cap C$.
(c) $B \cap C \cap D$. **(d)** $\cap(\cup\{B, C, D\})$.
(e) $A \sim B$. **(f)** $(A \sim B) \sim C$.
(g) $(A \sim B) \sim (C \sim D)$. **(h)** $((A \sim B) \sim C) \sim D$.

1.8.3. Let $A = \{\{1, 4\}, 3, 4, \{\{2\}\}, \{1, \{4\}\}\}$. Determine the sets:
(a) $A \sim 4$. **(b)** $A \sim 5$.
(c) $(\cup A) \sim 4$. **(d)** $(\cup\cup A) \sim 4$.

1.8.4. Determine the sets:
(a) $4 \sim 2$. **(b)** $\cap\{1, \{1\}\}$. **(c)** $\cap\{2, \{1\}\}$.

1.8.5. Let N ($\neq 0$) be a natural number. Determine $\cap N$.

1.8.6. Determine distinct sets A, B, and C such that $\cap\{A, B, C\} = \varnothing$, yet any two of these sets have a nonempty intersection.

†1.8.7. Prove that for all sets A and B the following are true:
(a) $A \subseteq A \cup B$.
(b) $A \cap B \subseteq A$.
(c) $\mathscr{P}(A) \cap \mathscr{P}(B) = \mathscr{P}(A \cap B)$.

†1.8.8. **(a)** Prove that for all sets A and B,

$$\mathscr{P}(A \sim B) \sim \{0\} \subseteq \mathscr{P}(A) \sim \mathscr{P}(B).$$

(b) Determine a necessary and sufficient condition for equality in (a).
(c) Prove (b).
(d) Prove that if $A \subseteq B$, then $\mathscr{P}(A) \sim \mathscr{P}(B) \in \mathscr{P}(A \sim B)$.

†1.8.9. Define $B \sim A$ without referring to a universe.

1.8.10. To which regions of a Venn diagram (Figure 1.8.1) do the sets in (1a) and in (4a) of Theorem 1.8.1 correspond?

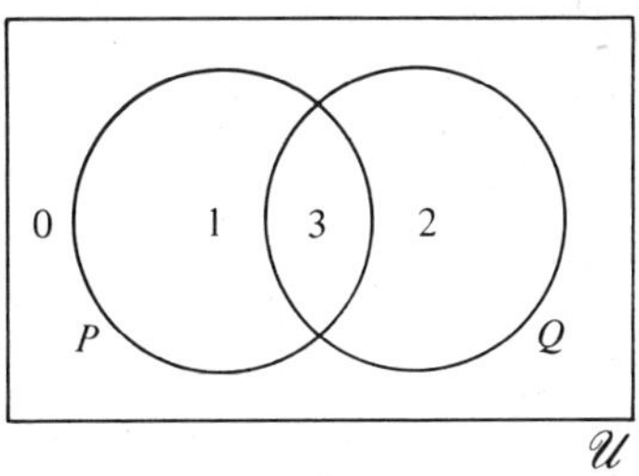

Fig. 1.8.1

1.8.11. To which regions of a Venn diagram (Figure 1.8.2) do the sets in (2b), (3a), and (5b) of Theorem 1.8.1 correspond?

1.8.12. In (3c) of Theorem 1.8.1, how do we know that the collection of $P \cup q$, for q in Q, is a set?

1.8.13. Prove Theorem 1.8.1, parts (3c), (3d), (4b), and (5c).

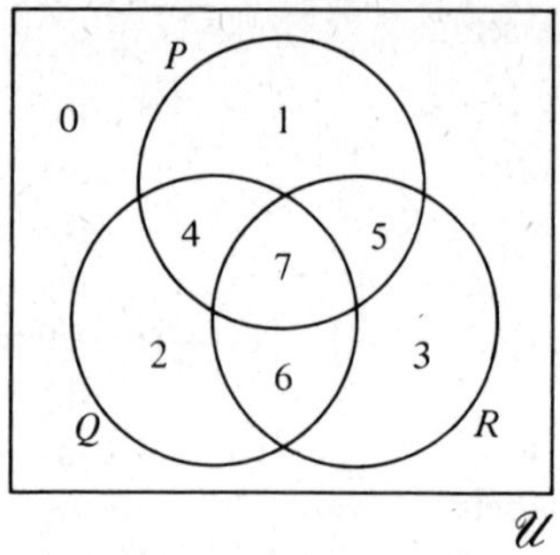

Fig. 1.8.2

†1.8.14. Prove the following **Generalization of De Morgan's Laws**: Suppose P is a subset of a universe. Then
(a) $\widetilde{\cup P} = \cap\{\tilde{p} : p \in P\}$, (b) $\widetilde{\cap P} = \cup\{\tilde{p} : p \in P\}$.

2

Relations

2.1 Ordered Pairs

Let a and b be sets. Without adding any new axioms, we can now introduce a set by means of which we will eventually express the notion "a precedes b." We define the **ordered pair** $\langle a, b\rangle$ to be "$\{\{a\}, \{a, b\}\}$".

Thus, ordered pairs are defined by means of three applications of the Pairing Axiom—we form the sets $\{a, a\}\,(= \{a\})$, $\{a, b\}$, and finally, $\{\{a\}, \{a, b\}\}$ $(= \langle a, b\rangle)$. Here, a is called the **first coordinate** and b the **second coordinate of** $\langle a, b\rangle$. Note that *if* $a = b$, *then* $\langle a, b\rangle = \langle b, a\rangle = \{\{a\}\}$, whereas *if* $a \neq b$, *then* $\langle a, b\rangle$ *and* $\langle b, a\rangle$ *are different*. Moreover, *if* a *and* b *are both ordinary and* $a \neq b$, *then the sets* $\{a, b\}$, $\langle a, b\rangle$, *and* $\langle b, a\rangle$ *are all different. In all cases* $\cup\langle a, b\rangle = \{a, b\}$.

EXAMPLE 2.1.1. $\langle 2, 4\rangle$ is the set $\{\{2\}, \{2, 4\}\}$; $\langle 2, 4\rangle \neq \{2, 4\}$; $\langle 2, 4\rangle \neq \{2, \{2, 4\}\}$.

Theorem 2.1.1. *For all sets a, b, c, and d,*

$$\langle a, b\rangle = \langle c, d\rangle \text{ iff both } a = c \text{ and } b = d.$$

PROOF: Suppose that $a = c$ and $b = d$; then $\{a\} = \{c\}$ and $\{a, b\} = \{c, d\}$. Thus, $\langle a, b\rangle = \{\{a\}, \{a, b\}\} = \{\{c\}, \{c, d\}\} = \langle c, d\rangle$.

Suppose that $\langle a, b\rangle = \langle c, d\rangle$; i.e., suppose that

$$\{\{a\}, \{a, b\}\} = \{\{c\}, \{c, d\}\}.$$

Then

$$\{a\} = \cap\{\{a\}, \{a, b\}\} = \cap\{\{c\}, \{c, d\}\} = \{c\};$$

hence, $a = c$. Similarly,

$$\{a, b\} = \cup\{\{a\}, \{a, b\}\} = \cup\{\{c\}, \{c, d\}\} = \{c, d\}.$$

If $a = b$ or $c = d$, then $a = b = c = d$. Otherwise,

$$\{b\} = \{a, b\} \sim \{a\} = \{c, d\} \sim \{c\} = \{d\} \qquad \text{and} \qquad b = d.$$

Exercises

2.1.1. Determine $\langle 0, 1\rangle \cap 3$.

2.1.2. Let a and b be distinct sets.

(a) Show that if a and b are ordinary, then $\{a, b\}$, $\langle a, b\rangle$, and $\langle b, a\rangle$ are all distinct.

(b) Which of the sets $\{a, b\}$, $\langle a, b\rangle$, and $\langle b, a\rangle$ are necessarily distinct from the others in each case?

(1) a is ordinary, but b is not?

(2) b is ordinary, but a is not?

(3) Neither a nor b is ordinary?

2.1.3. Let a and b be distinct ordinary sets. Some of the following sets are equal to others. Match all equal sets.

(1) $\langle a, b\rangle$. (2) $\{a, b\}$.

(3) $\langle a, \{b\}\rangle$. (4) $\{\{a, b\}\}$.

(5) $\{\{a\}, \{b, a\}\}$. (6) $\{\{a, b\}, \{b\}\}$.

(7) $\{\{a, b\}, \{a\}\}$. (8) $\langle b, a\rangle$.

(9) $\{b, a\}$. (10) $\{\{a\}, \{a, \{b\}\}\}$.

2.1.4. Determine an ordered pair equal to $\{\{\{a\}\}\}$.

2.1.5. Let a and b be sets. Determine

(a) $\cap\langle a, b\rangle$. **(b)** $\mathscr{P}(\langle a, b\rangle)$.

(c) $\langle a, b\rangle \cap \langle b, a\rangle$.

2.1.6. We know that the natural numbers $1(=\{0,0\})$ and $2(=\{0,1\})$ are (unordered) pairs.
(a) Show that no natural number is an ordered pair. (Used in Section 3.4.)
(b) Determine all subsets of ω that are ordered pairs.

★2.1.7. Use the Axiom of Pairing only once (together with other Axioms) to justify that if a and b are sets, then $\langle a, b\rangle$ is a set.

2.2 Cartesian Products

Let A and B be sets. The **Cartesian product of** A **and** B, written "$A \times B$", is defined to be the collection of ordered pairs $\langle a, b\rangle$ such that $a \in A$ and $b \in B$. As an exercise we show that *$A \times B$ is a set*. Also, *$A \times B = \varnothing$ iff either $A = \varnothing$ or $B = \varnothing$*. We write "$A^{②}$" for "$A \times A$".

The notion of Cartesian product is familiar to students of analytic geometry. There, we identify the points, P, of the plane with ordered pairs of real numbers, $\langle x, y\rangle$; we think of the (real) plane as $X \times Y$, where X and Y each represent the real numbers.

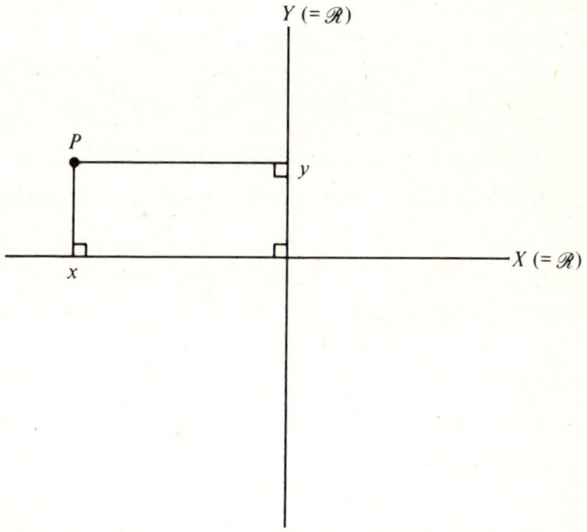

Fig. 2.2.1

If $s(x)$ is a formula that has x as its unique free variable, we will generally write "$\{\langle a, b\rangle \in A \times B : s\}$" or (when $A \times B$ is understood) "$\{\langle a, b\rangle : s\}$" instead of "$\left\{c \in A \times B : s\begin{pmatrix}x\\c\end{pmatrix}\right\}$".

EXAMPLE 2.2.1. Let a, b, c, and d be sets, let $A = \{a, b, c\}$, and let $B = \{a, c, d\}$.

$$A \times B = \{\langle a,a\rangle, \langle a,c\rangle, \langle a,d\rangle, \langle b,a\rangle, \langle b,c\rangle, \langle b,d\rangle, \langle c,a\rangle, \langle c,c\rangle, \langle c,d\rangle\}.$$

$$B \times A = \{\langle a,a\rangle, \langle a,b\rangle, \langle a,c\rangle, \langle c,a\rangle, \langle c,b\rangle, \langle c,c\rangle, \langle d,a\rangle, \langle d,b\rangle, \langle d,c\rangle\}.$$

EXAMPLE 2.2.2

$$\omega \times 2 = \{\langle 0,0\rangle, \langle 0,1\rangle, \langle 1,0\rangle, \langle 1,1\rangle, \langle 2,0\rangle, \langle 2,1\rangle, \ldots, \langle N,0\rangle, \langle N,1\rangle, \ldots\}.$$

$$2 \times \omega = \{\langle 0,0\rangle, \langle 1,0\rangle, \langle 0,1\rangle, \langle 1,1\rangle, \langle 0,2\rangle, \langle 1,2\rangle, \ldots, \langle 0,N\rangle, \langle 1,N\rangle, \ldots\}.$$

Theorem 2.2.1. The Principle of Double Induction. *Let A be a subset of $\omega^{②}$ with the following properties:*
(1) *For every N in ω, both $\langle 0, N\rangle \in A$ and $\langle N, 0\rangle \in A$.*
(2) *Whenever $\langle M, N\rangle \in A$, it follows that $\langle M^+, N^+\rangle \in A$.*
Then $A = \omega^{②}$.

(See Figure 2.2.2a.) Here, the assumption that $\langle M, N\rangle \in A$ [in (2)] is called the **inductive hypothesis.**

PROOF: Let $B = \{M \in \omega$: for every N in $\omega, \langle M, N\rangle \in A\}$. We utilize the Principle of Induction to show that $B = \omega$.

By assumption (1), $0 \in B$. Suppose that $M \in B$. Let

$$C_M = \{N \in \omega : \langle M^+, N\rangle \in A\}.$$

Then $0 \in C_M$, by (1). If N is an arbitrary nonzero natural number, then $N = K^+$ for some unique K in ω (Theorem 1.6.5). By the assumption that $M \in B$, it follows that $\langle M, K\rangle \in A$. By (2), $\langle M^+, K^+\rangle = \langle M^+, N\rangle \in A$, and thus $N \in C_M$. Therefore, $C_M = \omega$; hence, $M^+ \in B$. By induction, $B = \omega$. Thus, for every $\langle M, N\rangle$ in $\omega^{②}$, $\langle M, N\rangle \in A$. Because A is a subset of $\omega^{②}$, it follows that $A = \omega^{②}$.

We next consider a variant of the Principle of Double Induction that, although less interesting than Theorem 2.2.1, is nevertheless useful in obtaining other forms of this principle.

Theorem 2.2.2. *Let A be a subset of $\omega^{②}$ that has the following properties:*
(1) *For every N in ω, both $\langle 0, N\rangle \in A$ and $\langle N, 0\rangle \in A$.*
(2) *Whenever $\langle M, N\rangle \in A$, $\langle M^+, N\rangle \in A$ and $\langle M, N^+\rangle \in A$, it follows that $\langle M^+, N^+\rangle \in A$.*
Then $A = \omega^{②}$.

(See Figure 2.2.2b.) Theorem 2.2.2 can be proved by means of a minor modification of the proof of Theorem 2.2.1. We leave the details of this as an exercise.

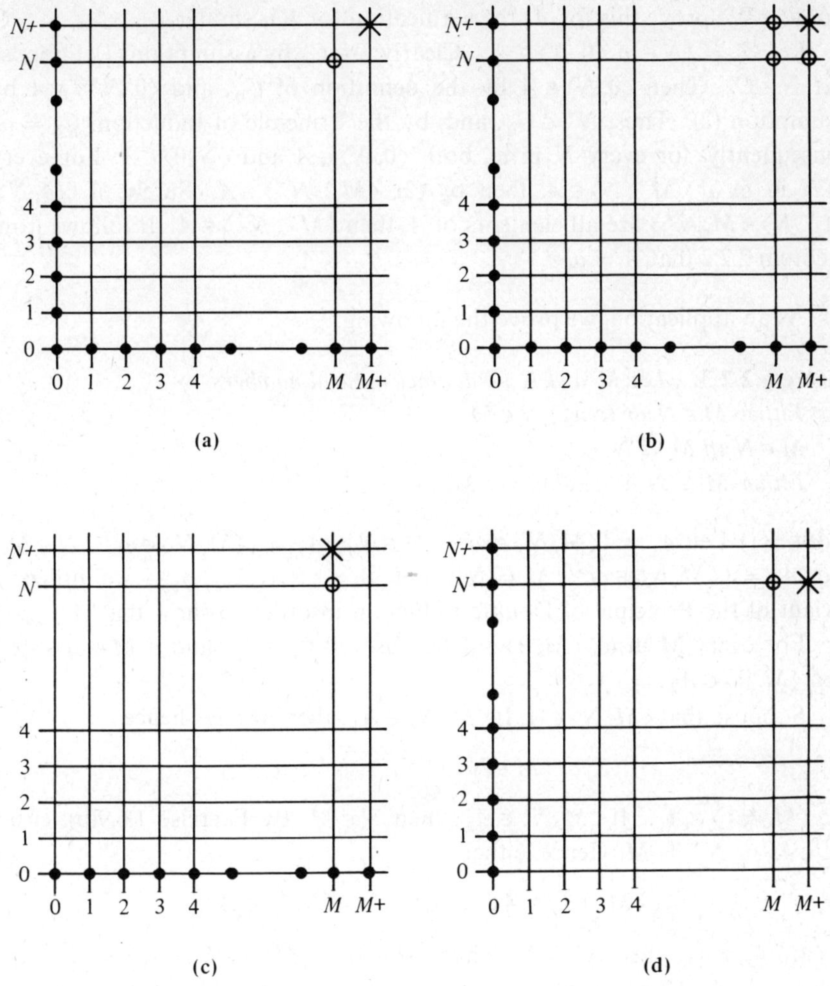

Fig. 2.2.2

Corollary. *The following are variants of the Principle of Double Induction: Let A be a subset of* $\omega^{②}$ *with the following properties:*

(1) *For every M in* ω, $\langle M, 0\rangle \in A$.

(2) *If* $\langle M, N\rangle \in A$, *then* $\langle M, N^+\rangle \in A$.

Let B be a subset of $\omega^{②}$ *with the following properties:*

(1′) *For every N in* ω, $\langle 0, N\rangle \in B$.

(2′) *If* $\langle M, N\rangle \in B$, *then* $\langle M^+, N\rangle \in B$.

Then $A = B = \omega^{②}$.

(See Figure 2.2.2c for the first variant and Figure 2.2.2d for the second.)

PROOF: We prove this for A; the verification for B is similar.

Let $C_0 = \{N \in \omega : \langle 0, N\rangle \in A\}$. Clearly, $0 \in C_0$ by assumption (1). Suppose that $N \in C_0$. Then $\langle 0, N\rangle \in A$ by the definition of C_0, and $\langle 0, N^+\rangle \in A$ by assumption (2). Thus, $N^+ \in C_0$, and, by the Principle of Induction, $C_0 = \omega$. Consequently, for every N in ω, both $\langle 0, N\rangle \in A$ and $\langle N, 0\rangle \in A$. For every M, N in ω, if $\langle M^+, N\rangle \in A$, then by (2), $\langle M^+, N^+\rangle \in A$. Surely, if $\langle M, N\rangle$, $\langle M^+, N\rangle, \langle M, N^+\rangle$ are all elements of A, then $\langle M^+, N^+\rangle \in A$. It follows from Theorem 2.2.2 that $A = \omega^{②}$.

As an application, we prove the following.

Theorem 2.2.3. *Let M and N be distinct natural numbers.*
(1) *Either $M \in N$ or (excl.) $N \in M$.*
(2) *$M \in N$ iff $M \subset N$.*
(3) *Either $M \subset N$ or (excl.) $N \subset M$.*

PROOF: (1) Let $A_1 = \{\langle M, N\rangle \in \omega^{②} : M \in N\}$, $A_2 = \{\langle M, N\rangle \in \omega^{②} : N \in M\}$, and $A_3 = \{\langle M, N\rangle \in \omega^{②} : M = N\}$. Let $A = \cup\{A_1, A_2, A_3\}$; we utilize a variant of the Principle of Double Induction in order to show that $A = \omega^{②}$.

For every M in ω, $\langle M, 0\rangle \in A$ because $\langle 0, 0\rangle \in A_3$; and if $M \in \omega \sim \{0\}$, then $\langle M, 0\rangle \in A_2$.

Suppose that $\langle M, N\rangle \in A$. If $\langle M, N\rangle \in A_1$, then $M \in N$; hence,

$$M \in N \cup \{N\} = N^+;$$

i.e., $\langle M, N^+\rangle \in A_1$. If $\langle M, N\rangle \in A_2$, then $N \in M$. By Exercise 1.6.9(b), either $N^+ \in M$ or $N^+ = M$. Hence, either

$$\langle M, N^+\rangle \in A_2 \qquad \text{or} \qquad \langle M, N^+\rangle \in A_3.$$

If $\langle M, N\rangle \in A_3$, then $M = N$. Thus, $M \in M \cup \{M\} = N \cup \{N\} = N^+$, and $\langle M, N^+\rangle \in A_1$. Consequently, $\langle M, N\rangle \in A$ implies $\langle M, N^+\rangle \in A$, and $A = \omega^{②}$.

Clearly, $A_1 \cap A_2 = \varnothing$, as otherwise, we would have natural numbers M and N for which $M \in N$ and $N \in M$. This would contradict Corollary 2 of Theorem 1.6.3.

(2) Let M and N be distinct natural numbers. Suppose that $M \in N$. Then $M \subseteq N$ because N is transitive, and $M \subset N$ because N is ordinary.

Suppose that $M \subset N$. Then $M \neq N$. Hence $M \in N$ or (excl.) $N \in M$. If $N \in M$, then $N \subset M$. Because this is not the case, $M \in N$.

(3) is a direct consequence of (1) and (2).

EXAMPLE 2.2.3. Consider the set $B = \{0, 1, \{1\}\}$. In Example 1.6.4, we utilized Theorem 1.6.5 to show that B is not a natural number. Now this follows because $2 \notin B$ and $B \notin 2$.

Exercises

†**2.2.1.** Show that if A and B are arbitrary sets, then $A \times B$ is also a set.

2.2.2. Let $A = 3$ and $B = 2$. Determine the following sets:

(a) $A \times B$. **(b)** $B \times A$.

2.2.3. Let $A = \{1, 3, 6, 9\}$. Determine the subset of $A^{②}$ consisting of all ordered pairs $\langle a, b\rangle$ in each case.

(a) $a \in b$. **(b)** $a = b$. **(c)** $a \in b$ or $b \in a$.

†**2.2.4.** **(a)** Prove that for arbitrary sets A and B, $A \cup B = \cup\cup(A \times B)$.

(b) Prove that for each set A, $A = \cup\cup(A^{②})$.

†**2.2.5.** Let A, B, C, and D be sets.

(a) Suppose that $A \times B = B \times A$. Under what conditions does it follow that $A = B$?

(b) Suppose that $A \times B = C \times D$. Under what conditions does it follow that $A = C$ and $B = D$?

(c) Suppose that $A \times A = B \times B$. Prove that $A = B$.

†**2.2.6.** Prove that for arbitrary sets A, B, and C the following are true:

(a) $A \times (B \cup C) = (A \times B) \cup (A \times C)$.

(b) $A \times (B \cap C) = (A \times B) \cap (A \times C)$.

2.2.7. Determine sets A, B, and C in each case.

(a) $A \cup (B \times C) \neq (A \cup B) \times (A \cup C)$.

(b) $(A \times B) \times C \neq A \times (B \times C)$.

†**2.2.8.** Prove that if $A\ (\neq \varnothing)$, B, and C are sets such that $A \times B = A \times C$, then $B = C$.

2.2.9. **(a)** Prove Theorem 2.2.2.

(b) Use Theorem 2.2.2 to prove Theorem 2.2.1.

2.2.10. Prove the following variant of the Principle of Double Induction: If A is any subset of $\omega^{②}$ that contains $\langle 0, 0\rangle$ and that contains $\langle M, N^+\rangle$ and $\langle M^+, N\rangle$ whenever it contains $\langle M, N\rangle$, then $A = \omega^{②}$. (Used in the proof of Exercise 3.2.2.)

2.3 Binary Relations

In analytic geometry we discuss graphs; these are depicted on the real plane. When we "draw a graph", we pictorially indicate a subcollection of $\mathscr{R}^{②}$, the Cartesian product of the reals with itself. The graph indicates which points of $\mathscr{R}^{②}$ are in the collection. Thus, the graph of the equation "$x^2 + y^2 = 25$" (Figure 2.3.1a) indicates a circle centered at the origin and of radius 5; the graph of the inequality "$x^2 + y^2 \leq 25$" (Figure 2.3.1b) indicates a disc centered at the origin and of radius 5 together with its boundary.

We now generalize this notion.

Let A and B be sets. A subset of $A \times B$ is called a **binary relation from** A **to (into)** B. A subset of $A^{②}$ is called a **binary relation on** A. We generally write

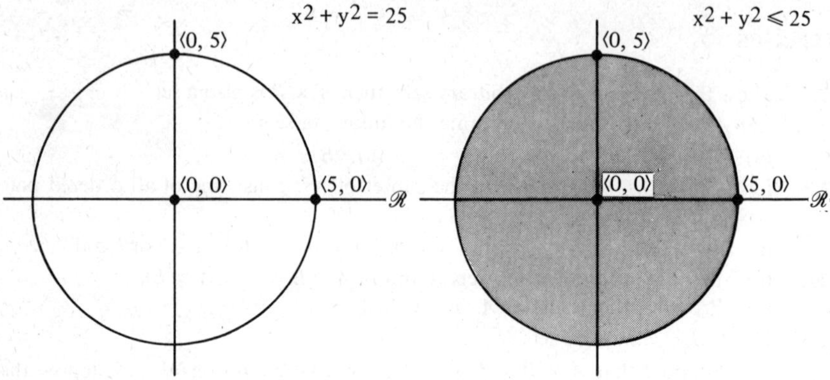

(a) (b)

Fig. 2.3.1

"relation" in lieu of "binary relation". If R is a relation on A, we speak of the **relational system** $\langle A, R\rangle$; if $C \subset A$, the subsystem $\langle C, R \cap C^{②}\rangle$ will often be denoted by "$\langle C, R\rangle$".

Relations arise in a natural way in life situations as well as in mathematics. For example, one common usage of the word "relation" indicates that among a group of people certain pairs have family connections. Specific relationships, such as "is the uncle of", designate ordered pairs of people.

In mathematics we may say that (the real number) a is the square of b. This indicates a relation from A to B, where A represents the collection of nonnegative real numbers and B, the collection of real numbers. Note that the same relation R can also be thought of as being on B because $R \subseteq A \times B \subseteq B^{②}$.

If X is the collection of circles in a plane, we can define a relation on X by saying that C_1 is related to C_2 provided that the two circles are concentric and that C_1 has the smaller radius.

Let R be a relation from A to B. We write "aRb" in case $\langle a, b\rangle \in R$; "$a\not{R}b$" stands for "$\langle a, b\rangle \notin R$". If aRb, we say that a **is R-related to** b and that b **is an R-relative of** a; also, a and b are called **R-comparable** if aRb or bRa. By the **domain of** R (dom R) we mean $\{a \in A : \text{for some } b,\ aRb\}$. By the **range of** R (range R) we mean $\{b \in B : \text{for some } a,\ aRb\}$. We let the **field of** R (field R) be dom $R \cup$ range R. Note that *dom $R \subseteq A$, range $R \subseteq B$, and field $R \subseteq A \cup B$. If dom $R \subseteq A'$ and range $R \subseteq B'$, then R is also a relation from A' to B'.*

Every set S of ordered pairs is a binary relation; we can define

$$\text{dom } S = \{a \in \cup\cup S : \text{for some } b,\ \langle a, b\rangle \in S\}$$

and

$$\text{range } S = \{b \in \cup\cup S : \text{for some } a,\ \langle a, b\rangle \in S\}.$$

Thus, S is a binary relation from dom S to range S or, for that matter, from A to B whenever A includes dom S and B includes range S.

EXAMPLE 2.3.1. Let $A = \omega$ and let $B = 3 = \{0, 1, 2\}$. Then each of the following are relations from A to B.

$$\varnothing;\ A \times B;$$

$$\begin{aligned} R_0 &= \{\langle N, 0\rangle : N \in \omega\} \\ &= \{\langle 0,0\rangle, \langle 1,0\rangle, \langle 2,0\rangle, \langle 3,0\rangle, \ldots\}; \\ R_1 &= \{\langle N, 1\rangle : N \in \omega\}; \\ R_2 &= \{\langle N, 2\rangle : N \in \omega\}; \\ R_3 &= \{\langle 0,0\rangle, \langle 2,1\rangle, \langle 5,1\rangle\}; \\ R_4 &= \{\langle 0,1\rangle, \langle 1,0\rangle, \langle 2,2\rangle, \langle 3,2\rangle\}. \end{aligned}$$

Dom $\varnothing$ = range $\varnothing = \varnothing$; dom $(A \times B)$ = dom R_0 = dom R_1 = dom R_2 = A; dom $R_3 = \{0, 2, 5\}$; dom $R_4 = 4$. Range $A \times B$ = range $R_4 = 3$; range $R_0 = 1$; range $R_1 = \{1\}$; range $R_2 = \{2\}$; range $R_3 = 2$.

EXAMPLE 2.3.2. Let $\mathscr{U}$ be a universe. Then each of the following is a relation on $\mathscr{U}$:

$$\begin{aligned} R_1 &= \{\langle A, B\rangle \in \mathscr{U}^{②} : A \in B\}. \\ R_2 &= \{\langle A, B\rangle \in \mathscr{U}^{②} : A \notin B\}. \\ R_3 &= \{\langle A, B\rangle \in \mathscr{U}^{②} : A = B\}. \\ R_4 &= \{\langle A, B\rangle \in \mathscr{U}^{②} : A \subset B\}. \end{aligned}$$

Note that $R_1 \cup R_2 = \mathscr{U}^{②}$ and $R_1 \cap R_2 = \varnothing$.

Let R be a relation from A to B. By the **inverse of** R, written "R^{-1}", we mean the collection of ordered pairs $\langle b, a\rangle$ for which $\langle a, b\rangle \in R$. Clearly, R^{-1} is a set; in fact, it is a relation from B to A. Moreover, $(R^{-1})^{-1} = R$ [Exercise 2.3.3(a)].

If R is a relation on the reals, then the graph of R^{-1} is obtained by rotating the graph of R about the line "$y = x$". (See Figure 2.3.2.)

Let R_1 be a relation from A to B and R_2 a relation from B to C. The collection of ordered pairs $\langle a, c\rangle$—such that there is some b in B for which both $\langle a, b\rangle \in R_1$ and $\langle b, c\rangle \in R_2$—is a relation from A to C [Exercise 2.3.1(b)]; it is known as the **composite relation of R_2 and R_1**—written "$R_2 \circ R_1$". We

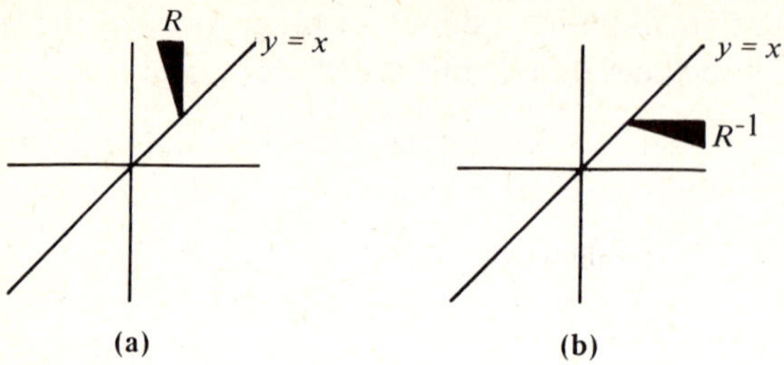

Fig. 2.3.2

note that $R_2 \circ R_1 = \varnothing$ *iff range* R_1 *and dom* R_2 *are disjoint*; in general, $R_2 \circ R_1 \neq R_1 \circ R_2$.

Among a group of men, "is the father of" is the inverse of "is the son of" and "is the brother of" is its own inverse; "is the paternal uncle of" is the composite relation $R_2 \circ R_1$, where R_1 denotes "is the father of" and R_2 denotes "is the brother of". What are $R_1 \circ R_2$, $R_1 \circ R_1$, $R_2 \circ R_2$, and $R_1^{-1} \circ R_2 \circ R_1$?

If R is a binary relation and X an arbitrary set, then $R''(X)$ is defined to be $\{b \in \text{range } R : \text{for some } a \text{ in } X, aRb\}$; *$R''(X)$ is thus the set of R-relatives of elements of X.*

$$R''(X) = \varnothing \text{ iff } X \cap \text{dom } R = \varnothing.$$

EXAMPLE 2.3.3. Let $R_1 = \{\langle N, N^+\rangle : N \in \omega\}$ and let

$$R_2 = \{\langle 2,0\rangle, \langle 0,1\rangle, \langle 0,2\rangle, \langle 5,3\rangle, \langle 4,4\rangle\}.$$

Then $R_1{}^{-1} = \{\langle N^+, N\rangle : N \in \omega\}$. Dom R_1 = range $R_1{}^{-1} = \omega$; range $R_1 = \text{dom } R_1{}^{-1} = \omega \sim \{0\}$. $R_2{}^{-1} = \{\langle 0,2\rangle, \langle 1,0\rangle, \langle 2,0\rangle, \langle 3,5\rangle, \langle 4,4\rangle\}$. Dom R_2 = range $R_2{}^{-1} = \{0, 2, 4, 5\}$; range $R_2 = \text{dom } R_2{}^{-1} = 5$.
$R_2 \circ R_1 = \{\langle 1,0\rangle, \langle 3,4\rangle, \langle 4,3\rangle\}$;
$R_1 \circ R_2 = \{\langle 2,1\rangle, \langle 0,2\rangle, \langle 0,3\rangle, \langle 5,4\rangle, \langle 4,5\rangle\}$.
$R_1''(4) = \{N^+ : N \in 4\} = \{1, 2, 3, 4\}$;
$R_2''(4) = \{N : \text{for some } M \text{ in } 4, \langle M, N\rangle \in R_2\} = \{0, 1, 2\} = 3$;
$(R_2 \circ R_1)''(4) = \{0, 4\}$; $(R_1 \circ R_2)''(4) = \{1, 2, 3\}$.

Exercises

†2.3.1. Let R_1 be relation from A to B and R_2 a relation from B to C. Prove the following:

(a) $R_1{}^{-1}$ is a relation from B to A.

(b) $R_2 \circ R_1$ is a relation from A to C.

2.3.2. Let $R_1 = \{\langle 0,0\rangle, \langle 2,1\rangle, \langle 2,2\rangle, \langle 4,0\rangle\}$ and let $R_2 = 2 \times 4$. Determine each set.

(a) dom R_1. **(b)** range R_1.
(c) field R_2. **(d)** $R_2 \circ R_1$.
(e) $R_1 \circ R_2$. **(f)** R_1^{-1}.
(g) $R_2^{-1} \circ R_1$. **(h)** $R_1''(4)$.

†2.3.3. Let R_1 and R_2 be binary relations. Prove each:

(a) $(R_1^{-1})^{-1} = R_1$. **(b)** $(R_2 \circ R_1)^{-1} = R_1^{-1} \circ R_2^{-1}$.
(c) $(R_1 \cup R_2)^{-1} = R_1^{-1} \cup R_2^{-1}$. **(d)** $(R_1 \cap R_2)^{-1} = R_1^{-1} \cap R_2^{-1}$.

2.3.4. Suppose that A is the set consisting of France, Russia, the United States, Spain, Portugal, and Canada. For $x, y \in A$, let xRy if and only if x borders on y. Which elements of $A^{②}$ are in R?

2.3.5. Let $A = 3$ and let $B = 7$. Which binary relations from A to B are described by the following graphic diagrams?

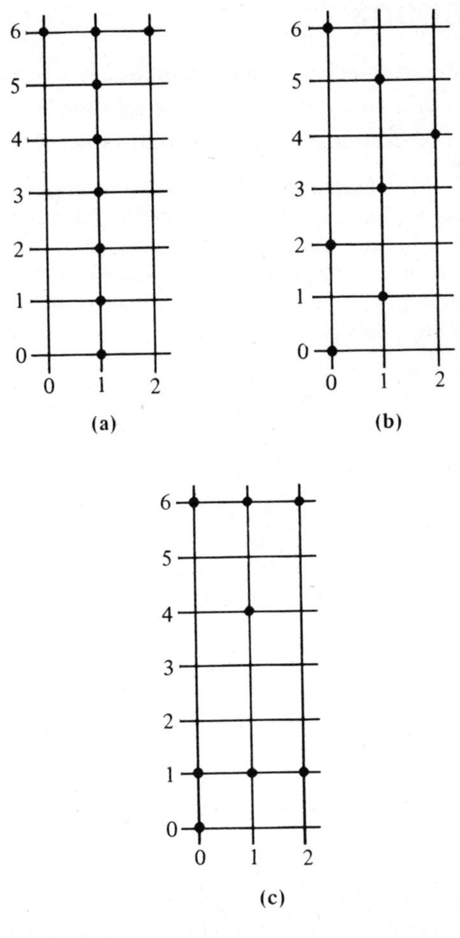

Fig. 2.3.3

2.3.6. Let $A = 5$ and let $B = 4$. Represent each of the following relations by means of a graphic diagram.

(a) $R_1 = \{\langle 1,3\rangle, \langle 2,2\rangle, \langle 2,3\rangle, \langle 4,0\rangle, \langle 4,1\rangle, \langle 4,2\rangle\}$.

(b) $R_2 = \{\langle 0,0\rangle, \langle 1,1\rangle, \langle 2,2\rangle, \langle 3,3\rangle, \langle 4,3\rangle\}$.

(c) $R_3 = \{\langle 0,0\rangle, \langle 1,0\rangle, \langle 2,0\rangle, \langle 3,0\rangle, \langle 4,0\rangle, \langle 0,1\rangle, \langle 0,2\rangle, \langle 0,3\rangle\}$.

2.3.7. Let A and B be sets and let R be a binary relation.

†(a) Prove that $R''(A \cup B) = R''(A) \cup R''(B)$.

†(b) Prove that $R''(A \cap B) \subseteq R''(A) \cap R''(B)$.

(c) Determine sets A_1 and B_1 and a relation R for which

$$R''(A_1 \cap B_1) \subset R''(A_1) \cap R''(B_1).$$

†2.3.8. Prove that if X and Y are sets for which $X \subseteq Y$, then for any relation R,

$$R''(X) \subseteq R''(Y).$$

2.4 Functions

A **function (correspondence, map, mapping)** is a relation F that is such that if aFb as well as aFb', then $b = b'$. Thus, a relation R is a function iff to any a in dom R there corresponds exactly one b in range R.

Functions are encountered frequently in elementary mathematics. A function is thought of as an association between two classes A and B by means of which a unique element of B corresponds to each element of A. For example, if both A and B represent the real numbers, then (in the usual notation) $f(x) = x^2 + 5$ and $g(x) = x - 3$ each indicate functions. These are depicted "graphically" on the (real) plane as follows:

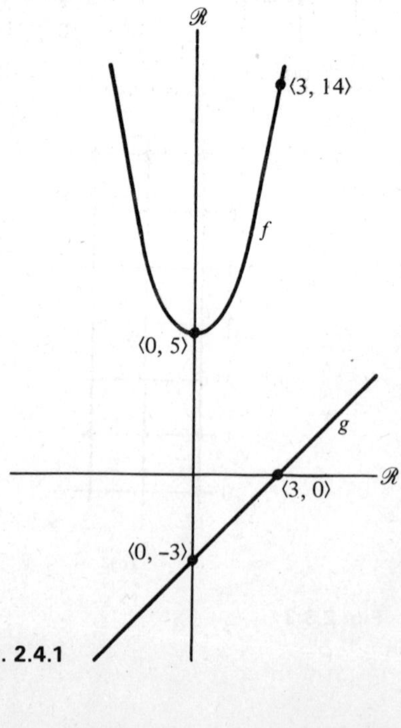

Fig. 2.4.1

A collection of points of the plane (a relation) represents a function iff each vertical line cuts it as most once. Thus, the relation specified by $x^2 + y^2 = 25$ is not a function. However, it can be broken up to yield two functions, h_1 and h_2, defined as follows:

$$h_1(x) = \sqrt{25 - x^2}, \ -5 \leq x \leq 5;$$
$$h_2(x) = -\sqrt{25 - x^2}, \ -5 < x < 5.$$

See Figure 2.4.2.

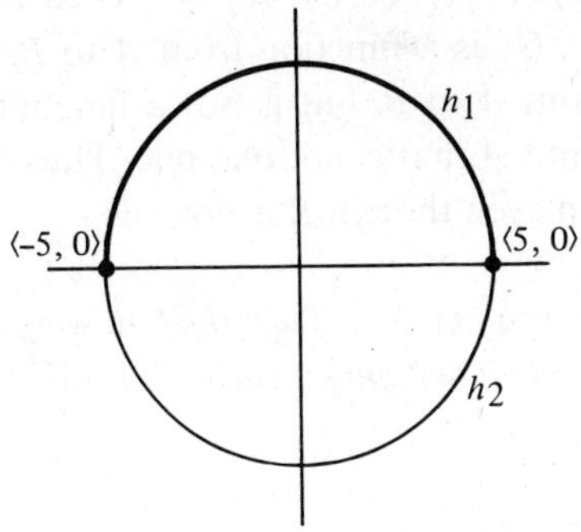

Fig. 2.4.2

A **function from** A **to (into)** B is a function whose domain is A and whose range is included in B. (Note that, according to this convention, if F is a function from A to B, then dom $F = A$, whereas if R is an arbitrary relation from A to B, dom R need only be a subset of A.) The notation, "$F : A \to B$" symbolizes that F is a function from A to B. Let $F : A \to B$ and let $x \in \text{dom } F$. Then x is called an **argument of** F. The unique y in B for which xFy is called the **value of** F **at** x or the **image of** x **under** F; it is written "$F(x)$". If all the values of F are equal, F is called a **constant function**.

Two functions are equal if they are equal as sets, i.e., if they contain the same ordered pairs.[1]

If $F : A \to B$ and $G : B \to C$, then the composite relation $G \circ F$ is a function from A into C (Exercise 2.4.4).

Let $F : A \to B$. Then F is called **one-one (injective)** if F has distinct values at distinct arguments. Moreover, F is **onto** B (F is **surjective**) if $B = \text{range } F$. The **identity function,** I_A, **on a set** A, by which we mean $\{\langle a, a\rangle : a \in A\}$, is an example of a one-one function of A onto itself. If F is a one-one function of A onto B, we say that F is a **one-one correspondence between** A **and** B (F is **bijective**). Alternatively, we say that A **is in one-one correspondence with** B or

[1] Algebraists often use a more stringent definition of "equality of functions". See (Mac Lane and Birkhoff, p. 4).

that A **and** B **are in one-one correspondence**. In this case we write "$A \approx B$"; otherwise, we write "$A \not\approx B$". (In the Corollary of Theorem 2.4.1 we will justify this usage.)

EXAMPLE 2.4.1. Let $A = \{0, 3, 4, 6\}$ and let $B = \{0, 2, 4, 6, 8\}$. Let
$F_1 = \{\langle 0,0\rangle, \langle 3,2\rangle, \langle 4,8\rangle, \langle 6,4\rangle\}$; let
$G_1 = \{\langle 0,0\rangle, \langle 3,2\rangle, \langle 4,2\rangle, \langle 6,6\rangle\}$; let
$F_2 = \{\langle 0,0\rangle, \langle 3,2\rangle, \langle 4,4\rangle, \langle 6,6\rangle, \langle 6,8\rangle\}$; let
$G_2 = \{\langle 0,0\rangle, \langle 2,3\rangle, \langle 4,4\rangle, \langle 6,6\rangle, \langle 8,6\rangle\}$. Then F_1 is a one-one function from A to B. Also, G_1 is a function from A to B, but G_1 is not one-one. F_2 is a relation from A to B, but is not a function. $G_2 = F_2^{-1}$ and is a function from B onto A that is not one-one. Thus the inverse of a relation R can be a function even though R is not one.

Theorem 2.4.1. *Let F be a function. Then the following sentences are equivalent (in the sense that one is true if and only if each of the others is true):*
(1) *F is one-one.*
(2) *F^{-1} is a function.*
(3) *F^{-1} is a one-one function.*

PROOF: Let $F : A \to B$. Then F is a set of ordered pairs $\langle a, b\rangle$, and for each a in A there is exactly one b in B for which $\langle a, b\rangle \in F$.

Clearly, (3) implies (2); thus, it suffices to show that (1) implies (3) and that (2) implies (1).

Suppose F is one-one. Then if $\langle a, b\rangle \in F$ and $\langle a', b\rangle \in F$, it follows that $a = a'$. Also, $F^{-1} = \{\langle b, a\rangle : \langle a, b\rangle \in F\}$. Thus, $\langle b, a\rangle \in F^{-1}$ and $\langle b, a'\rangle \in F^{-1}$ together imply that $a = a'$. Because a relation is a function iff it has this property, we know that F^{-1} is a function (from range F onto A). Moreover, if $\langle b, a\rangle \in F^{-1}$ and $\langle b', a\rangle \in F^{-1}$, then $\langle a, b\rangle \in F$ and $\langle a, b'\rangle \in F$. Thus, $b = b'$ because F is a function; hence, F^{-1} is one-one.

Suppose that F^{-1} is a function. Then, by the above, $F = (F^{-1})^{-1}$ is one-one.

Corollary. *If F is a one-one correspondence between A and B, then F^{-1} is a one-one correspondence between B and A. Thus, $A \approx B$ iff $B \approx A$.*

Let F be a function with domain X. Then if A is any subset of X, the function G with domain A, defined by $G(x) = F(x), x \in A$, is called the **restriction of** F **to** A, and is written "$F \upharpoonright A$". In this case F is called an **extension of** G **to** X.

Let $\mathscr{U}$ be a universe and let A be a subset of $\mathscr{U}$. We consider the function χ_A with domain $\mathscr{U}$ and range 2 defined by $\chi_A(x) = 1$ if $x \in A$ and $\chi_A(x) = 0$ if $a \in \tilde{A}$. We call χ_A the **characteristic function of** A. Characteristic functions enable us to discuss set membership in terms of functions.

EXAMPLE 2.4.2. Let $\mathscr{U} = \omega$ and let $A = \{1, 3, 5\}$. Then the function χ_A has domain ω and range 2, and is defined by

$$\chi_A(N) = \begin{cases} 1, & \text{if } N = 1, 3, \text{ or } 5; \\ 0, & \text{if } N \in \omega \sim \{1, 3, 5\}. \end{cases}$$

The function F with domain 6 and range 2—defined by

$$F(N) = \begin{cases} 1, & \text{if } N = 1, 3, \text{ or } 5; \\ 0, & \text{if } N = 0, 2, \text{ or } 4 \end{cases}$$

—is a restriction of χ_A to 6; hence, χ_A is an extension of F to ω.

Let $N \in \omega$. A set A is said to be an N-**element set** (or to **contain** N **elements**) if $N \approx A$. If F is a particular one-one correspondence between N and A, we can enumerate the elements of A according to the natural numbers to which they correspond under F. Thus, if a, b, c, and d are distinct sets and if $A = \{a, b, c, d\}$, then A is a 4-element set because there exists such a one-one function from 4 onto A—for example, $\{\langle 0, a\rangle, \langle 1, b\rangle, \langle 2, c\rangle, \langle 3, d\rangle\}$. With respect to this function, we write "$a = a_0$", "$b = a_1$", "$c = a_2$", "$d = a_3$". Each natural number N obviously contains N elements. Note that *$\varnothing$ is a one-one correspondence between $\varnothing$ and itself.*

A set A is called **finite** if for some N in ω, A is an N-element set; otherwise, A is said to be **infinite**.

Theorem 2.4.2. *Let $N \in \omega$.*
(1) *Suppose that $A \approx B$. Then A is an N-element set if and only if B is an N-element set.*
(2) *Two N-element sets are always in one-one correspondence.*

PROOF: We prove (1) and leave (2) as an exercise.

Let f be a one-one function of A onto B. Suppose that A is an N-element set. Then there exists a one-one mapping g of N onto A; clearly $f \circ g$ is a one-one mapping of N onto B [Exercise 2.4.4].

Similarly, if B is an N-element set, we can use the fact that f^{-1} is one-one function of B onto A in order to show that A is an N-element set.

Corollary. *If $A \approx B$, then both sets are finite or both are infinite.*

Many questions concerning finiteness and infinity are far more difficult than they appear to be. For example, intuitively, it is obvious that there is no one-one correspondence between an M-element set and an N-element set for distinct natural numbers M and N. The proof of this is considerably more complicated than that of Theorem 2.4.2. Moreover, other problems in this area require the addition of new axioms; they cannot be proved on the basis of the axioms thus far introduced. We will return to these matters in Chapter 4.

Exercises

2.4.1. Which of the following graphic diagrams of relations from 5 into itself represent functions? Which of these functions are one-one? Which are onto 5?

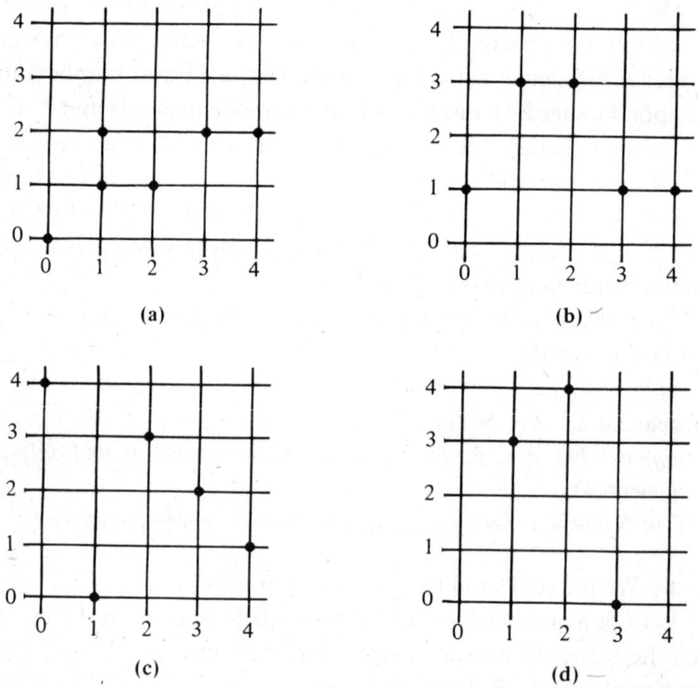

Fig. 2.4.3

2.4.2. Suppose that X is a set of people. Which of the following relations on X are necessarily functions?

(a) $\langle x, y\rangle \in R_1$ iff x is the brother of y.
(b) $\langle x, y\rangle \in R_2$ iff x is the father of y.
(c) $\langle x, y\rangle \in R_3$ iff y is the father of x.
(d) $\langle x, y\rangle \in R_4$ iff y is the grandfather of x.
(e) $\langle x, y\rangle \in R_5$ iff x is married to y.

2.4.3. Let $A = 4$, let $B = \{2, 5, 9, 10, 13\}$, and let $C = \{0, 2, 3\}$. Also, let $F_1 : A \to B$, $F_2 : B \to C$, $F_3 : B \to A$, $F_4 : C \to B$ be defined as follows:

$$F_1 = \{\langle 0, 2\rangle, \langle 1, 5\rangle, \langle 2, 2\rangle, \langle 3, 13\rangle\};$$

$$F_2 = \{\langle 2, 2\rangle, \langle 5, 0\rangle, \langle 9, 2\rangle, \langle 10, 3\rangle, \langle 13, 3\rangle\};$$

$$F_3 = \{\langle 2, 1\rangle, \langle 5, 2\rangle, \langle 9, 2\rangle, \langle 10, 3\rangle, \langle 13, 2\rangle\};$$

$$F_4 = \{\langle 0, 2\rangle, \langle 2, 2\rangle, \langle 3, 5\rangle\}.$$

(a) Determine $F_2 \circ F_1$. **(b)** Determine $(F_2 \circ F_1)^{-1}$.
(c) Determine $F_3 \circ F_4$. **(d)** Determine $F_1 \upharpoonright 3$.

2.4.4. Let $F : A \to B$ and $G : B \to C$.

†**(a)** Show that $G \circ F$ is a function from A into C.
†**(b)** Show that if F and G are one-one, so is $G \circ F$.
(c) Give an example in which F is one-one but $G \circ F$ is not one-one.
†**(d)** Show that if F and G are onto, so is $G \circ F$.
(e) Give an example in which F is onto but $G \circ F$ is not.

2.4.5. Let A and B be arbitrary sets and let F be a function. Prove (a) and (b).

†**(a)** $F''(\cup A) = \cup\{F''(B) : B \in A\}$.
†**(b)** $F''(\cap A) \subseteq \cap\{F''(B) : B \in A\}$.
(c) Determine sets A and B and a function F that are such that

$$F''(A \cap B) \subset F''(A) \cap F''(B).$$

†**(d)** What relationship holds between $F''(A \sim B)$ and $F''(A) \sim F''(B)$?

†**2.4.6.** Let A and B be arbitrary sets and let F be a function. Prove the following:

(a) $F^{-1}{}''(A \sim B) = F^{-1}{}''(A) \sim F^{-1}{}''(B)$.
(b) $F^{-1}{}''(\cup A) = \cup\{F^{-1}{}''(B) : B \in A\}$.
(c) $F^{-1}{}''(\cap A) = \cap\{F^{-1}{}''(B) : B \in A\}$.

2.4.7. Determine a function $F : \omega \to \omega$ which is **(a)** one-one but not onto, **(b)** onto but not one-one.

†**2.4.8.** Prove that composition of function is associative; i.e., if F, G, and H are functions, then $F \circ (G \circ H) = (F \circ G) \circ H$.

2.4.9. Prove the following properties of characteristic functions on a universe $\mathscr{U}$. Here, A and B vary over subsets of $\mathscr{U}$, and x varies over elements of $\mathscr{U}$.

†**(a)** $\chi_{A \cap B}(x) = \chi_A(x) \cap \chi_B(x)$.
†**(b)** $\chi_{A \cup B}(x) = \chi_A(x) \cup \chi_B(x)$.
†**(c)** $A \subseteq B$ iff for all x in $\mathscr{U}$, $\chi_A(x) \subseteq \chi_B(x)$.
(d) Is (c) still true if we replace both occurrences of "$\subseteq$" by "$\subset$"?
(e) In (a) determine sets, $\mathscr{U}$, A, and B for which $\chi_{A \cap B} \neq \chi_A \cap \chi_B$.
(f) In (b) determine sets $\mathscr{U}$, A, and B for which $\chi_{A \cup B} \neq \chi_A \cup \chi_B$.

In the next three problems, we utilize the familiar arithmetic operations—addition, multiplication, subtraction, division, and exponentiation, as well as the natural ordering, $\leq$, before they are introduced in the text.

2.4.10. How many relations are there on an N-element set?

2.4.11. Suppose that A is an N-element set, $N \in \omega$. How many one-one functions are there from N onto A?

2.4.12. Let X be an M-element set and Y an N-element set, $M, N \in \omega$. Let $\mathscr{F}(X, Y)$ be the set of functions $f: X \to Y$. How many elements are in $\mathscr{F}(X, Y)$? (Useful for Section 6.6.)

2.4.13. **(a)** Prove Theorem 2.4.2(2).

(b) Prove the corollary of Theorem 2.4.2.

†2.4.14. Let $F: A \to B, G: A' \to B'$, where F and G are each one-one correspondences and $A \cap A' \neq \varnothing$. Show that $F \cup G$ need not be a one-one correspondence between $A \cup A'$ and $B \cup B'$. What restrictions are needed to guarantee that $F \cup G$ will be one-one?

★2.4.15. A set $\mathscr{F}$ of functions is said to be **nested** if whenever $f_1, f_2 \in \mathscr{F}$, then either $f_1 \subseteq f_2$ or $f_2 \subseteq f_1$. Suppose that $\mathscr{F}$ is a nested set of functions. Prove the following:

(a) $\cup\mathscr{F}$ is a function.

(b) The collection of dom f, $f \in \mathscr{F}$, is a set and

$$\text{dom} \cup\mathscr{F} = \cup\{\text{dom}\, f : f \in \mathscr{F}\}.$$

(c) The collection of range f, $f \in \mathscr{F}$, is a set and

$$\text{range} \cup\mathscr{F} = \cup\{\text{range}\, f : f \in \mathscr{F}\}.$$

(d) If each f in $\mathscr{F}$ is one-one, then so is $\cup\mathscr{F}$.

(Used in Theorem 5.1.2.)

†2.4.16. Suppose that f and g are functions and suppose that for all x in dom g, $f \circ g(x) = x$.

(a) Show that g is one-one.

(b) Show that f need not be one-one.

2.4.17. Let F be a function. Show that $F = F_2 \circ F_1$, where F_1 is onto and F_2 is one-one.

2.5 Equivalence Relations

We now generalize the notion of set equality by abstracting some of its properties. For a given collection of (not necessarily distinct) sets, set equality is one means of identifying certain objects of the collection. We consider other relations on a set that effect an identification of related objects.

Let R be a (binary) relation on A. Then R is said to be **reflexive (on** A**)** if for each a in A, aRa. When the set A is understood, we simply write "**reflexive**". Clearly, *R is reflexive iff $I_A \subseteq R$.* Also, R is called **symmetric** if whenever aRb, then bRa. Thus, *R is symmetric iff $R = R^{-1}$.* Furthermore, R is said to be **transitive** if aRb and bRc together imply aRc. Hence, *R is transitive* iff $R \circ R \subseteq R$. A relation on A that is reflexive, symmetric, and transitive is known as an **equivalence relation on** A. Clearly, *$A^{②}$ and I_A are always equivalence relations on A.*

(Note that we have used the word "transitive" in two different senses:

(1) A set A is transitive if whenever $x \in A$, then $x \subseteq A$.
(2) A relation R is transitive if aRb and bRc together imply aRc.

Unfortunately, these two uses are not mutually exclusive because a relation is also a set. Unless otherwise specified, when we say that a relation is transitive, we mean in the sense of (2).

†EXAMPLE 2.5.1. In Chapter 3 we will introduce the set $\mathscr{Z}$ of integers as well as discuss the usual arithmetic concepts for integers. In number theory we say that the integer a **divides** the integer b provided there is an integer c for which $b = ac$. We write "$a|b$" for "a divides b." This relation on the nonzero integers is reflexive and transitive, but not symmetric. (We exclude zero because division by zero is usually undefined.) In terms of this relation we define "congruence, modulo m". Let m be a positive integer and let a and b be arbitrary integers. We say that ***a* is congruent to *b*, modulo *m***, symbolically, "$a \equiv b \pmod{m}$", iff $m|(a - b)$. Thus, $12 \equiv 2 \pmod{5}$ because 5 divides $12 - 2$. Similarly, $17 \equiv -1 \pmod{3}$ because 3 divides $17 - (-1)$.

It is not difficult to show that congruence, modulo m, is an equivalence relation on $\mathscr{Z}$. Clearly, $m|0$; thus, $a \equiv a \pmod{m}$ for each integer a, and congruence, modulo m, is reflexive. This relation is also symmetric. For if $a \equiv b \pmod{m}$, then $m|(a - b)$. Hence, $a - b = mc$ for some integer c; consequently, $b - a = m(-c)$. Thus, $m|(b - a)$ and $b \equiv a \pmod{m}$. Finally, we show that our relation is transitive. If $a \equiv b \pmod{m}$ and $b \equiv c \pmod{m}$, then for some integers d_1 and d_2, $a - b = md_1$ and $b - c \equiv = md_2$. Therefore, $a - c = m(d_1 + d_2)$. It follows that $a \equiv c \pmod{m}$.

Clearly, if R is reflexive on a set A, then R is transitive iff for *distinct* a, b, and c, aRb and bRc together imply aRc.

†EXAMPLE 2.5.2. Let $A = \{a, b, c, d\}$.

(1) Let $R_1 = I_A \cup \{\langle a, d\rangle, \langle d, a\rangle\}$.
Then R_1 is reflexive because $I_A \subseteq R_1$; also, R_1 is symmetric because $R_1^{-1} = R_1$. There are no three distinct elements x, y, z of A for which xRy and yRz. By the preceding remark, R_1 is transitive; hence, it is an equivalence relation on A.

(2) Let $R_2 = R_1 \cup \{\langle a, c\rangle, \langle c, a\rangle\}$.
Then R_2 is reflexive and symmetric. However, dR_2a and aR_2c; but $d\not{R}_2c$. Thus, R_2 is not transitive and, consequently, is not an equivalence relation on A.

†EXAMPLE 2.5.3. Let $A = 3 = \{0, 1, 2\}$. Let $R_1 = \{\langle 0, 0\rangle, \langle 2, 2\rangle\}$; $R_2 = \{\langle 0, 0\rangle, \langle 0, 1\rangle, \langle 0, 2\rangle\}$; $R_3 = \{\langle 0, 0\rangle, \langle 0, 1\rangle, \langle 1, 0\rangle\}$; $R_4 = \{\langle 0, 0\rangle, \langle 0, 1\rangle, \langle 1, 0\rangle, \langle 1, 1\rangle, \langle 2, 2\rangle$; $R_5 = \{\langle 0, 1\rangle, \langle 1, 2\rangle\}$; $R_6 = \{\langle 0, 1\rangle, \langle 1, 2\rangle, \langle 0, 2\rangle\}$. Then R_1 is an equivalence relation on $\{0, 2\}$ and R_4 is an equivalence relation on A. Also, R_2 is transitive; R_2 is neither reflexive ($\langle 1, 1\rangle \notin R_2$) nor symmetric ($\langle 1, 0\rangle \notin R_2$). Furthermore, R_3 is symmetric but is neither reflexive nor transitive; R_6 is transitive but is neither reflexive nor symmetric; R_5 has none of these properties.

A set X is a **partition of a set** $A(\neq \varnothing)$ if X is a set of pairwise disjoint nonempty sets and $\cup X = A$. The elements of a partition are called **cells**. Thus, if X is a partition of A, each element of A belongs to a unique cell. *For any nonempty set A, $\{A\}$ and $\{\{a\} : a \in A\}$ are each partitions of A.*

Let A be the collection of integers, let A_1 be the collection of odd integers, and let A_2 be the collection of even integers. Let X consist of A_1 and A_2. Then X is a partition of A. Furthermore, this partition is very closely related to congruence, modulo 2, which was discussed in Example 2.5.1. This connection is now explained.

Theorem 2.5.1. *Let A be a nonempty set.*

(1) *If R is an equivalence relation on A and if $a \in A$, let $[a] = \{x \in A : xRa\}$; $[a]$ is called the* **equivalence class of** *a. Then the collection of $[a]$ corresponding to a in A is a set; moreover, $\{[a] : a \in A\}$ is a partition of A, and is written "A/R".*

(2) *Conversely, if X is a partition of A we can define an equivalence relation R on A by letting aRb if and only if a and b belong to the same cell of X.*

(3) *There is a one-one correspondence between the set of equivalence relations on A and the set of partitions of A. In fact, the mapping from equivalence relations to partitions defined in part* (1) *is one-one onto and its inverse is the mapping of part* (2), *from partitions to equivalence relations.*

PROOF: (1) Let X be the collection of $[a]$ corresponding to elements a in A. Then X is a subcollection of $\mathscr{P}\mathscr{P}(A)$; a simple application of the Axiom of Subsets indicates that X is a set. Thus, $X = \{[a] : a \in A\}$. Each set $[a]$ is nonempty because $a \in [a]$; clearly, $X = \cup A$. We claim that if $[a], [b] \in X$, then either $[a]$ and $[b]$ are disjoint or they are equal. If $c \in [a] \cap [b]$, then cRa and cRb; hence, aRc (by symmetry) and aRb (by transitivity). Thus, if $x \in [a]$, then because xRa and aRb, it follows that xRb; i.e., $x \in [b]$. We have shown that $[a] \subseteq [b]$. Furthermore, aRb implies bRa, and we can similarly show that if $y \in [b]$, then $y \in [a]$. Consequently, $[b] \subseteq [a]$. Therefore, $[a]$ and $[b]$ are equal.

(2) Suppose that X is a partition of A and R is defined as in part (2) of the theorem. Then R is reflexive because $a \in [a]$ for each a in A. If aRb, then a and b

belong to the same cell of X, and hence bRa. Finally, if aRb and bRc, then a belongs to the unique cell to which b belongs; c must also belong to this cell. Thus, a and c belong to the same cell. It follows that aRc. Therefore, R is reflexive, symmetric, and transitive; i.e., R is an equivalence relation. Finally, dom $R = \cup X = A$.

(3) Let $\mathscr{E}_A$ be the set of equivalence relations on A and let $\mathscr{P}_A$ be the set of partitions of A. Let $F : \mathscr{E}_A \to \mathscr{P}_A$ be defined by $F(R) = A/R$ and let $G : \mathscr{P}_A \to \mathscr{E}_A$ be defined according to part (2). Then $G \circ F = I_{\mathscr{E}_A}$; i.e., $G = F^{-1}$. Because F and G are both functions, F is one-one, according to Theorem 2.4.1. For each partition X of A, $F(G(X)) = X$; hence, F is onto $\mathscr{P}_A$.

Corollary. *If R is an equivalence relation on a nonempty set A, then for each a in A, the cell of X/R to which a belongs is $[a]$.*

Example 2.5.4

(1) In Example 2.5.2(1), the partition corresponding to the equivalence relation R_1 is $\{\{a, d\}, \{b\}, \{c\}\}$.

(2) In Example 2.5.3, $\{0, 2\}/R_1 = \{\{0\}, \{2\}\}$ and $A/R_4 = \{\{0, 1\}, \{2\}\} = \{2, \{2\}\}$.

Example 2.5.5. Consider the partition $\{\{0, 4\}, \{1, 3, 5\}, \{2\}\}$ of 6. Let $R = \{\langle 0, 4\rangle, \langle 1, 3\rangle, \langle 1, 5\rangle, \langle 3, 5\rangle\}$. Then the equivalence relation corresponding to the given partition is $I_6 \cup R \cup R^{-1}$.

Exercises

2.5.1. Suppose that X is an arbitrary set of people. Let x and y vary over the individuals of X. Which of the following relations on X are reflexive? Which are symmetric? Which are transitive?

(a) x is a brother of y.

(b) x is a brother or (excl.) sister of y.

(c) x is a child of the mother of y.

(d) x is married to y.

(e) x is the father of y.

(f) x is a descendent of y.

2.5.2. Suppose that (in Euclidean geometry) the collection of all triangles on a plane is a set, A. Let $x, y \in A$. Which of the following relations on A are equivalence relations?

(a) xR_1y iff x is similar to y.

(b) xR_2y iff x is congruent to y.

(c) xR_3y iff x and y share a common side.
(d) xR_4y iff x and y have the same perimeter.

2.5.3. Let $A = 3$. Determine relations R_1, R_2, and R_3 on A satisfying the following conditions:

(a) R_1 is symmetric and transitive but not reflexive.
(b) R_2 is reflexive and symmetric but not transitive.
(c) R_3 is reflexive and transitive but not symmetric.

2.5.4. Let $A = \{1, 2, 3, 4\}$. Which of the following diagrams represent equivalence relations on A? Indicate the corresponding partitions. For the other diagrams indicate which properties of equivalence relations are lacking.

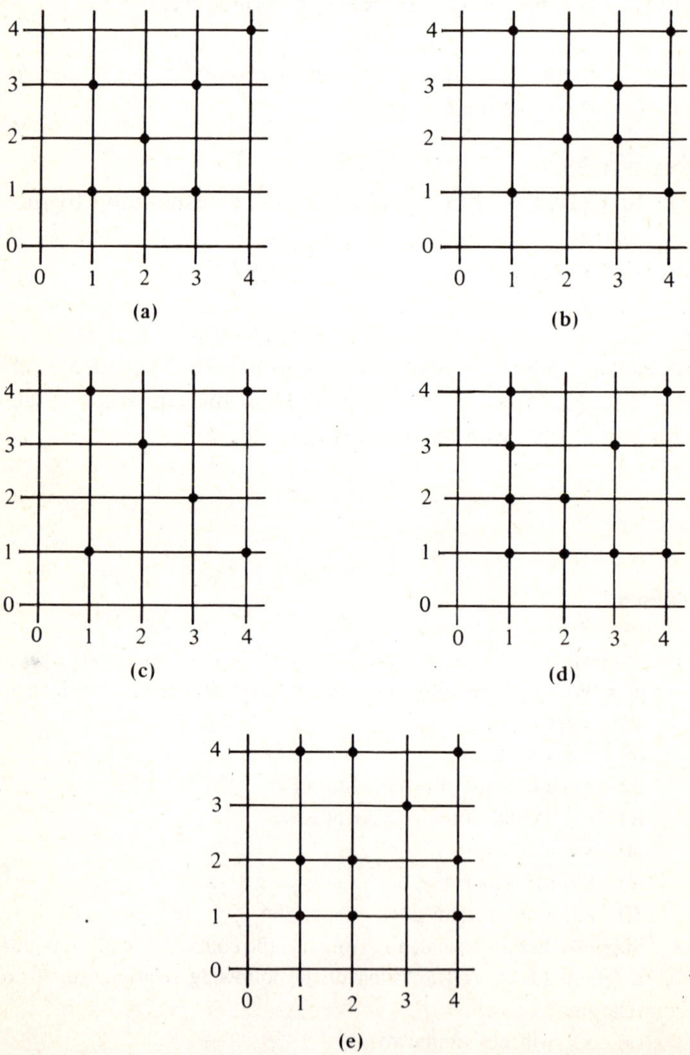

Fig. 2.5.1

2.5.5. Let $A = \{1, 2, 3, 4, 5, 6, 7\}$ and let $X = \{\{1, 2\}, \{3, 4, 5, 6\}, \{7\}\}$.
(a) Indicate the associated equivalence relation R_1 on A determined by X.
(b) Diagram R_1.

†**2.5.6.** **(a)** Let R be a relation that is symmetric and transitive. Show that R is an equivalence relation on dom R. Suppose that dom $R \subset A$. Show that R is not an equivalence relation on A.
(b) Let R be a reflexive and transitive relation on a set B. Show that $R \cap R^{-1}$ is an equivalence relation on B.

2.5.7. Determine all equivalence relations on 4.

2.5.8. Explain how equality of sets determines an equivalence relation on a universe, $\mathscr{U}$. Indicate the corresponding partition of $\mathscr{U}$.

2.5.9. Let R and S be equivalence relations on sets A and B, respectively. Is $R \cup S$ an equivalence relation on $A \cup B$?

2.5.10. Let $\mathscr{E}$ be a nonempty set of equivalence relations on a set A. Is $\cap\mathscr{E}$ an equivalence relation on A?

2.5.11. Let A be a set and let $B \subseteq A$. For X, Y in $\mathscr{P}(A)$, let XRY iff $X \cap B = Y \cap B$. Show that R is an equivalence relation.

2.5.12. Let X be a partition of a set A and let B be a nonempty subset of A. Show that $\{x \cap B : x \in X\} \sim \{\varnothing\}$ is a partition of B.

2.5.13. Suppose that the collection of real numbers is a set $\mathscr{R}$ and that the arithmetic operations have been defined. Consider the relation R on $\mathscr{R}^{②}$ defined by $\langle x_1, y_1\rangle R\langle x_2, y_2\rangle$ iff $x_1{}^2 + y_1{}^2 = x_2{}^2 + y_2{}^2$, $x_1, x_2, y_1, y_2 \in \mathscr{R}$.
(a) Show that R is an equivalence relation.
(b) Describe the corresponding partition geometrically.

2.5.14. Let $\mathscr{F}$ be the set of all functions $F : \omega \to \omega$. For F_1, F_2 in $\mathscr{F}$—define F_1RF_2 iff for some M in ω, $F_1(N) = F_2(N)$ for all N containing M. Show that R is an equivalence relation on $\mathscr{F}$. Indicate the corresponding partition of $\mathscr{F}$.

2.5.15. **(a)** How many partitions are there of a 3-element set?
(b) Of a 4-element set?
(c) Of a 5-element set?

***2.5.16.** Let P_N be the number of partitions and $\binom{N}{I}$ the number of I-element subsets of an N-element set. Show that for all N in ω, $P_{N^+} = \sum_{I=0}^{N} \binom{N}{I} \cdot P_I$. (See Section 3.2 for notation.)

2.6 Orderings

A relation R is said to be **antisymmetric** if xRy and yRx together imply $x = y$. Thus, *a relation R on a set A is antisymmetric if and only if $x, y \in A$, $x \neq y$, and xRy together imply $y \not R x$.*

Graphically, the *presence* of a point $P = \langle x, y\rangle$ off the "principal diagonal", $D = \{\langle x, x\rangle : x \in A\}$, implies the *absence* of the point $Q = \langle y, x\rangle$—the "mirror image of P in D".

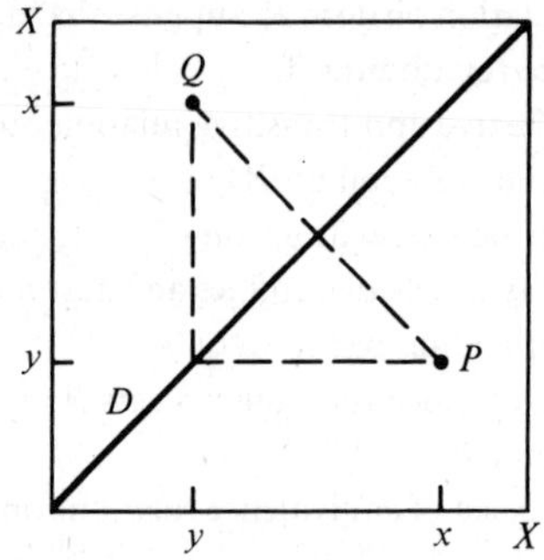

Fig. 2.6.1

†EXAMPLE 2.6.1. Let A be a set. By the relation $\in_A$, on A, we mean

$$\{\langle x, y\rangle \in A^{②} : x \in y\};$$

by the relation $\notin_A$, on A, we mean

$$\{\langle x, y\rangle \in A^{②} : x \notin y\}.$$

(1) If A is an arbitrary set, neither $\in_A$ not $\notin_A$ need be antisymmetric. For example, if a and b satisfy

$$a \in b \in a$$

and if

$$A = \{a, b, 1, \{2\}\},$$

then neither $\in_A$ nor $\notin_A$ is antisymmetric.

(2) For distinct natural numbers M and N,

$$\text{either } M \in N \text{ or (excl.) } N \in M.$$

Hence

$$\text{either } M \notin N \text{ or (excl.) } N \notin M.$$

Thus both $\in_\omega$ and $\notin_\omega$ are antisymmetric.

An equivalence relation on a set A was defined to be a reflexive, symmetric, and transitive relation on A. We now define a **partial ordering on** A to be a reflexive, antisymmetric, transitive relation on A. If R is a partial ordering on

A, $\langle A, R\rangle$ is called a **partially ordered system**; we also say that A **is partially ordered by** R.

†EXAMPLE 2.6.2. The relation "$a|b$" on the positive integers (see Example 2.5.1) is a partial ordering. It is not a partial ordering, however, on the nonzero integers because 1 and -1 divide one another.

If A is a nonempty finite set, a **chain diagram** can be employed to indicate a partial ordering on A. The elements of A are depicted in such a way that for distinct a, b in A, aRb iff there is a series of ascending directed line segments linking a to b.

†EXAMPLE 2.6.3. The following chain diagram represents a partial ordering R on a set $A = \{a, b, c, d, e, f, g\}$.

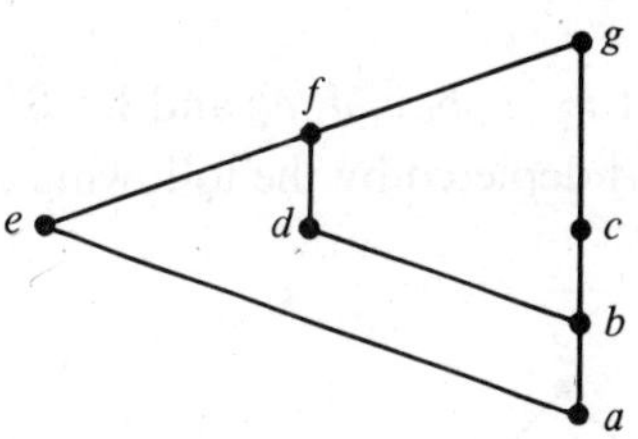

Fig. 2.6.2

Here, aRb, aRe, bRc, bRd, cRg, dRf, eRf, fRg; furthermore, xRx for each x in A (by reflexivity) and such relationships as aRc, aRf, dRg, etc., follow by transitivity. However, $c \not R f$ even though c can be linked to f because there is no series of *ascending* directed line segments linking c to f.

Let A be partially ordered by R and let B be any subset of A. Then a (in A) is called an **upper bound for** B if for every b in B, bRa. Also, a_0 is called the **least upper bound for** B if a_0 is an upper bound for B and if, in addition, a_0Ra for every upper bound a for B. Similarly, a (in A) is a **lower bound for** B if for every b in B, aRb; a_1 is called the **greatest lower bound for** B if a_1 is a lower bound for B and if, in addition, for every lower bound a for B, aRa_1. Note that these upper (lower) bounds for B can be in either B or in $A \sim B$. By the antisymmetry of R, *if the least upper bound* (*greatest lower bound*) *for B exists, it is unique.* If the greatest lower bound for B is an element of B, it is called the **first element of** B; if the least upper bound for B is in B, it is called the **last element of** B. The element b_0 (of B) is called a **maximal element of** B if b_0Rb and $b \in B$ together imply $b = b_0$. Similarly, b_1 (in B) is a **minimal element of** B if bRb_1 and $b \in B$ together imply

$b = b_1$. Surely, *a first* (*last*) *element of B is a minimal* (*maximal*) *element of B*. (We also speak of **upper bounds for** $\langle B, R\rangle$, **maximal elements of** $\langle B, R\rangle$, *etc.*)

EXAMPLE 2.6.4. Consider the set A and the partial ordering R of Example 2.6.3.

(1) Observe that g is the unique upper bound for A; hence, it is the least upper bound for A. Because $g \in A$, g is the last element of A; as such, it is a maximal element of A. Similarly, a is the greatest lower bound for A; consequently, it is the first element of A and is a minimal element of A.

(2) Let $B = \{b, c, d, f\}$. Then c and f are maximal elements of B, although neither is an upper bound for B; and g, in $A \sim B$, is the unique, hence least, upper bound for B. Therefore, B has no last element. Also, b is a lower bound for B; a, in $A \sim B$, is the only other lower bound for B. Therefore, b is the greatest lower bound for B; hence, b is the first element of B. As such, it is a minimal element of B.

EXAMPLE 2.6.5. Let $A = \{a, b, c, d, e\}$ and let $B = \{a, b, c\}$. Consider the partial orderings on A depicted by the following chain diagrams:

(1) R_1:

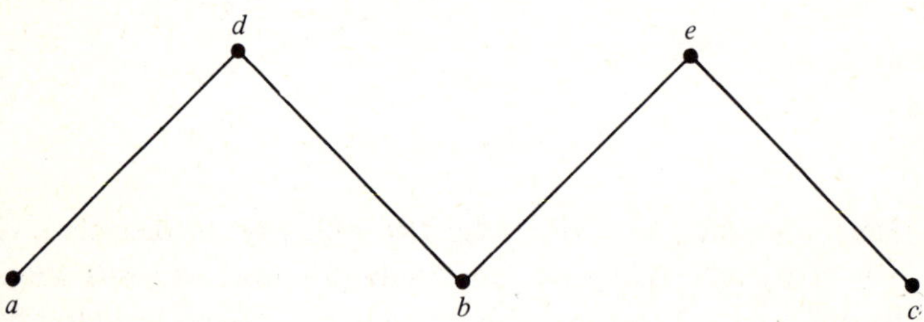

Fig. 2.6.3

Clearly, B has no upper bound with respect to R_1; hence, B has no last element. However, a, b, and c are all maximal elements of B.

(2) R_2:

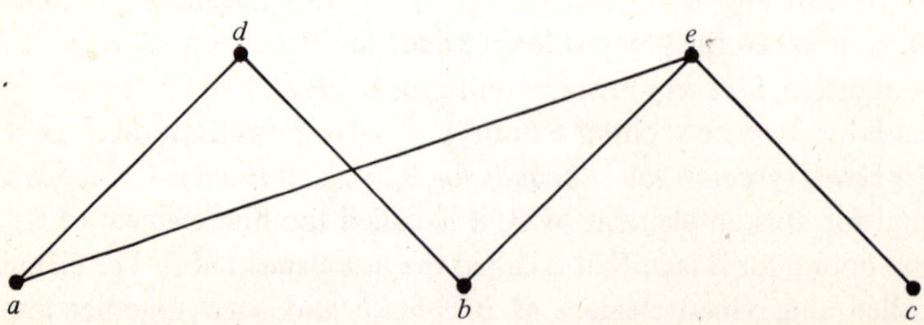

Fig. 2.6.4

Here, e is the unique upper bound for B with respect to R_2; hence, e is the least upper bound for B. Because $e \in A \sim B$, B has no last element. Again, a, b, and c are maximal elements of B.

(3) R_3:

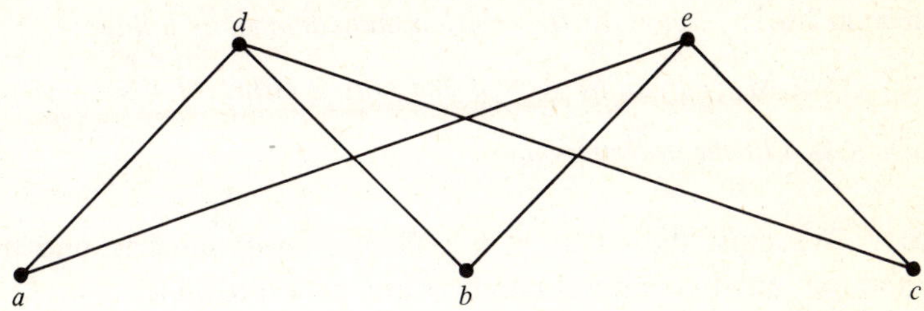

Fig. 2.6.5

Here, d and e are each upper bounds for B with respect to R_3. Neither is a least upper bound for B. (Again a, b, and c are maximal elements of B.)

A partial ordering R on A is called a **linear ordering (total ordering) on** A provided any two elements of A are R-comparable. In this case, $\langle A, R\rangle$ is called a **linearly (totally) ordered system**; we also say that A **is linearly (totally) ordered by** R. *If R is a linear ordering on A, then* by antisymmetry, *if x and y are distinct elements of A, xRy or* (*excl.*) *yRx*. For linear orderings the chain diagram representation can be a collection of points on a vertical line. *If R is a linear ordering, so is* R^{-1}; the chain diagram for R^{-1} is the diagram for R upside down. *Furthermore, let R be a linear ordering on A and also let* $B \subseteq A$. *It follows that if B has a maximal* (*minimal*) *element,* b_0, *then it is unique and it is the last* (*first*) *element of B* (Exercise 2.6.10).

†EXAMPLE 2.6.6. Let $\mathscr{U}$ be a universe. Then set inclusion, $\subseteq$, is a partial ordering on $\mathscr{P}(\mathscr{U})$ because for all sets A, B, and C, $A \subseteq A$; $A \subseteq B$ and $B \subseteq A$ together imply $A = B$; and from $A \subseteq B$ and $B \subseteq C$, it follows that $A \subseteq C$. In general, set inclusion is not a linear ordering on $\mathscr{U}$ because if X and Y are subsets of an arbitrary set $\mathscr{U}$, neither need be a subset of the other. If V is any subset of $\mathscr{P}(\mathscr{U})$, then $\cup V$ is the least upper bound and $\cap V$ the greatest lower bound, for V. In particular, $\mathscr{U}$ is the least upper bound and $\varnothing$ the greatest lower bound, for $\mathscr{P}(\mathscr{U})$.

†EXAMPLE 2.6.7. Suppose that the collection of points that lie on a horizontal line is a set, A. For $x, y \in A$, let xRy mean that either $x = y$ or x lies to the

left of y. Then R is a linear ordering on A. Each y in A is the least upper bound for the subset B of points that lie to the left of y as well as for $B \cup \{y\}$. Also, y is a maximal element of $B \cup \{y\}$. If z lies to the right of y, then z is an upper bound for B and for $B \cup \{y\}$.

Theorem 2.6.1. *Let $\leq$ be the relation on ω defined as follows:*

For M, N in ω, $M \leq N$ if and only if either $M \in N$ or $M = N$.

Then $\leq$ is a linear ordering on ω.

PROOF: We must show that $\leq$ is reflexive, antisymmetric, and transitive on ω, and that any two natural numbers are $\leq$-comparable.

Surely $N = N$; hence, $\leq$ is reflexive. By Theorem 2.2.3(1), we know that for distinct natural numbers M and N, either $M \in N$ or (excl.) $N \in M$. Consequently, for distinct M and N, $M \leq N$ or (excl.) $N \leq M$. Thus, $\leq$ is antisymmetric.

Suppose that $M \leq N$ and $N \leq R$. If either $M = N$ or $N = R$, then $M \leq R$. We need only consider the case $M \in N$, $N \in R$. But each natural number is transitive; thus, in this case, $M \in R$. Hence, $M \leq R$. Therefore, $\leq$ is transitive.

Corollary. *The inclusion relation $\subseteq$ is a linear ordering on ω.*

PROOF: By Theorem 2.2.3(2), $M \in N$ iff $M \subset N$, $M, N \in \omega$. It follows that $M \subseteq N$ iff $M \in N$ or $M = N$; i.e., $M \subseteq N$ iff $M \leq N$. Thus, $\subseteq$ also characterizes this linear ordering on ω.

We can define the relations $<$, $>$, and $\geq$ on ω in the usual way. Thus, $M < N$ iff $M \in N$ (i.e., iff $M \leq N$ and $M \neq N$); $M > N$ iff $N \in M$; $M \geq N$ iff $N \leq M$. As is customary, "$M < N \leq R$" means "$M < N$ and $N \leq R$"; other abbreviations of this sort will be employed. Henceforth, "ω_+" will stand for "$\omega \sim \{0\}$"; we will frequently write "$N \geq 1$" instead of "$N \in \omega_+$".

Exercises

†2.6.1. Let A be an arbitrary set and let I_A be the identity function on A.

(a) Prove that I_A is both symmetric and antisymmetric.

(b) Prove that a relation R is both symmetric and antisymmetric on A iff $R \subseteq I_A$.

2.6.2. Each of the following chain diagrams represents a partial ordering on 7. Represent each by a graphic diagram and determine the minimal, maximal, first, and last elements of 7 in each case.

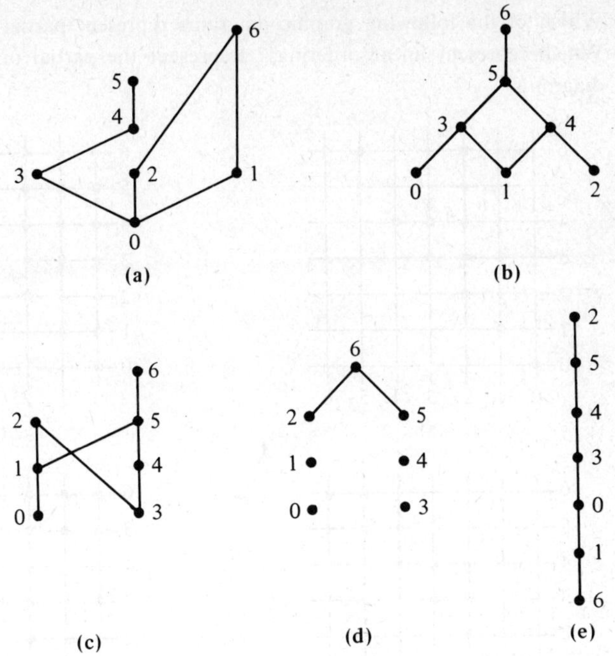

Fig. 2.6.6

2.6.3. Each of the following chain diagrams represents a partial ordering on 8. Represent each by a graphic diagram. Determine the minimal, maximal, first, and last elements of 8, of 5, in each case.

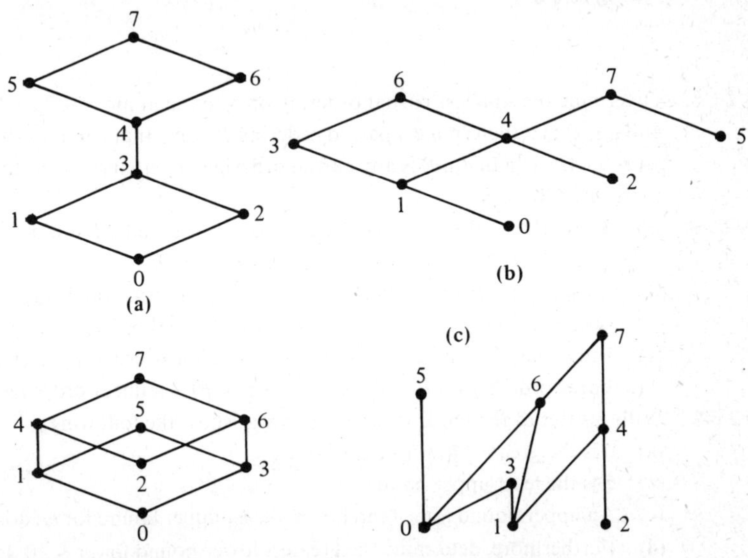

Fig. 2.6.7

2.6.4. Which of the following graphic diagrams represent partial orderings on 7? Which represent linear orderings? Represent the partial orderings by chain diagrams.

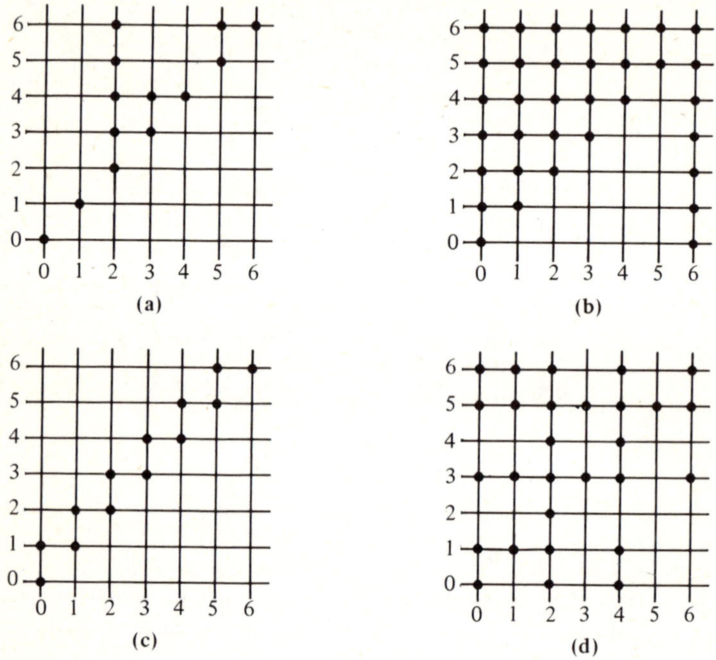

Fig. 2.6.8

2.6.5. Determine the smallest partial ordering on 5 that contains $\langle 0, 2\rangle$, $\langle 2, 3\rangle$, $\langle 2, 1\rangle$, and $\langle 3, 4\rangle$; i.e., determine a partial ordering R_0 on 5 that contains these ordered pairs and is such that if R is any partial ordering on 5 that contains these ordered pairs, then $R_0 \subseteq R$.

2.6.6. †**(a)** Prove that if R is a partial ordering on $A \sim B$ and if S is a partial ordering on B, then $R \cup S$ is a partial ordering on $A \cup B$.

(b) Show, by an example, that if R is a partial ordering on A and S is a partial ordering on B, then $R \cup S$ need not be a partial ordering on $A \cup B$.

†**(c)** Prove that if A and B are disjoint, if R is a linear ordering on A, and if S is a linear ordering on B, then $R \cup S \cup (A \times B)$ is a linear ordering on $A \cup B$.

2.6.7. With respect to the linear ordering $\leq$ on ω, show the following:

(a) 0 is the greatest lower bound for ω.

(b) 5 is the least upper bound for 6.

(c) An upper bound for ω (and hence a least upper bound for ω) does not exist.

(d) Furthermore, determine the greatest lower bound for $\omega \sim \{0, 1, 5, 6\}$.

2.6.8. List all partial orderings on 3; on 4. Which of these are linear orderings?

†2.6.9. **(a)** Prove that the inverse of a partial ordering on A is also a partial ordering on A.

(b) Prove that the inverse of a linear ordering on A is also a linear ordering on A.

2.6.10. †**(a)** Let $\langle A, R\rangle$ be a linearly ordered system. Prove that if the subset B of A has a maximal (minimal) element b_0, then b_0 is unique and is the last (first) element of B.

(b) Give an example of a partially ordered system $\langle A, R\rangle$ in which A has a unique maximal element, yet A has no least upper bound.

2.6.11. Suppose that we have the set of positive integers and the partial ordering "$a|b$" discussed in Example 2.6.2. What are the minimal and maximal elements?

†**2.6.12.** Let X be an arbitrary set. By the **inclusion relation on** X we mean the relation $\mathscr{I}_X$ on X defined by $A\mathscr{I}_X B$ iff $A \subseteq B$, $A, B \in X$. Show that $\mathscr{I}_X$ is a partial ordering on X.

†**2.6.13.** Let X be partially ordered by R, and let $A \subseteq X$. Show that an element x of X is the least upper bound for A iff x is the first element of the set of upper bounds for A.

2.7 Well-orderings

A linear ordering on A is called a **well-ordering on** A if every nonempty subset B of A contains a first element. If a well-ordering on A exists, A is said to be **well-orderable**; if R is a well-ordering on A, then A is said to be **well-ordered by** R, and $\langle A, R\rangle$ is called a **well-ordered system**.

Theorem 2.7.1. *The relation $\leq$ on ω is a well-ordering.*

PROOF: By Theorem 2.6.1, $\leq$ is a linear ordering on ω; we need only show that with respect to $\leq$, every nonempty subset of ω has a first element.

Let A be the set of all natural numbers N that have the property that whenever B is a subset of ω that contains some natural number M with $M \leq N$, then B has a first element. Thus, $0 \in A$ because if $0 \in B$ and $B \subseteq \omega$, then surely $0 \leq X$ for each X in B. Suppose that $N \in A$. Let B be a subset of ω that contains some natural number K with $K \leq N^+$. Then if $K \in B$ as well as $K \leq N$ for some natural number N, by the inductive hypothesis, B has a first element. Otherwise, N^+ is surely the first element of B. Thus, $N^+ \in A$, and, by induction, $A = \omega$.

The well-ordering $\leq$ rules out the second alternative discussed at the end of Section 1.6 *and guarantees that natural numbers conform to our intuitive notion about them. For if there were an "infinite descending $\in$-chain",*

$$\ldots \in L \in M \in N,$$

then the "set" of these natural numbers would lack a first element (*with respect to* $\leq$).

EXAMPLE 2.7.1. In Example 2.6.7, the linear ordering R on A is not a well-ordering. For any b in A, there is an a (in A) that lies to the left of b. Thus, A has no first element with respect to R.

Every subset of a well-orderable set is well-orderable (Exercise 2.7.2); *thus, every subset of* ω *is well-orderable.*

Every set that is in one-one correspondence with a well-orderable set is itself well-orderable. For if R_X is a well-ordering on X and if F is a one-one mapping of X onto Y, then the relation R_Y on Y, defined by $y_1 R_Y y_2$ iff $F^{-1}(y_1) R_X F^{-1}(y_2)$, is clearly a well-ordering on Y.

It follows from the preceding paragraph and from the fact that every natural number N (as a subset of ω) is well-orderable that *every finite set is well-orderable.* In fact, *if* A *is a finite set, then every linear ordering on* A *is a well-ordering* (Exercise 2.7.8).

In determining whether a given relation is a well-ordering, it is not necessary to test all the conditions for a linear ordering as well as the additional condition for a well-ordering. In fact, *a partial ordering* R *on a set* A *is a well-ordering iff every nonempty subset of* A *has a first element with respect to* R (Exercise 2.7.3). Moreover, *a relation* R *on a set* A *is a well-ordering iff every nonempty subset* B *of* A *contains a unique element that is* R*-related to every element of* B (Exercise 2.7.7).

We consider two more examples of infinite well-orderable sets.

†EXAMPLE 2.7.2. Consider $\omega^+ = \omega \cup \{\omega\}$. Let $\leq$ be the well-ordering on ω of Theorem 2.7.1, and let $R = \leq \cup \{\langle x, \omega \rangle : x \in \omega^+\}$. It is easily seen that R is a partial ordering on ω. If A is any nonempty subset of ω^+, then either $N \in A$ for some natural number N or (excl.) $A = \{\omega\}$. In the first case $A \sim \{\omega\}$ has a first element x_0 with respect to $\leq$; x_0 is also the first element with respect to R. In the second case ω is the first element of A (with respect to R). Thus, R is a well-ordering on ω^+.

†EXAMPLE 2.7.3. Consider

$$\omega \times 2 = \{\langle M, N \rangle : M \in \omega \text{ and } N \text{ is either 0 or 1}\}.$$

Define the relation R on X by $\langle M_1, N_1 \rangle R \langle M_2, N_2 \rangle$ if either $N_1 < N_2$ or if both $N_1 = N_2$ and $M_1 \leq M_2$. (In Figure 2.7.1, every point of the bottom row precedes every point of the top row; points of the same row are R-related to themselves and to points to their right.)

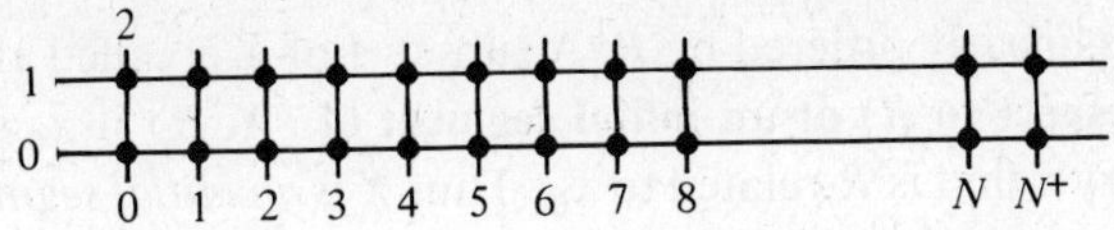

Fig. 2.7.1

It is not difficult to verify that R is a partial ordering. For any nonempty subset A of X, $A = A_0 \cup A_1$, where A_0 is the set of ordered pairs of A with second coordinate 0, and A_1 is the set of ordered pairs of A with second coordinate 1. The sets A_0 and A_1 are each in one-one correspondence with a subset of ω (the set of first coordinates in each case). Let R_I be the well-ordering induced in this manner (by $\leq$) on A_I, $I = 0, 1$. If $A_0 \neq \varnothing$, then the first element of A_0 (with respect to R_0) is the first element (with respect to R) of A. Otherwise, the first element of A_1 (with respect to R_1) is the first element of A.

A set can be well-ordered by one relation, yet fail to be well-ordered by another relation. For this reason we will often speak of well-ordered systems $\langle A, R\rangle$ rather than of well-orderable sets A.

†EXAMPLE 2.7.4. In Theorem 2.7.1, we showed that $\leq$ well-orders ω. However, the inverse relation, $\geq$, is a linear ordering on ω that is not a well-ordering:

$$\ldots \geq 4 \geq 3 \geq 2 \geq 1 \geq 0.$$

Clearly, ω has no first element with respect to this ordering.

If R is any relation on a set X and if $x \in X$, we define the **R-section determined by** x, written "$\text{sect}_R(x)$", to be $\{a \in X : aRx \text{ and } a \neq x\}$. *$\text{Sect}_R(x)$ is the set of elements of $X \sim \{x\}$ that are R-related to x.* We call A a **section of** X **(with respect to** R) if for some x in X, $A = \text{sect}_R(x)$. We also say that **the relational system** $\langle A, R\rangle$ **is a section of** $\langle X, R\rangle$. In particular, *if R well-orders X, then $\text{sect}_R(x)$ is empty iff x is the first element of X.*

EXAMPLE 2.7.5. Clearly, $\{0, 1, 2, \ldots, 9\}$ is a section with respect to the relation $\leq$ on ω; in fact, $\{0, 1, 2, \ldots, 9\} = \text{sect}_{\leq}(10)$. Observe that $\{0, 1, 2, \ldots, 9\}$ is also $\text{sect}_{\in_{11}}(10)$, where $\in_{11}$ is the relation $\{\langle x, y\rangle \in 11^{②} : x \in y\}$.

EXAMPLE 2.7.6. Let X be a set, let $A \subseteq X$, and let $\mathcal{I}_{\mathcal{P}(X)}$ be the inclusion relation on $\mathcal{P}(X)$. (See Exercise 2.6.12.) Then $\text{sect}\,\mathcal{I}_{\mathcal{P}(X)}(A) = \mathcal{P}(A) \sim \{A\}$, the set of proper subsets of A.

Let X be linearly ordered by R. A subset A of X is called an **initial segment of** X **(with respect to** R) or an **initial segment of** $\langle X, R\rangle$ if $x_0 \in A$ implies that $x \in A$ for every x that is R-related to x_0. Thus *X is an initial segment of itself with respect to any linear ordering on X*; all other initial segments of X (including $\varnothing$) are called **proper initial segments**. *Each section of X is a proper initial segment.*

Theorem 2.7.2. *A necessary and sufficient condition for a linear ordering R on X to be a well-ordering is for every proper initial segment A of X to be a section of X.*

PROOF: *Necessity*: Let R well-order X and let A be a proper initial segment of X. Then A is a proper subset of X; let x_0 be the first element of $X \sim A$. We claim that $A = \text{sect}_R(x_0)$. If aRx_0 and $a \neq x_0$, then $a \in A$ (because x_0 is the first element of $X \sim A$); thus, $\text{sect}_R(x_0) \subseteq A$. If $b \in A$ and cRb, then $c \in A$ (because A is an initial segment of X). In particular, not both $b \in A$ and x_0Rb (because $x_0 \notin A$). Thus, bRx_0 and $b \neq x_0$. Therefore, $b \in \text{sect}_R(x_0)$. It follows that $A = \text{sect}_R(x_0)$.

Sufficiency: Let R be a linear ordering on X that has the property that every proper initial segment of X (with respect to R) is a section of X. Let B be a nonempty subset of X and let $B' = \{x \in X \sim B : xRb \text{ for every } b \text{ in } B\}$. Possibly, B' is empty; in any case, if $x \in B'$ and yRx, surely $y \in B'$. Hence, B' is an initial segment of X. Moreover, it is a proper initial segment because $B \cap B' = \varnothing$. Thus, for some x_0 in X, $B' = \text{sect}_R(x_0)$. Clearly, $x_0 \notin \text{sect}_R(x_0)$. If for some b in B, bRx_0, then because $b \notin B'$, $b = x_0$. Thus, x_0Rb for every b in B; hence x_0 is the first element of B.

EXAMPLE 2.7.7. The linear ordering $\geq$ on ω (of Example 2.7.4) is not a well-ordering. Clearly, $\varnothing$ is a proper initial segment that is not a section of ω.

Theorem 2.7.3. The Principle of Transfinite Induction. *Let R be a well-ordering on X. Let A be a subset of X that is such that for each x in X, $\text{sect}_R(x) \subseteq A$ implies $x \in A$. Then $A = X$.*

PROOF: Suppose that A has this property and is a proper subset of X; then $X \sim A \neq \varnothing$. Let x_0 be the first element of $X \sim A$. It follows that $\text{sect}_R(x_0) \subseteq A$. But then $x_0 \in A$ Contradiction. Thus, $A = X$.

Corollary. **(Alternative Form of the Principle of Induction).** *Let B be a subset of ω that has the property that for each natural number N, whenever $N \subseteq B$ it follows that $N \in B$. Then $B = \omega$.*

PROOF: Observe that ω is well-ordered by $\subseteq$. For each N in ω, $\text{sect}_{\subseteq}(N) = N$. By the Principle of Transfinite Induction, if B is a subset of ω with the property that $N \subseteq B$ implies $N \in B$, then $B = \omega$.

The Principle of Induction enables us to prove that every natural number has a certain property; Transfinite Induction now provides a means of showing that every element of a given well-orderable set has some property in question.

Well-orderable sets will play a key role in our discussion of transfinite arithmetic (Chapter 5 and 6); the celebrated Well-Ordering Theorem will be the subject of Section 4.8.

Exercises

2.7.1. List all well-orderings on 3; on 4.

†2.7.2. Show that if R well-orders A and if $B \subseteq A$, then $R \cap B^{②}$ well-orders B.

***,†2.7.3.** Prove that a partial ordering R on a set A is a well-ordering iff every nonempty subset of A has a first element with respect to R.

†2.7.4. Show that if X and Y are well-orderable, then so are each of the following sets:

(a) $X \cap Y$. **(b)** $X \cup Y$. **(c)** $X \times Y$.

†2.7.5. **(a)** Let S be the relation on $\omega \times 2$ defined (in Example 2.7.3) by

$$\langle M_1, N_1 \rangle S \langle M_2, N_2 \rangle$$

if either $M_1 < M_2$ or if both $M_1 = M_2$ and $N_1 \leq N_2$. Is S a well-ordering?

(b) Show that $\omega^{②}$ is well-orderable.

2.7.6. Let X be linearly ordered by R. Show that with respect to R, (a) and (b) are true:

†(a) $\varnothing$ is an initial segment of X.

†(b) $\varnothing$ is a section of X iff X has a first element.

(c) Determine a set X and a linear ordering R on X for which (1) R is not a well-ordering, yet (2) $\varnothing$ is a section of X.

***,†2.7.7.** Let R be a relation on a set A. Show that R is a well-ordering iff every nonempty subset B of A contains a unique element that is R-related to every element of B.

†2.7.8. Prove that every linear ordering on a finite set is a well-ordering.

2.7.9. **(a)** Let $\langle X, R \rangle$ be a well-ordered system. Show that every element x_0, other than the last element (if it exists), of X has an "immediate successor"; i.e., there is some x_1 in $X \sim \{x_0\}$ such that $x_0 R x_1$, and if $x_0 R x_2$, then $x_1 R x_2$.

(b) Determine a well-ordered system $\langle X, R \rangle$ in which there is an element of X other than the first one that lacks an "immediate predecessor".

(This exercise will be useful for Chapter 5.)

2.8 Binary Operations

A **binary operation on a set** A is a function F from $A^{②}$ to A. Thus, a binary operation on A assigns an element of A to every ordered pair of elements of A. We generally write "$aFb = c$" instead of "$F(\langle a, b\rangle) = c$". Consequently, "$(aF_1b)F_2c$" stands for "$F_2(F_1(a, b), c)$" and "$(((aF_1b)F_2c)F_3(dF_4e))F_5f$" stands for "$F_5(F_3(F_2(F_1(a, b), c), F_4(d, e)), f)$".

†EXAMPLE 2.8.1. A binary operation, F, on 3 can be defined as follows:

$$0FN = NF0 = N, N = 0, 1, 2;$$

$$1F1 = 2; 1F2 = 2F1 = 0; 2F2 = 1.$$

What operation are we really describing?

We have already introduced various binary operations on arbitrary sets. Thus, if A and B are regarded as subsets of a universe $\mathscr{U}$, $\cup$, $\cap$, and $\sim$ are binary operations on $\mathscr{P}(\mathscr{U})$ because $A \cup B$, $A \cap B$, and $A \sim B$ are also subsets of $\mathscr{U}$. Another binary operation on $\mathscr{P}(\mathscr{U})$ is symmetric difference. If A and B are any two sets, we define their **symmetric difference**, $A \,\Delta\, B$, to be $(A \sim B) \cup (B \sim A)$. Clearly, if A and B are subsets of $\mathscr{U}$, so is $A \,\Delta\, B$.

EXAMPLE 2.8.2. $\{1, 3, 5, 7\} \,\Delta\, 5 = \{0, 2, 4, 5, 7\}$.

Complementation, on the other hand, is an example of a **unary operation on a set**, i.e., a function of a set into itself. If $A \subseteq \mathscr{U}$, then $\tilde{A} \subseteq \mathscr{U}$, as well; thus, complementation is a unary operation on $\mathscr{P}(\mathscr{U})$. The successor function on ω, $\{\langle N, N^+\rangle : N \in \omega\}$, is an example of a unary operation on ω.

EXAMPLE 2.8.3. A unary operation, G, on 3 can be defined as follows:

$$G(0) = 2; G(1) = 0; G(2) = 1.$$

Let F be a binary operation on A. Then F is said to be **associative** if for all a, b, c in A,

$$aF(bFc) = (aFb)Fc;$$

Furthermore, F is called **commutative** if for all a, b in A,

$$aFb = bFa.$$

If G is also a binary operation on A, then F is said to be **left distributive over** G if for all a, b, c in A,

$$aF(bGc) = (aFb)G(aFc);$$

and F is said to be **right distributive over** G if for all a, b, c in A,

$$(aGb)Fc = (aFc)G(bFc).$$

Finally, F is said to be **distributive over** G if it is both left and right distributive over G. *If F is commutative, then F is left distributive over G if and only if it is right distributive over G.* (Verify.) Theorem 1.8.1 indicates that for any set (universe) $\mathcal{U}$, union and intersection are each commutative and associative on $\mathcal{P}(\mathcal{U})$; furthermore, each of these operations is distributive over the other. If $\mathcal{U} \neq \varnothing$, relative complementation is neither commutative nor associative; however, intersection is distributive over relative complementation.

†Example 2.8.4. In Example 2.8.1, F is both commutative and associative on 3. Let G be the binary operation defined by

$$MGN = M; M, N = 0, 1, 2.$$

Then G is associative, but is noncommutative. F is distributive over G because F is commutative and because (for arbitrary M, N, R in 3)

$$MF(NGR) = MFN = (MFN)G(MFR).$$

However, G is right distributive over F, but is not left distributive over F. For

$$(MFN)GR = MFN = (MGR)F(NGR);$$

$$MG(NFR) = M, \text{ whereas } (MGN)F(MGR) = MFM \neq M, \text{ if } M = 1 \text{ or } 2.$$

Addition and multiplication of natural numbers are well-known examples of binary operations on ω; each of these will be defined (as sets) in Chapter 3, in accordance with our next major result:

Theorem 2.8.1. Definition by Induction. *Let X be a nonempty set, let $x_0 \in X$, and let $G : X \to X$. Then there is a unique function $F : \omega \to X$ that satisfies the following:*

(1) $F(0) = x_0$.

(2) *For all N in ω, $F(N^+) = G(F(N))$.*

[If we are given the function G and a particular element x_0 of X, we can define $F: \omega \to X$, inductively, by letting $F(0) = x_0$, $F(1) = G(x_0)$, $F(2) = G(G(x_0))$, and so forth.]

PROOF: Let $B = \{A : \langle 0, x_0\rangle \in A \in \mathscr{P}(\omega \times X)$ and for all $\langle N, x\rangle$ in $\omega \times X$, if $\langle N, x\rangle \in A$, then $\langle N^+, G(x)\rangle \in A\}$. Each A (in B) is a set of ordered pairs in $\omega \times X$; thus, A is a relation from ω to X. Clearly, $\omega \times X \in B$; thus, $B \neq \varnothing$. We claim that $\cap B$ is a function from ω to X.

We first show that $\cap B \in B$. Clearly, $\langle 0, x_0\rangle \in A$ for each A in B; hence, $\langle 0, x_0\rangle \in \cap B$. Each A in B is a subset of $\omega \times X$; hence, so is $\cap B$. If $\langle N, x\rangle \in \cap B$, then for each A in B, $\langle N, x\rangle \in A$. It follows that $\langle N^+, G(x)\rangle$ is in each A in B; hence $\langle N^+, G(x)\rangle \in \cap B$. Thus, $\cap B \in B$; moreover, for each A in B, $\cap B \subseteq A$.

Let $C = \{N \in \omega$: there is exactly one x in X such that $\langle N, x\rangle \in \cap B\}$.

Obviously, $\langle 0, x_0\rangle \in \cap B$. If for some $x_1 (\neq x_0)$ in X, $\langle 0, x_1\rangle \in \cap B$, then an argument similar to that of the second paragraph would show that $\cap B \sim \{\langle 0, x_1\rangle\}$ would be an element of B that is a proper subset of $\cap B$. Thus, x_0 is the unique element x of X for which $\langle 0, x\rangle \in \cap B$; hence, $0 \in C$.

Suppose that $N \in C$. Then $\langle N, x\rangle \in \cap B$ for some unique x in X. Because $\cap B \in B$, $\langle N^+, G(x)\rangle \in \cap B$. Suppose that $\langle N^+, x_2\rangle \in \cap B$ for some x_2 ($\neq G(x)$) in X. Again $\cap B \sim \{\langle N^+, x_2\rangle\}$ would be an element of B that is a proper subset of $\cap B$. Consequently, $G(x)$ is the unique element y of X for which $\langle N^+, y\rangle \in \cap B$; hence, $N^+ \in C$.

It follows, by induction, that $C = \omega$ and that for each natural number N there is precisely one x in X for which $\langle N, x\rangle \in \cap B$. Thus, $\cap B$ is a function from ω to X, as was to be shown.

If F is any function from ω to X that is such that $F(0) = x_0$ and such that for all N in ω, $F(N^+) = G(F(N))$, we let $D = \{N \in \omega : F(N) = (\cap B)(N)\}$. Then $0 \in D$ because $F(0) = (\cap B)(0) = x_0$. Suppose that $N \in D$. Then $F(N) = (\cap B)(N)$ and $F(N^+) = G(F(N)) = G((\cap B)(N)) = (\cap B)(N^+)$; hence, $N^+ \in D$. By induction, $D = \omega$, and the function $\cap B$ is uniquely characterized.

EXAMPLE 2.8.5. Let $X = \{a, b, c, d\}$; let $G(a) = c$, $G(b) = b$, $G(c) = G(d) = a$; let $x_0 = a$. Then F is defined by $F(0) = a$, $F(1) = c$, $F(2) = a$, $F(3) = c$, $F(4) = a$, $F(5) = c$, etc.

Exercises

2.8.1. Let A and B be subsets of a universe $\mathscr{U}$.

(a) Show that $A \,\Delta\, B = (A \cap \tilde{B}) \cup (B \cap \tilde{A})$.

(b) Show that $A \,\Delta\, B = (A \cup B) \sim (A \cap B)$.

(c) Draw a Venn Diagram for $\mathscr{U}$. To which regions does $A \,\Delta\, B$ correspond? (Useful for Section 6.8.)

2.8.2. Let A, B, and C be subsets of $\mathcal{U}$. Show that the symmetric difference operation is commutative, associative, and that intersection distributes over symmetric difference; i.e., show the following:

(a) $A \Delta B = B \Delta A$.

(b) $A \Delta (B \Delta C) = (A \Delta B) \Delta C$.

(c) $A \cap (B \Delta C) = (A \cap B) \Delta (A \cap C)$, $(A \Delta B) \cap C = (A \cap C) \Delta (B \cap C)$.

(Useful for Section 6.8.)

2.8.3. Let $A \subseteq \mathcal{U}$. Simplify each of the following expressions:

(a) $A \Delta \varnothing$. **(b)** $A \Delta \mathcal{U}$.

(c) $A \Delta A$. **(d)** $A \Delta \tilde{A}$.

2.8.4. Let A and B be subsets of $\mathcal{U}$. What relationship exists between A and B in each case?

(a) $A \Delta B = \varnothing$. **(b)** $A \Delta B = \mathcal{U}$.

(c) $(A \Delta B) \cup B = A$. **(d)** $A \cap (A \Delta B) = \varnothing$.

★2.8.5. Let X be a nonempty set and let F_1 and F_2 be binary operations on X. The "system" $\langle X, \langle F_1, F_2 \rangle\rangle$ is called a **ring** if conditions (1), (2), and (3) hold:

(1) F_1 is associative and commutative and for all x_1, x_2 in X, there is some y in X for which $x_1 F_1 y = x_2$. (We say that $\langle X, F_1 \rangle$ is a **commutative** or an **Abelian group**.)

(2) F_2 is associative.

(3) F_2 is distributive over F_1.

(a) Show that there is a *unique* z in X that is such that for all x in X, $xF_1 z = x$.

The *ring* is called **commutative** if F_2 is commutative; it is said to have an **identity** if for some e in $X \sim \{z\}$, $eF_2 x = xF_2 e = x$; it is called **Boolean** if it has an identity and if $xF_2 x = x$ for all x in X. (Note that the collection of integers under the usual definition of addition and multiplication yields an example of a non-Boolean commutative ring with an identity.)

(b) Show that for any universe $\mathcal{U}$, $\langle \mathcal{U}, \langle \Delta, \cap \rangle\rangle$ is a commutative Boolean ring with an identity. (Referred to in Exercises 3.2.16, 3.7.7, 3.10.14, and 3.12.8 and in Section 5.13.)

2.8.6. In Example 2.8.4, let $H(M) = 0$ for all M in 3.

(a) Is H commutative? **(b)** Is H associative?

(c) Is H distributive over F? Over G?

(d) Is F (is G) distributive over H?

2.8.7. Determine two unary operations on ω other than the successor operation.

2.8.8. Which of the following are binary operations on ω?

(a) $\{\langle\langle M, N\rangle, 0\rangle : M, N \in \omega\}$. **(b)** $\{\langle\langle M, 0\rangle, N\rangle : M, N \in \omega\}$.

(c) $\{\langle\langle M, N\rangle, M\rangle : M, N \in \omega\}$. **(d)** $\{\langle\langle M, N\rangle, N^+\rangle : M, N \in \omega\}$.

2.8.9. For each N in ω, how many binary operations are there on an N-element set?

★2.8.10. Prove that if X is a nonempty set, if $x_0 \in X$ and if $G: \mathscr{P}(X) \to X$, then there is a unique function F—with domain ω and range X—that satisfies (1) $F(0) = x_0$ and (2) $F(N^+) = G(\{F(N)\})$.

2.8.11. Let $X = \{a, b, c, d, e\}$ and let $G(a) = c$, $G(b) = d$, $G(c) = e$, $G(d) = a$, $G(e) = b$. Determine the function F—with domain ω and range X—whose existence is guaranteed by Definition by Induction in each case:
(a) $x_0 = a$. **(b)** $x_0 = b$.

2.8.12. Let $X = \{a, b, c, d, e\}$ and let $G(a) = c$, $G(b) = G(d) = d$, $G(c) = e$, $G(e) = a$. Determine the function $F : \omega \to X$ whose existence is guaranteed by Definition by Induction in each case:
(a) $x_0 = a$. **(b)** $x_0 = b$.

2.8.13. In the proof of Theorem 2.8.1, fill in the details of the argument that if $\langle 0, x_0 \rangle$ and $\langle 0, x_1 \rangle$ are both in $\cap B$ and if $x_0 \neq x_1$, then $\cap B \sim \{\langle 0, x_1 \rangle\} \in B$.

3

Real and Complex Numbers as Sets

3.1 Introduction

We have already introduced all the necessary concepts for defining the familiar real and complex numbers as sets. The purpose of this introductory section is twofold. For the unfamiliar reader we present a broad picture of how the ideas of the preceding chapters can be applied to construct various systems studied in mathematical analysis. On the other hand, undoubtedly many readers are already familiar with the general notions of this constructive approach, and will want to omit the details. Such readers need only read Sections 3.2 and 3.3 at this time before proceeding to Chapter 4. For these readers we outline the main concepts involved in the construction. The present section can be read before Section 4.5, if preferred. Adequate references to Sections 3.4–3.12 will be given subsequently.

Our starting point is the set, ω, of natural numbers. We use Definition by Induction to define addition and multiplication of natural numbers. We are unable to define the usual subtraction and division; at best, we can define a modified subtraction operation on ω.

An integer is defined to be an ordered pair of the form $\langle M, 0\rangle$ or $\langle 0, M\rangle$, where M is a natural number. The former is a "nonnegative integer" and is denoted by $\overline{M}$; the latter is a "nonpositive integer" and is denoted by "$-\overline{M}$". The set of integers is denoted by $\mathscr{Z}$.

Addition ($+_{\mathscr{Z}}$) and multiplication ($\cdot_{\mathscr{Z}}$) are defined on the set of integers in terms of the corresponding operations on ω. An operation $-_{\mathscr{Z}}$ is defined in terms of $+_{\mathscr{Z}}$; it expresses what one normally thinks of as subtraction of integers.

We introduce an equivalence relation, R_2, on $\mathscr{Z} \times (\mathscr{Z} \sim \{\overline{0}\})$, and define $\mathscr{Q}$, the set of rational numbers, as the corresponding set of equivalence classes. Briefly, if $\mathscr{M}_1$ and $\mathscr{M}_2$ are arbitrary integers and if $\mathscr{N}_1$ and $\mathscr{N}_2$ are nonzero integers, then

$$\langle \mathscr{M}_1, \mathscr{N}_1\rangle R_2 \langle \mathscr{M}_2, \mathscr{N}_2\rangle \text{ iff } \mathscr{M}_1 \cdot_{\mathscr{Z}} \mathscr{N}_2 = \mathscr{M}_2 \cdot_{\mathscr{Z}} \mathscr{N}_1.$$

We also sketch alternative approaches in which the integers are defined in terms of an equivalence relation, R_1, on $\mathscr{Z}$, and the rationals are defined as ordered pairs of integers.

All four of the usual operations—addition, multiplication, subtraction, and division—can be defined on $\mathscr{Q}$. However, "square roots" of "positive" rational numbers cannot always be extracted. Moreover, there will be bounded sets of rationals that have no least upper bound with respect to the "natural" ordering, $\leq_{\mathscr{Q}}$, on $\mathscr{Q}$; thus, there will be "gaps" in $\mathscr{Q}$. The real numbers, $\mathscr{R}$, will have neither of these defects. A real number (Dedekind cut) is defined to be a partition, $\{L, U\}$, of $\mathscr{Q}$ in which (with respect to $\leq_{\mathscr{Q}}$) L has no last element and in which each element of L precedes every element of U. Whereas there is no difficulty in extracting square roots of "nonnegative" real numbers, we are unable to apply this process to "negative" reals. This leads us to "extend" the set of real numbers to that of the complex numbers, $\mathscr{C}$; here, every element has a square root. A complex number is defined to be an ordered pair of real numbers.

In each of these stages the original set can be embedded in the new set in the sense that there exists a one-one mapping of the first set into the second that preserves the arithmetic operations and, except for the last stage, the "natural" order relation. Thus, the integers can be thought of as an extension of the natural numbers, the rational numbers as an extension of the integers (and, hence, of the natural numbers), etc.

We will utilize subscripts to indicate the set on which the various arithmetic operations and orderings are defined. Our convention will be to omit the subscript the first time a notion (such as addition or the natural ordering) is given; thus, there are no subscripts for the arithmetical operations and natural ordering on ω. Similarly, when division is defined on the rationals, we will write "$\mathscr{M}/\mathscr{N}$"; in contrast, we generally write "$/_{\mathscr{R}}$" and "$/_{\mathscr{C}}$" to indicate division of real and complex numbers, respectively. Each time we embed one set into another we will place a bar over the symbol for the embedded element. Thus "$\bar{0}$" will denote the integer zero, "$\bar{\bar{0}}$", the rational zero, etc. Finally, when the underlying set seems sufficiently clear from the context, we will sometimes omit these subscripts and bars.

3.2 Addition and Multiplication of Natural Numbers

In Definition by Induction (Theorem 2.8.1), let $X = \omega$ and let G be the successor function. Then for each M in ω there is a unique function F_M that is such that $F_M : \omega \to \omega$,

$$F_M(0) = M,$$

and

$$F_M(N^+) = G(F_M(N)) = F_M(N)^+ \qquad \text{for all } N \text{ in } \omega.$$

Let $+$ be the binary operation on ω defined by $M + N = F_M(N)$; i.e.,

$$+ = \{\langle\langle M, N\rangle, F_M(N)\rangle : M, N \in \omega\}.$$

†EXAMPLE 3.2.1. We show that for every natural number M,

$$M + 0 = 0 + M = M.$$

(1) By definition, $M + 0 = F_M(0) = M$.

(2) Let $A = \{M \in \omega : F_0(M) = M\}$. Then $F_0(0) = 0$ and $0 \in A$. Suppose that $K \in A$. Then $F_0(K) = K$; by definition, $F_0(K^+) = F_0(K)^+ = K^+$. Consequently, $K^+ \in A$; by the Principle of Induction, $A = \omega$. Thus, $0 + M = F_0(M) = M$ for every natural number M.

For each M in ω, let G_M be the unique operation on ω that satisfies

$$G_M(0) = 0$$

and

$$G_M(N^+) = F_M(G_M(N)) = M + G_M(N).$$

Let $\cdot$ be the binary operation on ω defined by $M \cdot N = G_M(N)$; i.e.,

$$\cdot = \{\langle\langle M, N\rangle, G_M(N)\rangle : M, N \in \omega\}.$$

[We will frequently write "MN" in lieu of "$M \cdot N$"; also, we will generally write "N^{++}" instead of "$(N^+)^+$".]

EXAMPLE 3.2.2

(1) $4 + 3 = F_4(3) = F_4(2)^+ = F_4(1)^{++} = F_4(0)^{+++} = 4^{+++} = 7.$

(2)
$$\begin{aligned} 4 \cdot 3 &= G_4(3) = F_4(G_4(2)) = F_4(F_4(G_4(1))) \\ &= F_4(F_4(F_4(G_4(0)))) = 4 + (4 + (4 + 0)) = 12. \end{aligned}$$

Addition and multiplication of natural numbers obey the following:

(1) *Commutativity:*

$$M + N = N + M, \; M \cdot N = N \cdot M.$$

(2) *Associativity:*

$$M + (N + P) = (M + N) + P, \; M \cdot (N \cdot P) = (M \cdot N) \cdot P.$$

Furthermore,

(3) *Multiplication distributes over addition:*

$$M \cdot (N + P) = (M \cdot N) + (M \cdot P), \; (M + N) \cdot P = (M \cdot P) + (N \cdot P).$$

(Here, $M, N, P \in \omega$.) The verification of these properties is left for the exercises. We write "$M + N + P$" for the sum of three natural numbers and "$M \cdot N \cdot P$" for the product. As is usual, we write "$M \cdot P + N \cdot P$" for "$(M \cdot P) + (N \cdot P)$", etc.

If $M_0, M_1, \ldots, M_N$ are natural numbers (not necessarily distinct), we write

$$\text{“}\sum_{K=0}^{N} M_K\text{” for “}M_0 + M_1 + \cdots + M_N\text{”}$$

and

$$\text{“}\prod_{K=0}^{N} M_K\text{” for “}M_0 \cdot M_1 \cdot \ldots \cdot M_N\text{”.}$$

We can also define **exponentation of natural numbers**, "$\exp(M, N)$", where M and N are not both 0, by induction (Exercise 3.2.10). We write "M^N" for "$\exp(M, N)$". We note that, in general, $M^{(N^R)} \neq (M^N)^R = M^{N \cdot R}$.

Neither subtraction nor division (in the usual sense) can be considered as binary operations on ω; each of these processes yields objects other than natural numbers. It is convenient to introduce a **modified subtraction on** ω as follows: Write

$$N \dot{-} M = \begin{cases} K, & \text{if there is a unique } K \text{ in } \omega \text{ such that } M + K = N; \\ 0, & \text{if } M > N. \end{cases}$$

These two cases are exhaustive and mutually exclusive—i.e., the first case is that in which $M \leq N$ (Exercise 3.2.6). It then follows that $N \dot{-} M = 0$ iff $M \geq N$.

Exercises

†3.2.1. Prove that for all natural numbers M, the following are true:
(a) $M + 1 = 1 + M = M^+$. **(b)** $M \cdot 1 = 1 \cdot M = M$.

†3.2.2. **(a)** Prove the associativity of addition of natural numbers; i.e., prove that for M, N, R in ω, $M + (N + R) = (M + N) + R$.
(b) Prove the commutativity of addition of natural numbers; i.e., prove that for M, N in ω, $M + N = N + M$.

†3.2.3. Prove that if $M, N, R \in \omega$ and $M + R = N + R$, then $M = N$.

†3.2.4. **(a)** Prove that $0 \cdot M = 0$ for all M in ω.
(b) Prove the commutativity of multiplication of natural numbers; i.e., prove that for M, N, in ω, $M \cdot N = N \cdot M$.
(c) Prove that multiplication distributes over addition:

$$M \cdot (N + R) = M \cdot N + M \cdot R;$$

$$(M + N) \cdot R = M \cdot R + N \cdot R \text{—for all } M, N, R \text{ in } \omega.$$

(d) Prove the associativity of multiplication of natural numbers; i.e., prove that for M, N, R in ω, $M \cdot (N \cdot R) = (M \cdot N) \cdot R$.

†3.2.5. Let $M, N, R \in \omega$.
(a) Prove that multiplication permits no "zero divisors"; i.e., prove that if $M \cdot N = 0$, then $M = 0$ or $N = 0$.
(b) Prove that if $M + N = 0$, then both M and N are zero.
(c) Prove that multiplicative "cancellation" holds: If $M \cdot N = M \cdot R$ and $M \neq 0$, then $N = R$.

†3.2.6. Prove that for M, N in ω, $M \leq N$ iff there exists a unique K in ω such that $M + K = N$.

†3.2.7. Let K, M, N be arbitrary natural numbers. Which of the following are true, in general?
(a) $K \cdot (N \dot{-} M) = (K \cdot N) \dot{-} (K \cdot M)$.
(b) $(N \dot{-} M) \dot{-} K = N \dot{-} (M \dot{-} K)$.
(c) $N \dot{-} (M + K) = (N \dot{-} M) \dot{-} K$.
(d) $M \dot{-} (M \dot{-} M^+) = M$.

†**3.2.8.** Show that if K, L, M, and N in ω are such that $K < L$ and $M \leq N$, then the following hold:

(a) $K + M < L + N$.

(b) $K \cdot M \leq L \cdot N$.

(c) If $N = K \cdot M > 0$ and $K \geq 2$, then $N > M$.

(d) $M \dot{-} L \leq N \dot{-} K$.

(e) $L \dot{-} K = M \dot{-} K$ implies $L = M$.

(f) When does equality hold in (b)? In (d)?

†**3.2.9.** Prove that if $C > B$ and $A + B = C + D$, then $A \dot{-} D = C \dot{-} B$.

†**3.2.10.** Define exponentiation of natural numbers by induction.

†**3.2.11.** Let $M, N, R \in \omega$. Show that exponentiation obeys the following rules:

(a) $M^{N+R} = M^N \cdot M^R$, if $M + NR > 0$.

(b) $(M \cdot N)^R = M^R \cdot N^R$, if $(M + R)(N + R) > 0$.

(c) $(M^N)^R = M^{N \cdot R}$, if $M + NR > 0$.

3.2.12. Suppose that H is defined by induction as follows:

$$H_M(0) = M,$$

$$H_M(N^+) = J_M(H_M(N)),$$

where

$$J_M(N) = \begin{cases} \exp(M, N), & \text{if } \langle M, N\rangle \in \omega^{②} \sim \{\langle 0, 0\rangle\}, \\ 1, & \text{if } M = N = 0; \end{cases}$$

$$H = \{\langle\langle M, N\rangle, H_M(N) > : M, N \in \omega\}.$$

Determine each:

(a) $H(1, 10)$. **(b)** $H(2, 1)$.

(c) $H(2, 3)$. **(d)** $H(3, 2)$.

3.2.13. Prove that the equation $3 \cdot X + 2 = 10$ has no solution in the natural numbers.

3.2.14. Prove that for all N in ω_+, $2 \cdot (1 + 2 + \cdots + N) = N \cdot (N + 1)$.

***3.2.15.** Let $K, N \in \omega$. Let $F(N) = 2^{N+2}$, $N \geq 0$. Show that for each $N \geq 0$ and for each $K \geq F(N)$,

(a) $K^N < 2^K$. **(b)** $(K + 1) \cdot 2^N < 2^{K \dot{-} 1}$.

(Used in Theorem 6.8.3.)

3.2.16. Is $\langle \omega, \langle +, \cdot \rangle\rangle$ a ring? (See Exercise 2.8.5.)

3.3 Fundamental Theorem of Arithmetic

Let $K, M, N \in \omega$. Then M is said to be a **factor of** N if for some K in ω, $N = K \cdot M$. For $N \geq 2$, N is said to be (a) **prime** if its only factors are 1 and N; if N is not (a) prime it is called (a) **composite**. According to Exercise 3.2.8(c), *if* $N = K \cdot M$ *and* $K \geq 2$, *then* $M < N$; clearly, *composites can be characterized* *by the fact that they are the natural numbers N for which* $N = K \cdot M$ *for* $2 \leq K < N$ *and* $2 \leq M < N$.

EXAMPLE 3.3.1

(1) 1, 2, 3, 4, 6, and 12 are (all) the factors of 12.
(2) Every natural number is a factor of 0.
(3) The first few primes are 2, 3, 5, 7, 11, and 13.
(4) 90, 91, 92, 93, 94, 95, and 96 are composites.

Theorem 3.3.1. The Fundamental Theorem of Arithmetic. *Let* $N \geq 2$. *Then* N *has a unique factorization of the form*

$$N = P_1^{E_1} \cdot P_2^{E_2} \cdot \ldots \cdot P_S^{E_S}, \tag{3.3.1}$$

where $P_1, P_2, \ldots, P_S$ *are primes,* $P_1 < P_2 < \cdots < P_S$, *and* $S, E_1, E_2, \ldots, E_S \in \omega_+$.

[A factorization of the form (3.3.1) is called a **prime factorization of** N.]

PROOF: We first consider the existence of a prime factorization, then its uniqueness. In each part we utilize (a variant of) the Principle of Induction.

Existence: Observe that 2 is already of the form (3.3.1) with $S = 1, P_1 = 2, E_1 = 1$. Similarly, every other prime is of the form (3.3.1).

Let N (> 2) be a composite and assume that for all N', $2 < N' < N$, N' has a prime factorization. Then $N = K \cdot M$, for $2 \leq K < N, 2 \leq M < N$. By the inductive hypothesis, $K = Q_1^{F_1} \cdot Q_2^{F_2} \cdot \ldots \cdot Q_T^{F_T}$, $M = R_1^{G_1} \cdot R_2^{G_2} \cdot \ldots \cdot R_U^{G_U}$, where $Q_1, Q_2, \ldots, Q_T, R_1, R_2, \ldots, R_U$ are primes, $Q_1 < Q_2 < \cdots < Q_T$, $R_1 < R_2 < \cdots < R_U$, and T, U, $F_1, F_2, \ldots, F_T, G_1, G_2, \ldots, G_U \in \omega_+$. Let $\hat{Q} = \{Q_1, Q_2, \ldots, Q_T\}$ and $\hat{R} = \{R_1, R_2, \ldots, R_U\}$; then $\hat{Q} \cup \hat{R}$ is an S-element set for some S in ω_+. Let $P_1 = \min \hat{Q} \cup \hat{R}$, let $P_2 = \min ((\hat{Q} \cup \hat{R}) \sim \{P_1\}), \ldots$, let $P_S = \max \hat{Q} \cup \hat{R}$; for $I = 1, 2, \ldots, S$, let

$$E_I = \begin{cases} F_J, & \text{if } P_I = Q_J \in \hat{Q} \sim \hat{R}; \\ G_K, & \text{if } P_I = R_K \in \hat{R} \sim \hat{Q}; \\ F_J + G_K, & \text{if } P_I = Q_J \in \hat{Q} \text{ and } P_I = R_K \in \hat{R}. \end{cases}$$

Then $N = P_1^{E_1} \cdot P_2^{E_2} \cdot \ldots \cdot P_S^{E_S}$ is of the form (3.3.1).

Uniqueness: If N is prime, its prime factorization is unique. Suppose that N is composite and that for all N', $2 \leq N' < N$, N' has a unique prime factorization. Suppose, further, that N has the distinct factorizations

$$N = P_1^{E_1} \cdot P_2^{E_2} \cdot \ldots \cdot P_S^{E_S} = Q_1^{F_1} \cdot Q_2^{F_2} \cdot \ldots \cdot Q_T^{F_T},$$

each of the form (3.3.1). Then $S > 1$ or $E_1 > 1$, and $T > 1$ or $F_1 > 1$. Moreover, P_1 cannot equal Q_1, as otherwise, $N = P_1 \cdot M$, where $M < N$. But then, in contradiction to the inductive hypothesis, M has the distinct factorizations

$$M = \tilde{P}_1^{\tilde{E}_1} \cdot \tilde{P}_2^{\tilde{E}_2} \cdot \ldots \cdot \tilde{P}_{\tilde{S}}^{\tilde{E}_{\tilde{S}}} = \tilde{Q}_1^{\tilde{F}_1} \cdot \tilde{Q}_2^{\tilde{F}_2} \cdot \ldots \cdot \tilde{Q}_{\tilde{T}}^{\tilde{F}_{\tilde{T}}}.$$

Here, in case $E_1 > 1$, $\tilde{E}_1 = E_1 \dot{-} 1$, whereas $\tilde{S}, \tilde{P}_1, \tilde{P}_2, \ldots, \tilde{P}_{\tilde{S}}, \tilde{E}_2, \ldots, \tilde{E}_{\tilde{S}}$ are, respectively, $S, P_1, P_2, \ldots, P_S, E_2, \ldots, E_S$; if $E_1 = 1$, then $\tilde{S} = S \dot{-} 1$ and $\tilde{P}_J = P_{J+1}$, $\tilde{E}_J = E_{J+1}$ for $J = 1, 2, \ldots, \tilde{S}$. A similar remark applies to the second factorization.

Suppose, arbitrarily, that $P_1 < Q_1$; then $1 \leq Q_1 \dot{-} P_1 < Q_1 < N$. Let

(3.3.2) $$L = N \dot{-} P_1 \cdot Q_1^{F_1 \dot{-} 1} \cdot Q_2^{F_2} \cdot Q_3^{F_3} \cdot \ldots \cdot Q_T^{F_T}.$$

Then

(3.3.2′) $$\begin{aligned} L &= Q_1^{F_1} \cdot Q_2^{F_2} \cdot \ldots \cdot Q_T^{F_T} \dot{-} P_1 \cdot Q_1^{F_1 \dot{-} 1} \cdot Q_2^{F_2} \cdot Q_3^{F_3} \cdot \ldots \cdot Q_T^{F_T} \\ &= (Q_1 \dot{-} P_1) \cdot (Q_1^{F_1 \dot{-} 1} \cdot Q_2^{F_2} \cdot Q_3^{F_3} \cdot \ldots \cdot Q_T^{F_T}). \end{aligned}$$

Furthermore,

(3.3.2″) $$\begin{aligned} L &= P_1^{E_1} \cdot P_2^{E_2} \cdot \ldots \cdot P_S^{E_S} \dot{-} P_1 \cdot Q_1^{F_1 \dot{-} 1} \cdot Q_2^{F_2} \cdot Q_3^{F_3} \cdot \ldots \cdot Q_T^{F_T} \\ &= P_1 \cdot (P_1^{E_1 \dot{-} 1} \cdot P_2^{E_2} \cdot \ldots \cdot P_S^{E_S} \dot{-} Q_1^{F_1 \dot{-} 1} \cdot Q_2^{F_2} \cdot Q_3^{F_3} \cdot \ldots \cdot Q_T^{F_T}). \end{aligned}$$

It follows from (3.3.2) and (3.3.2′) that $2 \leq L < N$ and, hence, that the prime factorization of L is unique. By (3.3.2″),

(3.3.3) P_1 is a prime factor of L.

Also, $Q_1 \dot{-} P_1$ is not 1, for otherwise, contrary to (3.3.3), (3.3.2′) would indicate that all the prime factors of L are in $\{Q_1, Q_2, \ldots, Q_T\}$. Thus $2 \leq Q_1 \dot{-} P_1 < N$ as well as $2 \leq Q_1^{F_1 \dot{-} 1} \cdot Q_2^{F_2} \cdot \ldots \cdot Q_T^{F_T} < N$. By the inductive hypothesis, each of the factors of L in (3.3.2′) (as well as L itself) has a unique prime factorization. Suppose V is the set of prime factors of $Q_1 \dot{-} P_1$. Then the set of prime factors of L is included in $V \cup \{Q_1, Q_2, \ldots, Q_T\}$. It follows from (3.3.3) that $P_1 \in V$. Thus, $Q_1 \dot{-} P_1 = K \cdot P_1$, $K \geq 1$, and $Q_1 = P_1 \cdot (K + 1)$. But this implies that Q_1 is composite. This contradiction establishes the uniqueness of the prime factorization of N.

EXAMPLE 3.3.2. The following are examples of prime factorizations:

(1) $17 = 17^1$. (2) $27 = 3^3$.
(3) $60 = 2^2 \cdot 3^1 \cdot 5^1$. (4) $288 = 2^5 \cdot 3^2$.
(5) $840 = 2^3 \cdot 3^1 \cdot 5^1 \cdot 7^1$.

Exercises

3.3.1. Classify each of the following natural numbers as either prime or composite. For each composite, indicate the prime factorization:

(a) 29. **(b)** 43. **(c)** 49. **(d)** 51.
(e) 53. **(f)** 57. **(g)** 61. **(h)** 63.

3.3.2. Determine the prime factorization of each:

(a) 96. **(b)** 124.
(c) 164. **(d)** 180.

3.3.3. Determine the prime factorization of each:

(a) 1176. **(b)** 1701.
(c) 2160. **(d)** 20,449.

3.3.4. Which of the following are equal?

(a) $3^7 \cdot 6^4$. **(b)** $2^4 \cdot 9^6$.
(c) $2^4 \cdot 3^{11}$. **(d)** $2^4 \cdot 3^1 \cdot 9^6$.

3.3.5. Two natural numbers, M and N, are said to be **relatively prime** if they have no prime factors in common. Determine all pairs of relatively prime numbers among the following:

(a) 3. **(b)** 8. **(c)** 12.
(d) 15. **(e)** 20.

(This definition is used in Exercise 4.1.12 and in Section 4.5; it is referred to in Section 3.6.)

3.3.6. Let $K, M, N, R \in \omega$. Suppose that $N = K \cdot M + R$, where $0 < R < M$. Prove that M is not a factor of N. (Used in Theorem 4.1.5.)

3.3.7. Let $M, R \in \omega_+$ and $N \in \omega$. We say that M **divides** N and that N **is a multiple of** M—if M is a factor of N. We write "$M|N$" in this case. (See Examples 2.5.1 and 2.6.2.) Prove the following:

(a) If P is a prime and $P|M \cdot N$, then $P|M$ or $P|N$.
(b) If R and M are relatively prime and $R|M \cdot N$, then $R|N$.

★3.3.8. Let M and N be natural numbers. We say that R is the **greatest common divisor of** M **and** N if R is the largest natural number that divides both M and N. We write R as "(M, N)". We say that S is the **least common multiple of** M **and** N if S is the smallest nonzero natural number that is a multiple of both M and N. We write S as "$[M, N]$".

(a) Show that if M and N are not both 0, then (M, N) exists.
(b) Show that if neither M nor N equals 0, then $[M, N]$ exists.
In the remaining parts assume that $M \geq 2$ and $N \geq 2$.
(c) Express (M, N) in terms of the prime factorizations of M and N.
(d) Express $[M, N]$ in terms of the prime factorizations of M and N.
(e) Determine the relationship between (M, N) and $[M, N]$.

★3.3.9. Let N be a nonzero natural number.

(a) Determine the number of divisors of N.
(b) Express the sum of the divisors of N.

★3.3.10. Show that for every nonzero natural number N there are N consecutive composites.

3.3.11. **(a)** Show that every composite N has a prime factor $P < \sqrt{N}$ (i.e., $P^2 < N$).
(b) Use (a) to determine all primes < 200.

★3.3.12. **(The Division Algorithm)** Let M ($\neq 0$) and N be natural numbers. Prove that there exist unique natural numbers K and R such that (1) $N = K \cdot M + R$ and (2) $0 \leq R < M$. (Referred to Exercise 3.4.14 and in Sections 5.7, 6.5, and 6.6.)

3.4 Integers

Let $\mathscr{Z} = (\omega \times 1) \cup (1 \times \omega_+)$; $\mathscr{Z}$ is called the **set of integers**. Thus, an integer is an ordered pair that is either of the form $\langle M, 0\rangle$ or $\langle 0, M\rangle$ for some natural number M. Theorem 2.1.1 guarantees that $\langle 0, 0\rangle$ is the only integer that admits both of these forms and that the natural number M in the representation of an integer is unique. Moreover, according to Exercise 2.1.6, an integer is not a natural number; i.e., $\mathscr{Z} \cap \omega = \varnothing$.

Let $M \in \omega$. We write "$\overline{M}$" for integers of the form "$\langle M, 0\rangle$" and "$-\overline{M}$" for those of the form "$\langle 0, M\rangle$". Thus, both "$\overline{0}$" and "$-\overline{0}$" represent "$\langle 0, 0\rangle$".

We proceed to define addition, $+_{\mathscr{Z}}$, multiplication, $\cdot_{\mathscr{Z}}$, and (a genuine) subtraction, $-_{\mathscr{Z}}$, on $\mathscr{Z}$. (The present definitions are due to M. Matzka.) Let $\langle M_1, N_1\rangle, \langle M_2, N_2\rangle \in \mathscr{Z}$. We define **addition of integers** by

$$\langle M_1, N_1\rangle +_{\mathscr{Z}} \langle M_2, N_2\rangle$$
$$= \langle (M_1 + M_2) \dot{-} (N_1 + N_2), (N_1 + N_2) \dot{-} (M_1 + M_2)\rangle$$

We define **multiplication of integers** by

$$\langle M_1, N_1\rangle \cdot_{\mathscr{Z}} \langle M_2, N_2\rangle = \langle M_1 \cdot M_2 + N_1 \cdot N_2, M_1 \cdot N_2 + M_2 \cdot N_1\rangle.$$

Thus, addition and multiplication of integers are defined in terms of the arithmetic of natural numbers. Observe that $M_1 = 0$ or $N_1 = 0$, and $M_2 = 0$ or $N_2 = 0$. Recall that for $N, M \in \omega$,

$$N \dot{-} M = \begin{cases} K, & \text{if there is a unique } K \text{ in } \omega \text{ such that } M + K = N; \\ 0, & \text{otherwise.} \end{cases}$$

The reader should verify that $+_{\mathscr{Z}}$ and $\cdot_{\mathscr{Z}}$ are binary operations on $\mathscr{Z}$—i.e., in all possible cases for M_1, N_1., M_2, and N_2, the sums and products are in $\mathscr{Z}$.

EXAMPLE 3.4.1

(1) $\overline{6} +_{\mathscr{Z}} \overline{4} = \langle 6,0\rangle +_{\mathscr{Z}} \langle 4,0\rangle = \langle (6+4) \dot{-} (0+0), (0+0) \dot{-} (6+4)\rangle$
$$= \langle 10 \dot{-} 0, 0 \dot{-} 10\rangle = \langle 10,0\rangle = \overline{10}.$$

(2) $6 +_{\mathscr{Z}} -\overline{4} = \langle 6,0\rangle +_{\mathscr{Z}} \langle 0,4\rangle$
$$= \langle (6+0) \dot{-} (0+4), (0+4) \dot{-} (6+0)\rangle$$
$$= \langle 6 \dot{-} 4, 4 \dot{-} 6\rangle = \langle 2,0\rangle = \overline{2}.$$

(3) $-\overline{6} +_{\mathscr{Z}} -\overline{4} = \langle 0,6\rangle +_{\mathscr{Z}} \langle 0,4\rangle$
$$= \langle (0+0) \dot{-} (6+4), (6+4) \dot{-} (0+0)\rangle$$
$$= \langle 0 \dot{-} 10, 10 \dot{-} 0\rangle = \langle 0,10\rangle = -\overline{10}.$$

(4) $\overline{0} +_{\mathscr{Z}} -\overline{6} = \langle 0,0\rangle +_{\mathscr{Z}} \langle 0,6\rangle$
$$= \langle (0+0) \dot{-} (0+6), (0+6) \dot{-} (0+0)\rangle$$
$$= \langle 0 \dot{-} 6, 6 \dot{-} 0\rangle = \langle 0,6\rangle = -\overline{6}.$$

EXAMPLE 3.4.2

(1) $\overline{6} \cdot_{\mathscr{Z}} \overline{4} = \langle 6,0\rangle \cdot_{\mathscr{Z}} \langle 4,0\rangle = \langle 6\cdot 4 + 0\cdot 0, 6\cdot 0 + 4\cdot 0\rangle$
$$= \langle 24,0\rangle = \overline{24}.$$

(2) $\overline{6} \cdot_{\mathscr{Z}} -\overline{4} = \langle 6,0\rangle \cdot_{\mathscr{Z}} \langle 0,4\rangle = \langle 6\cdot 0 + 0\cdot -4, 6\cdot 4 + 0\cdot 0\rangle$
$$= \langle 0,24\rangle = -\overline{24}.$$

(3) $-\overline{6} \cdot_{\mathscr{Z}} -\overline{4} = \langle 0,6\rangle \cdot_{\mathscr{Z}} \langle 0,4\rangle = \langle 0\cdot 0 + 6\cdot 4, 0\cdot 4 + 0\cdot 6\rangle$
$$= \langle 24,0\rangle = \overline{24}.$$

(4) $\overline{0} \cdot_{\mathscr{Z}} -\overline{6} = \langle 0,0\rangle \cdot_{\mathscr{Z}} \langle 0,6\rangle = \langle 0\cdot 0 + 0\cdot 6, 0\cdot 6 + 0\cdot 0\rangle$
$$= \langle 0,0\rangle = \overline{0}.$$

Let $M, M' \in \omega$; let $N, N' \in \omega_+$. It is easily established that addition of integers satisfies the following:

(1) $\overline{M} +_{\mathscr{Z}} \overline{M'} = \overline{M + M'}$.

(2) $\overline{M} +_{\mathscr{Z}} -\overline{N} = -\overline{N} +_{\mathscr{Z}} \overline{M} = \begin{cases} \overline{M \dot{-} N}, & \text{if } M \geq N, \\ -\overline{N \dot{-} M}, & \text{if } M < N. \end{cases}$

(3) $-\overline{N} +_{\mathscr{Z}} -\overline{N'} = -(\overline{N + N'})$.

Multiplication of integers satisfies the following:

(4) $\overline{M} \cdot_{\mathscr{Z}} \overline{M'} = \overline{M \cdot M'}$.

(5) $\overline{M} \cdot_{\mathscr{Z}} -\overline{N} = -\overline{N} \cdot_{\mathscr{Z}} \overline{M} = -\overline{M \cdot N}$.

(6) $-\overline{N} \cdot_{\mathscr{Z}} -\overline{N'} = \overline{N \cdot N'}$.

EXAMPLE 3.4.3

(1) $\overline{6} +_{\mathscr{Z}} \overline{4} = \overline{6 + 4} = \overline{10}; \overline{6} \cdot_{\mathscr{Z}} \overline{4} = \overline{6 \cdot 4} = \overline{24}$.

(2) $\overline{6} +_{\mathscr{Z}} -\overline{4} = \overline{6 \dot{-} 4} = \overline{2}; \overline{4} +_{\mathscr{Z}} -\overline{6} = -\overline{(6 \dot{-} 4)} = -\overline{2}$;

$\overline{6} \cdot_{\mathscr{Z}} -\overline{4} = \overline{4} \cdot_{\mathscr{Z}} -\overline{6} = -\overline{24}$.

(3) $-\overline{6} +_{\mathscr{Z}} -\overline{4} = -\overline{6 + 4} = -\overline{10}; -\overline{6} \cdot_{\mathscr{Z}} -\overline{4} = \overline{6 \cdot 4} = \overline{24}$.

(4) $\overline{0} +_{\mathscr{Z}} -\overline{6} = -\overline{(6 \dot{-} 0)} = -\overline{6}; \overline{0} \cdot_{\mathscr{Z}} -\overline{6} = -\overline{0 \cdot 6} = -\overline{0} (= \overline{0})$.

The definitions of these arithmetic operations on $\mathscr{Z}$ together with the properties of + and · (on ω) established in Chapter 2 immediately yield that *addition and multiplication of integers each obey commutativity and associativity*; furthermore, *multiplication distributes over addition.*

Thus, let $\mathscr{L}, \mathscr{M}, \mathscr{N} \in \mathscr{Z}$.

$$\mathscr{L} +_{\mathscr{Z}} \mathscr{M} = \mathscr{M} +_{\mathscr{Z}} \mathscr{L};\ \mathscr{L} \cdot_{\mathscr{Z}} \mathscr{M} = \mathscr{M} \cdot_{\mathscr{Z}} \mathscr{L}.$$

$$\mathscr{L} +_{\mathscr{Z}} (\mathscr{M} +_{\mathscr{Z}} \mathscr{N}) = (\mathscr{L} +_{\mathscr{Z}} \mathscr{M}) +_{\mathscr{Z}} \mathscr{N};$$

$$\mathscr{L} \cdot_{\mathscr{Z}} (\mathscr{M} \cdot_{\mathscr{Z}} \mathscr{N}) = (\mathscr{L} \cdot_{\mathscr{Z}} \mathscr{M}) \cdot_{\mathscr{Z}} \mathscr{N}.$$

$$\mathscr{L} \cdot_{\mathscr{Z}} (\mathscr{M} +_{\mathscr{Z}} \mathscr{N}) = \mathscr{L} \cdot_{\mathscr{Z}} \mathscr{M} +_{\mathscr{Z}} \mathscr{L} \cdot_{\mathscr{Z}} \mathscr{N};$$

$$(\mathscr{L} +_{\mathscr{Z}} \mathscr{M}) \cdot_{\mathscr{Z}} \mathscr{N} = \mathscr{L} \cdot_{\mathscr{Z}} \mathscr{N} +_{\mathscr{Z}} \mathscr{M} \cdot_{\mathscr{Z}} \mathscr{N}.$$

Clearly, $\overline{0}$ serves as an **additive identity** and $\overline{1}$ as a **multiplicative identity**—i.e., for any integer $\mathscr{M}$,

$$\mathscr{M} +_{\mathscr{Z}} \overline{0} = \overline{0} +_{\mathscr{Z}} \mathscr{M} = \mathscr{M};$$

$$\mathscr{M} \cdot_{\mathscr{Z}} \overline{1} = \overline{1} \cdot_{\mathscr{Z}} \mathscr{M} = \mathscr{M}.$$

It is an easy exercise to establish that $\overline{0}$ and $\overline{1}$ are the only integers that have these (respective) properties. Thus, *if $\mathscr{M}$ and $\mathscr{N}$ are arbitrary integers and $\mathscr{M} +_{\mathscr{Z}} \mathscr{N} = \mathscr{M}$, then $\mathscr{N}$ must be $\overline{0}$; if $\mathscr{K}$ is a nonzero integer and $\mathscr{K} \cdot_{\mathscr{Z}} \mathscr{N} = \mathscr{K}$, then $\mathscr{N}$ must be $\overline{1}$.*

Before considering subtraction of integers, we mention that **exponentiation** can be defined in terms of multiplication, by means of Definition by Induction, in essentially the same way as in Section 3.2. Thus, if $\mathscr{M}$ is any nonzero integer, we define

$$\mathscr{M}^0 = \overline{1}$$
$$\mathscr{M}^{N^+} = \mathscr{M}^N \cdot_{\mathscr{Z}} \mathscr{M}, \qquad N \in \omega;$$

furthermore, we define

$$\overline{0}^N = \overline{0}, \qquad N \in \omega_+.$$

Thus, exponentiation of integers, as we have defined it, is a mapping

$$\mathscr{E} : ((\mathscr{Z} \sim \{\overline{0}\}) \times \omega) \cup (\{\overline{0}\} \times \omega_+) \to \mathscr{Z}.$$

For $\mathscr{M} = \langle M, N \rangle \in \mathscr{Z}$, define "$-\mathscr{M}$" to be $\langle N, M \rangle$. Clearly,

$$\mathscr{M} +_{\mathscr{Z}} (-\mathscr{M}) = \overline{0}.$$

The integer $-\mathscr{M}$ is called the **additive inverse of** $\mathscr{M}$. Note that *for* $M \geq 0$, $-\overline{M}$ *is the additive inverse of* $\overline{M}$; hence, our present notation is consistent with our previous usage. Also, if $\mathscr{N}$ *is the additive inverse of* $\mathscr{M}$, *then* $\mathscr{M}$ *is the additive inverse of* $\mathscr{N}$; thus, *for* $N \geq 0$, $- -\overline{N} = \overline{N}$. We define **subtraction of integers** by

$$\mathscr{M} -_{\mathscr{Z}} \mathscr{N} = \mathscr{M} +_{\mathscr{Z}} -\mathscr{N}, \mathscr{M}, \mathscr{N} \in \mathscr{Z}.$$

Hence,

(3.4.1) $\qquad \mathscr{M} -_{\mathscr{Z}} \mathscr{N} = \mathscr{K}$ *if and only if* $\mathscr{K} +_{\mathscr{Z}} \mathscr{N} = \mathscr{M}$.

In fact, if $\mathscr{M} -_{\mathscr{Z}} \mathscr{N} = \mathscr{K}$, then

$$\begin{aligned}
\mathscr{K} +_{\mathscr{Z}} \mathscr{N} &= (\mathscr{M} -_{\mathscr{Z}} \mathscr{N}) +_{\mathscr{Z}} \mathscr{N} \\
&= (\mathscr{M} +_{\mathscr{Z}} -\mathscr{N}) +_{\mathscr{Z}} \mathscr{N} \\
&= \mathscr{M} +_{\mathscr{Z}} (-\mathscr{N} +_{\mathscr{Z}} \mathscr{N}) \\
&= \mathscr{M} +_{\mathscr{Z}} \overline{0} \\
&= \mathscr{M}.
\end{aligned}$$

Conversely, if $\mathscr{K} +_{\mathscr{Z}} \mathscr{N} = \mathscr{M}$, then

$$\begin{aligned}
\mathscr{M} -_{\mathscr{Z}} \mathscr{N} &= (\mathscr{K} +_{\mathscr{Z}} \mathscr{N}) -_{\mathscr{Z}} \mathscr{N} \\
&= (\mathscr{K} +_{\mathscr{Z}} \mathscr{N}) +_{\mathscr{Z}} -\mathscr{N} \\
&= \mathscr{K} +_{\mathscr{Z}} (\mathscr{N} +_{\mathscr{Z}} -\mathscr{N}) \\
&= \mathscr{K} +_{\mathscr{Z}} \overline{0} \\
&= \mathscr{K}.
\end{aligned}$$

EXAMPLE 3.4.4

(1) $\overline{8} -_{\mathscr{Z}} \overline{3} = \overline{5}$ because $\overline{5} +_{\mathscr{Z}} \overline{3} = \overline{8}$.
(2) $\overline{8} -_{\mathscr{Z}} -\overline{3} = \overline{11}$ because $\overline{11} +_{\mathscr{Z}} -\overline{3} = \overline{11 \dot{-} 3} = \overline{8}$.
(3) $-\overline{8} -_{\mathscr{Z}} -\overline{3} = -\overline{5}$ because $-\overline{5} +_{\mathscr{Z}} -\overline{3} = -\overline{5+3} = -\overline{8}$.

For any integers $\mathscr{M}$ and $\mathscr{N}$,

(1) $-(\mathscr{M} +_{\mathscr{Z}} \mathscr{N}) = (-\mathscr{M}) +_{\mathscr{Z}} (-\mathscr{N})$.
(2) $-(\mathscr{M} -_{\mathscr{Z}} \mathscr{N}) = \mathscr{N} -_{\mathscr{Z}} \mathscr{M}$.
(3) $-(\mathscr{M} \cdot_{\mathscr{Z}} \mathscr{N}) = (-\mathscr{M}) \cdot_{\mathscr{Z}} \mathscr{N} = \mathscr{M} \cdot_{\mathscr{Z}} (-\mathscr{N})$.

Here, (1) asserts that the additive inverse of the sum of two integers is the sum of their respective inverses.

We next consider the natural ordering of the integers; this is closely interrelated with the concept of "positiveness".

Every natural number N satisfies either (1) $N = 0$ or (excl.) (2) $N > 0$. In case (2), N is said to be positive. Clearly, the sum as well as the product of positive natural numbers are positive (Exercise 3.2.5).

An integer $\mathscr{M}$ **is** said to be **positive** if $\mathscr{M} = \overline{M}$ for some positive natural number M.

Theorem 3.4.1

(1) *The sum of two positive integers is positive.*
(2) *The product of two positive integers is positive.*
(3) **(Law of Trichotomy)** *For any integer $\mathscr{M}$, exactly one of the following holds:*
   (a) $\mathscr{M} = \overline{0}$.
   (b) *$\mathscr{M}$ is positive.*
   (c) *$-\mathscr{M}$ is positive.*

PROOF: [Parts (1) and (2) are left as an exercise.]

Part (3): For M_1, M_2, N_1, N_2 in ω,

$$\langle M_1, N_1 \rangle = \langle M_2, N_2 \rangle \text{ iff both } M_1 = M_2 \text{ and } N_1 = N_2.$$

It follows that $\mathscr{Z}$ is the disjoint union of the (nonempty) sets

$$\mathscr{Z}_0 = \{\langle 0, 0 \rangle\} = \{\overline{0}\};$$

$$\mathscr{Z}_+ = \{\langle M, 0 \rangle : M \in \omega_+\} = \{\overline{M} : M \in \omega_+\};$$

$$\mathscr{Z}_- = \{\langle 0, N \rangle : N \in \omega_+\} = \{-\overline{N} : N \in \omega_+\}.$$

The Law of Trichotomy follows from this partitioning of the integers.

An **integer** $\mathscr{M}$ **is negative** if case (c) of (3) holds. Thus, $\mathscr{Z}_+$ is the set of positive integers and $\mathscr{Z}_-$, the set of negative integers; $\mathscr{Z}_+ \cup \{\overline{0}\}$ is the set of nonnegative integers.

Let $\mathscr{M}$ and $\mathscr{N}$ be integers. We define "$\mathscr{M} <_{\mathscr{Z}} \mathscr{N}$" to mean that $\mathscr{N} -_{\mathscr{Z}} \mathscr{M}$ is positive. In terms of $<_{\mathscr{Z}}$ we can introduce the relations $\leq_{\mathscr{Z}}, >_{\mathscr{Z}}, \geq_{\mathscr{Z}}$ in the obvious manner. Thus "$\mathscr{M} \leq_{\mathscr{Z}} \mathscr{N}$" means "either $\mathscr{M} <_{\mathscr{Z}} \mathscr{N}$ or $\mathscr{M} = \mathscr{N}$". From the Law of Trichotomy it follows that *for any integers* $\mathscr{M}$ *and* $\mathscr{N}$, *exactly one of the relations*

$$\mathscr{M} <_{\mathscr{Z}} \mathscr{N}, \ \mathscr{M} = \mathscr{N}, \ \mathscr{M} >_{\mathscr{Z}} \mathscr{N}$$

holds. It is not difficult to establish that $\mathscr{M} \leq_{\mathscr{Z}} \mathscr{N}$ *if and only if for some* M, N *in* ω, *(at least) one of the following cases applies*:

(1) $\mathscr{M} = \overline{M}$, $\mathscr{N} = \overline{N}$, and $M \leq N$.
(2) $\mathscr{M} = -\overline{M}$, $\mathscr{N} = \overline{N}$.
(3) $\mathscr{M} = -\overline{M}$, $\mathscr{N} = -\overline{N}$, and $N \leq M$.

EXAMPLE 3.4.5

(1) $\overline{0} \leq_{\mathscr{Z}} \overline{4}$, $\overline{2} \leq_{\mathscr{Z}} \overline{2}$, by case (1).
(2) $-\overline{7} \leq_{\mathscr{Z}} \overline{1}$, $-\overline{3} \leq \overline{4}$, by case (2).
(3) $-\overline{8} \leq_{\mathscr{Z}} -\overline{7}$, $-\overline{1} \leq_{\mathscr{Z}} -\overline{1}$, by case (3).
(4) $-\overline{3} \nleq_{\mathscr{Z}} -\overline{5}$.

Theorem 3.4.2. $\leq_{\mathscr{Z}}$ *is a linear ordering on* $\mathscr{Z}$.

PROOF: The linear ordering $\leq$ on ω induces a linear ordering R on $\mathscr{Z}_+ \cup \{\overline{0}\}$:

$$\overline{M} R \overline{N} \text{ iff } M \leq N, \ M, N \in \omega.$$

Similarly, the linear ordering $\geq$ on ω_+ induces a linear ordering S on $\mathscr{Z}_-$:

$$-\overline{K} S \ -\overline{L} \text{ iff } K \geq L, K, L \in \omega_+.$$

Furthermore, $\mathscr{Z}_- \cap (\mathscr{Z}_+ \cup \{\overline{0}\}) = \varnothing$; according to Exercise 2.6.6(c),

$$R \cup S \cup (\mathscr{Z}_- \times (\mathscr{Z}_+ \cup \{\overline{0}\}))$$

is a linear ordering on $\mathscr{Z} = \mathscr{Z}_- \cup \mathscr{Z}_+ \cup \{\overline{0}\}$. Clearly,

$$\leq_{\mathscr{Z}} = R \cup S \cup (\mathscr{Z}_- \times (\mathscr{Z}_+ \cup \{\overline{0}\})).$$

It should be fairly obvious to the reader that $\leq_{\mathscr{Z}}$ *does not well-order* $\mathscr{Z}$.

An **isomorphism** is a one-one mapping of one structure into another that preserves the basic operations and, in some cases, the "natural" orderings. We next consider an isomorphic "embedding" of the natural numbers into the integers.

Theorem 3.4.3. *Let $F : \omega \to \mathscr{Z}$ be defined by $F(N) = \overline{N}$, $N \in \omega$. Then*
(1) *F is one-one.*
Furthermore, for all M, N in ω,
(2) $F(M + N) = F(M) +_{\mathscr{Z}} F(N)$.
(3) $F(M \cdot N) = F(M) \cdot_{\mathscr{Z}} F(N)$.
(4) $M < N$ *iff* $F(M) <_{\mathscr{Z}} F(N)$.

The proof is almost immediate.

Corollary. *The set of nonnegative integers is well-ordered by $\leq_{\mathscr{Z}}$.*

Thus, *the nonnegative numbers behave exactly as do the natural numbers with regard to addition, multiplication, and ordering (according to magnitude).* It follows that *the "purely arithmetic" theorems for natural numbers also hold for nonnegative integers.*

EXAMPLE 3.4.6. *The Fundamental Theorem of Arithmetic can be expressed in terms of nonnegative integers.* (See also Exercise 3.4.12.)

Let $M, N \in \omega$. The integer $\overline{M}$ is said to be a **factor of** $\overline{N}$ if for K in ω, $\overline{N} = \overline{K} \cdot_{\mathscr{Z}} \overline{M}$. For $\overline{N} \geq_{\mathscr{Z}} \overline{2}$, $\overline{N}$ is called a **$\mathscr{Z}_+$-prime** if its only factors are $\overline{1}$ and $\overline{N}$. *Let $\overline{N} \geq_{\mathscr{Z}} \overline{2}$. Then $\overline{N}$ has a unique factorization of the form*

$$\overline{N} = \overline{P}_1^{E_1} \cdot_{\mathscr{Z}} \overline{P}_2^{E_2} \cdot_{\mathscr{Z}} \cdots \cdot_{\mathscr{Z}} \overline{P}_S^{E_S},$$

where $S, E_1, E_2, \ldots, E_S \in \omega_+$, $\overline{P}_1, \overline{P}_2, \ldots, \overline{P}_S$ are $\mathscr{Z}_+$-primes, and

$$\overline{P}_1 <_{\mathscr{Z}} \overline{P}_2 <_{\mathscr{Z}} \cdots <_{\mathscr{Z}} \overline{P}_S.$$

Exercises

3.4.1. Which of the following are true?
- **(a)** $0 \in \mathscr{N}$ for some integer $\mathscr{N}$.
- **(b)** $1 \in \mathscr{N}$ for every nonnegative integer $\mathscr{N}$.
- **(c)** An integer contains either one or two elements.
- **(d)** $\cup\mathscr{Z} = \omega$.
- **(e)** $\cup\cup\mathscr{Z} = \omega$.

3.4.2. Use the definitions of addition and multiplication of integers to show each of the following:

(a) $\overline{7} +_{\mathscr{Z}} -\overline{10} = -\overline{3}$. (b) $-\overline{7} +_{\mathscr{Z}} \overline{10} = \overline{3}$.
(c) $\overline{7} \cdot_{\mathscr{Z}} -\overline{10} = -\overline{70}$. (d) $-\overline{7} \cdot_{\mathscr{Z}} -\overline{10} = \overline{70}$.
(e) $-\overline{7} +_{\mathscr{Z}} -\overline{10} \cdot_{\mathscr{Z}} \overline{3} = -\overline{37}$. (f) $(-\overline{7} +_{\mathscr{Z}} -\overline{10}) \cdot_{\mathscr{Z}} \overline{0} = \overline{0}$.

†3.4.3. Let $\mathscr{L}, \mathscr{M}, \mathscr{N} \in \mathscr{Z}$. Prove the following:
(a) $\mathscr{L} +_{\mathscr{Z}} \mathscr{M} = \mathscr{M} +_{\mathscr{Z}} \mathscr{L}$.
(b) $\mathscr{L} \cdot_{\mathscr{Z}} \mathscr{M} = \mathscr{M} \cdot_{\mathscr{Z}} \mathscr{L}$.
(c) $\mathscr{L} +_{\mathscr{Z}} (\mathscr{M} +_{\mathscr{Z}} \mathscr{N}) = (\mathscr{L} +_{\mathscr{Z}} \mathscr{M}) +_{\mathscr{Z}} \mathscr{N}$.
(d) $\mathscr{L} \cdot_{\mathscr{Z}} (\mathscr{M} \cdot_{\mathscr{Z}} \mathscr{N}) = (\mathscr{L} \cdot_{\mathscr{Z}} \mathscr{M}) \cdot_{\mathscr{Z}} \mathscr{N}$.
(e) $\mathscr{L} \cdot_{\mathscr{Z}} (\mathscr{M} +_{\mathscr{Z}} \mathscr{N}) = \mathscr{L} \cdot_{\mathscr{Z}} \mathscr{M} +_{\mathscr{Z}} \mathscr{L} \cdot_{\mathscr{Z}} \mathscr{N}$.
(f) $(\mathscr{L} +_{\mathscr{Z}} \mathscr{M}) \cdot_{\mathscr{Z}} \mathscr{N} = \mathscr{L} \cdot_{\mathscr{Z}} \mathscr{N} +_{\mathscr{Z}} \mathscr{M} \cdot_{\mathscr{Z}} \mathscr{N}$.

†3.4.4. Prove the following:
(a) If $\mathscr{M}$ and $\mathscr{N}$ are integers that satisfy $\mathscr{M} +_{\mathscr{Z}} \mathscr{N} = \mathscr{M}$, then $\mathscr{N} = \overline{0}$.
(b) If $\mathscr{K} \neq \overline{0}$ and $\mathscr{N}$ is an integer for which $\mathscr{K} \cdot_{\mathscr{Z}} \mathscr{N} = \mathscr{K}$, then $\mathscr{N} = \overline{1}$.

†3.4.5. Let $\mathscr{M}, \mathscr{N} \in \mathscr{Z}$. Prove that there is a unique integer $\mathscr{K}$ for which $\mathscr{M} +_{\mathscr{Z}} \mathscr{K} = \mathscr{N}$.

3.4.6. Use (3.4.1) to determine $\mathscr{M} -_{\mathscr{Z}} \mathscr{N}$ in each case.
(a) $\mathscr{M} = \overline{4}, \mathscr{N} = \overline{7}$.
(b) $\mathscr{M} = \overline{0}, \mathscr{N} = -\overline{4}$.
(c) $\mathscr{M} = -\overline{3}, \mathscr{N} = -\overline{5}$.
(d) $\mathscr{M} = -\overline{5}, \mathscr{N} = -\overline{3}$.

†3.4.7. Prove that for all integers $\mathscr{M}$ and $\mathscr{N}$ the following are true:
(a) $-(\mathscr{M} +_{\mathscr{Z}} \mathscr{N}) = (-\mathscr{M}) +_{\mathscr{Z}} (-\mathscr{N})$.
(b) $-(\mathscr{M} -_{\mathscr{Z}} \mathscr{N}) = \mathscr{N} -_{\mathscr{Z}} \mathscr{M}$.
(c) $-(\mathscr{M} \cdot_{\mathscr{Z}} \mathscr{N}) = (-\mathscr{M}) \cdot_{\mathscr{Z}} \mathscr{N} = \mathscr{M} \cdot_{\mathscr{Z}} (-\mathscr{N})$.

3.4.8. In Theorem 3.4.1, prove the following:
(a) The sum of two positive integers is positive.
(b) The product of two positive integers is positive.

†3.4.9. (a) Prove that multiplication of integers permits no "zero divisors"; thus, if $\mathscr{M}, \mathscr{N} \in \mathscr{Z}$ and $\mathscr{M} \cdot_{\mathscr{Z}} \mathscr{N} = \overline{0}$, prove that $\mathscr{M} = \overline{0}$ or $\mathscr{N} = \overline{0}$.
(b) Prove the following **Cancellation Law (for multiplication of integers)**:
For $\mathscr{L}, \mathscr{M}, \mathscr{N}$ in $\mathscr{Z}$, $\mathscr{L} \neq \overline{0}$ and $\mathscr{L} \cdot_{\mathscr{Z}} \mathscr{M} = \mathscr{L} \cdot_{\mathscr{Z}} \mathscr{N}$ implies $\mathscr{M} = \mathscr{N}$.

†3.4.10. Let $\mathscr{M}, \mathscr{N} \in \mathscr{Z}$. Prove that $\mathscr{M} \leq_{\mathscr{Z}} \mathscr{N}$ if and only if for some M, N in ω, (at least) one of the following holds:
(1) $\mathscr{M} = \overline{M}, \mathscr{N} = \overline{N}$, and $M \leq N$.
(2) $\mathscr{M} = -\overline{M}, \mathscr{N} = \overline{N}$.
(3) $\mathscr{M} = -\overline{M}, \mathscr{N} = -\overline{N}$, and $N \leq M$.

3.4.11. (a) Show that $\leq_{\mathscr{Z}}$ is not a well-ordering.
(b) Define a relation R on $\mathscr{Z}$ that well-orders $\mathscr{Z}$.

3.4.12. Modify the Fundamental Theorem of Arithmetic so that it now applies to negative as well as to positive integers.

3.4.13. Let X be a nonempty set and let F be a binary operation on X. We say that $\langle X, F\rangle$ is a **commutative** (or **Abelian**) **group** if F is associative and commutative, and if for every x_1, x_2 in X there is some y in X for which $x_1 F y = x_2$.
(a) Show that $\langle \mathscr{Z}, +_{\mathscr{Z}}\rangle$ is a commutative group.
(b) Show that $\langle \mathscr{Z} \sim \{\overline{0}\}, \cdot_{\mathscr{Z}}\rangle$ is a commutative group.

3.4.14. **(The Division Algorithm)** Suppose that $\mathscr{M}$ ($\neq \bar{0}$) and $\mathscr{N}$ are integers. Prove that there exist unique integers $\mathscr{K}$ and $\mathscr{L}$ such that (1) $\mathscr{N} = \mathscr{K} \cdot_{\mathscr{Z}} \mathscr{M} +_{\mathscr{Z}} \mathscr{L}$ and (2) $\bar{0} \leq_{\mathscr{Z}} \mathscr{L} <_{\mathscr{Z}} \operatorname{abs} \mathscr{M}$. Here,

$$\operatorname{abs} \mathscr{M} = \begin{cases} \mathscr{M}, & \text{if } \mathscr{M} \geq_{\mathscr{Z}} \bar{0}; \\ -\mathscr{M}, & \text{otherwise.} \end{cases}$$

(Referred to in Section 5.7; see also Exercise 3.3.12.)

3.5 Integers as Equivalence Classes

In lieu of the set $\mathscr{Z}$ that we have defined as the set of integers, many authors define integers in terms of an equivalence relation on $\omega^{②}$. Although we have not adopted this approach, we briefly consider this equivalence relation and the corresponding partition.

Thus we define a relation R_1 on $\omega^{②}$ as follows: Let $\langle M_1, N_1 \rangle$ and $\langle M_2, N_2 \rangle$ be any ordered pairs of natural numbers and let $\langle M_1, N_1 \rangle R_1 \langle M_2, N_2 \rangle$ iff $M_1 + N_2 = M_2 + N_1$.

Example 3.5.1

(1) $\langle 4, 1 \rangle R_1 \langle 12, 9 \rangle$ because $4 + 9 = 12 + 1$.
(2) $\langle 6, 6 \rangle R_1 \langle 3, 3 \rangle$.
(3) $\langle 6, 8 \rangle R_1 \langle 1, 3 \rangle$.
(4) $\langle 4, 2 \rangle \not{R}_1 \langle 8, 4 \rangle$.

Intuitively, we would like to consider pairs of natural numbers as equivalents if they have the same "differences". Unfortunately, these "differences" are not always defined within the set of natural numbers. For example, $\langle 3, 5 \rangle R_1 \langle 1, 3 \rangle$, but the usual subtraction "3 minus 5" or "1 minus 3" would take us out of the natural number system. R_1 expresses this relationship in terms of addition of natural numbers.

Note that *if* $\langle M_1, N_1 \rangle R_1 \langle M_2, N_2 \rangle$, *then* $M_1 \dot{-} N_1 = M_2 \dot{-} N_2$. Also, *if* $M_1 \geq N_1$, *then* $\langle M_1, N_1 \rangle R_1 \langle M_2, N_2 \rangle$ *iff* $M_1 \dot{-} N_1 = M_2 \dot{-} N_2$.

Theorem 3.5.1. *R_1 is an equivalence relation on ω.*

Proof: Let $M_I, N_I, I = 1, 2, 3$, be natural numbers.

(1) *Reflexivity*: $M_1 + N_1 = M_1 + N_1$; thus $\langle M_1, N_1 \rangle R_1 \langle M_1, N_1 \rangle$.

(2) *Symmetry*: Suppose that $\langle M_1, N_1\rangle R_1 \langle M_2, N_2\rangle$. Then $M_1 + N_2 = M_2 + N_1$; hence, $M_2 + N_1 = M_1 + N_2$. Thus,

$$\langle M_2, N_2\rangle R_1 \langle M_1, N_1\rangle.$$

(3) *Transitivity*: Suppose that

$$\langle M_1, N_1\rangle R_1 \langle M_2, N_2\rangle \text{ and } \langle M_2, N_2\rangle R_1 \langle M_3, N_3\rangle.$$

Then $M_1 + N_2 = M_2 + N_1$ and $M_2 + N_3 = M_3 + N_2$. It follows that

$$(M_1 + N_2) + (M_2 + N_3) = (M_2 + N_1) + (M_3 + N_2).$$

By the associativity and commutativity of addition (of natural numbers), this can be written as

$$(M_1 + N_3) + (N_2 + M_2) = (M_3 + N_1) + (N_2 + M_2).$$

By Exercise 3.2.3, $M_1 + N_3 = M_3 + N_1$; hence, $\langle M_1, N_1\rangle R_1 \langle M_3, N_3\rangle$.

Theorem 2.5.1(1) guarantees that $\omega^{②}/R_1$ is a set. We denote $\omega^{②}/R_1$ by $\mathscr{Z}^*$, and we call it the **set of (*)-integers**. Thus, a **(*)-integer** is an equivalence class corresponding to the relation R_1 on $\omega^{②}$.

Theorem 3.5.2. *$\mathscr{Z}$ is in one-one correspondence with $\mathscr{Z}^*$.*

PROOF: It suffices to show the following:

(1) Each equivalence class $[\langle M, N\rangle]$ (with respect to R_1) contains exactly one integer.

(2) Each integer is in exactly one such equivalence class.

(a) Let $M, N \in \omega$. If $M \geq N$, then $\overline{M \dot{-} N} = \langle M \dot{-} N, 0\rangle \in [\langle M, N\rangle]$ because $(M \dot{-} N) + N = M = M + 0$. Suppose that

$$\overline{K} = \langle K, 0\rangle \in [\langle M, N\rangle]$$

for K in ω. Then $K + N = M + 0 = M$. It follows that

$$K = M \dot{-} N.$$

Let $L \in \omega_+$, and consider $-\overline{L} = \langle 0, L\rangle$; we claim that $-\overline{L} \notin [\langle M, N\rangle]$. Otherwise, $N = 0 + N = M + L$, and this would contradict the assumption that $M \geq N$.

Similarly, if $M < N$, then $-\overline{N \dot{-} M}$ is the unique integer in $[\langle M, N\rangle]$. (The details are left as an exercise.)

(b) Each nonnegative integer $\langle M, 0\rangle$ is in $[\langle M, 0\rangle]$ and each negative integer $\langle 0, N\rangle$ is in $[\langle 0, N\rangle]$.

By part (1), we see that

$$\mathscr{Z}^* = \{[\langle M,0\rangle] : M \in \omega\} \cup \{[\langle 0, N\rangle] : N \in \omega_+\}.$$

If $\langle K,0\rangle \in [\langle M,0\rangle]$, then

$$K = K + 0 = M + 0 = M.$$

If $\langle 0, N\rangle \in [\langle M,0\rangle]$, then $M + N = 0$; hence, $M = N = 0$. Similarly, one shows that $\langle 0, N\rangle$ is the unique integer in $[\langle 0, N\rangle]$.

Corollary. *Let $F^* : \mathscr{Z}^* \to \mathscr{Z}$ be defined by*

$$F^*([\langle M,0\rangle]) = \langle M,0\rangle,\ M \in \omega$$

and

$$F^*([\langle 0,N\rangle]) = \langle 0,N\rangle,\ N \in \omega_+.$$

Then F^* is a one-one mapping of $\mathscr{Z}^*$ onto $\mathscr{Z}$.

EXAMPLE 3.5.2

(1) $[\langle 14, 3\rangle] = [\langle 11, 0\rangle]$; thus, $\overline{11} \in [\langle 14, 3\rangle]$.
(2) $[\langle 5, 5\rangle] = [\langle 0, 0\rangle]$.
(3) $F^*([\langle 3, 14\rangle]) = F^*([\langle 0, 11\rangle]) = -\overline{11}$.

The mapping F^* of the corollary enables us to define the arithmetic operations and (natural) ordering on $\mathscr{Z}^*$ in terms of the definitions given for $\mathscr{Z}$. Thus, let $y^*, z^* \in \mathscr{Z}^*$; define

"$y^* +_{\mathscr{Z}^*} z^*$" to be "$F^{*-1}(F^*(y^*) +_{\mathscr{Z}} F^*(z^*))$";

"$y^* -_{\mathscr{Z}^*} z^*$" to be "$F^{*-1}(F^*(y^*) -_{\mathscr{Z}} F^*(z^*))$";

"$y^* \cdot_{\mathscr{Z}^*} z^*$" to be "$F^{*-1}(F^*(y^*) \cdot_{\mathscr{Z}} F^*(z^*))$";

let

$$y^* \leq_{\mathscr{Z}^*} z^* \text{ iff } F^*(y^*) \leq_{\mathscr{Z}} F^*(z^*).$$

EXAMPLE 3.5.3

(1) $[\langle 3, 0\rangle] +_{\mathscr{Z}^*} [\langle 0, 5\rangle] = F^{*-1}(\langle 3, 0\rangle +_{\mathscr{Z}} \langle 0, 5\rangle)$
$= F^{*-1}(\overline{3} +_{\mathscr{Z}} -\overline{5}) = F^{*-1}(-(5 \dot{-} 3)) = F^{*-1}(\langle 0, 2\rangle) = [\langle 0, 2\rangle]$.
(2) $[\langle 0, 7\rangle] -_{\mathscr{Z}^*} [\langle 0, 4\rangle] = [\langle 0, 3\rangle]$ because $-\overline{7} -_{\mathscr{Z}} -\overline{4} = -\overline{3}$.
(3) $[\langle 5, 2\rangle] \cdot_{\mathscr{Z}^*} [\langle 1, 4\rangle] = [\langle 0, 9\rangle]$ because $\langle 5, 2\rangle R_1 \langle 3, 0\rangle$, $\langle 1, 4\rangle R_1 \langle 0, 3\rangle$, and $\langle 3, 0\rangle \cdot_{\mathscr{Z}} \langle 0, 3\rangle = \overline{3} \cdot_{\mathscr{Z}} -\overline{3} = -\overline{9}$.
(4) $[\langle 2, 12\rangle] <_{\mathscr{Z}^*} [\langle 2, 1\rangle]$ because $-\overline{10} <_{\mathscr{Z}} \overline{1}$.

The **positive elements of** $\mathscr{Z}^*$ are those equivalence classes $[\langle M, 0\rangle]$ for $M > 0$; the **negative elements of** $\mathscr{Z}^*$ are the equivalence classes $[\langle 0, N\rangle]$ for $N > 0$. Thus, the **Law of Trichotomy** can be phrased: *For any element* z^* *of* $\mathscr{Z}^*$, *exactly one of the following holds true*:

(1) $\langle 0, 0\rangle \in z^*$.
(2) $\langle M, 0\rangle \in z^*$ *for some* M *in* ω_+.
(3) $\langle 0, N\rangle \in z^*$ *for some* N *in* ω_+.

We note that instead of defining the above operations and ordering on $\mathscr{Z}^*$ in terms of those on $\mathscr{Z}$, we could have defined these directly in terms of the operations on ω. (We do not recommend this approach.) For example, addition on $\mathscr{Z}^*$ could be defined by

$$[\langle M_1, N_1\rangle] \boxplus [\langle M_2, N_2\rangle] = [\langle M_1 + M_2, N_1 + N_2\rangle];$$

multiplication could be defined by

$$[\langle M_1, N_1\rangle] \boxdot [\langle M_2, N_2\rangle] = [\langle M_1M_2 + N_1N_2, M_1N_2 + M_2N_1\rangle].$$

[These definitions are motivated by the need to "preserve the differences"; for example, the second definition can be thought of as expressing the product:

$$\begin{array}{l} M_1 - N_1 \\ \underline{M_2 - N_2} \\ M_1M_2 - (M_1N_2 + M_2N_1) + N_1N_2.] \end{array}$$

However, it is a rather tedious task to show that these operations are well-defined in the sense that they do not depend on the particular elements chosen from the equivalence classes. It is conceivable that there might be natural numbers M_I, N_I, $1 \leq I \leq 4$, such that

$$\langle M_1, N_1\rangle R_1 \langle M_3, N_3\rangle, \langle M_2, N_2\rangle R_1 \langle M_4, N_4\rangle,$$

yet

$$\langle M_1 + M_2, N_1 + N_2\rangle \not R_1 \langle M_3 + M_4, N_3 + N_4\rangle.$$

We leave it as an exercise to show that this (and the corresponding statement for multiplication) can never occur. Also, one must verify that $\boxplus = +_{\mathscr{Z}^*}$ and $\boxdot = \cdot_{\mathscr{Z}^*}$.

Exercises

3.5.1. Determine M and N:

(a) $\langle 8, 4\rangle R_1 \langle M, 7\rangle$.

(b) $\langle 3, 10\rangle R_1 \langle M, 7\rangle$.

(c) $\langle 5, 9\rangle R_1 \langle M, N\rangle$ and $\langle M, 2\rangle R_1 \langle 6, 1\rangle$.

3.5.2. Determine the largest subset of $\omega^{②}$ consisting of elements $\langle M, N\rangle$ for which each of the following is true:

(a) $\langle M, N\rangle R_1 \langle N, M\rangle$. **(b)** $\langle M, N\rangle R_1 \langle M, M\rangle$.

(c) $\langle M, M + N\rangle R_1 \langle 1, N^+\rangle$. **(d)** $\langle M + N, M\rangle R_1 \langle M \cdot N, M\rangle$.

3.5.3. In the proof of Theorem 3.5.2, show that if $M < N$, then $\overline{N \dot{-} M}$ is the unique integer in $[\langle M, N\rangle]$.

3.5.4. Define $-_{\mathscr{Z}^*}$ in terms of $+_{\mathscr{Z}^*}$; show that this definition is equivalent to the one given in the text.

3.5.5. Show the following:

(a) $[\langle 5, 1\rangle] +_{\mathscr{Z}^*} [\langle 2, 5\rangle] = [\langle 1, 0\rangle]$.

(b) $[\langle 5, 1\rangle] \cdot_{\mathscr{Z}^*} [\langle 2, 5\rangle] = [\langle 0, 12\rangle]$.

(c) $[\langle 5, 1\rangle] -_{\mathscr{Z}^*} [\langle 2, 5\rangle] = [\langle 7, 0\rangle]$.

3.5.6. Determine M:

(a) $[\langle 2, 7\rangle] +_{\mathscr{Z}^*} [\langle 5, 3\rangle] = [\langle 7, M\rangle]$.

(b) $[\langle 2 + M, M^2\rangle] +_{\mathscr{Z}^*} [\langle 10, 1\rangle] = [\langle 4, 5\rangle]$.

(c) $[\langle 4, 1\rangle] \cdot_{\mathscr{Z}^*} [\langle 5, 6\rangle] = [\langle M, 6\rangle]$.

(d) $[\langle 2, M\rangle] \cdot_{\mathscr{Z}^*} [\langle M, 1\rangle] = [\langle 1, 7\rangle]$.

(e) $[\langle 7, 8\rangle] -_{\mathscr{Z}^*} [\langle 5, 7\rangle] = [\langle 1, M\rangle]$.

(f) $[\langle M, M^2\rangle] -_{\mathscr{Z}^*} [\langle 3, 5\rangle] = [\langle 2, 6\rangle]$.

3.5.7. With respect to $+_{\mathscr{Z}^*}$, what is the additive inverse of $[\langle M, N\rangle]$?

3.5.8. Show that an element z^* of $\mathscr{Z}^*$ is positive iff for every $\langle M, N\rangle$ in z^*, $M > N$.

***3.5.9.** Without referring to the operations on $\mathscr{Z}$, show that the operations $\boxplus$ and $\boxdot$ on $\mathscr{Z}^*$ are well-defined. Thus for natural numbers $M_I, N_I, 1 \le I \le 4$, that are such that

$$\langle M_1, N_1\rangle R_1 \langle M_3, N_3\rangle, \langle M_2, N_2\rangle R_1 \langle M_4, N_4\rangle,$$

show the following:

(a) $\langle M_1 + M_2, N_1 + N_2\rangle R_1 \langle M_3 + M_4, N_3 + N_4\rangle$.

(b) $\langle M_1 M_2 + N_1 N_2, M_1 N_2 + M_2 N_1\rangle R_1 \langle M_3 M_4 + N_3 N_4, M_3 N_4 + M_4 N_3\rangle$.

(c) Prove that $\boxplus = +_{\mathscr{Z}^*}$ and $\boxdot = \cdot_{\mathscr{Z}^*}$.

3.6 Rational Numbers

There was a very natural way to define the integers as certain ordered pairs of natural numbers. In this way the usual operations and ordering on the integers could easy be defined, and the basic properties of the integers could quickly be established.

It would not be difficult to define a rational number as an ordered pair of integers. To this end we could modify the notion of "relatively prime natural numbers" (Exercise 3.3.5) so as to apply to integers. Thus, let $\mathscr{M} \in \mathscr{Z}$ and $\mathscr{N} \in \mathscr{Z}_+$; for unique natural numbers M and N ($\neq 0$), $\mathscr{N} = \overline{N}$ and either

$\mathscr{M} = \overline{M}$ or $\mathscr{M} = -\overline{M}$. Define $\langle \mathscr{M}, \mathscr{N} \rangle$ to be **relatively prime** if M and N are relatively prime. We could then define a rational number as a relatively prime *ordered* pair $\langle \mathscr{M}, \mathscr{N} \rangle$ in $\mathscr{Z} \times \mathscr{Z}_+$. For example, $\langle \overline{2}, \overline{3} \rangle$, $\langle \overline{3}, \overline{2} \rangle$, $\langle -\overline{3}, \overline{2} \rangle$, $\langle \overline{0}, \overline{1} \rangle$ would all be rational numbers according to this definition; $\langle \overline{3}, -\overline{2} \rangle$, $\langle \overline{3}, \overline{0} \rangle$, $\langle \overline{0}, \overline{4} \rangle$, $\langle \overline{2}, \overline{4} \rangle$ would not be.

The difficulty with this approach is that each time we define an arithmetic operation or establish a basic property of rationals, we have to worry about "dividing out" common factors. This is by no means an insurmountable difficulty, but it proves to be somewhat cumbersome! We choose, instead, to define rational numbers by considering an equivalence relation.

Let R_2 be the following relation on $\mathscr{Z} \times (\mathscr{Z} \sim \{\overline{0}\})$:

$$\langle \mathscr{M}_1, \mathscr{N}_1 \rangle R_2 \langle \mathscr{M}_2, \mathscr{N}_2 \rangle \text{ iff } \mathscr{M}_1 \cdot_{\mathscr{Z}} \mathscr{N}_2 = \mathscr{M}_2 \cdot_{\mathscr{Z}} \mathscr{N}_1.$$

Thus, R_2 relates ordered pairs, the first coordinate of each of which can be any integer, whereas the second coordinate must be a nonzero integer. Just as R_1 served as a means of introducing the notions of negative numbers and subtraction, the relation R_2 will be utilized to introduce the notions of fractions, rational numbers, and division.

EXAMPLE 3.6.1

(1) $\langle \overline{6}, \overline{2} \rangle R_2 \langle \overline{3}, \overline{1} \rangle$ because $\overline{6} \cdot_{\mathscr{Z}} \overline{1} = \overline{3} \cdot_{\mathscr{Z}} \overline{2}$.
(2) $\langle \overline{4}, -\overline{10} \rangle R_2 \langle -\overline{2}, \overline{5} \rangle$ because $\overline{4} \cdot_{\mathscr{Z}} \overline{5} = (-\overline{2}) \cdot_{\mathscr{Z}} (-\overline{10})$.
(3) $\langle \overline{10}, \overline{6} \rangle \not R_2 \langle \overline{6}, \overline{2} \rangle$ because $\overline{10} \cdot_{\mathscr{Z}} \overline{2} \neq \overline{6} \cdot_{\mathscr{Z}} \overline{6}$. Note that $\langle 10, 6 \rangle R_1 \langle 6, 2 \rangle$.

Theorem 3.6.1. *R_2 is an equivalence relation on $\mathscr{Z} \times (\mathscr{Z} \sim \{\overline{0}\})$.*

PROOF: Let $\mathscr{M}_I$ be integers and $\mathscr{N}_I$ be nonzero integers, $I = 1, 2, 3$.

(1) *Reflexivity*: By the reflexivity of equality, $\mathscr{M}_1 \cdot_{\mathscr{Z}} \mathscr{N}_1 = \mathscr{M}_1 \cdot_{\mathscr{Z}} \mathscr{N}_1$; thus $\langle \mathscr{M}_1, \mathscr{N}_1 \rangle R_2 \langle \mathscr{M}_1, \mathscr{N}_1 \rangle$.

(2) *Symmetry*: Suppose that $\langle \mathscr{M}_1, \mathscr{N}_1 \rangle R_2 \langle \mathscr{M}_2, \mathscr{N}_2 \rangle$. Then $\mathscr{M}_1 \cdot_{\mathscr{Z}} \mathscr{N}_2 = \mathscr{M}_2 \cdot_{\mathscr{Z}} \mathscr{N}_1$. By the symmetry of equality, $\mathscr{M}_2 \cdot_{\mathscr{Z}} \mathscr{N}_1 = \mathscr{M}_1 \cdot_{\mathscr{Z}} \mathscr{N}_2$. Thus,

$$\langle \mathscr{M}_2, \mathscr{N}_2 \rangle R_2 \langle \mathscr{M}_1, \mathscr{N}_1 \rangle.$$

(3) *Transitivity*: Suppose that

$$\langle \mathscr{M}_1, \mathscr{N}_1 \rangle R_2 \langle \mathscr{M}_2, \mathscr{N}_2 \rangle \quad \text{and} \quad \langle \mathscr{M}_2, \mathscr{N}_2 \rangle R_2 \langle \mathscr{M}_3, \mathscr{N}_3 \rangle.$$

Therefore,

$$\mathscr{M}_1 \cdot_{\mathscr{Z}} \mathscr{N}_2 = \mathscr{M}_2 \cdot_{\mathscr{Z}} \mathscr{N}_1 \quad \text{and} \quad \mathscr{M}_2 \cdot_{\mathscr{Z}} \mathscr{N}_3 = \mathscr{M}_3 \cdot_{\mathscr{Z}} \mathscr{N}_2.$$

If $\mathcal{M}_2 = \overline{0}$, then because $\mathcal{N}_2 \neq \overline{0}$,

$$\mathcal{M}_1 = \mathcal{M}_3 = \overline{0} = \mathcal{M}_1 \cdot_{\mathscr{Z}} \mathcal{N}_3 = \mathcal{M}_3 \cdot_{\mathscr{Z}} \mathcal{N}_1.$$

Consequently, in this case $\langle \mathcal{M}_1, \mathcal{N}_1 \rangle R_2 \langle \mathcal{M}_3, \mathcal{N}_3 \rangle$. If $\mathcal{M}_2 \neq \overline{0}$, then

$$\begin{aligned}(\mathcal{M}_1 \cdot_{\mathscr{Z}} \mathcal{N}_3) \cdot_{\mathscr{Z}} (\mathcal{M}_2 \cdot_{\mathscr{Z}} \mathcal{N}_2) &= (\mathcal{M}_1 \cdot_{\mathscr{Z}} \mathcal{N}_2) \cdot_{\mathscr{Z}} (\mathcal{M}_2 \cdot_{\mathscr{Z}} \mathcal{N}_3) \\ &= (\mathcal{M}_2 \cdot_{\mathscr{Z}} \mathcal{N}_1) \cdot_{\mathscr{Z}} (\mathcal{M}_3 \cdot_{\mathscr{Z}} \mathcal{N}_2) \\ &= (\mathcal{M}_3 \cdot_{\mathscr{Z}} \mathcal{N}_1) \cdot_{\mathscr{Z}} (\mathcal{M}_2 \cdot_{\mathscr{Z}} \mathcal{N}_2),\end{aligned}$$

and, because $\mathcal{M}_2 \cdot_{\mathscr{Z}} \mathcal{N}_2 \neq \overline{0}$, it follows that $\mathcal{M}_1 \cdot_{\mathscr{Z}} \mathcal{N}_3 = \mathcal{M}_3 \cdot_{\mathscr{Z}} \mathcal{N}_1$. Again, $\langle \mathcal{M}_1, \mathcal{N}_1 \rangle R_2 \langle \mathcal{M}_3, \mathcal{N}_3 \rangle$.

The relation R_2 expresses our need to consider as equivalent those ordered pairs of integers that, intuitively, express the same quotient. For in elementary arithmetic we say that for nonzero numbers y and y', x/y and x'/y' are equivalent fractions iff $xy' = x'y$. Thus, three-sixths is said to be equivalent to one-half. In our description of numbers as sets, it is these equivalence classes of integers that represent the notion of division.

We define a **rational (number)** as an equivalence class corresponding to the relation R_2 on $\mathscr{Z} \times (\mathscr{Z} \sim \{\overline{0}\})$. The set of rational numbers is denoted by $\mathscr{Q}$. Each element of a rational number is called a **fraction**. Thus, a fraction is an ordered pair; the **numerator** of the fraction is defined to be its first coordinate—the **denominator**, its second coordinate.

EXAMPLE 3.6.2. The set representing "three fifths" is

$$\alpha = \{\langle \mathcal{M}, \mathcal{N} \rangle : \mathcal{M} \in \mathscr{Z}, \mathcal{N} \in \mathscr{Z} \sim \{\overline{0}\}, \text{ and } \mathcal{M} \cdot_{\mathscr{Z}} \overline{5} = \overline{3} \cdot_{\mathscr{Z}} \mathcal{N}\}.$$

The numerator and denominator of a fraction in α are in the ratio 3 : 5. Thus, $\{\langle \overline{3}, \overline{5} \rangle, \langle -\overline{3}, -\overline{5} \rangle, \langle \overline{9}, \overline{15} \rangle\} \subset \alpha$.

Note that in the above example, had we allowed $\overline{0}$ as a possible value for $\mathcal{N}$, then $\langle \overline{0}, \overline{0} \rangle$ would have been an element of α. But we want α to consist of those fractions that are "reducible" to three fifths.

Exercises

3.6.1. Fill in:

(a) $\langle \overline{8}, \overline{3} \rangle R_2 \langle -\overline{16}, \mathcal{N} \rangle$.

(b) $\langle \overline{3}, \mathcal{N} \rangle R_2 \langle -1, 3 \rangle$.

(c) $\langle \mathcal{M}^2, \overline{5} \rangle R_2 \langle \overline{10}, \overline{2} \rangle$.

(d) $\langle \mathcal{M}^2, \overline{5} \rangle R_2 \langle \mathcal{M}, \overline{1} \rangle$.

(e) $\langle \overline{0}, \overline{4} \rangle R_2 \langle \overline{0}, \mathcal{N} \rangle$.

3.6.2. Partition the following set of ordered pairs of integers according to the relation R_2: $\{\langle \bar{4}, \bar{2}\rangle, \langle \bar{3}, \bar{4}\rangle, \langle \bar{1}, -\bar{1}\rangle, \langle -\bar{1}, \bar{1}\rangle, \langle \bar{1}, \bar{2}\rangle, \langle -\bar{1}, \bar{2}\rangle, \langle \bar{6}, \bar{3}\rangle, \langle \bar{6}, \bar{8}\rangle, \langle -\bar{2}, -\bar{1}\rangle\}$.

3.6.3. For $M, N\ (\neq 0) \in \omega$, let $[\langle M, N\rangle]$ refer to R_1 and let $[\langle \bar{M}, \bar{N}\rangle]$ refer to R_2. Which of the following are true?

(a) $\langle 4, 2\rangle \in [\langle 2, 1\rangle]$. **(b)** $\langle 4, 2\rangle \in [\langle \bar{2}, \bar{1}\rangle]$.
(c) $\langle \bar{4}, \bar{2}\rangle \in [\langle \bar{2}, \bar{1}\rangle]$. **(d)** $\langle 3, 2\rangle \in [\langle 2, 1\rangle]$.
(e) $\langle 3, 2\rangle \in [\langle \bar{2}, \bar{1}\rangle]$. **(f)** $\langle \bar{3}, \bar{2}\rangle \in [\langle \bar{2}, \bar{1}\rangle]$.

3.6.4. Give a detailed explanation of the equalities in the proof of the transitivity of R_2.

3.6.5. Suppose that S_2 is the relation on $\mathscr{Z}^{\textcircled{2}}$ defined by $\langle \mathscr{M}_1, \mathscr{N}_1\rangle S_2 \langle \mathscr{M}_2, \mathscr{N}_2\rangle$ iff $\mathscr{M}_1 \cdot_{\mathscr{Z}} \mathscr{N}_2 = \mathscr{N}_1 \cdot_{\mathscr{Z}} \mathscr{M}_2$. Is S_2 an equivalence relation?

3.6.6. Let $M_1, M_2, N_1, N_2 \in \omega$. Define the relation $\bar{R}_1$ on $\mathscr{Z}^{\textcircled{2}}$ by $\langle \bar{M}_1, \bar{N}_1\rangle \bar{R}_1 \langle \bar{M}_2, \bar{N}_2\rangle$ iff $\langle M_1, N_1\rangle R_1 \langle M_2, N_2\rangle$. Determine $\bar{R}_1 \cap R_2$.

3.6.7. Show that every rational number contains exactly one relatively prime fraction. (Used in Section 4.5.)

3.6.8. (For readers familiar with matrices.) Show that the collection of all 2×2 nonsingular matrices whose entries are rational is a set. (First define a matrix as a set.)

3.7 Arithmetic of Rational Numbers

We now let "$[\mathscr{M}/\mathscr{N}]$" denote the rational number "$[\langle \mathscr{M}, \mathscr{N}\rangle]$" (corresponding to the relation R_2 on $\mathscr{Z} \times (\mathscr{Z} \sim \{\bar{0}\})$. Moreover, in the present section we let "$\mathscr{K}/\mathscr{L}$" denote a fraction, $\langle \mathscr{K}, \mathscr{L}\rangle$, in $[\mathscr{M}/\mathscr{N}]$. In other words, the numerator $\mathscr{K}$ and the denominator $\mathscr{L}$ of such a fraction satisfy $\mathscr{K} \cdot_{\mathscr{Z}} \mathscr{N} = \mathscr{M} \cdot_{\mathscr{Z}} \mathscr{L}$. We will further simplify our notation and write

$$\text{“}[\mathscr{M}/\mathscr{N}]\text{” instead of “}[\bar{M}/\bar{N}]\text{”}$$

and

$$\text{“}M/N\text{” instead of “}\bar{M}/\bar{N}\text{”,}$$

$M, N\ (\neq 0)$ in ω. Thus, "[3/5]" denotes the rational number "$[\langle \bar{3}, \bar{5}\rangle]$" and "6/10" denotes a fraction in [3/5]. We write

$$\text{“}[-M/N]\text{” for “}[(-\bar{M})/\bar{N}]\text{”,}$$

$$\text{“}[M/-N]\text{” for “}[\bar{M}/(-\bar{N})]\text{”,}$$

$$\text{“}[-M/-N]\text{” for “}[(-\bar{M})/(-\bar{N})]\text{”,}$$

and we similarly abbreviate the associated fractions.

The arithmetic operations on $\mathscr{Q}$, which we now consider, correspond to the familiar notions of adding, multiplying, subtracting, and dividing fractions. The definitions we give for rational numbers $[\mathscr{M}/\mathscr{N}]$ are set-theoretic counterparts of the usual rules governing the arithmetic of the fractions $\mathscr{M}/\mathscr{N}$.

For $I = 1, 2$, let $\mathscr{M}_I$ be integers and $\mathscr{N}_I$ be nonzero integers. We define **addition, $+_{\mathscr{Q}}$, subtraction, $-_{\mathscr{Q}}$, and multiplication, $\cdot_{\mathscr{Q}}$, of rational numbers** as follows:

$$[\mathscr{M}_1/\mathscr{N}_1] +_{\mathscr{Q}} [\mathscr{M}_2/\mathscr{N}_2] = [\mathscr{M}_1 \cdot_{\mathscr{Z}} \mathscr{N}_2 +_{\mathscr{Z}} \mathscr{M}_2 \cdot_{\mathscr{Z}} \mathscr{N}_1/\mathscr{N}_1 \cdot_{\mathscr{Z}} \mathscr{N}_2];$$

$$[\mathscr{M}_1/\mathscr{N}_1] -_{\mathscr{Q}} [\mathscr{M}_2/\mathscr{N}_2] = [\mathscr{M}_1 \cdot_{\mathscr{Z}} \mathscr{N}_2 -_{\mathscr{Z}} \mathscr{M}_2 \cdot_{\mathscr{Z}} \mathscr{N}_1/\mathscr{N}_1 \cdot_{\mathscr{Z}} \mathscr{N}_2];$$

$$[\mathscr{M}_1/\mathscr{N}_1] \cdot_{\mathscr{Q}} [\mathscr{M}_2/\mathscr{N}_2] = [\mathscr{M}_1 \cdot_{\mathscr{Z}} \mathscr{M}_2/\mathscr{N}_1 \cdot_{\mathscr{Z}} \mathscr{N}_2].$$

Theorem 3.7.1. *Addition, subtraction, and multiplication of rational numbers are well-defined. Thus, if* $\mathscr{M}_I, \mathscr{N}_I (\neq \overline{0})$, $1 \leq I \leq 4$, *are integers for which* $\langle \mathscr{M}_1, \mathscr{N}_1 \rangle R_2 \langle \mathscr{M}_3, \mathscr{N}_3 \rangle$ *and* $\langle \mathscr{M}_2, \mathscr{N}_2 \rangle R_2 \langle \mathscr{M}_4, \mathscr{N}_4 \rangle$, *then the following are true:*

(1) $\langle \mathscr{M}_1 \cdot_{\mathscr{Z}} \mathscr{N}_2 +_{\mathscr{Z}} \mathscr{M}_2 \cdot_{\mathscr{Z}} \mathscr{N}_1, \mathscr{N}_1 \cdot_{\mathscr{Z}} \mathscr{N}_2 \rangle R_2$
$\langle \mathscr{M}_3 \cdot_{\mathscr{Z}} \mathscr{N}_4 +_{\mathscr{Z}} \mathscr{M}_4 \cdot_{\mathscr{Z}} \mathscr{N}_3, \mathscr{N}_3 \cdot_{\mathscr{Z}} \mathscr{N}_4 \rangle$.

(2) $\langle \mathscr{M}_1 \cdot_{\mathscr{Z}} \mathscr{N}_2 -_{\mathscr{Z}} \mathscr{M}_2 \cdot_{\mathscr{Z}} \mathscr{N}_1, \mathscr{N}_1 \cdot_{\mathscr{Z}} \mathscr{N}_2 \rangle R_2$
$\langle \mathscr{M}_3 \cdot_{\mathscr{Z}} \mathscr{N}_4 -_{\mathscr{Z}} \mathscr{M}_4 \cdot_{\mathscr{Z}} \mathscr{N}_3, \mathscr{N}_3 \cdot_{\mathscr{Z}} \mathscr{N}_4 \rangle$.

(3) $\langle \mathscr{M}_1 \cdot_{\mathscr{Z}} \mathscr{M}_2, \mathscr{N}_1 \cdot_{\mathscr{Z}} \mathscr{N}_2 \rangle R_2 \langle \mathscr{M}_3 \cdot_{\mathscr{Z}} \mathscr{M}_4, \mathscr{N}_3 \cdot_{\mathscr{Z}} \mathscr{N}_4 \rangle$.

PROOF: We prove only part (1). (The addition and multiplication operations here are on the integers; thus the subscript "$\mathscr{Z}$" is understood.)

According to the definition of R_2, we must show that

$$(\mathscr{M}_1 \cdot \mathscr{N}_2 + \mathscr{M}_2 \cdot \mathscr{N}_1) \cdot (\mathscr{N}_3 \cdot \mathscr{N}_4) = (\mathscr{M}_3 \cdot \mathscr{N}_4 + \mathscr{M}_4 \cdot \mathscr{N}_3) \cdot (\mathscr{N}_1 \cdot \mathscr{N}_2).$$

By hypothesis, $\langle \mathscr{M}_1, \mathscr{N}_1 \rangle R_2 \langle \mathscr{M}_3, \mathscr{N}_3 \rangle$ and $\langle \mathscr{M}_2, \mathscr{N}_2 \rangle R_2 \langle \mathscr{M}_4, \mathscr{N}_4 \rangle$; hence, $\mathscr{M}_1 \cdot \mathscr{N}_3 = \mathscr{M}_3 \cdot \mathscr{N}_1$ and $\mathscr{M}_2 \cdot \mathscr{N}_4 = \mathscr{M}_4 \cdot \mathscr{N}_2$. Thus

$$\begin{aligned}
(\mathscr{M}_1 \cdot \mathscr{N}_2 + \mathscr{M}_2 \cdot \mathscr{N}_1) \cdot (\mathscr{N}_3 \cdot \mathscr{N}_4) &= (\mathscr{M}_1 \cdot \mathscr{N}_2) \cdot (\mathscr{N}_3 \cdot \mathscr{N}_4) + (\mathscr{M}_2 \cdot \mathscr{N}_1) \cdot (\mathscr{N}_3 \cdot \mathscr{N}_4) \\
&= (\mathscr{M}_1 \cdot \mathscr{N}_3) \cdot (\mathscr{N}_2 \cdot \mathscr{N}_4) + (\mathscr{M}_2 \cdot \mathscr{N}_4) \cdot (\mathscr{N}_1 \cdot \mathscr{N}_3) \\
&= (\mathscr{M}_3 \cdot \mathscr{N}_1) \cdot (\mathscr{N}_2 \cdot \mathscr{N}_4) + (\mathscr{M}_4 \cdot \mathscr{N}_2) \cdot (\mathscr{N}_1 \cdot \mathscr{N}_3) \\
&= (\mathscr{N}_1 \cdot \mathscr{N}_2) \cdot (\mathscr{M}_3 \cdot \mathscr{N}_4) + (\mathscr{N}_1 \cdot \mathscr{N}_2) \cdot (\mathscr{M}_4 \cdot \mathscr{N}_3) \\
&= (\mathscr{N}_1 \cdot \mathscr{N}_2) \cdot (\mathscr{M}_3 \cdot \mathscr{N}_4 + \mathscr{M}_4 \cdot \mathscr{N}_3) \\
&= (\mathscr{M}_3 \cdot \mathscr{N}_4 + \mathscr{M}_4 \cdot \mathscr{N}_3) \cdot (\mathscr{N}_1 \cdot \mathscr{N}_2).
\end{aligned}$$

EXAMPLE 3.7.1. $\langle \overline{4}, \overline{8} \rangle R_2 \langle \overline{1}, \overline{2} \rangle$; $\langle \overline{7}, -\overline{4} \rangle R_2 \langle -\overline{21}, \overline{12} \rangle$. Thus,

(1) $\langle \overline{40}, -\overline{32} \rangle = \langle \overline{4} \cdot_{\mathscr{Z}} (-\overline{4}) +_{\mathscr{Z}} \overline{7} \cdot_{\mathscr{Z}} \overline{8}, \overline{8} \cdot_{\mathscr{Z}} (-\overline{4}) \rangle$
$R_2 \langle \overline{1} \cdot_{\mathscr{Z}} \overline{12} +_{\mathscr{Z}} (-\overline{21}) \cdot_{\mathscr{Z}} \overline{2}, \overline{2} \cdot_{\mathscr{Z}} \overline{12} \rangle = \langle -\overline{30}, \overline{24} \rangle$.

Otherwise stated,

$$[40/-32] = [4/8] +_{\mathscr{Q}} [7/-4] = [1/2] +_{\mathscr{Q}} [-21/12] = [-30/24].$$

(2) $\langle -\overline{72}, -\overline{32}\rangle = \langle \overline{4}\cdot(-\overline{4}) - \overline{7}\cdot\overline{8}, \overline{8}\cdot(-\overline{4})\rangle$
$R_2\langle \overline{1}\cdot\overline{12} - (-\overline{21})\cdot\overline{2}, \overline{2}\cdot\overline{12}\rangle = \langle \overline{54}, \overline{24}\rangle$.

(3) $\langle \overline{28}, -\overline{32}\rangle = \langle \overline{4}\cdot\overline{7}, \overline{8}\cdot(-\overline{4})\rangle R_2 \langle \overline{1}\cdot(-\overline{21}), \overline{2}\cdot\overline{12}\rangle = \langle -\overline{21}, \overline{24}\rangle$.

In this chapter, we use lower-case Greek letters to denote rational numbers.

Theorem 3.7.2. *Addition and multiplication of rational numbers are commutative and associative; moreover, multiplication distributes with respect to addition as well as with respect to subtraction. Thus, for $\alpha, \beta, \gamma \in \mathscr{Q}$, the following are true:*

(1) $\alpha +_{\mathscr{Q}} \beta = \beta +_{\mathscr{Q}} \alpha$, $\alpha \cdot_{\mathscr{Q}} \beta = \beta \cdot_{\mathscr{Q}} \alpha$.
(2) $\alpha +_{\mathscr{Q}} (\beta +_{\mathscr{Q}} \gamma) = (\alpha +_{\mathscr{Q}} \beta) +_{\mathscr{Q}} \gamma$, $\alpha \cdot_{\mathscr{Q}} (\beta \cdot_{\mathscr{Q}} \gamma) = (\alpha \cdot_{\mathscr{Q}} \beta) \cdot_{\mathscr{Q}} \gamma$.
(3) $\alpha \cdot_{\mathscr{Q}} (\beta +_{\mathscr{Q}} \gamma) = \alpha \cdot_{\mathscr{Q}} \beta +_{\mathscr{Q}} \alpha \cdot_{\mathscr{Q}} \gamma$.
(4) $\alpha \cdot_{\mathscr{Q}} (\beta -_{\mathscr{Q}} \gamma) = \alpha \cdot_{\mathscr{Q}} \beta -_{\mathscr{Q}} \alpha \cdot_{\mathscr{Q}} \gamma$.

(The proof of Theorem 3.7.2 is left as an exercise.)

We next embed the integers in the rationals. Define

$$G : \mathscr{Z} \to \mathscr{Q} \text{ by } G(\mathscr{M}) = [\mathscr{M}/\overline{1}].$$

It is easily verified that G *is a one–one mapping that preserves sums, differences, and products; i.e., if $\mathscr{M}, \mathscr{N} \in \mathscr{Z}$, then*

$$G(\mathscr{M} +_{\mathscr{Z}} \mathscr{N}) = G(\mathscr{M}) +_{\mathscr{Q}} G(\mathscr{N});$$

$$G(\mathscr{M} -_{\mathscr{Z}} \mathscr{N}) = G(\mathscr{M}) -_{\mathscr{Q}} G(\mathscr{N});$$

$$G(\mathscr{M} \cdot_{\mathscr{Z}} \mathscr{N}) = G(\mathscr{M}) \cdot_{\mathscr{Q}} G(\mathscr{N}).$$

Moreover, G *preserves additive and multiplicative identities in the sense that for all α in $\mathscr{Q}$,*

$$\alpha +_{\mathscr{Q}} G(\overline{0}) = G(\overline{0}) +_{\mathscr{Q}} \alpha = \alpha;$$

$$\alpha \cdot_{\mathscr{Q}} G(\overline{1}) = G(\overline{1}) \cdot_{\mathscr{Q}} \alpha = \alpha.$$

We note that

$$G(\overline{0}) = [0/1] \qquad \textit{and} \qquad G(\overline{1}) = [1/1].$$

For every α in $\mathscr{Q}$ there is a unique β in $\mathscr{Q}$ such that $\alpha +_{\mathscr{Q}} \beta = \beta +_{\mathscr{Q}} \alpha = [0/1]$. In fact, if α is $[\mathscr{M}/\mathscr{N}]$, *then β is* $[-\mathscr{M}/\mathscr{N}]$, and β is called the **additive inverse of** α. *For every α, the additive inverse of α is the difference,* $[0/1] -_{\mathscr{Q}} \alpha$. We agree to write β as "$-_{\mathscr{Q}} \alpha$".

We let "$\overline{\mathscr{M}}$" stand for "$G(\mathscr{M})$", $\mathscr{M} \in \mathscr{Z}$. Thus, $-_{\mathscr{Q}} \overline{\mathscr{M}} = \overline{-\mathscr{M}}$, $\mathscr{M} \in \mathscr{Z}$. Also, "$\overline{\overline{M}}$" stands for $G(\overline{M})$, $M \in \omega$. In particular, $\overline{\overline{0}}$ and $\overline{\overline{1}}$ are, respectively, the additive and multiplicative identities of $\mathscr{Q}$.

The new operation that holds in the rationals is division. The equation "$M \cdot X = N$", $M \neq 0$, did not always have a solution in the set of natural numbers; neither did the equation "$\mathcal{M} \cdot_{\mathcal{Z}} \mathcal{X} = \mathcal{N}$", $\mathcal{M} \neq \bar{0}$, have a solution in the set of integers.

Theorem 3.7.3. *Let α $(\neq \bar{\bar{0}})$, $\beta \in \mathcal{Q}$. Then there is a unique γ in $\mathcal{Q}$ for which $\alpha \cdot_{\mathcal{Q}} \gamma = \gamma \cdot_{\mathcal{Q}} \alpha = \beta$.*

PROOF: Let $\alpha = [\mathcal{M}_1/\mathcal{N}_1]$ and $\beta = [\mathcal{M}_2/\mathcal{N}_2]$, $\mathcal{M}_1, \mathcal{M}_2, \mathcal{N}_1$ $(\neq \bar{0})$, $\mathcal{N}_2$ $(\neq \bar{0}) \in \mathcal{Z}$. Then $\mathcal{M}_1 \neq \bar{0}$, for otherwise, α would equal $\bar{\bar{0}}$. Because there are no zero divisors in $\mathcal{Z}$, $\mathcal{M}_1 \cdot_{\mathcal{Z}} \mathcal{N}_2 \neq \bar{0}$. Hence, $[\mathcal{M}_2 \cdot_{\mathcal{Z}} \mathcal{N}_1/\mathcal{M}_1 \cdot_{\mathcal{Z}} \mathcal{N}_2] \in \mathcal{Q}$ and

$$\begin{aligned}
&[\mathcal{M}_1/\mathcal{N}_1] \cdot_{\mathcal{Q}} [\mathcal{M}_2 \cdot_{\mathcal{Z}} \mathcal{N}_1/\mathcal{M}_1 \cdot_{\mathcal{Z}} \mathcal{N}_2] \\
&\quad = [\mathcal{M}_1 \cdot_{\mathcal{Z}} (\mathcal{M}_2 \cdot_{\mathcal{Z}} \mathcal{N}_1)/\mathcal{N}_1 \cdot_{\mathcal{Z}} (\mathcal{M}_1 \cdot_{\mathcal{Z}} \mathcal{N}_2)] \\
&\quad = [\mathcal{M}_2 \cdot_{\mathcal{Z}} (\mathcal{M}_1 \cdot \mathcal{N}_1)/\mathcal{N}_2 \cdot_{\mathcal{Z}} (\mathcal{M}_1 \cdot \mathcal{N}_1)] \\
&\quad = [\mathcal{M}_2/\mathcal{N}_2].
\end{aligned}$$

Thus, if $\gamma = [\mathcal{M}_2 \cdot_{\mathcal{Z}} \mathcal{N}_1/\mathcal{M}_1 \cdot_{\mathcal{Z}} \mathcal{N}_2]$, then $\alpha \cdot_{\mathcal{Q}} \gamma = \beta$; by multiplicative commutativity in $\mathcal{Q}$, $\gamma \cdot_{\mathcal{Q}} \alpha = \beta$.

Suppose that $\mathcal{K}, \mathcal{L}$ $(\neq \bar{0}) \in \mathcal{Z}$ are such that $[\mathcal{M}_1/\mathcal{N}_1] \cdot_{\mathcal{Q}} [\mathcal{K}/\mathcal{L}] = [\mathcal{M}_2/\mathcal{N}_2]$. Then $(\mathcal{M}_1 \cdot_{\mathcal{Z}} \mathcal{K}) \cdot_{\mathcal{Z}} \mathcal{N}_2 = \mathcal{M}_2 \cdot_{\mathcal{Z}} (\mathcal{N}_1 \cdot_{\mathcal{Z}} \mathcal{L})$. By multiplicative associativity and commutativity in $\mathcal{Z}$, $\mathcal{K} \cdot_{\mathcal{Z}} (\mathcal{M}_1 \cdot_{\mathcal{Z}} \mathcal{N}_2) = (\mathcal{M}_2 \cdot_{\mathcal{Z}} \mathcal{N}_1) \cdot_{\mathcal{Z}} \mathcal{L}$. Hence,

$$\langle \mathcal{K}, \mathcal{L} \rangle R_2 \langle \mathcal{M}_2 \cdot_{\mathcal{Z}} \mathcal{N}_1, \mathcal{M}_1 \cdot \mathcal{N}_2 \rangle;$$

consequently, $[\mathcal{K}/\mathcal{L}] = \gamma$.

We write "γ" as "β/α", and we call γ the **quotient of** β **and** α. Thus, $\gamma = \beta/\alpha$ *iff* $\alpha \cdot_{\mathcal{Q}} \gamma = \beta$. Clearly, *if* $\beta \neq \bar{\bar{0}}$, *then* $\gamma = \beta/\alpha$ *iff* $\alpha = \beta/\gamma$.

Corollary. *Let α $(\neq \bar{\bar{0}}) \in \mathcal{Q}$. Then there is a unique δ in $\mathcal{Q}$ such that*

$$\alpha \cdot_{\mathcal{Q}} \delta = \delta \cdot_{\mathcal{Q}} \alpha = \bar{\bar{1}}.$$

The rational number δ $(= \bar{\bar{1}}/\alpha)$ is called the **multiplicative inverse of** α. *Thus, every nonzero rational number α has a unique multiplicative inverse.*

EXAMPLE 3.7.2. Let $\beta = \bar{\bar{4}}$, $\alpha = \bar{\bar{3}}$. It follows that $\beta/\alpha = \bar{\bar{4}}/\bar{\bar{3}} = [4/3]$ because $[3/1] \cdot_{\mathcal{Z}} [4/3] = [4/1]$. Note that $4/3 \in \bar{\bar{4}}/\bar{\bar{3}}$.

EXAMPLE 3.7.3

(1) Suppose that $[3/7] \cdot_{\mathcal{Q}} \gamma = [-2/5]$. Then $\gamma = [(-\bar{2} \cdot_{\mathcal{Z}} \bar{7})/(\bar{3} \cdot_{\mathcal{Z}} \bar{5})] = [-14/15]$.

(2) $[8/3]/[4/7] = [14/3]$, because the unique solution of $[4/7] \cdot_{\mathcal{Q}} \gamma = [8/3]$ is $[(\bar{8} \cdot_{\mathcal{Z}} \bar{7})/(\bar{4} \cdot_{\mathcal{Z}} \bar{3})] = [56/12] = [14/3]$.

The following well-known laws, relating quotients of rationals to sums, differences, and products, are easily established. (We omit the subscript "$\mathscr{Q}$" here and in the ensuing exercises.)

Let $\alpha, \gamma \in \mathscr{Q}$ *and* $\beta, \delta \in \mathscr{Q} \sim \{\bar{\bar{0}}\}$.

(1) $$(\alpha/\beta) + (\gamma/\delta) = (\alpha \cdot \delta + \beta \cdot \gamma)/(\beta \cdot \delta);$$

$$(\alpha/\beta) - (\gamma/\delta) = (\alpha \cdot \delta - \beta \cdot \gamma)/(\beta \cdot \delta).$$

(2) $$(\alpha/\beta) \cdot (\gamma/\delta) = (\alpha \cdot \gamma)/(\beta \cdot \delta).$$

(3) $$(\beta/\delta) \cdot (\delta/\beta) = \bar{\bar{1}}.$$

(4) $$(-\alpha)/\beta = \alpha/(-\beta) = -(\alpha/\beta).$$

(5) $$\alpha/\beta = \gamma/\delta \text{ iff } \alpha \cdot \delta = \beta \cdot \gamma.$$

Because of the equalities of (4), we agree to omit parentheses in such expressions.

Finally, we would like to define negative as well as positive powers of nonzero rationals. Thus, we essentially define a function $\mathscr{E}_{\mathscr{Z}}$, where

$$\mathscr{E}_{\mathscr{Q}} : (((\mathscr{Q} \sim \{\bar{\bar{0}}\}) \times \mathscr{Z}) \cup (\{\bar{\bar{0}}\} \times \mathscr{Z}_+)) \to \mathscr{Q}.$$

For a nonzero rational α, we let

$$\alpha^{\bar{0}} = \bar{\bar{1}},$$

and for N in ω,

$$\alpha^{\overline{N^+}} = \alpha^{\bar{N}} \cdot_{\mathscr{Q}} \alpha \qquad \text{and} \qquad \alpha^{-\bar{N}} = 1/\alpha^{\bar{N}};$$

we define

$$\bar{\bar{0}}^{\bar{N}} = \bar{0}, \qquad N \in \omega_+.$$

We generally omit the bars and write "α^N" or "α^{-N}". In particular, α^{-1} is clearly the multiplicative inverse of α; i.e., $\alpha^{-1} = \bar{\bar{1}}/\alpha$. It is easily seen that

$$(\alpha^{-1})^{-1} = \alpha$$

and that

$$(\alpha \cdot \beta)^{-1} = \alpha^{-1}\beta^{-1}.$$

Exercises

3.7.1. **(a)** Determine $[4/3] + [3/-1]$.
(b) Determine $[4/3] - [3/-1]$.
(c) Determine $[4/3] \cdot [3/-1]$.
(d) Determine $[4/3]/[3/-1]$.

3.7.2. **(a)** Determine $[8/3] + [-1/7]$.
(b) Determine $[8/3] - [-1/7]$.
(c) Determine $[8/3] \cdot [-1/7]$.
(d) Determine $[8/3]/[-1/7]$.

3.7.3. Prove Theorem 3.7.2. Thus, show that for α, β, γ in $\mathscr{Q}$, the following are true:
(a) $\alpha + \beta = \beta + \alpha$, $\alpha \cdot \beta = \beta \cdot \alpha$.
(b) $\alpha + (\beta + \gamma) = (\alpha + \beta) + \gamma$, $\alpha \cdot (\beta \cdot \gamma) = (\alpha \cdot \beta) \cdot \gamma$.
(c) $\alpha \cdot (\beta + \gamma) = \alpha \cdot \beta + \alpha \cdot \gamma$.
(d) $\alpha \cdot (\beta - \gamma) = \alpha \cdot \beta - \alpha \cdot \gamma$.

†3.7.4. Prove that if $G : \mathscr{Z} \to \mathscr{Q}$ is defined by $G(\mathscr{M}) = [\mathscr{M}/\bar{1}]$, then for $\mathscr{M}, \mathscr{N} \in \mathscr{Z}$, $\alpha \in \mathscr{Q}$, the following are true:
(a) $G(\mathscr{M} +_{\mathscr{Z}} \mathscr{N}) = G(\mathscr{M}\varepsilon +_{\mathscr{Q}} G(\mathscr{N})$.
(b) $G(\mathscr{M} -_{\mathscr{Z}} \mathscr{N}) = G(\mathscr{M}) -_{\mathscr{Q}} G(\mathscr{N})$.
(c) $G(\mathscr{M} \cdot_{\mathscr{Z}} \mathscr{N}) = G(\mathscr{M}) \cdot_{\mathscr{Q}} G(\mathscr{N})$.
(d) $\alpha +_{\mathscr{Q}} G(\bar{0}) = G(\bar{0}) +_{\mathscr{Q}} \alpha = \alpha$.
(e) $\alpha \cdot_{\mathscr{Q}} G(\bar{1}) = G(\bar{1}) \cdot_{\mathscr{Q}} \alpha = \alpha$.
(f) If $G(\mathscr{M}) = G(\mathscr{N})$, then $\mathscr{M} = \mathscr{N}$.

†3.7.5. Prove that $\mathscr{Q}$ has no zero divisors.

†3.7.6. Let $\alpha, \gamma \in \mathscr{Q}$ and $\beta, \delta \in \mathscr{Q} \sim \{\overline{\overline{0}}\}$. Prove each of the following:
(a) $(\alpha/\beta) + (\gamma/\delta) = (\alpha \cdot \delta + \beta \cdot \gamma)/(\beta \cdot \delta)$, $(\alpha/\beta) - (\gamma/\delta) = (\alpha \cdot \delta - \beta \cdot \gamma)/(\beta \cdot \delta)$.
(b) $(\alpha/\beta) \cdot (\gamma/\delta) = (\alpha \cdot \gamma)/(\beta \cdot \delta)$.
(c) $(\alpha/\beta) \cdot (\beta/\alpha) = \overline{\overline{1}}$, $\alpha \neq \overline{\overline{0}}$.
(d) $(-\alpha)/\beta = \alpha/(-\beta) = -(\alpha/\beta)$, $\alpha \cdot (\gamma/\delta) = (\alpha \cdot \gamma)/\delta$.
(e) $\alpha/\beta = \gamma/\delta$ iff $\alpha \cdot \delta = \beta \cdot \gamma$.

3.7.7. (See Example 2.8.5 for terminology.) Let $\langle X, \langle F_1, F_2 \rangle\rangle$ be a commutative ring with an identity. $\langle X, \langle F_1, F_2 \rangle\rangle$ is called a **field** if for each x_1 in $X \sim \{z\}$ there is an x_2 in X that satisfies $x_1 F_2 x_2 = e$. Show that $\langle \mathscr{Q}, \langle +_{\mathscr{Q}}, \cdot_{\mathscr{Q}} \rangle\rangle$ is a field. (Referred to in Exercises 3.10.14 and 3.12.8.)

★3.7.8. Let $\alpha \in \mathscr{Q} \sim \{\bar{0}, \bar{1}, -\bar{1}\}$. Prove that α can be expressed uniquely in the form

$$\alpha = \varepsilon \cdot_{\mathscr{Q}} \overline{\overline{P_1}}^{\mathscr{E}_1} \cdot_{\mathscr{Q}} \overline{\overline{P_2}}^{\mathscr{E}_2} \cdot_{\mathscr{Q}} \cdots \cdot_{\mathscr{Q}} \overline{\overline{P_S}}^{\mathscr{E}_S},$$

where $\varepsilon = \overline{\overline{1}}$ or $-\overline{\overline{1}}$, $P_1, P_2, \ldots, P_S$ are primes, $P_1 < P_2 < \cdots < P_S$, $S \in \omega_+$, and $\mathscr{E}_1, \mathscr{E}_2, \ldots, \mathscr{E}_S$ are nonzero integers. (Useful for Section 4.5.)

3.8 Ordering the Rational Numbers

We now introduce the "natural" order relation on the set of rationals. As was the case with the integers, we begin by defining the "positive" elements.

We revert to our original notation for fractions.

A **rational number** α **is** said to be **positive** provided there is some $\langle \mathscr{M}, \mathscr{N} \rangle$ in α for which $\mathscr{M} \cdot_{\mathscr{Z}} \mathscr{N}$ is a positive integer. A **rational number** α **is negative** if there is *some* $\langle \mathscr{M}, \mathscr{N} \rangle$ in α for which $\mathscr{M} \cdot_{\mathscr{Z}} \mathscr{N}$ is a negative integer.

EXAMPLE 3.8.1

(1) $[5/2]$ is positive because $\langle\overline{5},\overline{2}\rangle \in [5/2]$ and $\overline{5} \cdot_{\mathscr{Z}} \overline{2}$ is a positive integer.
(2) $[-3/-1]$ and $[-1/-3]$ are both positive.
(3) $[-2/1]$ and $[1/-2]$ are both negative.
(4) $[0/2]$ is neither positive nor negative.

Theorem 3.8.1

(1) *A rational number α is positive if and only if for* every $\langle\mathscr{M},\mathscr{N}\rangle$ in α, $\mathscr{M} \cdot_{\mathscr{Z}} \mathscr{N}$ *is a positive integer.*
(2) *A rational number α is negative if and only if for* every $\langle\mathscr{M},\mathscr{N}\rangle$ *in α,* $\mathscr{M} \cdot_{\mathscr{Z}} \mathscr{N}$ *is a negative integer.*

PROOF: We prove only part (1). Suppose that α is a positive rational. Then there are integers $\mathscr{M}_1$ and $\mathscr{N}_1$ such that $\mathscr{M}_1 \cdot \mathscr{N}_1$ is positive and $\langle\mathscr{M}_1,\mathscr{N}_1\rangle \in \alpha$. Let $\mathscr{M}$ and $\mathscr{N}$ be any integers for which $\langle\mathscr{M},\mathscr{N}\rangle \in \alpha$. Then $\langle\mathscr{M},\mathscr{N}\rangle R_2 \langle\mathscr{M}_1,\mathscr{N}_1\rangle$; hence $\mathscr{M} \cdot \mathscr{N}_1 = \mathscr{N} \cdot \mathscr{M}_1$. Neither $\mathscr{M}$ nor $\mathscr{N}$ can be $\overline{0}$. Moreover, $\mathscr{M} \cdot \mathscr{N}_1^2 = \mathscr{N} \cdot (\mathscr{M}_1 \cdot \mathscr{N}_1)$. Clearly, $\mathscr{N}_1^2$ and $\mathscr{M}_1 \cdot \mathscr{N}_1$ are each positive. Because multiplication by a positive integer preserves positiveness (respectively, negativeness), $\mathscr{M}$ and $\mathscr{N}$ are either simultaneously positive or (*excl.*) negative. Thus, $\mathscr{M} \cdot \mathscr{N}$ is positive.

Conversely, suppose that for every $\langle\mathscr{M},\mathscr{N}\rangle$ in α, $\mathscr{M} \cdot \mathscr{N}$ is a positive integer. We know that a rational number is a nonempty set of ordered pairs, $\langle\mathscr{M},\mathscr{N}\rangle$, for $\mathscr{M}$ in $\mathscr{Z}$ and $\mathscr{N}$ in $\mathscr{Z} \sim \{\overline{0}\}$. Thus, there exist integer $\mathscr{M}_1$ and $\mathscr{N}_1$ such that $\langle\mathscr{M}_1,\mathscr{N}_1\rangle \in \alpha$ and, consequently, $\mathscr{M}_1 \cdot \mathscr{N}_1$ is a positive integer. Hence, α is a positive rational.

Corollary. *The* **Law of Trichotomy** *holds for the rationals. Thus, for every rational number α, exactly one of the following is true:*

$$(1)\ \alpha = \overline{\overline{0}}; \qquad (2)\ \alpha \textit{ is positive}; \qquad (3)\ \alpha \textit{ is negative.}$$

PROOF: We use the theorem together with the Law of Trichotomy for $\mathscr{Z}$. Let α be any rational. Then α is a nonempty set of ordered pairs, $\langle\mathscr{M},\mathscr{N}\rangle$ for $\mathscr{M}$ in $\mathscr{Z}$ and $\mathscr{N}$ in $\mathscr{Z} \sim \{\overline{0}\}$. If for some $\langle\mathscr{M},\mathscr{N}\rangle$ in α, $\mathscr{M} \cdot \mathscr{N}$ is positive (respectively, negative), then for every $\langle\mathscr{M},\mathscr{N}\rangle$ in α, $\mathscr{M} \cdot \mathscr{N}$ is positive (respectively, negative). It follows that if for some $\langle\mathscr{M},\mathscr{N}\rangle$ in α, $\mathscr{M} \cdot \mathscr{N} = \overline{0}$, then every $\langle\mathscr{M},\mathscr{N}\rangle$ in α must have this property. These three possibilities are mutually exclusive and exhaustive.

We leave it to the reader to establish that *the sum, product, and quotient of two positive rationals are positive.* Also, *the sum of two negative rationals is negative,* whereas *the product and quotient of two negative rationals is positive.*

Let $\alpha, \beta \in \mathscr{Q}$. We define "$\alpha <_{\mathscr{Q}} \beta$" to mean that $\beta -_{\mathscr{Q}} \alpha$ is positive. Thus, *for any $\langle \mathscr{M}_1, \mathscr{N}_1 \rangle$ in α and $\langle \mathscr{M}_2, \mathscr{N}_2 \rangle$ in β, $\alpha <_{\mathscr{Q}} \beta$ iff*

$$(\mathscr{M}_2 \cdot_{\mathscr{Z}} \mathscr{N}_1 - \mathscr{M}_1 \cdot_{\mathscr{Z}} \mathscr{N}_2) \cdot_{\mathscr{Z}} (\mathscr{N}_1 \cdot_{\mathscr{Z}} \mathscr{N}_2) >_{\mathscr{Z}} \bar{0}.$$

We can always choose representative fractions $\langle \mathscr{M}_1, \mathscr{N}_1 \rangle$ in α and $\langle \mathscr{M}_2, \mathscr{N}_2 \rangle$ in β for which $\mathscr{N}_1 >_{\mathscr{Z}} \bar{0}$ and $\mathscr{N}_2 >_{\mathscr{Z}} \bar{0}$; with reference to such representatives, $\alpha <_{\mathscr{Q}} \beta$ iff $\mathscr{M}_2 \cdot_{\mathscr{Z}} \mathscr{N}_1 >_{\mathscr{Z}} \mathscr{M}_1 \cdot_{\mathscr{Z}} \mathscr{N}_2$.

We define "$\leq_{\mathscr{Q}}$", "$>_{\mathscr{Q}}$", and "$\geq_{\mathscr{Q}}$" in the usual manner; for instance, $\alpha \leq_{\mathscr{Q}} \beta$ iff either $\alpha <_{\mathscr{Q}} \beta$ or $\alpha = \beta$. Once again, the Law of Trichotomy implies that *for any two rationals α and β, exactly one of the relations $\alpha <_{\mathscr{Q}} \beta$, $a = \beta$, and $\alpha >_{\mathscr{Q}} \beta$ holds.*

Theorem 3.8.2. *$\leq_{\mathscr{Q}}$ is a linear ordering of $\mathscr{Q}$.*

(We leave the proof as an exercise.)

In Section 3.7 we considered a one-one function G that is such that $G : \mathscr{Z} \to \mathscr{Q}$ and $G(\mathscr{M}) = \overline{\mathscr{M}}$, $\mathscr{M} \in \mathscr{Z}$. This function G not only preserves the arithmetic operations of $\mathscr{Z}$, but it preserves order as well. Thus

$$\mathscr{M} <_{\mathscr{Z}} \mathscr{N} \text{ iff } \overline{\mathscr{M}} <_{\mathscr{Q}} \overline{\mathscr{N}}$$

and

$$\mathscr{M} \leq_{\mathscr{Z}} \mathscr{N} \text{ iff } \overline{\mathscr{M}} \leq_{\mathscr{Q}} \overline{\mathscr{N}}.$$

Consequently, *G maps the integers isomorphically into the rationals.* It follows that *the set of rationals cannot be well-ordered by $\leq_{\mathscr{Q}}$*, for if it could be, then $G''(\mathscr{Z})$, as a subset of $\mathscr{Q}$, would automatically be well-ordered by $\leq_{\mathscr{Q}}$. But this would imply that $\mathscr{Z}$ is well-ordered by $\leq_{\mathscr{Z}}$.

Let F be the isomorphism of ω into $\mathscr{Z}$ discussed in Theorem 3.4.3. Then $G \circ F : \omega \to \mathscr{Q}$; the composite function $G \circ F$ is clearly one-one and preserves sums, products, and order; thus, *the natural numbers can be embedded isomorphically in the rationals.*

The ordering $\leq_{\mathscr{Q}}$ possesses one very important characteristic not shared by either of the orderings $\leq$ or $\leq_{\mathscr{Z}}$.

Let R be an antisymmetric, transitive relation whose field is a set X. Then X **is** said to be **dense with respect to** R if whenever a and b are distinct elements of X for which aRb, there is an element c in $X \sim \{a, b\}$ such that aRc and cRb. Clearly, *X is dense with respect to R iff X is dense with respect to $R \cup \mathscr{I}_X$*. If X is dense with respect to a linear ordering R, then R is said to **densely order** X.

EXAMPLE 3.8.2. Suppose that the collection of points that lie on a horizontal line is a set X and that for x, y in X, xRy means that x lies to the left of y.

(Note that here, in contrast with the relation of Example 2.6.7, we do not assume reflexivity.) Then R densely orders X. We are utilizing the intuitive notion that between any two distinct points there lies a third point.

EXAMPLE 3.8.3. Let X be any nonempty set and consider the proper inclusion relation, $\subset$, on $\mathscr{P}(X)$. Then X is not dense with respect to $\subset$. For if A is any proper subset of X and if $a \in X \sim A$, then $A \subset A \cup \{a\}$. However, there is no Y in $\mathscr{P}(X)$ for which $A \subset Y \subset A \cup \{a\}$ holds true.

Clearly, $\leq$ does not densely order ω, for there is no natural number N other than 0 or 1 for which $0 \leq N \leq 1$. Similarly, $\leq_{\mathscr{Z}}$ does not densely order the integers.

Theorem 3.8.3. *The linear ordering $\leq_{\mathscr{Q}}$ densely orders the rational numbers.*

PROOF: We need only show that if α and β are any two distinct rationals that satisfy $\alpha \leq \beta$ (and hence $\alpha < \beta$), then there exists γ in $\mathscr{Q} \sim \{\alpha, \beta\}$ for which $\alpha \leq \gamma \leq \beta$. Let $\gamma = (\alpha + \beta)/2$. Then

$$\gamma - \alpha = \frac{\alpha + \beta}{2} - \alpha = \frac{\alpha + \beta - 2\cdot\alpha}{2} = \frac{\beta - \alpha}{2}.$$

Because $\beta - \alpha$ and 2 are both positive, so is $\gamma - \alpha$. Hence, $\alpha \leq \gamma$. Similarly, we can show $\gamma \leq \beta$. Moreover, $\gamma \neq \alpha$ and $\gamma \neq \beta$. For example, if $\gamma = \alpha$, then $\alpha + \beta$ would equal $2\cdot\alpha(= \alpha + \alpha)$. But then β would equal α, which is a contradiction.

Exercises

3.8.1. Classify the following rationals as (1) zero, (2) positive, or (3) negative:

(a) $[3/5]$. **(b)** $[0/7]$.
(c) $[1/-2]$. **(d)** $[-1/-4]$.
(e) $[-1/5]$. **(f)** $[0/-1]$.
(g) $[3/5] + [1/-2]$. **(h)** $[3/5]\cdot[1/-2]$.
(i) $[3/5] - [1/-2]$. **(j)** $[3/5]/[1/-2]$.

3.8.2. Order the following rationals according to $<_{\mathscr{Q}}$:

(a) $[7/-4]$. **(b)** $[-1/-2]$.
(c) $[0/-3]$. **(d)** $[-2/4]$.
(e) $[7/10]$.

†3.8.3. Let α and β be positive rationals and let γ and δ be negative rationals. Prove that each of (a)–(e) is positive:

(a) $\alpha + \beta$. **(b)** $\alpha\cdot\beta$.
(c) α/β. **(d)** $\gamma\cdot\delta$.
(e) γ/δ. **(f)** Prove that $\gamma + \delta$ is negative.

3.8.4. Prove Theorem 3.8.2; i.e., prove that $\leq_{\mathscr{Q}}$ is a linear ordering.

†3.8.5. Prove that for all $\mathscr{M}$, $\mathscr{N}$ in $\mathscr{Z}$, the following are true:

(a) $\mathscr{M} <_{\mathscr{Z}} \mathscr{N}$ iff $\overline{\mathscr{M}} <_{\mathscr{Q}} \overline{\mathscr{N}}$.

(b) $\mathscr{M} \leq_{\mathscr{Z}} \mathscr{N}$ iff $\overline{\mathscr{M}} \leq_{\mathscr{Q}} \overline{\mathscr{N}}$.

†3.8.6. Prove that if α and β are nonzero rationals such that $\alpha < \beta$, then $\bar{\bar{1}}/\beta < \bar{\bar{1}}/\alpha$.

†3.8.7. **(a)** Prove that if $\alpha < \beta$, there is some γ_1 in $\mathscr{Q}$ for which $\alpha + \gamma_1 > \beta$.

(b) Prove that if α is nonzero and $\alpha < \beta$, there is some γ_2 in $\mathscr{Q}$ for which $\alpha \cdot \gamma_2 > \beta$.

(Thus, $\mathscr{Q}$ possesses the "**Archimedian Property**" with respect to both addition and multiplication.)

†3.8.8. Prove that if $\alpha >_{\mathscr{Q}} \bar{\bar{0}}$, then there is some $N \geq 1$ such that $[1/N] <_{\mathscr{Q}} \alpha$.

†3.8.9. Let $\alpha, \beta, \gamma, \delta, \varepsilon \in \mathscr{Q}$ and suppose that $\beta < \gamma$, $0 < \delta$, and $\varepsilon < 0$.

(a) Show that $\alpha + \beta < \alpha + \gamma$.

(b) What can be said about $\alpha \cdot \beta$ and $\alpha \cdot \gamma$?

(c) Show that $\delta \cdot \beta < \delta \cdot \gamma$.

(d) Show that $\varepsilon \cdot \beta > \varepsilon \cdot \gamma$.

†3.8.10. Suppose that X is dense with respect to R and that a and b are distinct elements of X that are such that aRb. Prove that $\{x \in X : aRx \text{ and } xRb\}$ is infinite.

3.8.11. Suppose that X is a nonempty finite set. Let X be dense with respect to a partial ordering R. What can be said about R?

3.8.12. For $I = 1, 2, 3, 4$, which of the following subsets S_I of $\mathscr{Q}$ are densely ordered by $\leq_{\mathscr{Q}} \upharpoonright S_I$?

(a) $S_1 = \{\text{positive rationals}\}$.

(b) $S_2 = \{\text{nonzero rationals}\}$.

(c) $S_3 = \{\alpha \in \mathscr{Q} : \alpha = [1/N] : N \in \omega_+\}$.

(d) $S_4 = \{\alpha \in \mathscr{Q} : \bar{\bar{0}} < \alpha < \bar{\bar{1}}\}$.

3.9 Real Numbers as Dedekind Cuts

We begin by considering a major defect in the arithmetic of the rationals: not every rational number has a square root.

Theorem 3.9.1. *There is no α in $\mathscr{Q}$ for which $\alpha^2 = \bar{\bar{2}}$.*

PROOF: Suppose that such a rational, α, exists. Because $\alpha^2 = (-\alpha)^2$ and $\bar{\bar{0}}^2 = \bar{\bar{0}} \neq \bar{\bar{2}}$, we can suppose that α is positive. Therefore, α can be written in the form $[M/N]$, where $M \geq 1$, $N \geq 1$. Consequently,

$$[2/1] = \bar{\bar{2}} = \alpha^2 = [M/N] \cdot_{\mathscr{Q}} [M/N] = [\overline{M^2}/\overline{N^2}].$$

Thus, $\bar{2} \cdot_{\mathscr{Z}} \overline{N^2} = \overline{M^2}$, and hence $2 \cdot N^2 = M^2$. Let

$$K = 2 \cdot N^2 = M^2. \tag{3.9.1}$$

Surely, the prime factor 2 appears an even number of times (possibly not at all) in the prime factorization of each of the numbers M^2 and N^2. But then (3.9.1) implies that K has two distinct prime factorizations. (Contradiction.)

We have seen that there are numbers that are "missing" from the set $\mathcal{Q}$, despite the denseness of $\mathcal{Q}$. Thus, were we to arrange the elements of $\mathcal{Q}$ on a straight line according to the ordering $\leq_{\mathcal{Q}}$, there would be "gaps" in the line corresponding to such numbers as "$\sqrt{\overline{\overline{2}}}$". There are infinitely many such gaps; in fact, the proof of Theorem 3.9.1 could easily be modified to show that there is no rational number α such that $\alpha^2 = \overline{\overline{P}}$, where $P(\in\omega)$ is any prime. Our definition of real numbers will correspond to the intuitive notion of "cutting the real line" both at the rational numbers and at these "irrational gaps". For this reason, this process of defining the real numbers is called "the method of **Dedekind cuts**" (after Richard Dedekind, 1831–1916).

A **real number** is a partition $\{L, U\}$ of $\mathcal{Q}$ that is such that with respect to $\leq_{\mathcal{Q}}$, L has no last element and L is an initial segment of $\mathcal{Q}$. *A real number is a set; and the collection of real numbers is a set* (Exercise 3.9.1), which is denoted by $\mathcal{R}$.

The definition of a real number $\{L, U\}$ immediately implies that

(3.9.2) $L \neq \varnothing, U \neq \varnothing$;
(3.9.3)$_a$ $L \cup U = \mathcal{Q}$;
(3.9.3)$_b$ $L \cap U = \varnothing$.

Moreover, *let α and β be arbitrary rationals.*

(3.9.4)$_a$ *If $\alpha \in L$ and $\beta \in U$, then $\alpha < \beta$.*
(3.9.4)$_b$ *If $\alpha \in L$ and $\beta < \alpha$, then $\beta \in L$.*
(3.9.4)$_c$ *If $\alpha \in U$ and $\beta > \alpha$, then $\beta \in U$.*
(3.9.5) *For every α_1 in L there is an α_2 in L that satisfies $\alpha_1 < \alpha_2$.*

In fact, *conditions* (3.9.2), (3.9.3), (3.9.4), and (3.9.5) *serve as an equivalent definition of a real number* (Exercise 3.9.5). Thus, any set that satisfies these conditions is a real number. Finally, note that *a real number is completely determined by a nonempty proper initial segment of $\mathcal{Q}$ that has no last element.*

EXAMPLE 3.9.1

(1) If L is the set of negative rationals and U is the set consisting of $\overline{\overline{0}}$ together with the positive rationals, then $\{L, U\}$ is a real number.
(2) If L' consists of the negative rationals and U' consists of the positive rationals, then $\{L', U'\}$ is not a real number because $\overline{\overline{0}} \notin L' \cup U'$.

(3) If $L'' = L' \cup \{\bar{\bar{0}}\}$, then $\{L'', U'\}$ is not a real number because L'' has a last element, $\bar{\bar{0}}$.

The set U may or may not have a first element. We first consider the case in which U has a first element.

Theorem 3.9.2. *For α in $\mathcal{Q}$, let $H(\alpha) = \{L_\alpha, U_\alpha\}$, where $L_\alpha = \{\beta \in \mathcal{Q} : \beta < \alpha\}$ and $U_\alpha = \{\beta \in \mathcal{Q} : \beta \geq \alpha\}$. Then H is a one-one mapping of $\mathcal{Q}$ into $\mathcal{R}$.*

PROOF: Let $\alpha \in \mathcal{Q}$ and consider $H(\alpha) = \{L_\alpha, U_\alpha\}$; we first show that H_α is, indeed, a real number. By the Law of Trichotomy for $\mathcal{Q}$, $L_\alpha \cup U_\alpha = \mathcal{Q}$ and $L_\alpha \cap U_\alpha = \varnothing$. Moreover, $\alpha \in U$ and $\alpha - \bar{\bar{1}} \in L_\alpha$; hence, neither of these sets is empty. Thus, $\{L_\alpha, U_\alpha\}$ is a partition of $\mathcal{Q}$. If $\beta_1 \in L_\alpha$ and $\beta_2 < \beta_1$, then by the transitivity of $<_{\mathcal{Q}}$, $\beta_2 < \alpha$; thus, $\beta_2 \in L_\alpha$, and L_α is an initial segment of $\mathcal{Q}$. Finally, L_α has no last element. For if β is any element of L_α, the denseness of $\mathcal{Q}$ with respect to $\leq_{\mathcal{Q}}$ guarantees that there is some γ in $\mathcal{Q}$ for which $\beta < \gamma < \alpha$. Hence, $\gamma \in L_\alpha$ and $\gamma > \beta$.

We now show that H is one-one. Suppose that α_1 and α_2 are distinct elements of $\mathcal{Q}$. Then, by the Law of Trichotomy, one of these, say α_1, is the smaller. It follows that $\alpha_1 \in U_{\alpha_1}$; but also $\alpha_1 \in L_{\alpha_2}$. Thus, $\alpha_1 \notin U_{\alpha_2} = \mathcal{Q} \sim L_{\alpha_2}$ and $U_{\alpha_1} \neq U_{\alpha_2}$. Surely, $U_{\alpha_1} \neq L_{\alpha_2}$ because $\alpha_2 \in U_{\alpha_1} \sim L_{\alpha_2}$. Thus,

$$H(\alpha_1) = \{L_{\alpha_1}, U_{\alpha_1}\} \neq \{L_{\alpha_2}, U_{\alpha_2}\} = H(\alpha_2).$$

Corollary. *Every real number $\{L, U\}$ in which U has a first element is in the range of H. In fact, if α is the first element of U, then $\{L, U\} = H(\alpha)$. Thus, every real number of this sort corresponds to a unique rational number.*

Following previous notational practice, we write "$\bar{\alpha}$" for $H(\alpha)$, $\alpha \in \mathcal{Q}$. In particular, for natural numbers N, $\bar{\bar{\bar{N}}} = H(\bar{\bar{N}})$.

In the remainder of this section and in the next one, when there is no likelihood of confusion, we will generally omit bars, subscripts, and square brackets when denoting rational numbers. Thus, "3/4" will denote the rational number formerly indicated by "[3/4]". (We will have no further need to denote fractions in this manner.) We will continue to use natural numbers as subscripts and exponents; otherwise, if nothing is said to the contrary, the numbers referred to will be rational.

EXAMPLE 3.9.2. Let $\alpha = 3/4$. Thus, $L_{3/4} = \{\beta \in \mathcal{Q} : \beta < 3/4\}$ and $U_{3/4} = \{\beta \in \mathcal{Q} : \beta \geq 3/4\}$. The real number $H(3/4)$ corresponding to 3/4 is

$$\{L_{3/4}, U_{3/4}\}.$$

We now give an example of a real number that is not in the range of H, and hence does not correspond (under H) to a rational number. Such real numbers are called **irrational.**

†EXAMPLE 3.9.3. Let $L_1 = \{\text{negative rationals}\}$, $L_2 = \{\alpha \in \mathcal{Q} : \alpha^2 < 2\}$, and let $L = L_1 \cup L_2$ and $U = \mathcal{Q} \sim L$. Thus,

$$U = \{\alpha \in \mathcal{Q} : \alpha \text{ is positive and } \alpha^2 > 2\}.$$

We claim that $\{L, U\}$ is a real number. Clearly, $L \cup U = \mathcal{Q}$ and $L \cap U = \varnothing$. Also, $0 \in L_2 \subseteq L$ and $2 \in U$ (because $2^2 = 4 > 2$); hence, $\{L, U\}$ is a partition of $\mathcal{Q}$. Suppose that $\alpha \in L$ and $\beta \in U$. If $\alpha \in L_1$, then α is negative and β is positive; thus, $\alpha < \beta$. If $\alpha \in L_2 \sim L_1$, then $\alpha^2 < 2 < \beta^2$. By the transitivity of $<_{\mathcal{Q}}$, $\alpha^2 < \beta^2$. Thus, $\beta^2 - \alpha^2 = (\beta - \alpha)(\beta + \alpha)$ is positive. Because α and β are both positive, so is $\beta + \alpha$. Consequently, $\beta - \alpha$ is also positive, and $\beta > \alpha$. Therefore, L is an initial segment of $\mathcal{Q}$.

We must finally show that L has no last element; hence, for each γ in L we must find some δ in L for which $\gamma < \delta$. If $\gamma \in L_1$, then $\gamma < 0$ and $0 \in L$. Consider any γ in $L_2 \sim L_1$, and let $\varepsilon = 2 - \gamma^2$. Then ε is positive, as is $\varepsilon/(2\gamma + 1)$; hence, by Exercise 3.8.8, there is an M such that $M \geq 1$ and $1/M < \varepsilon/(2\gamma + 1)$. Thus, $(2\gamma + 1)/M < \varepsilon$, and

$$\left(\gamma + \frac{1}{M}\right)^2 = \gamma^2 + \frac{2\gamma}{M} + \frac{1}{M^2} \leq \gamma^2 + \frac{2\gamma + 1}{M} < \gamma^2 + \varepsilon = 2.$$

We therefore let $\delta = \gamma + 1/M$. This completes the proof that $\{L, U\}$ is a real number.

A modification of the argument in the preceding paragraph shows that U has no first element. Here, we consider an arbitrary positive rational ρ for which $\rho^2 > 2$, and we let "$\delta(= \rho^2 - 2)$" replace "ε". The remainder of this proof is left as an exercise.

Thus, $\{L, U\}$ is not in the range of H; we denote $\{L, U\}$ by "$\sqrt{2}$"; we justify this notation later.

The next theorem shows that there is no gulf separating the cells L and U of a real number $\{L, U\}$.

Theorem 3.9.3. *Let $\{L, U\}$ be a real number. For every nonzero natural number N, there is an α in L and a β in U such that $\beta - \alpha = 1/N$.*

PROOF: Let N be a nonzero natural number. Because $L \neq \varnothing$, we can let α_1 be an arbitrary element of L. If $\delta \in L$ and if $\gamma < \delta$, then $\gamma \in L$. Thus for some nonzero natural number K, $\alpha_1 + K/N \in U$, as, otherwise, U would be empty. Let K_N be the smallest nonzero natural number K for which $\alpha_1 + K/N \in U$. Let $\beta = \alpha_1 + K_N/N$ and let $\alpha = \alpha_1 + (K_N \dot{-} 1)/N$.

Corollary. *If $\{L, U\}$ is a real number, then for every nonzero natural number M, there is an α in L and a β in U such that $\beta - \alpha < 1/M$.*

PROOF: For each such M, let $N = M^+$ in the theorem.

Exercises

†3.9.1. Show that $\mathcal{R}$, the collection of real numbers, is a set.

3.9.2. Let P be any prime. Show that the equation $\alpha^2 = \overline{\overline{P}}$ has no rational root. (Useful for Section 4.6.)

3.9.3. Determine the set of all natural numbers N for which $\alpha^2 = \overline{\overline{N}}$ has a rational root.

†3.9.4. Let $\{L, U\}$ be a real number. Prove that conditions (3.9.4) and (3.9.5) hold. Thus, show the following:

(3.9.4)$_a$ If $\alpha \in L$ and $\beta \in U$, then $\alpha < \beta$.
(3.9.4)$_b$ If $\alpha_1 \in L$ and $\alpha_2 < \alpha_1$, then $\alpha_2 \in L$.
(3.9.4)$_c$ If $\beta_1 \in U$ and $\beta_2 > \beta_1$, then $\beta_2 \in U$.
(3.9.5) If $\alpha_1 \in L$, then there is some α_2 in L such that $\alpha_1 < \alpha_2$.

†3.9.5. Let L and U be subsets of $\mathcal{Q}$.

(a) Prove that if L and U satisfy conditions (3.9.2), (3.9.3)$_a$, (3.9.4)$_a$, and (3.9.5) (for real numbers), then $\{L, U\}$ is a real number.

(b) Suppose that L and U satisfy both parts of condition (3.9.3). Then show that they satisfy all parts of condition (3.9.4) or none of these.

3.9.6. Let $\{L, U\}$ be a real number and suppose that the rational number 3 is in L. Which of the following statements (about rationals) necessarily hold?

(a) $0 \in L$. **(b)** $4 \in L$.
(c) $2 \notin U$. **(d)** $10 \in U$.
(e) For some α in $\mathcal{Q}$, $\alpha > 3$ and $\alpha \in L$. **(f)** For some β in $\mathcal{Q}$, $\beta > 3$ and $\beta \in U$.

3.9.7. In Example 3.9.3, show that U has no first element.

3.10 Arithmetic and Ordering of Real Numbers

We now define the arithmetic operations and the (natural) order relation on $\mathcal{R}$. The reader should bear in mind that the definitions to be given hold whether or not the real numbers involved correspond (under the mapping H of Theorem 3.9.2) to rational numbers.

For the remainder of the section let $\{L_I, U_I\}$, $I = 1, 2$, be real numbers. We first define **addition**:

$$\{L_1, U_1\} +_{\mathscr{R}} \{L_2, U_2\} = \{L, U\},$$

where

$$L = \{\alpha +_{\mathscr{Q}} \beta : \alpha \in L_1, \beta \in L_2\} \text{ and } U = \mathscr{Q} \sim L.$$

We must show that $\{L, U\}$ is actually a real number. Note that the problem of proving addition to be "well-defined" is different than in the case of rational addition in that the summands are no longer equivalence classes.

Theorem 3.10.1. *Addition of real numbers is well-defined. Thus, if $\{L_1, U_1\}$, $\{L_2, U_2\}$, and $\{L, U\}$ are defined as above, then $\{L, U\}$ is a real number.*

PROOF: (The numbers we discuss are all rational; thus, the operations and ordering refer to $\mathscr{Q}$.)

Because $L_1 \neq \varnothing$ and $L_2 \neq \varnothing$, if α_0 and β_0 are arbitrary elements of L_1 and L_2, respectively, then, by definition, $\alpha_0 + \beta_0 \in L$. If $\rho \in L$ and $\sigma < \rho$, then $\rho = \alpha_1 + \beta_1$ for some α_1 in L_1 and β_1 in L_2. Thus, $\beta_1 - (\rho - \sigma) \in L_2$ and $\sigma(= \alpha_1 + (\beta_1 - (\rho - \sigma))) \in L$. We have shown that L is a nonempty initial segment of $\mathscr{Q}$.

If $\gamma \in U_1$ and $\delta \in U_2$, then for every α in L_1 and β in L_2, $\alpha < \gamma$ and $\beta < \delta$; hence, $\alpha + \beta < \gamma + \delta$. Thus, $\gamma + \delta \in U$. Clearly, $\{L, U\}$ is a partition of $\mathscr{Q}$, and we must finally show that L has no last element.

Indeed, for any ρ in L, $\rho = \alpha_1 + \beta_1$ for some α_1 in L_1 and β_1 in L_2. There are α_2 in L_1 and β_2 in L_2 such that $\alpha_1 < \alpha_2$ and $\beta_1 < \alpha_2$, because neither L_1 nor L_2 has a last element. It follows that $\alpha_2 + \beta_2 \in L$ and $\rho = \alpha_1 + \beta_1 < \alpha_2 + \beta_2$.

EXAMPLE 3.10.1. Let $L_1 = \{\alpha \in \mathscr{Q} : \alpha < 1\}$, $U_1 = \{\alpha \in \mathscr{Q} : \alpha \geq 1\}$, $L_2 = \{\alpha \in \mathscr{Q} : \alpha < 5\}$, $U_2 = \{\alpha \in \mathscr{Q} : \alpha \geq 5\}$, $L_3 = \{\alpha \in \mathscr{Q} : \alpha \text{ is negative or } a^2 < 5\}$, $U_3 = \mathscr{Q} \sim L_2$. (See Figure 3.10.1a.)

(1) Let $\{L, U\} = \{L_1, U_1\} +_{\mathscr{R}} \{L_2, U_2\}$; then $L = \{\alpha \in \mathscr{Q} : \alpha < 6\}$ and $U = \mathscr{Q} \sim L$. In fact, if $\beta < 6$, $\beta = 6 - \gamma$ for some positive rational γ. Then $\beta = \beta_1 + \beta_2$, where $\beta_1 = 1 - \gamma/2 \in L_1$ and $\beta_2 = 5 - \gamma/2 \in L_2$. Conversely, if $\delta = \delta_1 + \delta_2$, where $\delta_1 \in L_1$ and $\delta_2 \in L_2$, then $\delta_1 < 1$ and $\delta_2 < 5$. Hence, $\delta < 6$. (See Figure 3.10.1b.)

(2) Let $\{L^*, U^*\} = \{L_1, U_1\} +_{\mathscr{R}} \{L_3, U_3\}$. In the light of part (1), we would like to say that $L^* = \{\alpha \in \mathscr{Q} : \alpha < 1 + \sqrt{5}\}$. Of course, this makes no sense because $1 + \sqrt{5} \notin \mathscr{L}$. But from this point of view, for the number α in question,

$$\alpha - 1 < \sqrt{5}$$

and, provided that $\alpha - 1 > 0$,

$$(\alpha - 1)^2 < 5.$$

Thus, let

$$L' = \{\alpha \in \mathcal{Q} : \alpha \leq 1\}, L'' = \{\alpha \in \mathcal{Q} : (\alpha - 1)^2 < 5\},$$
$$L^* = L' \cup L'', \text{ and } U^* = \mathcal{Q} \sim L^*.$$

(See Figure 3.10.1c.)

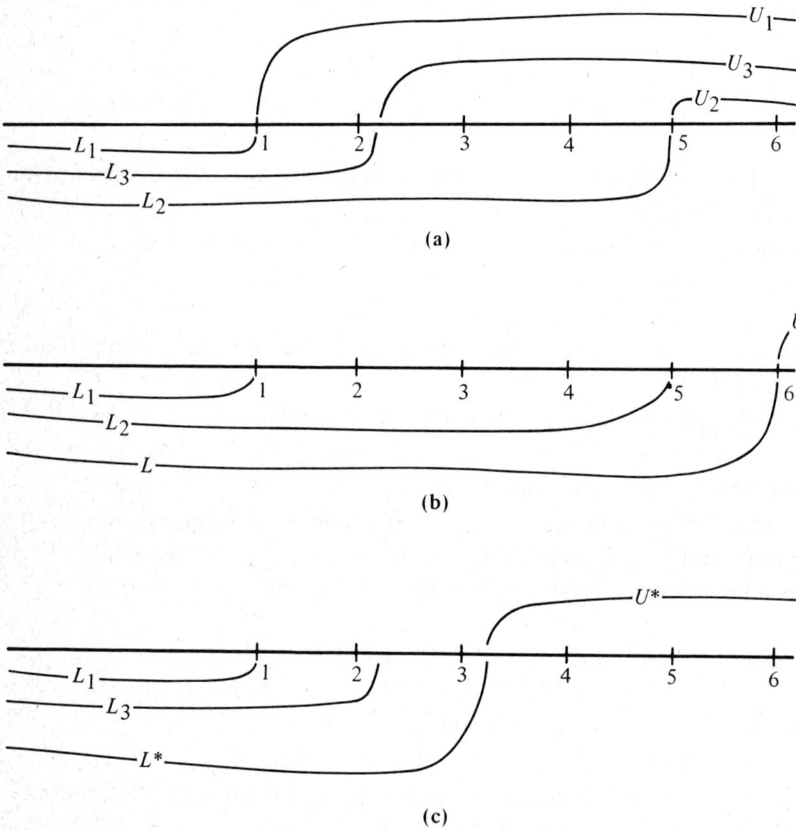

Fig. 3.10.1

When we cannot describe the sum of two real numbers as in part (1) of Example 3.10.1, we often describe the sum $\{L, U\}$ of $\{L_1, U_1\}$ and $\{L_2, U_2\}$ simply by means of the definition. Thus, we say that

$$L = \{\alpha + \beta \in \mathcal{Q} : \alpha \in L_1 \text{ and } \beta \in L_2\}$$

and $U = \mathcal{Q} \sim L$.

We will define subtraction, $-_{\mathscr{R}}$, in terms of $+_{\mathscr{R}}$. Before doing so, we define a unary operation on $\mathscr{R}$ that will associate with each real number its **additive inverse**.

Let $\{L, U\}$ be any real number. Define

$$L^- = \{\alpha \in \mathscr{Q} : \text{there is some } \beta \text{ in } U \text{ for which } \alpha < -\beta\},$$

and let $U^- = \mathscr{Q} \sim L^-$.

EXAMPLE 3.10.2

(1) If $L_1 = \{\alpha \in \mathscr{Q} : \alpha < 2\}$ and $U_1 = \{\alpha \in \mathscr{Q} : \alpha \geq 2\}$, then we claim that $L_1^- = \{\alpha \in \mathscr{Q} : \alpha < -2\}$ and $U_1^- = \{\alpha \in \mathscr{Q} : \alpha \geq -2\}$. (See Figure 3.10.2a.) For if $\beta \in U_1$, then $2 \leq \beta$, and thus $-\beta \leq -2$. Hence, if $\alpha < -\beta$ for some β in U_1, then $\alpha < -2$. Conversely, if $\alpha < -2$, then, by the denseness of $\mathscr{Q}$, there is some γ such that $\alpha < \gamma < -2$. Therefore, $2 < -\gamma$, and $-\gamma \in U_1$. Because $\alpha < -(-\gamma) = \gamma$, $\alpha \in L_1^-$.

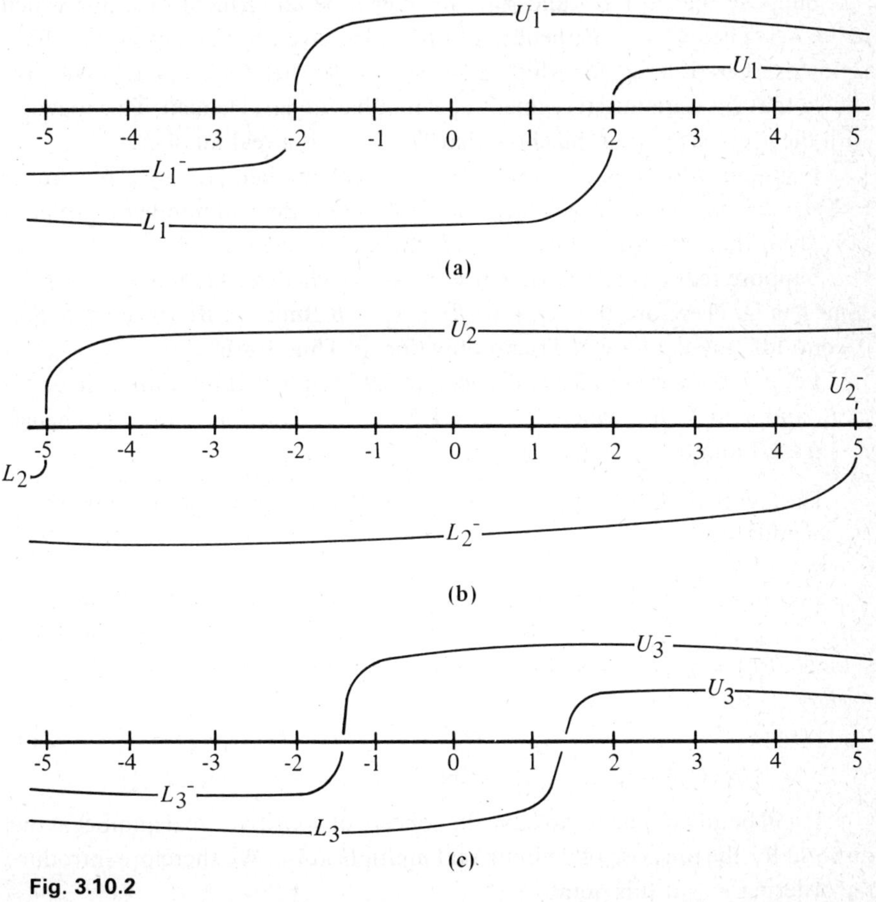

Fig. 3.10.2

(2) If $L_2 = \{\alpha \in \mathcal{Q} : \alpha < -5\}$ and $U_2 = \{\alpha \in \mathcal{Q} : \alpha \geq -5\}$, then $L_1^- = \{\alpha \in \mathcal{Q} : \alpha < 5\}$ and $U_2^- = \{\alpha \in \mathcal{Q} : \alpha \geq 5\}$. (See Figure 3.10.2b.)

(3) If $L_3 = \{\alpha \in \mathcal{Q} : \alpha$ is negative or $\alpha^2 < 2\}$ and $U_3 = \{\alpha \in \mathcal{Q} : \alpha$ is positive and $\alpha^2 > 2\}$, then $L_3^- = \{\alpha \in \mathcal{Q} : \alpha$ is negative and $\alpha^2 > 2\}$ and $U_3^- = \{\alpha \in \mathcal{Q} : \alpha^2 < 2$ or α is positive$\}$. (See Figure 3.10.2c.)

Theorem 3.10.2. *For every real number $\{L, U\}$, the corresponding $\{L^-, U^-\}$ is a real number that has the property that*

$$\{L, U\} +_{\mathcal{R}} \{L^-, U^-\} = \{L^-, U^-\} +_{\mathcal{R}} \{L, U\} = \overset{\equiv}{0}.$$

PROOF: Let $\{L, U\}$ be a real number and let L^- and U^- be defined as above. Let $\alpha \in L$ and let $\beta \in U$. Then $-\beta - 1 \in L^-$. Also, $\alpha < \beta$; hence, $-\alpha > -\beta$. Because β was arbitrarily chosen from U, it follows that $-\alpha \in U^-$. From the definition of U^-, it now follows that $\{L^-, U^-\}$ is a partition of $\mathcal{Q}$.

Suppose that $\alpha_1 \in L^-$ and $\alpha_2 < \alpha_1$. Let β be an element of U for which $\alpha_1 < -\beta$. Then $\alpha_2 < -\beta$; hence, $\alpha_2 \in L^-$. Moreover, by the denseness of the rationals, there is an α_3 for which $\alpha_1 < \alpha_3 < -\beta$. Therefore, $\alpha_3 \in L^-$. We have shown that L^- is an initial segment of $\mathcal{Q}$ that has no last element. This together with the previous paragraph show that $\{L^-, U^-\}$ is a real number.

Consequently, $\{L, U\} +_{\mathcal{R}} \{L^-, U^-\}$ is a real number, $\{L^*, U^*\}$. According to Theorem 3.9.2, in order to show that $\{L^*, U^*\}$ is the real number 0, we need only show that the rational number 0 is the first element of U^*.

Suppose that 0 is in L^*. Then $0 = \alpha_1 + \alpha_2$, where $\alpha_1 \in L$ and $\alpha_2 < -\beta$ for some β in U. Therefore, $0 < \alpha_1 + (-\beta) = \alpha_1 - \beta$. But $\alpha_1 < \beta$; hence, $\alpha_1 - \beta < 0$, contradicting the Law of Trichotomy (for $\mathcal{Q}$). Thus, $0 \in U^*$.

Let $\rho < 0$. Exercise 3.8.8 and Theorem 3.9.3 imply that there are N in ω_+, μ in L, and ν in U for which $\nu - \mu = 1/N < -\rho$. Thus, $\rho - \mu < -\nu$; hence, $\rho - \mu \in L^-$ and $\rho = \mu + (\rho - \mu) \in L^*$.

The fact that $\{L^-, U^-\} +_{\mathcal{R}} \{L, U\} = \overset{\equiv}{0}$ will follow from the commutativity of real addition.

In conformity with our previous convention, we write "$-_{\mathcal{R}}\{L, U\}$" in lieu of "$\{L^-, U^-\}$", and we generally drop the subscript "$\mathcal{R}$". In this notation, $-(-\{L, U\}) = \{L, U\}$, and for $\alpha \in \mathcal{Q}$ and H as in Theorem 3.9.2, $H(-\alpha) = -H(\alpha)$.

Define

"$\{L_1, U_1\} -_{\mathcal{R}} \{L_2, U_2\}$" to be "$\{L_1, U_1\} +_{\mathcal{R}} (-\{L_2, U_2\})$".

It will be advantageous to have the concept of "positive" real number at our disposal for the purpose of defining real multiplication. We therefore introduce the ordering $\leq_{\mathcal{R}}$ at this point.

A **real number** $\{L, U\}$ **is** said to be **positive** if there is some positive rational in L; $\{L, U\}$ **is** called **negative** if some negative rational is in U. It is an immediate consequence of this definition that *$\{L, U\}$ is positive iff* $\overline{\overline{0}} \in L$; hence, *the sum of two positive real numbers is positive.* Moreover, *$\{L, U\}$ is positive iff $-\{L, U\}$ is negative.* We define

"$\{L_1, U_1\} <_{\mathscr{R}} \{L_2, U_2\}$" to mean "$\{L_2, U_2\} -_{\mathscr{R}} \{L_1, U_1\}$ is positive."

$\leq_{\mathscr{R}}, >_{\mathscr{R}}, \geq_{\mathscr{R}}$ are defined in the usual manner. Clearly, the **Law of Trichotomy** holds for $\mathscr{R}$: *For every real number $\{L, U\}$ exactly one of the following is true:*

(1) $\{L, U\} = \overline{\overline{0}}$.
(2) $\{L, U\}$ *is positive.*
(3) $\{L, U\}$ *is negative.*

Equivalently, *if $\{L_1, U_1\}$ and $\{L_2, U_2\}$ are any real numbers, then exactly one of the following holds*:

(1) $\{L_1, U_1\} = \{L_2, U_2\}$.
(2) $\{L_1, U_1\} <_{\mathscr{R}} \{L_2, U_2\}$.
(3) $\{L_1, U_1\} >_{\mathscr{R}} \{L_2, U_2\}$.

The following theorem shows that *set inclusion characterizes the ordering* $\leq_{\mathscr{R}}$; we recall that (in a somewhat different sense) inclusion also characterized the ordering $\leq$ on the natural numbers. (Note that this was not the case for either of the orderings $\leq_{\mathscr{Z}}$ or $\leq_{\mathscr{Q}}$.)

Theorem 3.10.3. *Let $\{L_1, U_1\}$ and $\{L_2, U_2\}$ be real numbers. Then $\{L_1, U_1\} <_{\mathscr{R}} \{L_2, U_2\}$ if and only if $L_1 \subset L_2$.*

PROOF: Let $\{L^*, U^*\} = \{L_2, U_2\} -_{\mathscr{R}} \{L_1, U_1\}$. We must show that $\{L^*, U^*\}$ is positive iff $L_1 \subset L_2$ or, equivalently, that the rational 0 is in L^* iff $L_1 \subset L_2$.

Suppose that $0 \in L^*$. Then there are α_2 in L_2 and β in U_1 such that $0 = \alpha_2 + \alpha_1$ and $\alpha_1 < -\beta$. We claim that $\alpha_2 \in L_2 \sim L_1$. Otherwise, $\alpha_2 \in L_1$; hence, $\alpha_2 < \beta$ because $\beta \in U_1$. Therefore, $0 = \alpha_2 + \alpha_1 < \beta + (-\beta) = 0$. This contradiction establishes our assertion. If $\gamma \in L_1$, then $\gamma < \alpha_2$, and by condition $(3.9.4)_b$, $\gamma \in L_2$. We have shown that $L_1 \subset L_2$.

Suppose that $L_1 \subset L_2$. Let $\delta \in L_2 \sim L_1 = L_2 \cap U_1$. Then by condition (3.9.5), there is an ε in L_2 such that $\delta < \varepsilon$. Thus, $-\varepsilon < -\delta$; hence, $-\varepsilon \in L_1^-$ and $0 = \varepsilon + (-\varepsilon) \in L^*$.

Corollary. *For any real numbers* $\{L_1, U_1\}$ *and* $\{L_2, U_2\}$, $\{L_1, U_1\} <_{\mathscr{R}} \{L_2, U_2\}$ *if and only if* $U_2 \subset U_1$.

We define the **product of** $\{L_1, U_1\}$ **and** $\{L_2, U_2\}$ by cases as follows:

(1) $\{L_1, U_1\} \cdot_{\mathscr{R}} \{L_2, U_2\} = \overset{\equiv}{0}$ if either $\{L_1, U_1\}$ or $\{L_2, U_2\}$ is $\overset{\equiv}{0}$.

(2) Suppose that $\{L_1, U_1\}$ and $\{L_2, U_2\}$ are both positive. Let L' be the set of nonpositive rationals and let

$$L'' = \{\gamma \in \mathscr{Q}: \text{there are positive rationals } \alpha \text{ and } \beta \text{ such that } \alpha \in L_1, \beta \in L_2, \text{ and } \gamma = \alpha \cdot \beta\}.$$

Let $L = L' \cup L''$, $U = \mathscr{Q} \sim L$, and define

$$\{L_1, U_1\} \cdot_{\mathscr{R}} \{L_2, U_2\} = \{L, U\}.$$

(3) If $\{L_1, U_1\}$ and $\{L_2, U_2\}$ are both negative, then $-\{L_1, U_1\}$ and $-\{L_2, U_2\}$ are positive; we define

$$\{L_1, U_1\} \cdot_{\mathscr{R}} \{L_2, U_2\} = (-\{L_1, U_1\}) \cdot_{\mathscr{R}} (-\{L_2, U_2\}).$$

(4) If $\{L_1, U_1\}$ is positive and $\{L_2, U_2\}$ is negative, let

$$\{L_1, U_1\} \cdot_{\mathscr{R}} \{L_2, U_2\} = -(\{L_1, U_1\} \cdot_{\mathscr{R}} (-\{L_2, U_2\})).$$

(5) If $\{L_1, U_1\}$ is negative and $\{L_2, U_2\}$ is positive, let

$$\{L_1, U_1\} \cdot_{\mathscr{R}} \{L_2, U_2\} = -((-\{L_1, U_1\}) \cdot_{\mathscr{R}} \{L_2, U_2\}).$$

We leave as an exercise the verification that *multiplication of real numbers is well-defined* (in the sense that the product of two reals is real). Clearly, the only interesting case is (2); here the proof is somewhat similar to that of Theorem 3.10.1. *The product of two positive real numbers is positive*; moreover, *there are no zero divisors.*

EXAMPLE 3.10.3. Let $L_0 = \{\text{negative rationals}\}$, $L_1 = \{\alpha \in \mathscr{Q} : \alpha < 2\}$, $L_2 = \{\alpha \in \mathscr{Q} : \alpha \text{ is negative or } \alpha^2 < 7\}$, $L_3 = \{\alpha \in \mathscr{Q} : \alpha < -3\}$, $L_4 = \{\alpha \in \mathscr{Q} : \alpha < -5\}$, and for $I = 1, 2, 3, 4$, let $U_I = \mathscr{Q} \sim L_I$.

(1) $\{L_0, U_0\} \cdot_{\mathscr{R}} \{L_I, R_I\} = \{L_0, U_0\} = \overset{\equiv}{0}$, $I = 0, 1, 2, 3, 4$.

(2) $\{L_1, U_1\} \cdot_{\mathscr{R}} \{L_2, U_2\} = \{L, U\}$, where

$$L = \{\alpha \in \mathscr{Q} : \alpha \text{ is negative or } \alpha^2 < 28\},$$

and $U = \mathscr{Q} \sim L$.

(3) $\{L_2, U_2\} \cdot_{\mathscr{R}} \{L_3, U_3\} = -(\{L_2, U_2\} \cdot_{\mathscr{R}} (-\{L_3, U_3\})) = -\{L^*, U^*\}$, where $L^* = \{\alpha \in \mathscr{Q} : \alpha \text{ is negative or } \alpha^2 < 63\}$ and $U^* = \mathscr{Q} \sim L^*$. This

product can also be expressed as $\{L^{**}, U^{**}\}$, where

$$L^{**} = \{\alpha \in \mathcal{Q} : \alpha \text{ is negative and } \alpha^2 > 63\}$$

and $U^{**} = \mathcal{Q} \sim L^{**}$.

(4) $\{L_3, U_3\} \cdot_{\mathcal{R}} \{L_4, U_4\} = (-\{L_3, U_3\}) \cdot_{\mathcal{R}} (-\{L_4, U_4\})$
$= \{\{\alpha \in \mathcal{Q} : \alpha < 3\}, \{\alpha \in \mathcal{Q} : \alpha \geq 3\}\} \cdot_{\mathcal{R}} \{\{\alpha \in \mathcal{Q} : \alpha < 5\}, \{\alpha \in \mathcal{Q} : \alpha \geq 5\}\}$
$= \{\{\alpha \in \mathcal{Q} : \alpha < 15\}, \{\alpha \in \mathcal{Q} : \alpha \geq 15\}\} = \overline{\overline{\overline{15}}}$.

It is easy to show that $\overline{\overline{\overline{1}}}$ *serves as a multiplicative identity in* $\mathcal{R}$.

Exponentiation can be defined with rational (or even real) exponents (Exercise 3.10.8). We will omit bars in writing these exponents. The following theorem does not depend on the definition of exponentiation; however, we utilize the usual notation for "multiplicative inverses".

Theorem 3.10.4. *For every nonzero real number* $\{L, U\}$ *there exists a real number, to be denoted by* "$\{L, U\}^{-1}$", *with the property that*

$$\{L, U\} \cdot_{\mathcal{R}} \{L, U\}^{-1} = \{L, U\}^{-1} \cdot_{\mathcal{R}} \{L, U\} = \overline{\overline{\overline{1}}}.$$

PROOF: If $\{L, U\}$ is positive, let L' be the set of nonpositive rationals; let $L'' = \{\alpha \in \mathcal{Q}$: there is some β in L for which $\beta > 0$ and $\alpha = 1/\beta\}$. Let $L_1 = L' \cup L''$, let $U_1 = \mathcal{Q} \sim L_1$, and let $\{L, U\}^{-1} = \{L_1, U_1\}$.

If $\{L, U\}$ is negative, let $\{L, U\}^{-1} = -((-\{L, U\})^{-1})$.

If $\{L_2, U_2\} \neq \overline{\overline{\overline{0}}}$, define

"$\{L_1, U_1\} /_{\mathcal{R}} \{L_2, U_2\}$" to be "$\{L_1, U_1\} \cdot_{\mathcal{R}} (\{L_2, U_2\}^{-1})$".

We prove the uniqueness of $\{L, U\}^{-1}$ in Exercise 3.10.4(d).

Many of the properties of the arithmetical operations and of the (linear) ordering, $\leq_{\mathcal{R}}$, on the reals are the same as the corresponding ones for the rationals; we discuss some of these in the exercises. We mention only that $\leq_{\mathcal{R}}$ *densely orders* $\mathcal{R}$. Moreover, *the mapping H of* $\mathcal{Q}$ *into* $\mathcal{R}$ (defined in the preceding section) *is an isomorphism in the sense that H preserves the basic operations on* $\mathcal{Q}$ *and the ordering* $\leq_{\mathcal{Q}}$.

We began our discussion of the reals with the observation that not every positive rational number had a square root. One would hope that for real numbers this situation has been remedied—and, indeed, this is so!

Theorem 3.10.5. *Let* $\{L, U\}$ *be a positive real number. Then there is a positve real number* $\{L_1, U_1\}$ *such that* $\{L_1, U_1\}^2 = \{L, U\}$.

PROOF: Suppose that $\{L, U\}$ is a positive real number. Let L' be the nonpositive rationals and let $L'' = \{\beta : \beta \text{ is a positive rational and } \beta^2 \in L\}$. Let $L_1 = L' \cup L''$ and let $U_1 = \mathcal{Q} \sim L_1$. Then $0 \in L_1$ and $\varnothing \neq \{\alpha \in U : \alpha > 1\} \subset U_1$; hence, $\{L_1, U_1\}$ is a partition of $\mathcal{Q}$.

In order to show that L_1 is an initial segment of $\mathcal{Q}$, it suffices to show that for every α in L'' and β in U_1, $\alpha < \beta$. Suppose, on the contrary, there exist α_1 in L'' and β_1 in U_1 such that $\alpha_1 \geq \beta_1$. Then $\alpha_1{}^2 \geq \beta_1{}^2$. Because L is an initial segment of $\mathcal{Q}$ and $\alpha_1{}^2 \in L$, $\beta_1{}^2 \in L$. But $\beta_1 \in U_1 \subseteq \{\text{positive rationals}\}$. It follows that $\beta_1 \in L'' \subseteq L_1$; this contradicts the disjointness of L_1 and U_1.

To complete the proof that $\{L_1, U_1\}$ is a real number, we must show that L_1 has no last element. We first note that L'' is nonempty. In fact, because $\{L, U\}$ is a positive real, there is a positive rational, α_2, in L. Choose a nonzero natural number N such that (in $\mathcal{Q}$) $1/N^2 < \alpha_2$. Thus, $1/N \in L''$. [We note that the presence of $1/N$ in L'' $(\subset L_1)$ shows that $\{L_1, U_1\}$ is a positive real.] Because each element of L' precedes every element of L'' (in the ordering $\leq_{\mathcal{Q}}$), it suffices to show that L'' has no last element.

Let $\alpha \in L''$. Then $\alpha > 0$ and $\alpha^2 \in L$. Because L has no last element, we can choose β in L such that $\alpha^2 < \beta$. Therefore, $\varepsilon = \beta - \alpha^2$ is positive, as is $\varepsilon/(2\cdot\alpha + 1)$. Choose a nonzero natural number M such that $1/M < \varepsilon/(2\alpha + 1)$. Hence,

$$\left(\alpha + \frac{1}{M}\right)^2 = \alpha^2 + 2\cdot\frac{\alpha}{M} + \frac{1}{M^2} \leq \alpha^2 + \frac{2\cdot\alpha + 1}{M} < \alpha^2 + \varepsilon = \beta.$$

Because $0 < \alpha < \alpha + 1/M$, $\alpha + 1/M \in L''$, and L'' has no last element.

The verification that $\{L_1, U_1\}^2 = \{L, U\}$ is an exercise.

Note that the technique employed in the final paragraph was essentially the same as that of Example 3.9.3.

Corollary. *If $\{L, U\}$ is any positive real number and if $\{L_1, U_1\}^2 = \{L, U\}$, then $(-\{L_1, U_1\})^2 = \{L, U\}$.*

Thus, *every positive real number has at least 2 distinct square roots;* Exercise 3.10.10 asserts that, in fact, there are exactly 2 square roots. Moreover, $\overset{\equiv}{0}$ is its own unique square root.

The set of real numbers possesses another important property that is lacking in the rationals. Recall that if R is a relation on a set X and if $x_0 \in X$, $Y \subseteq X$, and aRx_0 for every a in Y, then x_0 is said to be an upper bound for Y. It is possible for a nonempty set of rationals to have an upper bound (with respect to $\leq_{\mathcal{Q}}$) and yet to lack a least upper bound. For example, if Y is the set of rationals whose square is less than 2, then 2 is certainly an upper bound for Y. But there is no first rational whose square is greater than or equal to 2.

Theorem 3.10.6. *Every nonempty set of real numbers that has an upper bound (with respect to $\leq_{\mathscr{R}}$) has a least upper bound.*

PROOF: Let X be a nonempty set of reals for which an upper bound for X—$\{L^*, U^*\}$—exists. Let $L^\# = \cup\{L : \{L, U\} \in X\}$ and let $U^\# = \mathscr{Q} \sim L^\#$. We claim first that $\{L^\#, U^\#\}$ is a real.

For each $\{L, U\}$ in X, $L \subseteq L^\#$; hence, $L^\#$ is nonempty. By Theorem 3.10.3, $L \subseteq L^*$ for each $\{L, U\}$ in X; therefore, $L^\# \subseteq L^*$. Consequently, $U^* \subseteq U^\#$, and $U^\# \neq \varnothing$. It follows that $\{L^\#, U^\#\}$ is a partition of $\mathscr{Q}$.

By the Generalization of De Morgan's Laws (Exercise 1.8.14), $U^\# = \cap\{U : \{L, U\} \in X\}$; hence, if $\gamma \in L^\#$ and $\delta \in U^\#$, then for some $\{L, U\}$ in X, $\gamma \in L$ and $\delta \in U$. Thus, $\gamma < \delta$, and it follows that $L^\#$ is an initial segment of $\mathscr{Q}$.

If $L^\#$ were to have a last element, then so would some L, contradicting our assumption that each $\{L, U\}$ is a real. This completes the proof that $\{L^\#, U^\#\}$ is a real number.

Let $\{L, U\}$ be an arbitrary element of X and let $\{L^*, U^*\}$ be an arbitrary upper bound for X. Then $L \subseteq L^\# \subseteq L^*$; hence, by Theorem 3.10.3,

$$\{L, U\} \leq_{\mathscr{R}} \{L^\#, U^\#\} \leq_{\mathscr{R}} \{L^*, U^*\}.$$

Consequently, $\{L^\#, U^\#\}$ is the least upper bound for X.

Corollary. *Every nonempty set of real numbers that has a lower bound (with respect to $\leq_{\mathscr{R}}$) has a greatest lower bound.*

Exercises

3.10.1. Let $L_1 = \{\alpha \in \mathscr{Q} : \alpha < -4\}$, $L_2 = \{\alpha \in \mathscr{Q} : \alpha \text{ is negative or } \alpha^2 < 3\}$, and $L_3 = \{\alpha \in \mathscr{Q} : \alpha < 5\}$. For $I = 1, 2, 3$, let $\Gamma_I = \{L_I, \mathscr{Q} \sim L_I\}$. Determine each:

(a) $\Gamma_1 +_{\mathscr{R}} \Gamma_2$.
(b) $\Gamma_1 +_{\mathscr{R}} \Gamma_2 +_{\mathscr{R}} \Gamma_3$.
(c) $\Gamma_1 -_{\mathscr{R}} \Gamma_3$.
(d) $\Gamma_2 \cdot_{\mathscr{R}} \Gamma_3$.
(e) $\Gamma_3 /_{\mathscr{R}} \Gamma_1$.

†3.10.2. Prove that if $\{L_1, U_1\}$ and $\{L_2, U_2\}$ are distinct real numbers, then $\{L_1^-, U_2^-\} \neq \{L_2^-, U_2^-\}$.

***3.10.3.** **†(a)** Prove that multiplication of real numbers is well-defined. Thus, show that if $\{L_1, U_1\}$ and $\{L_2, U_2\}$ are positive real numbers, if

$$L = \{\gamma \in \mathscr{Q} : \gamma \text{ is nonpositive or } \gamma = \alpha \cdot \beta \text{ for some } \alpha \text{ in } L_1 \text{ and } \beta \text{ in } L_2\},$$

and if $U = \mathscr{Q} \sim L$, then $\{L, U\}$ is a real number.

(b) Suppose that

$$U^* = \{\delta \in \mathscr{Q} : \delta = \rho \cdot \sigma \text{ for some } \rho \text{ in } U_1 \text{ and some } \sigma \text{ in } U_2\}$$

and $L^* = \mathscr{Q} \sim U^*$. Show that $\{L^*, U^*\}$ is not necessarily a real number.

†3.10.4. Let $\{L_I, U_I\}$ be real numbers for $I = 1, 2, 3$.

(a) Prove that $\{L_2, U_2\} <_{\mathscr{R}} \{L_3, U_3)$ iff

$$\{L_1, U_1\} +_{\mathscr{R}} \{L_2, U_2\} <_{\mathscr{R}} \{L_1, U_1\} +_{\mathscr{R}} \{L_3, U_3\}.$$

(b) Prove that if $\{L_1, U_1\} +_{\mathscr{R}} \{L_2, U_2\} = \overset{\equiv}{0}$, then $\{L_2, U_2\} = -\{L_1, U_1\}$.

(c) Prove that if $\{L_1, U_1\} >_{\mathscr{R}} \overset{\equiv}{0}$, then $\{L_2, U_2\} <_{\mathscr{R}} \{L_3, U_3\}$ iff

$$\{L_1, U_1\} \cdot_{\mathscr{R}} \{L_2, U_2\} <_{\mathscr{R}} \{L_1, U_1\} \cdot_{\mathscr{R}} \{L_3, U_3\}.$$

(d) Prove that if $\{L_1, U_1\} \cdot_{\mathscr{R}} \{L_2, U_2\} = \overset{\equiv}{1}$, then $\{L_2, U_2\} = \{L_1, U_1\}^{-1}$.

†3.10.5. Prove that addition and multiplication of real numbers are each associative and commutative and that multiplication distributes over both addition and subtraction.

†3.10.6. Prove that $\mathscr{R}$ is dense with respect to $\leq_{\mathscr{R}}$. In fact, prove that if $\{L_1, U_1\} <_{\mathscr{R}} \{L_2, U_2\}$, then the following are true:

(a) There is a rational α such that $\{L_1, U_1\} <_{\mathscr{R}} \bar{\alpha} <_{\mathscr{R}} \{L_2, U_2\}$.

(b) There is an irrational number $\{L_3, U_3\}$ such that

$$\{L_1, U_1\} <_{\mathscr{R}} \{L_3, U_3\} <_{\mathscr{R}} \{L_2, U_2\}.$$

†3.10.7. Show that the mapping H of $\mathscr{Q}$ into $\mathscr{R}$, defined by $H(\alpha) = \bar{\alpha}$, is an isomorphism. Thus, show that for α, β in $\mathscr{Q}$, the following are true:

(a) $H(\alpha +_{\mathscr{Q}} \beta) = H(\alpha) +_{\mathscr{R}} H(\beta)$.

(b) $H(\alpha \cdot_{\mathscr{Q}} \beta) = H(\alpha) \cdot_{\mathscr{R}} H(\beta)$.

(c) $\alpha \leq_{\mathscr{Q}} \beta$ iff $H(\alpha) \leq_{\mathscr{R}} H(\beta)$.

★,†3.10.8. Define exponentiation of real numbers

(a) with rational exponents.

(b) (for students with a calculus background) with real exponents.

3.10.9. In Theorem 3.10.5, show that $\{L_1, U_1\}^2 = \{L, U\}$.

3.10.10. Prove that every positive real number has exactly 2 square roots.

3.10.11. Determine the least upper bound and the greatest lower bound of each of the following subsets of $\mathscr{R}$, whenever these exist:

(a) $\{\overset{\equiv}{0}, \overset{\equiv}{1}, \sqrt{\overset{\equiv}{2}}, \overset{\equiv}{2}\}$. **(b)** $\{\overset{\equiv}{N} : N \in \omega\}$.

(c) $\{\overset{\equiv}{1} /_{\mathscr{R}} \overset{\equiv}{N} : N \geq 1\}$. **(d)** $\{(\overset{\equiv}{N} -_{\mathscr{R}} \overset{\equiv}{1}) /_{\mathscr{R}} (\overset{\equiv}{N} +_{\mathscr{R}} \overset{\equiv}{1}) : N \in \omega\}$.

3.10.12. Prove the corollary of Theorem 3.10.6. Thus, show that every nonempty set of reals that has a lower bound (with respect to $\leq_{\mathscr{R}}$) has a greatest lower bound.

†3.10.13 Prove that $\mathscr{R}$ has the following **Archimedean Property**: For any positive real numbers $\{L_1, U_1\}, \{L_2, U_2\}$ such that $\{L_1, U_1\} <_{\mathscr{R}} \{L_2, U_2\}$, there exists N for which $N \geq 1$ and $\overset{\equiv}{N} \cdot_{\mathscr{R}} \{L_1, U_1\} >_{\mathscr{R}} \{L_2, U_2\}$.

3.10.14. (See Example 2.8.5 and Exercise 3.7.7 for terminology.) Prove that $\langle \mathscr{R}, \langle +_{\mathscr{R}}, \cdot_{\mathscr{R}} \rangle \rangle$ is a field.

★3.10.15. **(a)** (**Dedekind's Theorem**). Let $\{\mathscr{L}, \mathscr{U}\}$ be a partition of $\mathscr{R}$ that is such that $\mathscr{L}$ is an initial segment of $\mathscr{R}$. Show that either $\mathscr{L}$ has a largest element or (excl.) $\mathscr{U}$ has a smallest element. (Useful for Section 6.2.)

(b) Use Dedekind's Theorem to prove the least upper bound property (Theorem 3.10.6).

★3.10.16. An **open interval** is a set of real numbers of the form $\{\Gamma \in \mathscr{R} : \Gamma_1 <_{\mathscr{R}} \Gamma <_{\mathscr{R}} \Gamma_2\}$, for $\Gamma_1, \Gamma_2 \in \mathscr{R}$. Let $S \subseteq \mathscr{R}$. A real number Γ_0 is called an **accumulation point of** S is every open interval containing Γ_0 contains points of $S \sim \{\Gamma_0\}$. Also, S is said to be **bounded** if for some $\Gamma_1 \in \mathscr{R}_+$, $S \subseteq \{\Gamma \in \mathscr{R};\ -\Gamma_1 <_{\mathscr{R}} \Gamma <_{\mathscr{R}} \Gamma_1\}$. (**Bolzano–Weierstrass Theorem**): Show that every infinite bounded set of real numbers has an accumulation point. (Useful for Section 5.12; referred to in Exercise 3.12.9.)

3.11 Complex Numbers: Intuitive Approach

Before defining the complex numbers as sets, we pause to consider the problem on an intuitive level. The reader with a working knowledge of elementary complex arithmetic should proceed directly to Section 3.12.

In extending the rational numbers to the reals, we began by observing that the equation $x^2 = \overset{\equiv}{2}$ has no rational solution. We then introduced the real numbers as Dedekind cuts and proved that every nonnegative real number has a (real) square root. However, this statement is clearly false if we omit the adjective "nonnegative". In fact, the starting point for investigations of "complex numbers" is the adjunction of a so-called "imaginary" root of the equation $x^2 +_R \overset{\equiv}{1} = \overset{\equiv}{0}$. We let i be this root, and define an (extended) multiplication in such a way that $i^2 = -\overset{\equiv}{1}$; if multiplication is to obey the usual rules, it will then follow that $i^3 = -i$, where $-i$ is a second adjunction to the real numbers (and is an additive inverse of i). Moreover, $i^4 = \overset{\equiv}{1}$, and for natural numbers $K \geq 1$ and $0 \leq L \leq 3$, $i^{4K+L} = i^L$.

In fact, from a naïve point of view, we could extend real arithmetical operations to "complex operations", which we write temporarily as $\oplus, \odot, \ominus, \oslash$, so that the "complex expressions"

$$\Gamma \quad \oplus \quad \Delta \odot i, \Gamma, \qquad \Delta \in \mathscr{R},$$

satisfy the following conditions:

Let $\Gamma_1, \Gamma_2, \Delta_1$, and Δ_2 be real numbers.

(1a) $(\Gamma_1 \oplus \overset{\equiv}{0}\odot i) \oplus (\Gamma_2 \oplus \overset{\equiv}{0}\odot i) = (\Gamma_1 +_{\mathscr{R}} \Gamma_2) \oplus \overset{\equiv}{0}\odot i$;

(1b) $(\Gamma_1 \oplus \overset{\equiv}{0}\odot i) \ominus (\Gamma_2 \oplus \overset{\equiv}{0}\odot i) = (\Gamma_1 -_{\mathscr{R}} \Gamma_2) \oplus \overset{\equiv}{0}\odot i$;

(1c) $(\Gamma_1 \oplus \overset{\equiv}{0}\odot i) \odot (\Gamma_2 \oplus \overset{\equiv}{0}\odot i) = (\Gamma_1 \cdot_{\mathscr{R}} \Gamma_2) \oplus \overset{\equiv}{0}\odot i$;

(1d) $(\Gamma_1 \oplus \overset{\equiv}{0}\odot i) \oslash (\Gamma_2 \oplus \overset{\equiv}{0}\odot i) = (\Gamma_1 /_{\mathscr{R}} \Gamma_2) \oplus \overset{\equiv}{0}\odot i$,

provided $\Gamma_2 \neq \overset{\equiv}{0}$.

(2a) $(\overset{\equiv}{0} \oplus \Delta_1\odot i) \oplus (\overset{\equiv}{0} \oplus \Delta_2\odot i) = \overset{\equiv}{0} \oplus (\Delta_1 +_{\mathscr{R}} \Delta_2)i$;

(2b) $(\overset{\equiv}{0} \oplus \Delta_1\odot i) \ominus (\overset{\equiv}{0} \oplus \Delta_2\odot i) = \overset{\equiv}{0} \oplus (\Delta_1 -_{\mathscr{R}} \Delta_2)i$;

(2c) $(\overset{\equiv}{0} \oplus \Delta_1\odot i) \odot (\overset{\equiv}{0} \oplus \Delta_2\odot i) = -_{\mathscr{R}}(\Delta_1 \cdot_{\mathscr{R}} \Delta_2) \oplus \overset{\equiv}{0}\odot i$;

(2d) $(\overset{\equiv}{0} \oplus \Delta_1\odot i) \oslash (\overset{\equiv}{0} \oplus \Delta_2\odot i) = (\Delta_1 /_{\mathscr{R}} \Delta_2) \oplus \overset{\equiv}{0}\odot i$,

provided $\Delta_2 \neq \overset{\equiv}{0}$.

(3a) $(\Gamma_1 \oplus \overset{\equiv}{0}\odot i) \oplus (\overset{\equiv}{0} \oplus \Delta_2\odot i) = \Gamma_1 \oplus \Delta_2\odot i$;

(3b) $(\Gamma_1 \oplus \overset{\equiv}{0}\odot i) \ominus (\overset{\equiv}{0} \oplus \Delta_2\odot i) = \Gamma_1 \ominus \Delta_2\odot i$;

(3c) $(\Gamma_1 \oplus \overset{\equiv}{0}\odot i) \odot (\overset{\equiv}{0} \oplus \Delta_2\odot i) = \overset{\equiv}{0} \oplus (\Gamma_1 \cdot_{\mathscr{R}} \Delta_2)\odot i$;

(3d) $(\Gamma_1 \oplus \overset{\equiv}{0}\odot i) \oslash (\overset{\equiv}{0} \oplus \Delta_2\odot i) = \overset{\equiv}{0} \oplus -_{\mathscr{R}}(\Gamma_1 /_{\mathscr{R}} \Delta_2)\odot i$,

provided $\Delta_2 \neq \overset{\equiv}{0}$.

(4a) $(\overset{\equiv}{0} \oplus \Delta_1\odot i) \oplus (\Gamma_2 \oplus \overset{\equiv}{0}\odot i) = \Gamma_2 \oplus \Delta_1\odot i$;

(4b) $(\overset{\equiv}{0} \oplus \Delta_1\odot i) \ominus (\Gamma_2 \oplus \overset{\equiv}{0}\odot i) = -_{\mathscr{R}}\Gamma_2 \oplus \Delta_1\odot i$;

(4c) $(\overset{\equiv}{0} \oplus \Delta_1\odot i) \odot (\Gamma_2 \oplus \overset{\equiv}{0}\odot i) = \overset{\equiv}{0} \oplus (\Delta_1 \cdot_{\mathscr{R}} \Gamma_2)\odot i$;

(4d) $(\overset{\equiv}{0} \oplus \Delta_1\odot i) \oslash (\Gamma_2 \oplus \overset{\equiv}{0}\odot i) = \overset{\equiv}{0} \oplus (\Delta_1 /_{\mathscr{R}} \Gamma_2)\odot i$,

provided $\Gamma_2 \neq \overset{\equiv}{0}$.

What we have in mind are the following "operations":

Addition:

$$(\Gamma_1 \oplus \Delta_1\odot i) \oplus (\Gamma_2 \oplus \Delta_2\odot i) = (\Gamma_1 +_{\mathscr{R}} \Gamma_2) \oplus (\Delta_1 +_{\mathscr{R}} \Delta_2)\odot i.$$

Subtraction:

$$(\Gamma_1 \oplus \Delta_1\odot i) \ominus (\Gamma_2 \oplus \Delta_2\odot i) = (\Gamma_1 -_{\mathscr{R}} \Gamma_2) \oplus (\Delta_1 -_{\mathscr{R}} \Delta_2)\odot i.$$

Multiplication:

$$(\Gamma_1 \quad \oplus \quad \Delta_1 \odot i) \odot (\Gamma_2 \quad \oplus \quad \Delta_2 \odot i)$$
$$= (\Gamma_1 \cdot_{\mathscr{R}} \Gamma_2 -_{\mathscr{R}} \Delta_1 \cdot_{\mathscr{R}} \Delta_2) \oplus (\Gamma_1 \cdot_{\mathscr{R}} \Delta_2 +_{\mathscr{R}} \Gamma_2 \cdot_{\mathscr{R}} \Delta_1) \odot i.$$

Division:

$$(\Gamma_1 \quad \oplus \quad \Delta_1 \odot i) \oslash (\Gamma_2 \quad \oplus \quad \Delta_2 \odot i)$$
$$= (\Gamma_1 \cdot_{\mathscr{R}} \Gamma_2 +_{\mathscr{R}} \Delta_1 \cdot_{\mathscr{R}} \Delta_2) /_{\mathscr{R}} ({\Gamma_2}^2 +_{\mathscr{R}} {\Delta_2}^2)$$
$$\oplus ((\Gamma_2 \cdot_{\mathscr{R}} \Delta_1 -_{\mathscr{R}} \Gamma_1 \cdot_{\mathscr{R}} \Delta_2) /_{\mathscr{R}} ({\Gamma_2}^2 +_{\mathscr{R}} {\Delta_2}^2)) \odot i, \qquad {\Gamma_2}^2 +_{\mathscr{R}} {\Delta_2}^2 \neq \overset{\equiv}{0}.$$

The "definition" of multiplication is motivated by the concern that i^2 $(= i \odot i)$ is to be $-\overset{\equiv}{1}$; we need only consider the four terms involved in multiplying

$$\begin{array}{ccc} \Gamma_1 & \oplus & \Delta_1 \odot i \\ \Gamma_2 & \oplus & \Delta_2 \odot i \\ \hline \end{array}$$

by the "usual" rules of elementary algebra.

Division is to be the inverse operation to multiplication; if Γ_2 and Δ_2 are not both zero, then

$$(\Gamma_1 \quad \oplus \quad \Delta_1 \odot i) \oslash (\Gamma_2 \quad \oplus \quad \Delta_2 \odot i) = \Gamma_3 \quad \oplus \quad \Delta_3 \odot i$$

iff

(3.11.1) $$(\Gamma_2 \quad \oplus \quad \Delta_2 \odot i) \odot (\Gamma_3 \quad \oplus \quad \Delta_3 \odot i) = \Gamma_1 \quad \oplus \quad \Delta_1 \odot i.$$

We use the above "definition" of multiplication and the fact that two expressions, $\rho_1 \quad \oplus \quad \sigma_1 \odot i$ and $\rho_2 \quad \oplus \quad \sigma_2 \odot i$, are to be considered equal iff both

$$\rho_1 = \rho_2 \qquad \text{and} \qquad \sigma_1 = \sigma_2.$$

Thus, equation (3.11.1) yields the following two equations (over $\mathscr{R}$):

(3.11.2) $$\begin{aligned} \Gamma_2 \cdot \Gamma_3 - \Delta_2 \cdot \Delta_3 &= \Gamma_1; \\ \Gamma_2 \cdot \Delta_3 + \Gamma_3 \cdot \Delta_2 &= \Delta_1. \end{aligned}$$

We regard (3.11.2) as a system of two equations in the two unknowns Γ_3 and Δ_3 in which the real constants $\Gamma_1, \Gamma_2, \Delta_1$, and Δ_2 appear as indicated. We solve for Γ_3 and Δ_3 and obtain

$$\Gamma_3 = (\Gamma_1 \cdot_{\mathscr{R}} \Gamma_2 +_{\mathscr{R}} \Delta_1 \cdot_{\mathscr{R}} \Delta_2) /_{\mathscr{R}} ({\Gamma_2}^2 +_{\mathscr{R}} {\Delta_2}^2)$$

and

$$\Delta_3 = (\Gamma_2 \cdot_{\mathscr{R}} \Delta_1 -_{\mathscr{R}} \Gamma_1 \cdot_{\mathscr{R}} \Delta_2) /_{\mathscr{R}} ({\Gamma_2}^2 +_{\mathscr{R}} {\Delta_2}^2).$$

We agree to write

$$\text{“}\Gamma\text{” instead of “}\Gamma \quad \overset{\equiv}{\oplus} \quad 0 \odot i\text{”},$$

$$\text{“}\Delta i\text{” instead of “}0 \quad \overset{\equiv}{\oplus} \quad \Delta \odot i\text{”}, \qquad \Delta \neq \overset{\equiv}{0},$$

and

$$\text{“}i\text{” instead of “}\overset{\equiv}{1} \odot i\text{”}.$$

"It is easily verified that any expression "$\Gamma \quad \overset{\equiv}{\oplus} \quad \Delta \odot i$" can now be written as "$\Gamma \oplus \Delta i$" and that if $-\Delta$ is the (real) additive inverse of Δ, then

$$\Gamma \oplus (-\Delta)i = \Gamma \ominus \Delta i.$$

(Henceforth, we omit bars and the subscript "$\mathscr{R}$".)

EXAMPLE 3.11.1. $5 \ominus 2i = 5 \oplus (-2)i = (5 \oplus 0i) \oplus (0 \oplus (-2)i)$, by the above conventions.

EXAMPLE 3.11.2

(1) $(5 \ominus 2i) \oplus (7 \oplus 3i) = (5 + 7) \oplus (-2 + 3)i = 12 \oplus i.$
(2) $(5 \ominus 2i) \ominus (7 \oplus 3i) = (5 - 7) \oplus (-2 - 3)i = -2 \ominus 5i.$
(3) $(5 \ominus 2i) \odot (7 \oplus 3i) = (5 \cdot 7 - (-2) \cdot 3) \oplus (5 \cdot 3 + 7 \cdot (-2))i = 41 \oplus i.$
(4) $(5 \ominus 2i) \oslash (7 \oplus 3i) = (5 \cdot 7 + (-2) \cdot 3)/(7^2 + 3^2)$
$\oplus ((7 \cdot (-2) - 5 \cdot 3)/(7^2 + 3^2))i = 1/2 \ominus (1/2)i = 1/2 \odot (1 \ominus i).$

Exercises

3.11.1. Determine each:
(a) $(2 \oplus i) \odot (3i)$. **(b)** $12 \odot (6 \ominus i)$.
(c) $(24 \oplus 2i) \oslash 8$. **(d)** $5 \oslash (10 \ominus 5i)$.

3.11.2. Determine each:
(a) $(6 \oplus 3i) \ominus (2 \ominus 2i)$. **(b)** $(10 \ominus i) \odot (1 \oplus 2i)$.
(c) $(1 \oplus i) \oslash (1 \ominus i)$. **(d)** $(2 \oplus 5i) \oslash (1 \ominus i)$.

3.11.3. Determine the square roots of each:
(a) -3. **(b)** $-1 \oplus \sqrt{3}i$. **(c)** $5 \oplus 2i$.

The next three problems utilize various concepts from elementary mathematics.

3.11.4. With reference to an expression $\Gamma \oplus \Delta i$, $\Gamma, \Delta \in \mathscr{R}$, not both zero, define $P = \sqrt{\Gamma^2 + \Delta^2}$. The real number P is called the **modulus** of $\Gamma \oplus \Delta i$. Let Φ be any real number that is such that $\tan(\Delta/\Gamma) = \Phi$ if $\Gamma \neq 0$, and let $\Phi = \mathscr{M}\pi/2$ for some odd "integer" $\mathscr{M}$, if $\Gamma = 0$.

(a) Show that $\Gamma \oplus \Delta i = P \odot (\cos\Phi \oplus \sin\Phi i)$.

(b) Show that $(\Gamma \oplus \Delta i)^N = P^N \odot (\cos(N\Phi) \oplus (\sin(N\Phi))i)$ for each N in ω (**de Moivre's Formula**).

(c) Use (b) to extract square roots and cube roots. (Useful for Exercises 3.12.3 and 3.12.4.)

3.11.5. The expression $\Gamma \ominus \Delta i$ is called the **complex conjugate of** $\Gamma \oplus \Delta i$; it is denoted by "Conj $(\Gamma \oplus \Delta i)$". Show the following:

(a) Conj (Conj $(\Gamma \oplus \Delta i)) = \Gamma \oplus \Delta i$.

(b) The modulus of Conj $(\Gamma \oplus \Delta i)$ is that of $\Gamma \oplus \Delta i$.

(c) If $\Gamma \oplus \Delta i = P \odot (\cos\Phi \oplus (\sin\Phi)i)$, then Conj$(\Gamma \oplus \Delta i) = P \odot (\cos\Phi \ominus (\sin\Phi)i)$. (Useful for Exercise 3.12.3.)

***3.11.6.** Consider $\Gamma_1 \oplus \Delta_1 i$ and $\Gamma_2 \oplus \Delta_2 i$. Suppose Γ_2 and Δ_2 are not both zero. How do we know that division, as defined in the text, always yields exactly one quotient?

3.12 Complex Numbers as Sets

The transition from real to complex numbers (as sets) is the simplest of the four extensions we consider in this chapter. In fact, the properties noted in the preceding section clearly suggest that complex numbers can be defined as ordered pairs of real numbers.

A **complex number** $\mathscr{A}$ is an ordered pair $\langle\Gamma, \Delta\rangle$ of real numbers; the first coordinate, Γ, is called the **real part of** $\mathscr{A}$ and the second coordinate, Δ, is the **imaginary part of** $\mathscr{A}$. Note that *the imaginary part of a complex number, as well as the real part, is a real number.*

With the model of Section 3.11 before us we proceed to define the arithmetic operations on $\mathscr{C}$, the set of complex numbers:

$$\langle\Gamma_1, \Delta_1\rangle +_{\mathscr{C}} \langle\Gamma_2, \Delta_2\rangle = \langle\Gamma_1 +_{\mathscr{R}} \Gamma_2, \Delta_1 +_{\mathscr{R}} \Delta_2\rangle;$$

$$\langle\Gamma_1, \Delta_1\rangle -_{\mathscr{C}} \langle\Gamma_2, \Delta_2\rangle = \langle\Gamma_1 -_{\mathscr{R}} \Gamma_2, \Delta_1 -_{\mathscr{R}} \Delta_2\rangle;$$

$$\langle\Gamma_1, \Delta_1\rangle \cdot_{\mathscr{C}} \langle\Gamma_2, \Delta_2\rangle = \langle\Gamma_1 \cdot_{\mathscr{R}} \Gamma_2 -_{\mathscr{R}} \Delta_1 \cdot_{\mathscr{R}} \Delta_2, \Gamma_1 \cdot_{\mathscr{R}} \Delta_2 +_{\mathscr{R}} \Gamma_2 \cdot_{\mathscr{R}} \Delta_1\rangle;$$

$$\langle\Gamma_1, \Delta_1\rangle /_{\mathscr{C}} \langle\Gamma_2, \Delta_2\rangle = \langle(\Gamma_1 \cdot_{\mathscr{R}} \Gamma_2 +_{\mathscr{R}} \Delta_1 \cdot_{\mathscr{R}} \Delta_2) /_{\mathscr{R}} (\Gamma_2{}^2 +_{\mathscr{R}} \Delta_2{}^2), (\Gamma_2 \cdot_{\mathscr{R}} \Delta_1 -_{\mathscr{R}} \Gamma_1 \cdot_{\mathscr{R}} \Delta_2) /_{\mathscr{R}} (\Gamma_2{}^2 +_{\mathscr{R}} \Delta_2{}^2)\rangle,$$

Γ_2 and Δ_2 not both $\overset{\equiv}{0}$.

Note that the familiar "parallelogram law" of vector addition and subtraction can easily be recognized in $+_{\mathscr{C}}$ and $-_{\mathscr{C}}$; we can, of course, identify vectors with points of the plane—hence with ordered pairs of real numbers. The rules for $+_{\mathscr{C}}$ and $-_{\mathscr{C}}$ are the same as for addition and subtraction of planar vectors.

The relationship between the intuitive approach and the set-theoretic approach can be expressed by means of the following table:

Intuitive approach	Set-theoretic approach
$\oplus$	$+_{\mathscr{C}}$
$\ominus$	$-_{\mathscr{C}}$
$\odot$	$\cdot_{\mathscr{C}}$
$\oslash$	$/_{\mathscr{C}}$
$\Gamma \oplus \Delta i$	$\langle \Gamma, \Delta \rangle$

EXAMPLE 3.12.1

(1) $\langle 5, 3 \rangle \cdot_{\mathscr{C}} \langle -2, 4 \rangle = \langle 5 \cdot_{\mathscr{R}} (-2) -_{\mathscr{R}} 3 \cdot_{\mathscr{R}} 4, 5 \cdot_{\mathscr{R}} 4 +_{\mathscr{R}} (-2) \cdot_{\mathscr{R}} 3 \rangle$

$$= \langle -22, 14 \rangle.$$

Note that the complex number $\langle -22, 14 \rangle$ is not equivalent to $\langle -11, 7 \rangle$; in fact, the former is twice the latter (i.e., $\langle -22, 14 \rangle = \langle 2, 0 \rangle \cdot_{\mathscr{C}} \langle -11, 7 \rangle$).

(2) $(5 \oplus 3i) \odot (-2 \oplus 4i)$

$$= (5 \cdot_{\mathscr{R}} (-2) -_{\mathscr{R}} 3 \cdot_{\mathscr{R}} 4) \oplus (5 \cdot_{\mathscr{R}} 4 +_{\mathscr{R}} (-2) \cdot_{\mathscr{R}} 3)i$$

$$= -22 \oplus 14i.$$

We mention the following two theorems, whose proofs are left as routine exercises.

Theorem 3.12.1. *Let* $\mathscr{A}_1 = \langle \Gamma_1, \Delta_1 \rangle$, $\mathscr{A}_2 = \langle \Gamma_2, \Delta_2 \rangle$, *and* $\mathscr{A}_3 = \langle \Gamma_3, \Delta_3 \rangle$ *be arbitrary complex numbers.*

(1) *Complex addition and multiplication are each commutative and associative; moreover, multiplication distributes over addition:*

(a) $\mathscr{A}_1 +_{\mathscr{C}} \mathscr{A}_2 = \mathscr{A}_2 +_{\mathscr{C}} \mathscr{A}_1$.
(b) $\mathscr{A}_1 +_{\mathscr{C}} (\mathscr{A}_2 +_{\mathscr{C}} \mathscr{A}_3) = (\mathscr{A}_1 +_{\mathscr{C}} \mathscr{A}_2) +_{\mathscr{C}} \mathscr{A}_3$.
(c) $\mathscr{A}_1 \cdot_{\mathscr{C}} \mathscr{A}_2 = \mathscr{A}_2 \cdot_{\mathscr{C}} \mathscr{A}_1$.
(d) $\mathscr{A}_1 \cdot_{\mathscr{C}} (\mathscr{A}_2 \cdot_{\mathscr{C}} \mathscr{A}_3) = (\mathscr{A}_1 \cdot_{\mathscr{C}} \mathscr{A}_2) \cdot_{\mathscr{C}} \mathscr{A}_3$.
(e) $\mathscr{A}_1 \cdot_{\mathscr{C}} (\mathscr{A}_2 +_{\mathscr{C}} \mathscr{A}_3) = \mathscr{A}_1 \cdot_{\mathscr{C}} \mathscr{A}_2 +_{\mathscr{C}} \mathscr{A}_1 \cdot_{\mathscr{C}} \mathscr{A}_3$.
(f) $(\mathscr{A}_1 +_{\mathscr{C}} \mathscr{A}_2) \cdot_{\mathscr{C}} \mathscr{A}_3 = \mathscr{A}_1 \cdot_{\mathscr{C}} \mathscr{A}_3 +_{\mathscr{C}} \mathscr{A}_2 \cdot_{\mathscr{C}} \mathscr{A}_3$.

(2) $\langle \overset{\equiv}{0}, \overset{\equiv}{0} \rangle$ *is the unique additive identity;* $\langle \overset{\equiv}{1}, \overset{\equiv}{0} \rangle$ *is the unique multiplicative identity:*

(a) $\langle \Gamma_1, \Delta_1 \rangle +_{\mathscr{C}} \langle \overset{\equiv}{0}, \overset{\equiv}{0} \rangle = \langle \overset{\equiv}{0}, \overset{\equiv}{0} \rangle +_{\mathscr{C}} \langle \Gamma_1, \Delta_1 \rangle = \langle \Gamma_1, \Delta_1 \rangle.$

(b) *If* $\langle \Gamma_1, \Delta_1 \rangle +_{\mathscr{C}} \langle \Gamma_2, \Delta_2 \rangle = \langle \Gamma_1, \Delta_1 \rangle$, *then* $\langle \Gamma_2, \Delta_2 \rangle = \langle \overset{\equiv}{0}, \overset{\equiv}{0} \rangle.$

(c) $\langle \Gamma_1, \Delta_1 \rangle \cdot_{\mathscr{C}} \langle \overset{\equiv}{1}, \overset{\equiv}{0} \rangle = \langle \overset{\equiv}{1}, \overset{\equiv}{0} \rangle \cdot_{\mathscr{C}} \langle \Gamma_1, \Delta_1 \rangle = \langle \Gamma_1, \Delta_1 \rangle.$

(d) *If* $\langle \Gamma_1, \Delta_1 \rangle \cdot_{\mathscr{C}} \langle \Gamma_2, \Delta_2 \rangle = \langle \Gamma_1, \Delta_1 \rangle$ *and if* $\langle \Gamma_1, \Delta_1 \rangle \neq \langle 0, 0 \rangle$, *then* $\langle \Gamma_2, \Delta_2 \rangle = \langle \overset{\equiv}{1}, \overset{\equiv}{0} \rangle.$

(3) *For each complex number* $\langle \Gamma_1, \Delta_1 \rangle$, $\langle -\Gamma_1, -\Delta_1 \rangle$ *is its unique additive inverse:*

(a) $\langle \Gamma_1, \Delta_1 \rangle +_{\mathscr{C}} \langle -\Gamma_1, -\Delta_1 \rangle = \langle -\Gamma_1, -\Delta_1 \rangle +_{\mathscr{C}} \langle \Gamma_1, \Delta_1 \rangle = \langle \overset{\equiv}{0}, \overset{\equiv}{0} \rangle.$

(b) *If* $\langle \Gamma_1, \Delta_1 \rangle +_{\mathscr{C}} \langle \Gamma_2, \Delta_2 \rangle = \langle \Gamma_2, \Delta_2 \rangle +_{\mathscr{C}} \langle \Gamma_1, \Delta_1 \rangle = \langle \overset{\equiv}{0}, \overset{\equiv}{0} \rangle$, *then* $\langle \Gamma_2, \Delta_2 \rangle = \langle -\Gamma_1, -\Delta_1 \rangle.$
For each nonzero complex number $\langle \Gamma_1, \Delta_1 \rangle$, *its unique multiplicative inverse is* $\langle \Gamma_1 /_{\mathscr{R}} (\Gamma_1^2 +_{\mathscr{R}} \Delta_1^2), -\Delta_1 /_{\mathscr{R}} (\Gamma_1^2 +_{\mathscr{R}} \Delta_1^2) \rangle$:

(c) $\langle \Gamma_1, \Delta_1 \rangle \cdot_{\mathscr{C}} \langle \Gamma_1 /_{\mathscr{R}} (\Gamma_1^2 +_{\mathscr{R}} \Delta_1^2), -\Delta_1 /_{\mathscr{R}} (\Gamma_1^2 +_{\mathscr{R}} \Delta_1^2) \rangle = \langle \overset{\equiv}{1}, \overset{\equiv}{0} \rangle.$

(d) *If* $\langle \Gamma_1, \Delta_1 \rangle \cdot_{\mathscr{C}} \langle \Gamma_2, \Delta_2 \rangle = \langle \overset{\equiv}{1}, \overset{\equiv}{0} \rangle$, *then*

$$\langle \Gamma_2, \Delta_2 \rangle = \langle \Gamma_1 /_{\mathscr{R}} (\Gamma_1^2 +_{\mathscr{R}} \Delta_1^2), -\Delta_1 /_{\mathscr{R}} (\Gamma_1^2 +_{\mathscr{R}} \Delta_1^2) \rangle.$$

Thus,

$$\langle 0, 1 \rangle \cdot_{\mathscr{C}} \langle 0, 1 \rangle = \langle 0 \cdot_{\mathscr{R}} 0 -_{\mathscr{R}} 1 \cdot_{\mathscr{R}} 1, 0 \cdot_{\mathscr{R}} 1 +_{\mathscr{R}} 0 \cdot_{\mathscr{R}} 1 \rangle = \langle -1, 0 \rangle;$$

because $\langle -1, 0 \rangle$ corresponds to the real number -1, clearly $\langle 0, 1 \rangle$ plays the role of i.

Theorem 3.12.2. *Let* $J : \mathscr{R} \to \mathscr{C}$ *be defined by* $J(\Gamma) = \langle \Gamma, \overset{\equiv}{0} \rangle$, *for* $\Gamma \in \mathscr{R}$. *Then* J *is an isomorphism of the real numbers into the complex numbers:*

(1) $J(\Gamma +_{\mathscr{R}} \Delta) = J(\Gamma) +_{\mathscr{C}} J(\Delta) = \langle \Gamma +_{\mathscr{R}} \Delta, \overset{\equiv}{0} \rangle.$

(2) $J(\Gamma \cdot_{\mathscr{R}} \Delta) = J(\Gamma) \cdot_{\mathscr{C}} J(\Delta) = \langle \Gamma \cdot_{\mathscr{R}} \Delta, \overset{\equiv}{0} \rangle.$

In line with our previous conventions we write $\overline{\Gamma}$ for $J(\Gamma)$; thus, for N in ω, $\overset{\equiv}{\mathrm{N}} \in \mathscr{C}$. The reader will notice that we omitted the usual mention of the orderings in the above theorem. This is because there is no "natural" ordering of the complex numbers according to magnitude. There are various ways of ordering the set $\mathscr{C}$ of complex numbers. For example, we can order $\mathscr{C}$ **lexicographically**; i.e., we can define

$$\langle \Gamma_1, \Delta_1 \rangle R \langle \Gamma_2, \Delta_2 \rangle \text{ — iff } \Gamma_1 <_{\mathscr{R}} \Gamma_2, \text{or (excl.) } \Gamma_1 = \Gamma_2 \text{ and } \Delta_1 \leq_{\mathscr{R}} \Delta_2.$$

However, there is no method of ordering the complex numbers in terms of which we can partition $\mathscr{C}$ into cells of "positive" numbers, "negative" numbers, and zero. For we would want these to be subject to the Law of Trichotomy, and we would want sums and products of positive numbers to be positive. Squares of nonzero numbers should be positive. But $\langle 0, 1 \rangle^2 = \langle -1, 0 \rangle$, and it is fairly easily seen that because of the isomorphism expressed in Theorem 3.12.2, $\langle -1, 0 \rangle$ would have to be negative.

In each of the extensions we have considered in this chapter there were always new equations that could be solved in the extended system, but not in the original one. Thus, $x + 1 = 0$ had no solution in the natural numbers, but $x +_{\mathscr{I}} \overline{1} = \overline{0}$ was solvable in the integers; $\overline{2} \cdot_{\mathscr{I}} x - \overline{1} = \overline{0}$ had no integer solution, but $\overset{=}{2} \cdot_{\mathscr{Q}} x -_{\mathscr{Q}} \overset{=}{1} = \overset{=}{0}$ had a rational solution; $x^2 -_{\mathscr{Q}} \overset{=}{5} = \overset{=}{0}$ was unsolvable in the rationals, but its real counterpart had solutions. Now we have "adjoined" the "square roots of $-\overset{\equiv}{1}$" in extending the reals to the complex numbers. Moreover, it is easily seen that (with the appropriate definition of exponentiation) for any N in ω,

$$\langle \overset{\equiv}{0}, \sqrt{\overset{\equiv}{N}} \rangle^2 = \langle -\overset{\equiv}{N}, \overset{\equiv}{0} \rangle;$$

thus, "square roots" of all negatives have been adjoined in the process. The natural question at this point would be whether there is any further extension of $\mathscr{C}$ that contains the roots of equations with complex coefficients that are unsolvable in $\mathscr{C}$. It turns out that we need never seek such an extension of $\mathscr{C}$.

Fundamental Theorem of Algebra: *Every polynomial*

$$\mathscr{A}_N \cdot_{\mathscr{C}} x^N +_{\mathscr{C}} \mathscr{A}_{N-1} \cdot_{\mathscr{C}} x^{N-1} +_{\mathscr{C}} \cdots +_{\mathscr{C}} \mathscr{A}_1 \cdot_{\mathscr{C}} x +_{\mathscr{C}} \mathscr{A}_0,$$

in which $N \geq 1$ *and* $\mathscr{A}_0, \mathscr{A}_1, \ldots, \mathscr{A}_N (\neq \overset{\equiv}{0})$ *are complex, has a complex root.*

Moreover, it can be shown that *all the roots of such a polynomial are complex*, so that from the standpoint of adjoining roots, we have reached the end of the line! The proof of the Fundamental Theorem of Algebra involves analytic (rather than algebraic) techniques and is usually considered in a course in complex variables.

Exercises

Except when used as subscripts and as exponents and when otherwise indicated, the numbers N refer to the isomorphic images of natural numbers in $\mathscr{R}$. Thus, "8" stands for "$\overset{\equiv}{8}$".

3.12.1. Determine each of the following:

(a) $\langle 7, 4\rangle \cdot_{\mathscr{C}} \langle 0, 1\rangle$.
(b) $\langle 8, 3\rangle -_{\mathscr{C}} \langle 1/2, 1\rangle$.
(c) $\langle 9, \sqrt{2}\rangle +_{\mathscr{C}} \langle \sqrt{2}, \sqrt{3}\rangle$.
(d) $\langle -6, 5\rangle /_{\mathscr{C}} \langle 1, -2\rangle$.

3.12.2. Determine all complex numbers satisfying $(\mathscr{A} - \overset{\equiv}{2})^2 + \overset{\equiv}{3} = \overset{\equiv}{0}$.

3.12.3. Let $\mathscr{A} = \langle \Gamma, \Delta\rangle$ be a complex number.

(a) Define the modulus of $\mathscr{A}$, the conjugate of $\mathscr{A}$.
(b) Express $\mathscr{A}$ in the form $P \cdot_{\mathscr{C}} \langle \cos\theta, \sin\theta\rangle$, where P is the modulus of $\mathscr{A}$.

3.12.4. Let $N \in \omega$. A complex number $\mathscr{A}$ is called an Nth **root of unity** if $\mathscr{A}^N = \overset{\equiv}{1}$.

(a) Prove that the sum of all the distinct Nth roots of unity is zero.
(b) What do the Nth roots of unity represent geometrically?

3.12.5. Prove the following parts of Theorem 3.12.1:
Let $\Gamma_1, \Gamma_2, \Delta_1, \Delta_2 \in \mathscr{R}$.

(2)(c) $\langle \Gamma_1, \Delta_1\rangle \cdot_{\mathscr{C}} \langle 1, 0\rangle = \langle 1, 0\rangle \cdot_{\mathscr{C}} \langle \Gamma_1, \Delta_1\rangle = \langle \Gamma_1, \Delta_1\rangle$.

(2)(d) If $\langle \Gamma_1, \Delta_1\rangle \cdot_{\mathscr{C}} \langle \Gamma_2, \Delta_2\rangle = \langle \Gamma_1, \Delta_1\rangle$ for $\langle \Gamma_1, \Delta_1\rangle$ in $\mathscr{C} \sim \{\overset{\equiv}{0}\}$, then $\Gamma_2 = 1$ and $\Delta_2 = 0$.

(3)(c) and (d) If $\Gamma_1{}^2 +_{\mathscr{R}} \Delta_1{}^2 >_{\mathscr{R}} 0$, then

$$\langle \Gamma_1, \Delta_1\rangle \cdot_{\mathscr{C}} \langle \Gamma_1 /_{\mathscr{R}} (\Gamma_1{}^2 +_{\mathscr{R}} \Delta_1{}^2), -\Delta_1 /_{\mathscr{R}} (\Gamma_1{}^2 +_{\mathscr{R}} \Delta_1{}^2)\rangle = \langle 1, 0\rangle,$$

and if $\langle \Gamma_1, \Delta_1\rangle \cdot_{\mathscr{C}} \langle \Gamma_2, \Delta_2\rangle = \langle 1, 0\rangle$, then

$$\langle \Gamma_2, \Delta_2\rangle = \langle \Gamma_1 /_{\mathscr{R}} (\Gamma_1{}^2 +_{\mathscr{R}} \Delta_1{}^2), -\Delta_1 /_{\mathscr{R}} (\Gamma_1{}^2 + \Delta_1{}^2)\rangle.$$

3.12.6. Prove Theorem 3.12.2. Thus, prove that if $J : \mathscr{R} \to \mathscr{C}$ is defined by $J(\Gamma) = \langle \Gamma, 0\rangle$, then the following are true:

(a) $J(\Gamma +_{\mathscr{R}} \Delta) = J(\Gamma) +_{\mathscr{C}} J(\Delta) = (\Gamma +_{\mathscr{R}} \Delta, 0)$.
(b) $J(\Gamma \cdot_{\mathscr{R}} \Delta) = J(\Gamma) \cdot_{\mathscr{C}} J(\Delta) = (\Gamma \cdot_{\mathscr{R}} \Delta, 0)$.

3.12.7. Let $f : \mathscr{R} \to \mathscr{C}$ be defined by $f(\Gamma) = \langle 0, -\Gamma\rangle$, for $\Gamma \in \mathscr{R}$. Is f an isomorphism?

3.12.8. (See Examples 2.8.5 and 3.7.7 for terminology.) Prove that $\langle C, \langle +_c, \cdot_c\rangle\rangle$ is a field.

★3.12.9. State and prove the complex analogue of the Bolzano–Weierstrass Theorem (Exercise 3.10.16). Begin by defining the relevant concepts for the complex case.

★3.12.10. Consider the sets S_N discussed in Section 1.7.

(a) Determine N so that $C \in S_N$.
(b) Let $f : C \to C$. Determine M so that $f \in S_M$.

4

Denumerability and Choice

4.1 Finite Sets; Infinite Sets

Recall that a set A is **finite** if for some N in ω, A is an N-element set; otherwise, A is **infinite**. Thus, A is finite if and only if there is a natural number N such that there exists a one-one function of N onto A. If A is finite and nonempty and if F is a particular one-one mapping of N onto A, we write

$$A = \{a_0, a_1, \ldots, a_{N \dot{-} 1}\};$$

here, we have enumerated the elements of A according to the elements of N to which they correspond under F.

Theorem 4.1.1. *A subset of a finite set is finite. In fact, if $N \in \omega$, if A is an N-element set, and if $B \subset A$, then B is an M-element set for $M < N$.*

PROOF: It clearly suffices to prove the second assertion; the proof is by induction.

If A is a 0-element set, then $A = \varnothing$ and A has no proper subsets.

Suppose that for $N \geq 0$ every proper subset of an N-element set has M elements for $M < N$. Let A be an N^+-element set, let f be a one-one mapping of N^+ onto A, and let $B \subset A$.

If $f(N) \in B$, then $B \sim \{f(N)\}$ must be a proper subset of $A \sim \{f(N)\}$; because $A \sim \{f(N)\}$ is an N-element set, $B \sim \{f(N)\}$ is a K-element set for $K < N$. Thus, B is a K^+-element set and $K^+ < N^+$.

If $f(N) \notin B$, then either $B = A \sim \{f(N)\}$ and thus is an N-element set, or else B is a proper subset of the N-element set $A \sim \{f(N)\}$. It then follows that B is an L-element set for $L < N < N^+$.

We next consider a question raised at the end of Section 2.4. Can a set be simultaneously an M-element set and an N-element set for distinct M and N?

Theorem 4.1.2. *Let $M, N \in \omega$. Then M and N are in one-one correspondence if and only if $M = N$.*

PROOF: If $M = N$, I_M is an appropriate one-one function of M onto N. Let

$$A_1 = \{\langle M, M\rangle : M \in \omega\}(=I_\omega)$$

and let

$$A_2 = \{\langle M, N\rangle \in \omega^{②} : \text{there is no one-one mapping of } M \text{ onto } N\}.$$

[Clearly, the defining condition for A_2 could be written in terms of a formula (of set theory) with one free variable.] Let $A = A_1 \cup A_2$. We know that $A_1 \cap A_2 = \varnothing$; we must show that $A = \omega^{②}$. Clearly, $\langle 0, 0\rangle \in A_1$; moreover, $\langle 0, N\rangle$ and $\langle N, 0\rangle$ are in A_2 for every $N(\neq 0)$ in ω. Suppose that $\langle M, N\rangle \in A$. If $\langle M, N\rangle \in A_1$, then $M = N$; hence, $M^+ = N^+$, and $\langle M^+, N^+\rangle \in A_1 \subset A$. Suppose that $\langle M, N\rangle \in A_2$ and suppose that there is a one-one mapping F of M^+ onto N^+. If $F(M) = N$, then let $G = F \upharpoonright M$; G is a one-one mapping of M onto N. If $F(M) \neq N$, define $G: M \to N$ to be such that

$$G(K) = \begin{cases} F(M), & \text{if } F(K) = N; \\ F(K), & \text{otherwise.} \end{cases}$$

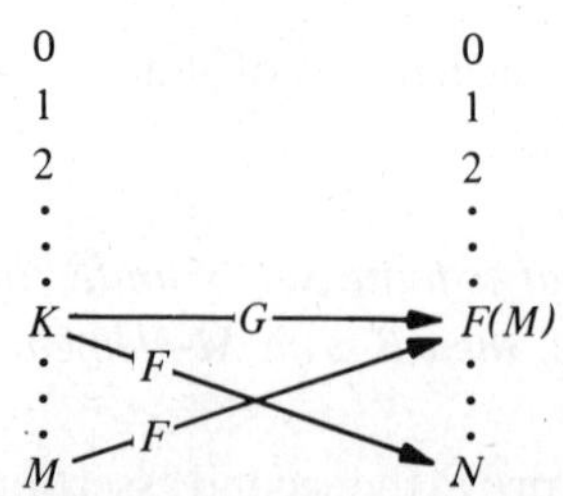

Fig. 4.1.1

Then, in either case, G is a one-one mapping of M onto N, contradicting the assumption that $\langle M, N\rangle \in A_2$. Thus, $\langle M, N\rangle \in A_2$ implies $\langle M^+, N^+\rangle \in A_2$; hence, $\langle M, N\rangle \in A$ implies $\langle M^+, N^+\rangle \in A$. By the Principle of Double Induction, $A = \omega^{②}$.

Corollary 1. *For distinct natural numbers M and N, an M-element set cannot be in one-one correspondence with an N-element set.*

(The proof is left as an exercise.)

Corollary 2. *ω is infinite.*

PROOF: Suppose that ω is finite. Then for some N in ω, there exists a one-one map F of N onto ω. Consequently, F^{-1} maps ω one-one onto N, and $F^{-1} \upharpoonright N$ maps N one-one onto a proper subset of itself. But, according to Theorem 4.1.1, a proper subset of N contains M elements for $M < N$; this contradicts Corollary 1.

A set is called **Dedekind infinite** if it is in one-one correspondence with a proper subset of itself; otherwise, it is called **Dedekind finite**.

Theorem 4.1.3
(1) *ω is Dedekind infinite.*
(2) *Every natural number is Dedekind finite.*

PROOF: (1) Let $F : \omega \to \omega$ be such that $F(N) = N^+$. By Theorem 1.6.5 and Peano Postulates (2) and (3), F is a function of ω onto ω_+. Also, F is one-one because $M^+ = N^+$ implies $M = N$ [by Peano Postulate (4)].

(2) This follows from Theorem 4.1.1 and Corollary 1 of Theorem 4.1.2.

Corollary. *Every finite set is Dedekind finite.*

(The proof of this corollary is left as an exercise. The converse of this corollary will be considered in the next section.)

Let A be an N-element set and let $\{a_0, a_1, \ldots, a_{N \dot{-} 1}\}$ be a specific enumeration of the elements of A. Two subsets of A are distinct iff there is some a_J, $J < N$, for which a_J is an element of one of these sets but not of the other. Each subset B of A is therefore determined by selecting the true sentences from among

$$a_0 \in B, a_0 \notin B, a_1 \in B, a_1 \notin B; \ldots; a_{N \dot{-} 1} \in B, a_{N \dot{-} 1} \notin B.$$

Clearly, there are 2^N different possibilities for B because for each a_J, either $a_J \in B$ or $a_J \notin B$; moreover, if $J \neq K, J, K \in N$, the choices for a_J and a_K are mutually independent. Thus, there are 2^N distinct subsets of an N-element set, $N \geq 1$. We can extend this to cover the case, in which $N = 0$; for then $A = \varnothing$, and A is the only subset of itself. (The name "power set" is derived from this property.) The following theorem has thus been proved.

Theorem 4.1.4. *If A is finite, then $\mathscr{P}(A)$ is finite; in fact, if A is an N-element set, then $\mathscr{P}(A)$ is a 2^N-element set.*

EXAMPLE 4.1.1

(1) Let $A = \{a, b, c\}$. Then $\mathscr{P}(A)$ has 8 ($= 2^3$) elements. (See Example 1.7.1.)

(2) Let $B = \{a, b, c, d\}$. Then $\mathscr{P}(B)$ consists of the following 16 ($= 2^4$) subsets of B:

$$\varnothing; \{a\}, \{b\}, \{c\}, \{d\}; \{a, b\}, \{a, c\}, \{a, d\}, \{b, c\}, \{b, d\}, \{c, d\};$$
$$\{a, b, c\}, \{a, b, d\}, \{a, c, d\}, \{b, c, d\}; B.$$

A set A is called **denumerable** if there exists a one-one correspondence between ω and A. If A is denumerable and if F is some such one-one correspondence, we can **enumerate the elements of** A according to the natural numbers to which they correspond under F. We write $A = \{a_0, a_1, a_2, \ldots, a_N, \ldots\}$. In this case we also speak of an **enumeration of** A. Clearly, a set A is denumerable iff there exists a one-one correspondence between ω_+ and A; with respect to any such correspondence we write $A = \{a_1, a_2, a_3, \ldots, a_N, \ldots\}$. *Every subset of a denumerable set is either finite or denumerable* (Exercise 4.1.11).

Theorem 4.1.5. *The set of primes is denumerable.*

PROOF: Suppose that there were a finite number, N, of primes; call these $P_1, P_2, \ldots, P_N$. Let $M = P_1 \cdot P_2 \cdot \ldots \cdot P_N + 1$. Then for each J, $1 \leq J \leq N$, there is a K_J such that $M = K_J \cdot P_J + 1 > P_J$ and $K_J \geq 1$. Hence, by Exercise 3.3.6, none of the primes $P_1, P_2, \ldots, P_N$ can be a factor of M. Therefore, either M is itself a prime or else M has a prime factor distinct from $P_1, P_2, \ldots, P_N$. In either case we have contradicted the assumption that there was a finite number, N, of primes. Thus, the set of primes is an infinite subset of ω. As such, it must be denumerable.

A denumerable set is infinite, as is any set that includes a denumerable set; however, not every infinite set is denumerable. Note that the power set of a finite set A contains "more elements" than does A. It may not be too surprising, then, that $\mathscr{P}(\omega)$ turns out to be a nondenumerable (infinite) set. To prove this we must use an argument similar to the one used to produce the Russell Paradox.

Theorem 4.1.6. *$\mathscr{P}(\omega)$ is a nondenumerable infinite set.*

PROOF: Let $A = \{\{N\} : N \in \omega\}$. Then $A \subseteq \mathscr{P}(\omega)$. There is a one-one correspondence between ω and A defined by $F(N) = \{N\}$; hence, A is infinite. Every set that includes an infinite set must be infinite; thus, $\mathscr{P}(\omega)$ is infinite.

Suppose that $\mathscr{P}(\omega)$ is denumerable; let G be a one-one mapping of ω onto $\mathscr{P}(\omega)$. Let $A = \{N \in \omega : N \notin G(N)\}$; $A \in \mathscr{P}(\omega)$. Thus, there is some M in ω such that $G(M) = A$. Suppose that $M \in A$; then $M \notin G(M) = A$. This is a contradiction. Thus, $M \notin A$. However, $A = G(M)$; hence, $M \notin G(M)$. Consequently, $M \in A$, by the definition of A. Again there is a contradiction. Thus, our original assumption, that $\mathscr{P}(\omega)$ is denumerable, is false, and $\mathscr{P}(\omega)$ is a nondenumerable infinite set.

In contrast with this result, we will show (in Exercise 4.3.4) that *the set of finite subsets of a denumerable set is denumerable.*

The argument of Theorem 4.1.6 can be generalized; we leave the verification of the following theorem as an exercise.

Theorem 4.1.7. Cantor's Theorem. *A set A cannot be in one-one correspondence with $\mathscr{P}(A)$.*

Exercises

4.1.1. Let $M, N, K \in \omega$, let A be an M-element set, and let B be an N-element set. Prove that $A \cap B$ is a K-element set, for $K \in \omega$, iff $A \cup B$ is an $(M + N) \dot{-} K$-element set.

4.1.2. Let A and B be N-element sets, $N \in \omega$, and let $F : A \to B$. Prove that F is one-one iff it is onto.

***4.1.3.** Consider the relation $\mathscr{I}_{\mathscr{P}(X)}$ induced by inclusion on $\mathscr{P}(X)$. Let $B \in Y \in \mathscr{P}\mathscr{P}(X)$. Then B is an $\mathscr{I}_{\mathscr{P}(X)}$-minimal element iff whenever $A \in Y$, then $A \not\subset B$. Prove that if a set X is finite, then every nonempty subset of $\mathscr{P}(X)$ contains an $\mathscr{I}_{\mathscr{P}(X)}$-minimal element. (The converse of this statement will be considered in the next section.)

4.1.4. **(a)** Prove Corollary 1 of Theorem 4.1.2.
(b) Prove the Corollary of Theorem 4.1.3.

4.1.5. Suppose that A is finite. Which of the following sets must also be finite? Prove the finiteness of the set in question or indicate a counterexample.
(a) $A^{②}$. **(b)** $\cap A$. **(c)** $\cup A$.
(d) The set of binary relations on A.

4.1.6. Show that if X is infinite, then $\cup X$ is infinite.

†4.1.7. Show that every denumerable set is infinite.

4.1.8. Characterize denumerable subsets of ω in terms of the well-ordering $\leq$.

4.1.9. Show that a set X is denumerable iff for each N in ω there exists a one-one correspondence between $\omega \sim N$ and X.

4.1.10. †**(a)** Prove that every denumerable set is well-orderable.

(b) Prove that every denumerable set can be linearly ordered by a relation that is not a well-ordering.

†**4.1.11.** Prove that an infinite subset of a denumerable set is denumerable.

4.1.12. For M in ω_+ let P_M be the set of natural numbers that are relatively prime to M. Show that each P_M is denumerable. (Used in Section 4.3.)

4.1.13. Show that the set of composites is denumerable.

***4.1.14.** Prove that a set X is finite iff there is a relation R on X such that both R and R^{-1} well-order X.

4.1.15. Prove Cantor's Theorem.

4.2 Axiom of Choice

We would like to demonstrate that denumerable sets are the "smallest" infinite sets in the sense that every infinite set includes a denumerable subset; in order to do so, we must consider another axiom, first introduced by Zermelo and Russell.

Axiom of Choice: *Let X be any nonempty set of nonempty sets. Then there is a function F mapping X into $\cup X$ with the property that $F(Y) \in Y$ for each Y in X.*

Such a function F will be called a **choice function on** X.

Let X as well as the elements of X be nonempty sets. Then $\cup X$ is a nonempty set, as is $X \times \cup X$.

The function F—mentioned in the statement of the Axiom of Choice—is a subcollection of $X \times \cup X$. It is natural to seek a formula $s(x)$ that is such that F is the collection of all ordered pairs $a(= \langle x_1, x_2 \rangle)$ in $X \times \cup X$ for which $s\begin{pmatrix} x \\ a \end{pmatrix}$ is true. Were we able to find such a formula, we could then guarantee the existence of a choice function F on X by means of the Axiom of Subsets. In fact, it is known that we cannot find such a formula for an arbitrary set X. (However, the method of demonstrating this nonexistence is beyond the scope of this book.)

Note that for finite sets X or for sets of sets of natural numbers, we need not resort to the Axiom of Choice to guarantee the existence of such a function; in these cases choice functions exist by virtue of the preceding axioms (Exercises 4.2.1 and 4.2.2).

EXAMPLE 4.2.1. Let a, b, c, d, e, and f be sets and let $A = \{a, b, c, d\}$, $B = \{a, c, e\}$, $C = \{b, d\}$, $D = \{e\}$, $E = \{a, f\}$, and $X = \{A, B, C, D, E\}$. Then $\{\langle A, a\rangle, \langle B, a\rangle, \langle C, b\rangle, \langle D, e\rangle, \langle E, a\rangle\}$ is a choice function on X. Another choice function on X is $\{\langle A, b\rangle, \langle B, a\rangle, \langle C, b\rangle, \langle D, e\rangle, \langle E, f\rangle\}$.

EXAMPLE 4.2.2. Let $X = \{\omega \sim \{N\} : N \in \omega\}$. Then

$$\{\langle \omega \sim \{N\}, N^+\rangle : N \in \omega\} \quad \text{and} \quad \{\langle \omega \sim \{N\}, 0\rangle : N \in \omega\} \cup \{\langle \omega_+, 1\rangle\}$$

are each choice functions on X.

The Axiom of Choice has stirred up considerable controversy. It is perhaps the least intuitive of the axioms we have discussed. Moreover, it is "nonconstructive" in that it asserts the existence of a function without indicating how to construct such a function. For these reasons many mathematicians and philosophers have objected to using the Axiom of Choice. Nevertheless, a large part of traditional mathematics would not be true without assuming this axiom or some equivalent form of it (see Sections 4.7, 4.8, 5.13, and 6.9) to be true. It has been proven by Kurt Gödel (1939) and Paul Cohen (1964) that the Axiom of Choice is "consistent with and independent of the other axioms of set theory" (including two more to be introduced in Chapters 5 and 6). What is meant by this is that if the other axioms of set theory do not lead to a contradiction, then neither does the system of set theory obtained by adding as an axiom either the Axiom of Choice or its negation. Furthermore, neither the Axiom of Choice nor its negation can be proved from the other axioms of set theory.[1]

We may add that it is an article of faith among most mathematicians that these axioms are consistent in this sense. As yet no such contradiction has been deduced; although it is conceivable that a contradiction might be derived sometime in the future, this is believed to be highly unlikely. Gödel's famous Incompleteness Theorem (1931) shows that the consistency of most interesting systems can only be demonstrated using more powerful systems. Thus, we could not prove that our system of set theory is consistent within the system itself, even if we went to the trouble of formulating our underlying logical framework more precisely. (See the further discussion in Section 6.12.)

We now consider two theorems that require the Axiom of Choice. (In this connection, see Exercise 4.2.8.) In stating such theorems, the symbol "(A.C.)" will stand for the statement "Assume the Axiom of Choice (among our axioms)."

A "naïve proof" that every infinite set X includes a denumerable subset A might run along these lines. Suppose that $X \neq \varnothing$; let $a\ (= a_0) \in X$. Then $X \sim \{a_0\} \neq \varnothing$; otherwise X would be a 1-element set. Let $a_1 \in X \sim \{a_0\}$. Similarly, let $a_2 \in X \sim \{a_0, a_1\}$, $a_3 \in X \sim \{a_0, a_1, a_2\}, \ldots, a_N \in X \sim \{a_0, a_1, a_2, \ldots, a_{N-1}\}, \ldots$. Let $A = \{a_0, a_1, a_2, \ldots, a_N, \ldots\}$.

[1] These consistency and independence questions are studied in *formal* treatments of set theory.

The difficulty with this argument is that it takes infinitely many statements, each dependent on the preceding one, to produce the set A. We cannot legitimize the sethood of A on the basis of the Axiom of Subsets.

Theorem 4.2.1. (*A.C.*) *Every infinite set has a denumerable subset.*

PROOF: Let X be an infinite set and let G^* be a choice function on $\mathscr{P}(X) \sim \{\varnothing\}$. Let $G : \mathscr{P}(X) \sim \{\varnothing\} \to \mathscr{P}(X) \sim \{\varnothing\}$ be defined by $G(A) = A \sim \{G^*(A)\}$, for every nonempty subset A of X. By Theorem 2.8.1 (Definition by Induction), there is a unique function F that is such that $F : \omega \to \mathscr{P}(X) \sim \{\varnothing\}$, $F(0) = G(X) = X \sim \{G^*(X)\}$, and for all N in ω, $F(N^+) = G(F(N)) = F(N) \sim \{G^*(F(N))\}$. Hence,

$$X \supset F(0) \supset F(1) \supset \cdots \supset F(N) \supset F(N^+) \supset \cdots ;$$

moreover, $X \sim F(0)$ is a unit set, as is $F(N^+) \sim F(N)$ for each N in ω. Thus, $X \sim \cup \text{ Range } F$ is a denumerable subset of X.

Theorem 4.2.2. (*A.C.*) *Every infinite set is Dedekind infinite.*

PROOF: Let X be an infinite set. Then, by Theorem 4.2.1, X includes a denumerable subset, A. We can enumerate A (in accordance with some one-one correspondence with ω) as $\{a_0, a_1, a_2, \ldots, a_N, \ldots\}$. Let $B = A \sim \{a_0\}$, and let f be the one-one correspondence between A and its proper subset B given by $f(a_N) = a_{N^+}$, $N \in \omega$. Then $f \cup I_{X \sim A}$ is a one-one correspondence between X and its proper subset $B \cup (X \sim A) = X \sim \{a_0\}$.

Corollary 1. (*A.C.*) *A set is infinite if and only if it is Dedekind infinite.*

Corollary 2. (*A.C.*) *A set is finite if and only if it is Dedekind finite.*

Exercises

4.2.1. Let X be a nonempty set of nonempty subsets of natural numbers. Without using the Axiom of Choice (or any equivalent), show that a choice function on X exists; i.e., show that there is a function F that is such that $F : X \to \cup X$ and $F(Y) \in Y$ for each Y in X.

4.2.2. Let X be a nonempty finite set of nonempty sets. Without using the Axiom of Choice, show that there exists a choice function on X.

†**4.2.3.** (A.C.) Prove that if A is infinite and if B is denumerable, then $A \approx A \cup B$.

4.2.4. (A.C.) Let X be a linearly orderable infinite set. Show that there exists a linear ordering on X that is not a well-ordering. (See Exercise 2.7.8.)

★**4.2.5.** (A.C.) Prove that a set is finite iff every nonempty subset of $\mathscr{P}(X)$ contains an $\mathscr{I}_{\mathscr{P}(X)}$-minimal element. (See Exercise 4.1.3.)

★**4.2.6.** Let $N \in \omega_+$. Let X be an infinite set and let $\mathscr{P}^{(N)}(X)$ be the set of N-element subsets of X.

(a) (A.C.) **Ramsey's Pigeonhole Problem** is to prove that if $\{C_1, C_2\}$ is an arbitrary partition of $\mathscr{P}^{(N)}(X)$, then there is an infinite subset Y of X that is such that either $\mathscr{P}^{(N)}(Y) \subseteq C_1$ or $\mathscr{P}^{(N)}(Y) \subseteq C_2$.

(b) (A.C.) Prove that for $R \geq 2$, if $\{C_1, C_2, \ldots, C_R\}$ is an arbitrary partition of $\mathscr{P}^{(N)}(X)$, then there is an infinite subset Z of X that is such that $\mathscr{P}^{(N)}(Z) \subseteq C_I$ for some I, $1 \leq I \leq R$.

4.2.7. For each N in ω_+, let $[N]$ be the following **Axiom of Choice for** N**-element sets**: Let X be a nonempty set of N-element sets. Then there is a function F mapping X into $\cup X$ with the property that $F(Y) \in Y$ for each Y in X. Prove the following:

(a) The Axiom of Choice implies $[N]$ for all N in ω_+.

(b) $[1]$ holds, even without the Axiom of Choice.

★**(c)** $[2]$ implies $[4]$.

★**(d)** For all K, N in ω_+, $[K \cdot N]$ implies $[N]$.

(Referred to in Section 5.13.)

★**4.2.8.** The **Axiom of Dependent Choices** is the following statement: Let X be any nonempty set on which there is a relation R with the property that for each x in X, there is some $y (\neq x)$ for which xRy. Then there is a denumerable subset $\{x_0, x_1, x_2, \ldots\}$ of X for which $x_N R x_{N^+}$ for all N in ω.

(a) Show that the Axiom of Choice implies the Axiom of Dependent Choices.

(b) Show the Axiom of Dependent Choices can be used in lieu of the Axiom of Choice to prove Theorems 4.2.1 and 4.2.2.

(Used in the proof of Exercise 6.1.12.)

4.3 Denumerable Unions, Intersections, and Products

We use the adjective "**countable**" to describe a set that is either finite or denumerable. Every natural number, as well as the set of natural numbers, is countable. An **uncountable** set is thus a nondenumerable infinite set.

Theorem 4.3.1. (*A.C.*)

(1) *The union of a nonempty countable set of denumerable sets is denumerable.*

(2) *The union of a countable set of countable sets is countable.*

PROOF: (1) We consider the case in which X is denumerable; the case in which X is finite (but nonempty) is left as an exercise. Let X be a denumerable set; thus, $X = \{A_0, A_1, A_2, \ldots, A_N, \ldots\}$ with respect to a particular one-one correspondence between ω and X.

For each N in ω, W_N, the collection one-one correspondences between ω and A_N, is a nonempty set; W, the collection of W_N, $N \in \omega$, is also a set. By the Axiom of Choice, there is a function F defined on W that is such that $F(W_N) \in W_N$, for each N in ω; thus $F(W_N)$ enumerates the elements of A_N as $\{a_{N,1}, a_{N,2}, a_{N,3}, \ldots, a_{N,M}, \ldots\}$. Consequently,

$$\cup X = \{a_{0,0}, a_{0,1}, a_{1,0}, a_{0,2}, a_{1,1}, a_{2,0}, a_{0,3}, a_{1,2}, a_{2,1}, a_{3,0}, a_{0,4}, \ldots\};$$

therefore, we can order the elements of $\cup X$ according to the following considerations:

(a) If $N + M < K + L$, we let $a_{N,M}$ precede $a_{K,L}$.
(b) If $N + M = K + L$, we let $a_{N,M}$ precede $a_{K,L}$ iff $N < K$.
(c) We delete those elements that have already appeared.

(The elements of X need not be pairwise disjoint.) In this manner we obtain a one-one function G of ω onto $\cup X$. (This enumeration can be defined, precisely, by Definition by Induction.)

(2) The argument here is similar for the most part. In some cases $\cup X$ will be finite. For example, if X is finite and if all the A_N are finite, then dom G will be a natural number.

The method utilized to enumerate $\cup X$ is known as the **Cauchy diagonal method.**

In Chapter 6 we give another proof of this theorem by means of "transfinite arithmetic".

There are pairwise disjoint denumerable sets X of denumerable sets A_N for which we do not require the Axiom of Choice in order to guarantee the denumerability of $\cup X$ (Exercise 4.3.5).

Note that we required the Axiom of Choice in the proof of Theorem 4.3.1 to select a specific well-ordering on each A_N. However, if there is a "natural" well-ordering on A_N—for example, an ordering by magnitude—then the Axiom of Choice is unnecessary for the specific set X in question.

In case X ($\neq \varnothing$) is an N-element set, we can select a particular enumeration of the elements of X. Thus $X = \{A_0, A_1, \ldots, A_{N \dot{-} 1}\}$; we write "$\bigcup_{J=0}^{N \dot{-} 1} A_J$" as well as "$A_0 \cup A_1 \cup \cdots \cup A_{N \dot{-} 1}$" for "$\cup X$", and "$\bigcap_{J=0}^{N \dot{-} 1} A_J$" or "$A_0 \cap A_1 \cap \cdots \cap$

$A_{N \dot{-} 1}$" for "$\cap X$". If X is denumerable and if we know of a specific one-one correspondence with ω, then (with respect to this correspondence)

$$X = \{A_0, A_1, \ldots, A_N, \ldots\};$$

we write "$\bigcup_{J \in \omega} A_J$" for "$\cup X$" and "$\bigcap_{J \in \omega} A_J$" for "$\cap X$".

The following theorem can be proved on the basis of Theorem 4.3.1; however, its proof does not really require the Axiom of Choice. These matters will be considered in the exercises.

Theorem 4.3.2
(1) *Let A and B be denumerable sets. Then $A \times B$ is denumerable.*
(2) *Let A and B be countable sets. Then $A \times B$ is countable.*

Exercises

†4.3.1. Without using the Axiom of Choice, prove that every countable set is well-orderable.

4.3.2. Determine a denumable set X, each element of which is a denumerable set A_J, that satisfies the specified condition:
- **(a)** $\cap X$ is empty.
- **(b)** $\cap X$ is an K-element set, $K \in \omega_+$.
- **(c)** $\cap X$ is denumerable.

4.3.3. Consider the proof of Theorem 4.3.1.
- **(a)** Show that each W_N is a set.
- **(b)** Show that W is a set.
- **(c)** Use Definition by Induction to define the enumeration of the elements of $\cup X$.
- **(d)** Prove Theorem 4.3.1(1) for the case in which X is finite but nonempty.

†4.3.4. Prove that the set of finite subsets of a denumerable set is denumerable—
- **(a)** as a corollary of Theorem 4.3.1,
- **(b)** without using the Axiom of Choice.

4.3.5. Show that ω is the union of a denumerable set of pairwise disjoint denumerable sets.

4.3.6. Show that every denumerable set is the union of a denumerable set of pairwise disjoint denumerable sets.

4.3.7. **(a)** Let A and B each be denumerable pairwise disjoint sets. With respect to particular one-one correspondences with ω, let $A = \{a_0, a_1, a_2, \ldots, a_N, \ldots\}$ and let $B = \{b_0, b_1, b_2, \ldots, b_N, \ldots\}$. Suppose that for each N in ω, a_N is in one-one correspondence with b_N. Show that $\cup A$ is in one-one correspondence with $\cup B$.
- **(b)** Is the Axiom of Choice used in (a)?

4.3.8. Prove Theorem 4.3.2(1). Thus, prove that if A and B are each denumerable sets, then $A \times B$ is also denumerable.

(a) First use Theorem 4.3.1 (and, hence, the Axiom of Choice) to prove this.

(b) Now prove this without using the Axiom of Choice.

4.4 Sequences, N-tuples, and N-place Relations

Let X be a nonempty set. Suppose that $F : N \to X$, where $N \in \omega$; such a function is called an N-**sequence (in** X). We write "$[X_0, X_1, \ldots, X_{N \dot{-} 1}]$" or "$[F_0, F_1, \ldots, F_{N \dot{-} 1}]$" to indicate the correspondence. Similarly, a function F with domain ω and range X is called a **denumerable sequence (in** X) and is denoted by "$[X_0, X_1, \ldots, X_N, \ldots]$" or by "$[F_0, F_1, \ldots, F_N, \ldots]$". N-sequences and denumerable sequences are collectively called **sequences**. N-sequences are N-element sets (of ordered pairs) and denumerable sequences are denumerable sets. However, the range of an N-sequence, $N \geq 1$, can be an M-element subset of X, $1 \leq M \leq N$, and the range of a denumerable sequence can be any nonempty countable subset of X.

Strictly speaking, "$[X_0, X_1, \ldots, X_{N \dot{-} 1}]$" or "$[F_0, F_1, \ldots, F_{N \dot{-} 1}]$" denotes the set "$\{\langle 0, F(0)\rangle, \langle 1, F(1)\rangle, \ldots, \langle N \dot{-} 1, F(N \dot{-} 1)\rangle\}$"; a similar remark applies to denumerable sequences.

EXAMPLE 4.4.1. Let $X = \omega$.

(1) Each of the following is a 5-sequence:

(a) $[0, 1, 2, 3, 4] = \{\langle 0, 0\rangle, \langle 1, 1\rangle, \langle 2, 2\rangle, \langle 3, 3\rangle, \langle 4, 4\rangle\}$;

(b) $[2, 4, 6, 4, 8] = \{\langle 0, 2\rangle, \langle 1, 4\rangle, \langle 2, 6\rangle, \langle 3, 4\rangle, \langle 4, 8\rangle\}$;

(c) $[3, 3, 3, 3, 3] = \{\langle 0, 3\rangle, \langle 1, 3\rangle, \langle 2, 3\rangle, \langle 3, 3\rangle, \langle 4, 3\rangle\}$.

(2) Each of the following is a denumerable sequence:

(a) $[0, 2, 4, \ldots, 2N, \ldots] = \{X : X = \langle N, 2N\rangle \text{ for some } N \text{ in } \omega\}$;

(b) $[0, 1, 0, 1, 0, 1, \ldots, 0, 1, \ldots] = \{X : X = \langle 2N, 0\rangle \text{ for some } N \text{ in } \omega\} \cup \{X : X = \langle 2N + 1, 1\rangle \text{ for some } N \text{ in } \omega\}$.

Let $x, y, x_1, x_2, \ldots$ be sets. We have already defined the ordered pair $\langle x, y\rangle$ to be the (unordered) pair $\{\{x\}, \{x, y\}\}$; we utilize this definition to define, inductively, **ordered** N-**tuples** for every $N \geq 1$. Let "$\langle x_1\rangle$" be "x_1" and for $K \geq 1$, define "$\langle x_1, x_2, \ldots, x_K, x_{K^+}\rangle$" to be "$\langle\langle x_1, x_2, \ldots, x_K\rangle, x_{K^+}\rangle$". This definition preserves the meaning of $\langle x_1, x_2\rangle$. Thus, an ordered 2-tuple is simply an ordered pair. Ordered 3-tuples are called **ordered triples**. For $1 \leq I \leq K$, we say that x_I is the Ith **coordinate of** $\langle x_1, x_2, \ldots, x_K\rangle$.

†EXAMPLE 4.4.2

$$\langle x_1, x_2, x_3\rangle = \langle\langle x_1, x_2\rangle, x_3\rangle = \{\{\langle x_1, x_2\rangle\}, \{\langle x_1, x_2\rangle, x_3\}\}$$
$$= \{\{\{x_1\}, \{x_1, x_2\}\}, \{\{\{x_1\}, \{x_1, x_2\}\}, x_3\}\};$$

in particular, $\langle x, x, x\rangle = \{\{\{x\}\}, \{\{\{x\}\}, x\}\}$.

Theorem 4.4.1. *Let $N \geq 1$. Then $\langle x_1, x_2, \ldots, x_N\rangle = \langle y_1, y_2, \ldots, y_N\rangle$ if and only if all the following equalities hold: $x_1 = y_1, x_2 = y_2, \ldots,$ and $x_N = y_N$.*

PROOF: For $N = 1$, this is obvious, because $\langle x_1\rangle = x_1$. Suppose that this is true for $N = K$. Then $\langle x_1, x_2, \ldots, x_K\rangle = \langle y_1, y_2, \ldots, y_K\rangle$ iff $x_1 = y_1$, $x_2 = y_2, \ldots, x_K = y_K$. By Theorem 2.1.1,

$$\langle\langle x_1, x_2, \ldots, x_K\rangle, x_{K+1}\rangle = \langle\langle y_1, y_2, \ldots, y_K\rangle, y_{K+1}\rangle$$

iff both $\langle x_1, x_2, \ldots, x_K\rangle = \langle y_1, y_2, \ldots, y_K\rangle$ and $x_{K+1} = y_{K+1}$. Thus, by the definition of an ordered $(K + 1)$-tuple together with the inductive hypothesis,

$$\langle x_1, x_2, \ldots, x_K, x_{K+1}\rangle = \langle y_1, y_2, \ldots, y_K, y_{K+1}\rangle$$

iff $x_1 = y_1, x_2 = y_2, \ldots, x_K = y_K$, and $x_{K+1} = y_{K+1}$. The theorem follows by induction.

Let $N \geq 1$. With each N-tuple $\langle x_1, x_2, \ldots, x_N\rangle$ there is associated the unique N-sequence $[x_1, x_2, \ldots, x_N]$, and vice versa. Each ordered N-tuple, for $N > 1$, is an ordered pair; hence it is either a unit set or a 2-element set. The associated N-sequence is an N-element set whose elements are ordered pairs. It will be convenient to refer to ordered N-tuples also as **sequences**; the notation "$\langle x_1, x_2, \ldots, x_N\rangle$" as opposed to "$[x_1, x_2, \ldots, x_N]$" will clarify the intention. Suppose that $1 \leq K \leq N$. We call x_K a **term** of each of the sequences $\langle x_1, x_2, \ldots, x_N\rangle$ and $[x_1, x_2, \ldots, x_N]$. We also say that x_K **appears in a sequence** M **times** if M of the terms are x_K. For arbitrary sets $x_1, x_2, \ldots, x_N$, we will sometimes say that x_K **appears among** $x_1, x_2, \ldots, x_N$ M **times** if x_K appears in the sequence $\langle x_1, x_2, \ldots, x_N\rangle$ M times.

If $\langle x_1, x_2, \ldots, x_N\rangle$ (or $[x_1, x_2, \ldots, x_N]$) is a sequence, then for each M, $1 \leq M \leq N$, $\langle x_1, x_2, \ldots, x_M\rangle$ (or $[x_1, x_2, \ldots, x_M]$) will be called an **initial segment of** $\langle x_1, x_2, \ldots, x_N\rangle$ (or **of** $[x_1, x_2, \ldots, x_N]$).

We extend some of the previous concepts—Cartesian product, relation, operation—to cover the case of N sets for $N > 2$. Let $A_1, A_2, \ldots, A_N$ be sets. By the N-**fold Cartesian product**, $A_1 \times A_2 \times \cdots \times A_N$, we mean

$$\{\langle a_1, a_2, \ldots, a_N\rangle : a_1 \in A_1, a_2 \in A_2, \ldots, a_N \in A_N\};$$

if $A_1 = A_2 = \cdots = A_N = A$, we write "$A^{(N)}$" instead of "$A \times A \times \cdots \times A$". An N-place relation is a subset of $A_1 \times A_2 \times \cdots \times A_N$ for some sets $A_1, A_2, \ldots,$

A_N; in particular, if $A_1 = A_2 = \cdots = A_N = A$, we speak of an N-**place relation on** A. Thus, an $(N + 1)$-place relation F with domain $A_1 \times A_2 \times \cdots \times A_N$ that has the property that $\langle a_1, a_2, \ldots, a_N\rangle Fb$ and $\langle a_1, a_2, \ldots, a_N\rangle Fb'$ together imply that $b = b'$ is a function; we call F a **function of** N **arguments** and write "$F(a_1, a_2, \ldots, a_N)$" instead of "$F(\langle a_1, a_2, \ldots, a_N\rangle)$".

†EXAMPLE 4.4.3. Let $A_1 = A_2 = A_3 = \omega$.

(1) The following are 3-place relations on ω:

(a) $\{\langle M, N, R\rangle : M, N, R \in \omega \text{ and } M + N + R = 2\}$

$$= \{\langle 0,0,2\rangle, \langle 0,2,0\rangle, \langle 2,0,0\rangle, \langle 0,1,1\rangle, \langle 1,0,1\rangle, \langle 1,1,0\rangle\}.$$

(b) $\{\langle M, M, M\rangle : M \in \omega\} = \{\langle 0,0,0\rangle, \langle 1,1,1\rangle, \langle 2,2,2\rangle, \ldots\}$.

(c) $\omega^{③}$.

(d) $\varnothing$.

(2) The following are functions of 3 arguments from $\omega^{③}$ to ω:

(a) $F_1 = \{\langle M, N, R, M + N + R\rangle : M, N, R \in \omega\}$

—i.e., $F_1(M, N, R) = M + N + R$ for each $\langle M, N, R\rangle \in \omega^{③}$.

(b) $F_2 = \{\langle M, N, R, M \cdot N \cdot R\rangle : M, N, R \in \omega\}$

—i.e., $F_2(M, N, R) = M \cdot N \cdot R$ for each $\langle M, N, R\rangle \in \omega^{③}$.

(c) $F_3 = \{\langle M, N, R, 0\rangle : M, N, R \in \omega\}$

—i.e., $F_3(M, N, R) = 0$ for each $\langle M, N, R\rangle \in \omega^{③}$. Here, F_3 is a constant function.

For N in ω, an N-**place operation on** A is defined to be a function F with domain $A^{Ⓝ}$ and range A—i.e., a function that assigns an element of A to each ordered N-tuple of elements of A. Each of the functions F_1, F_2, F_3 of Example 4.4.3 is a 3-place operation on ω.

Exercises

4.4.1. Which of the following 3-place relations are functions? Determine the domain and range of each such function.

(a) $\{\langle 1,7,2\rangle, \langle 1,4,7\rangle, \langle 2,4,1\rangle\}$.

(b) $\{\langle 0,0,0\rangle\}$.

(c) $\{\langle 1,0,0\rangle, \langle 0,1,0\rangle, \langle 0,1,1\rangle, \langle 1,0,1\rangle, \langle 1,1,0\rangle, \langle 1,1,1\rangle\}$.

(d) $\{\langle M, 0, M\rangle : M \in \omega\}$.

(e) $\{\langle 0, M, M\rangle : M \in \omega\}$.
(f) $\{\langle 0, 0, M\rangle : M \in \omega\}$.

4.4.2. For each $K \geq 0$ and $N \geq 1$, how many N-place relations are there on a K-element set?

4.4.3. For $N > 2$, formulate a "Principle of N-fold Induction".

4.5 Denumerability of Rational Numbers

The sets that were constructed in Chapter 3—the integers, the rationals, the reals, and the complex numbers—are infinite sets. We now show that the first two of these are denumerable.

According to Theorem 4.3.2, $\omega^{②}$ is denumerable. Thus $\mathscr{Z}$, as an infinite subset of a denumerable set, is also denumerable (Exercise 4.1.11).

Exercise 3.6.7 indicates that $\mathscr{Q}$ is in one-one correspondence with $\mathscr{F}$, the set of relatively prime fractions. In fact, $\mathscr{F} \subseteq \mathscr{Z}^{②}$; hence, we can modify the above argument to obtain the denumerability of $\mathscr{Q}$.

We now consider proofs of the denumerability of $\mathscr{Z}$ and of $\mathscr{Q}$ in which we explicitly indicate the one-one mappings of ω onto each of these sets.

Every integer can be represented uniquely as either $\overline{N}$ for N in ω or (excl.) as $-\overline{N}$ for N in ω_+. Moreover, every natural member is of the form $2K$ or (excl.) $2K + 1$ for some unique K in ω. Define $G : \omega \to \mathscr{Z}$ as follows: If N is of the form $2K$, let $G(N) = \overline{K}$; if N is of the form $2K + 1$, let $G(N) = -\overline{K^+}$. Clearly, G is a one-one correspondence of ω onto $\mathscr{Z}$. The first few values of G are

$$0, -1, 1, -2, 2, -3, 3.$$

We next sketch the enumeration of $\mathscr{F}$, the set of relatively prime fractions. Consider the fractions $\langle \mathscr{M}, \mathscr{N}\rangle$ in $\mathscr{F}$ in which $\mathscr{M} >_{\mathscr{Z}} 0$; each such fraction is of the form $\langle \overline{M}, \overline{N}\rangle$ for $M, N \in \omega_+$. First list all such fractions in which $M + N = 2$—then those for which $M + N$ is successively $3, 4, 5, \ldots$. Order the (relative prime) fractions for which $M + N = K$ according to the increasing magnitude of M. (Note that not all the $K \dot{-} 1$ fractions for which $M + N = K$ are relative prime.) The first few fractions in our enumeration are

$$\langle 1, 1\rangle, \langle 1, 2\rangle, \langle 2, 1\rangle, \langle 1, 3\rangle, \langle 3, 1\rangle, \langle 1, 4\rangle, \langle 2, 3\rangle, \langle 3, 2\rangle, \langle 4, 1\rangle, \langle 1, 5\rangle, \langle 5, 1\rangle.$$

Next, intersperse the $\langle \mathscr{M}, \mathscr{N}\rangle$ in $\mathscr{F}$ for which $\mathscr{M} \leq_{\mathscr{Z}} 0$ in the above enumeration as we previously did in the indicated enumeration of $\mathscr{Z}$. Let F_1 be this enumeration of $\mathscr{F}$ and let F_2 be the one-one mapping of $\mathscr{F}$ onto $\mathscr{Q}$ whose existence is indicated in Exercise 3.6.7. Then $F_2 \circ F_1$ is an enumeration of $\mathscr{Q}$.

Freilich exhibited an explicit one-one correspondence between ω and $\mathcal{Q}$. Let $f : \omega \to \mathcal{Z}$ be defined by

$$f(N) = \begin{cases} \overline{K}, & \text{if } N = 2K; \\ -\overline{L}, & \text{if } N = 2L - 1. \end{cases}$$

Clearly, f is one-one onto $\mathcal{Z}$. Now define $g : \mathcal{Z} \to \mathcal{Q}$ by

$$g(\mathcal{M}) = \begin{cases} \overline{\overline{P_1}}^{f(N_1)} \cdot_{\mathcal{Q}} \overline{\overline{P_2}}^{f(N_2)} \cdot_{\mathcal{Q}} \cdots \cdot_{\mathcal{Q}} \overline{\overline{P_K}}^{f(N_K)}, & \\ \quad \text{if } \mathcal{M} = \overline{M} \geq_{\mathcal{Z}} \overline{2} \text{ and } M = P_1^{N_1} \cdot P_2^{N_2} \cdot \ldots \cdot P_K^{N_K} & \\ \quad \text{is the prime factorization of } M \text{ (Theorem 3.3.1)}; & \\ \overline{\overline{1}}, & \text{if } \mathcal{M} = \overline{1}; \\ \overline{\overline{0}}, & \text{if } \mathcal{M} = \overline{0}; \\ -_{\mathcal{Q}}\, g(-\mathcal{M}), & \text{if } \mathcal{M} <_{\mathcal{Z}} \overline{0}. \end{cases}$$

Then g is one-one onto $\mathcal{Q}$ (Exercise 4.5.2) and $h = g \circ f$ is the desired one-one correspondence between ω and $\mathcal{Q}$.

Exercises

4.5.1. Give a detailed account of the first enumeration of $\mathcal{Q}$.

★4.5.2. In Freilich's "denumerability formula", show that g is one-one onto $\mathcal{Q}$.

4.5.3. Prove that if $\{L, U\}$ is a real number, then L and U are each denumerable.

4.5.4. Show that the set of real numbers of the form

$$\bar{\gamma}_1 + \bar{\gamma}_2\sqrt{\overline{\overline{\overline{2}}}} + \bar{\gamma}_3\sqrt{\overline{\overline{\overline{3}}}}, \gamma_1, \gamma_2, \gamma_3 \in \mathcal{Q},$$

is denumerable.

★4.5.5. A real number Γ is said to be **algebraic** if for some $N \geq 1$ there exist natural numbers $A_0, A_1, A_2, \ldots, A_N (\neq 0)$ for which

$$\overline{\overline{\overline{A_N}}}\Gamma^N + \overline{\overline{\overline{A_{N \dot{-} 1}}}}\Gamma^{N \dot{-} 1} + \cdots + \overline{\overline{\overline{A_1}}}\Gamma + \overline{\overline{\overline{A_0}}} = \overline{\overline{\overline{0}}}.$$

Thus, for rational numbers α, $\bar{\alpha}$ is algebraic because $\bar{\alpha} = \overline{\overline{\overline{M}}}/\overline{\overline{\overline{N}}}$ for $M, N (\neq 0)$ in ω; hence, $-\overline{\overline{\overline{M}}} + \overline{\overline{\overline{N}}}\bar{\alpha} = \overline{\overline{\overline{0}}}$. Also, such irrationals as $\Gamma_1 = \sqrt{\overline{\overline{\overline{2}}}}$, $\Gamma_2 = \sqrt[5]{\overline{\overline{\overline{17}}}}$, satisfy the respective equations

$$-\overline{\overline{\overline{2}}} + \Gamma_1{}^2 = \overline{\overline{\overline{0}}}, \qquad -\overline{\overline{\overline{17}}} + \Gamma_2{}^5 = \overline{\overline{\overline{0}}}.$$

Show that the set of all algebraic numbers is denumerable.

4.6 Nondenumerability of Real Numbers

The most direct method of demonstrating the denumerability of a set X is the exhibition of a one-one correspondence between some known denumerable set and X. In contrast with this procedure, in order to prove the nondenumerability of an infinite set Y, we must show that no one-one correspondence could exist between a denumerable set and Y. In fact, we generally begin by assuming that such a one-one correspondence actually exists, and we show that this supposition leads to a contradiction. [In Theorem 4.1.6 we used this method to show that $\mathscr{P}(\omega)$ is nondenumerable.]

We will be concerned with various sets of numbers. Although in some places we could speak about sets of natural numbers and sets of rationals, we will generally discuss the isomorphic images of these sets in $\mathscr{R}$. (An exception is made in the proof of Theorem 4.6.1, where we will be concerned with a real number $\{L, U\}$, i.e., with a partition of $\mathscr{Q}$.) We will refer to these respective isomorphic images as the $\mathscr{R}$-natural numbers and the $\mathscr{R}$-rationals. As we have previously remarked, there is no difficulty in extending purely arithmetic theorems about ω (or $\mathscr{Q}$) so as to apply to $\mathscr{R}$-natural numbers (or $\mathscr{R}$-rationals). Thus, all the arithmetic operations and the natural ordering we discuss refer to $\mathscr{R}$. Consequently, we can drop the subscript "$\mathscr{R}$" and (except in one proof) the bars in our notation. (Strictly speaking, when we write an expression such as "d_N", the subscript "N" refers to a natural number; we could just as well utilize $\mathscr{R}$-natural numbers.

Our first objective will be to show that the set of real numbers $\mathbf{U}$, defined by

$$\mathbf{U} = \{\Gamma \in \mathscr{R} : 0 < \Gamma \leq 1\},$$

is nondenumerable. To this end, we define a set by means of which we can represent nonterminating decimal expansions of elements of $\mathbf{U}$.

We let $\mathscr{T}$ be the set of $\mathscr{R}$-natural numbers < 10. Consider all denumerable sequences $d = \langle d_0, d_1, d_2, \ldots, d_N, \ldots \rangle$ that are such that

(4.6.1) $d_0 = 0$

and for each $N = 1, 2, 3, \ldots,$

(4.6.2) $d_N \in \mathscr{T},$

and

(4.6.3) there is a natural number $M, M \geq N$, such that $d_M > 0$.

Recall that "d_N" denotes "$d(N)$", $N \geq 0$. Because d_0 is automatically 0 for each such sequence d, we usually omit explicit mention of this. *The collection of all such sequences is a set* (Exercise 4.6.1), which we denote by $\mathscr{D}$.

Let $d\,(= \langle d_0, d_1, d_2, \ldots, d_K, \ldots\rangle) \in \mathscr{D}$. We write "$\delta_N$" for the $\mathscr{R}$-rational number "$\sum_{K=1}^{N} d_K/10^K$", $N \geq 1$. In effect, each d in $\mathscr{D}$ determines a sequence δ defined by $\delta(0) = 0$; $\delta(N) = \delta_N$, $N \geq 1$. A sequence δ determined by such a sequence d in D will represent a "nonterminating decimal expansion" of a real number in **U**; we will say that δ **corresponds to** d.

EXAMPLE 4.6.1. Consider the sequence d defined by

$$d_N = d(N) = \begin{cases} 1, & \text{if } N \text{ is odd;} \\ 2, & \text{if } N\,(\geq 2) \text{ is even.} \end{cases}$$

Then

$$\delta_1 = d_1/10^1 = 1/10.$$

$$\delta_2 = d_1/10^1 + d_2/10^2 = 1/10 + 2/100 = 12/100\,(= 3/25).$$

$$\delta_3 = d_1/10^1 + d_2/10^2 + d_3/10^3 = 1/10 + 2/100 + 1/1000 = 121/1000.$$

EXAMPLE 4.6.2

(1) The sequence d of $\mathscr{D}$ that will represent the real number 1/3 will be the sequence defined by $d(N) = 3$ for all N in ω_+. Note that the numbers $\delta_1\,(= 3/10)$, $\delta_2\,(= 33/100)$, $\delta_3\,(= 333/1000)$, etc., "approach" 1/3.

(2) The sequence d of $\mathscr{D}$ that will represent 1/2 will be defined by $d_1 = 4$ and $d_N = 9$ for $N \geq 2$. Here, the rational numbers δ_N "approach" 1/2. Note that requirement (4.6.3) for sequences d of $\mathscr{D}$ specifically eliminates the sequence e defined by $e(1) = 5$ and $e(N) = 0$ for $N \geq 2$. Thus, condition (4.6.3) for membership in $\mathscr{D}$ assures that only one sequence in $\mathscr{D}$ will represent an element of **U**.

(3) The sequence d of $\mathscr{D}$ that will represent $\sqrt{2}/2$ (approximately "1.414/2") will have $d_1 = 7$, $d_2 = 0$, and $d_3 = 7$.

We can define an order relation $\leq_{\mathscr{D}}$ on $\mathscr{D}$ "lexicographically". Thus, if d and d' are arbitrary elements of $\mathscr{D}$, we define $d <_{\mathscr{D}} d'$ iff for some $N = 1, 2, 3, \ldots$, $d_K = d'_K$ for $K = 0, 1, 2, \ldots, N-1$ and $d_N < d'_N$; $d \leq_{\mathscr{D}} d'$ if either $d <_{\mathscr{D}} d'$ or $d = d'$. We leave as an exercise verification that $\leq_{\mathscr{D}}$ is a linear ordering on $\mathscr{D}$.

Suppose that d and d' are in $\mathscr{D}$ *with* $d <_{\mathscr{D}} d'$, *and let* δ *and* δ' *be the respective corresponding sequences. Let* N *be the smallest number for which* $d_N < d'_N$. *Then* it is important to note the following:

(1) $\delta_K = \delta'_K$ *for* $K = 0, 1, 2, \ldots, N - 1$.
(2) $\delta_K < \delta'_K$ *for* $K = N, N + 1, N + 2, \ldots$.

In fact,

(3) $\delta_L < \delta'_M$ *for all* $L = 0, 1, 2, \ldots,$ *and for all* $M = N, N + 1, N + 2, \ldots$.

EXAMPLE 4.6.3

(1) Let $d = \langle 0, 4, 1, 8, 3, 6, \ldots \rangle$ and $d' = \langle 0, 4, 1, 9, 0, 5, \ldots \rangle$. Then $d <_{\mathscr{D}} d'$ because $d_K = d'_K$ for $K = 0, 1, 2$, but $d_3 = 8 < 9 = d'_3$. Furthermore,

$$\delta_1 = \delta'_1 = 4/10;$$

$$\delta_2 = \delta'_2 = 41/100;$$

$$\delta_3 = 418/1000 < \delta'_3 = 419/1000;$$

and for all $L = 0, 1, 2, \ldots$ and $M = 4, 5, 6, \ldots,$

$$\delta_L < 4184/10{,}000, \qquad \delta'_M \geq 4190/10{,}000.$$

(2) $\langle 0, 0, 9, 9, 9, 9, \ldots \rangle <_{\mathscr{D}} \langle 0, 1, 0, 1, 0, 1, \ldots \rangle$.

Theorem 4.6.1. $\mathscr{D}$ *is in one-one correspondence with* $\mathbf{U}$.

PROOF (SKETCH): Let $f : \mathscr{D} \to \mathbf{U}$ be defined as follows. For d in $\mathscr{D}$, let δ be the corresponding sequence. Let

$$f(d) = \{L, U\}, \text{ where } U = \{\alpha \in \mathscr{Q} : \text{for every } N = 1, 2, 3, \ldots, \delta_N < \bar{\alpha}\},$$

and let $L = \mathscr{Q} \sim U$. [We leave as an exercise the verification that each $f(d)$ is in $\mathbf{U}$.] We show that f maps $\mathscr{D}$ one-one onto $\mathbf{U}$.

Suppose that d and d' are distinct sequences of $\mathscr{D}$. Let $f(d) = \{L, U\}$ and $f(d') = \{L', U'\}$; we first show that $\{L, U\} \neq \{L', U'\}$. Let H be the isomorphism defined in Sections 3.9 and 3.10) of $\mathscr{Q}$ onto the set of $\mathscr{R}$-rationals. Clearly, it suffices to show that $\{H''(L), H''(U)\} \neq \{H''(L'), H''(U')\}$; in fact, it suffices to show that $H''(U) \neq H''(U')$. We may suppose that $d <_{\mathscr{D}} d'$. Let K be the smallest number for which $d_K \neq d'_K$. Then $d_K < d'_K$. Let δ and δ' be the sequences corresponding to d and d', respectively. Then $\delta_N < \delta'_K$ for $N = 1, 2, 3, \ldots$; hence, $\delta'_K \in H''(U)$. However, for some M with $M > K$, $d'_M > 0$; therefore, $\delta'_K < \delta'_M$ and $\delta'_K \notin H''(U')$. Thus, $\{H''(L), H''(U)\} \neq \{H''(L'), H''(U')\}$.

Now let $\{L, U\}$ be any real number in **U**. We must find a sequence d of $\mathscr{D}$ for which $f(d) = \{L, U\}$. By Theorem 3.9.3. and its Corollary, for each $N \geq 1$ there is an α in L and a β in U for which $\beta - \alpha \leq 1/N$. Because L has no last element we can define, by induction, α_N (in L) and β_N (in U), $N \geq 1$, such that for each such N,

$$0 < \overline{\alpha_N} < \overline{\alpha_{N^+}} \text{ and } \overline{\beta_N} - \overline{\alpha_N} \leq 1/N. \tag{4.6.4}$$

It is intuitively clear (and not difficult to prove) that if ρ and σ are $\mathscr{R}$-rationals in **U**, with $\rho < \sigma$, then there exists a K, $K \geq 1$, and an $\mathscr{R}$-rational δ_K—of the form $\sum_{I=1}^{K} e_I/10^I$, where $e_I \in \mathscr{T}$ for $I = 1, 2, \ldots, K$—such that $\rho \leq \delta_K < \sigma$. Thus, we can define δ_N, $N \geq 1$, such that

$$\alpha_N \leq \delta_N < \alpha_{N^+}. \tag{4.6.5}$$

The δ_N determine a sequence d in $\mathscr{D}$ defined as follows: Let

$d_1 = \max\{d \in \mathscr{T} : d/10 < \delta_N \text{ for some } N = 1, 2, \ldots\}$. Suppose that $d_1, d_2, \ldots, d_K$ have already been defined. Let

$$d_{K^+} = \max\{d \in \mathscr{T} : \sum_{I=1}^{K} d_I/10^I + d/10^{K^+} < \delta_N \text{ for some } N = 1, 2, \ldots\}.$$

(In Exercise 4.6.4 we will show that such a sequence d exists by virtue of Definition by Induction.) By means of the definitions of f and of d, as well as the inequalities (4.6.4) and (4.6.5), one verifies that $f(d) = \{L, U\}$.

Theorem 4.6.2. *$\mathscr{D}$ is nondenumerable.*

PROOF: Suppose that $\mathscr{D}$ is denumerable. Then there exists a one-one correspondence between ω_+ and $\mathscr{D}$. With respect to any such one-one correspondence g, we can enumerate *all* the sequences d of $\mathscr{D}$; we let $d^{(I)}$ be $g(I)$, $I = 0, 1, 2, \ldots$. Thus, each $d^{(I)}$ is of the form $\langle 0, d^{(I)}{}_1, d^{(I)}{}_2, \ldots, d^{(I)}{}_N, \ldots \rangle$. In order to obtain a contradiction, we show that this enumeration, in fact, leaves out some sequence of $\mathscr{D}$. We produce such an omitted sequence by means of **Cantor's Diagonal Process.** [Recall the proofs of Cantor's Theorem (4.1.7) and of Theorem 4.1.6.] Thus,

$$d^{(1)} = \langle 0, d^{(1)}{}_1, d^{(1)}{}_2, d^{(1)}{}_3, \ldots, d^{(1)}{}_N, \ldots \rangle$$
$$d^{(2)} = \langle 0, d^{(2)}{}_1, d^{(2)}{}_2, d^{(2)}{}_3, \ldots, d^{(2)}{}_N, \ldots \rangle$$
$$d^{(3)} = \langle 0, d^{(3)}{}_1, d^{(3)}{}_2, d^{(3)}{}_3, \ldots, d^{(3)}{}_N, \ldots \rangle$$
$$\cdots$$
$$d^{(N)} = \langle 0, d^{(N)}{}_1, d^{(N)}{}_2, d^{(N)}{}_3, \ldots, d^{(N)}{}_N, \ldots \rangle.$$

Call the numbers $d^{(N)}{}_N$, $N = 1, 2, 3, \ldots$, the **diagonal entries**; hence, the diagonal entry of $d^{(1)}$ is $d^{(1)}{}_1$, and of $d^{(2)}$ is $d^{(2)}{}_2$. Define d^* as follows:

$$d^*{}_0 = 0 \text{ and for } N \geq 1, d^*{}_N = \begin{cases} 1, & \text{if } d^{(N)}{}_N \neq 1; \\ 2, & \text{if } d^{(N)}{}_N = 1. \end{cases}$$

Thus, we have defined d^* in such a way that for $N = 1, 2, 3, \ldots$, each $d^*{}_N$ differs from the diagonal entry of $d^{(N)}$. Because disagreement at a single entry is sufficient to assure that two sequences are distinct, we see that $d^* \neq d^{(1)}$, $d^* \neq d^{(2)}, \ldots, d^* \neq d^{(N)}, \ldots$. Consequently, d^* is not listed among the elements of $\mathscr{D}$. But surely $d^* \in \mathscr{D}$ because $d^*{}_0 = 0$ and for $N = 1, 2, 3, \ldots, d^*{}_N$ is 1 or 2. We have contradicted the assumption that there exists a one-one correspondence between ω_+ and $\mathscr{D}$. Thus, $\mathscr{D}$ is an infinite nondenumerable set.

Corollary. **U** *is nondenumerable.*

EXAMPLE 4.6.4. We illustrate the proof of Theorem 4.6.2. Suppose that the enumeration (determined by g) begins as follows:

$$d^{(1)} = \langle 0, 3, 4, 7, 0, 1, 8, \ldots \rangle,$$

$$d^{(2)} = \langle 0, 2, 1, 5, 0, 8, 9, \ldots \rangle,$$

$$d^{(3)} = \langle 0, 0, 1, 2, 6, 9, 9, \ldots \rangle,$$

$$d^{(4)} = \langle 0, 6, 3, 6, 0, 6, 5, \ldots \rangle,$$

$$d^{(5)} = \langle 0, 1, 1, 1, 1, 1, 1, \ldots \rangle.$$

Then $d^{(1)}{}_1 = 3$, $d^{(2)}{}_2 = 1$, $d^{(3)}{}_3 = 2$, $d^{(4)}{}_4 = 0$, and $d^{(5)}{}_5 = 1$. Consequently, d^* is the sequence that begins $\langle 0, 1, 2, 1, 1, 2, \ldots \rangle$.

Theorem 4.6.3. *$\mathscr{R}$ is in one-one correspondence with* **U** *(and hence with $\mathscr{D}$). Thus, $\mathscr{R}$ is nondenumerable.*

PROOF: Let $\mathbf{U}^* = \mathbf{U} \sim \{1\}$. We define a function $h_1 : \mathbf{U}^* \to \mathscr{R}$ as follows:

$$h_1(\Gamma) = \begin{cases} (1/(2\Gamma)) - 1, & \text{if } 0 < \Gamma \leq 1/2; \\ 1 - 1/(2 - 2\Gamma), & \text{if } 1/2 < \Gamma < 1. \end{cases}$$

Then it is a routine algebraic exercise to establish that h_1 is a one-one correspondence between $\mathbf{U}^*$ and $\mathscr{R}$.

Next, the function $h_2 : \mathbf{U} \to \mathbf{U}^*$—defined by

$$h_2(1/N) = 1/N^+, \qquad N = 1, 2, 3, \ldots ;$$

$$h_2(\Gamma) = \Gamma, \qquad \text{if } \Gamma \neq 1/N \text{ for any } N = 1, 2, 3, \ldots$$

—is one-one onto $\mathbf{U}^*$.

Finally, $h_1 \circ h_2 : \mathbf{U} \to \mathscr{R}$ is the desired one-one correspondence. (The function h_1 is "graphed" in Figure 4.6.1.)

Corollary. *$\mathscr{R}$ is nondenumerable.*

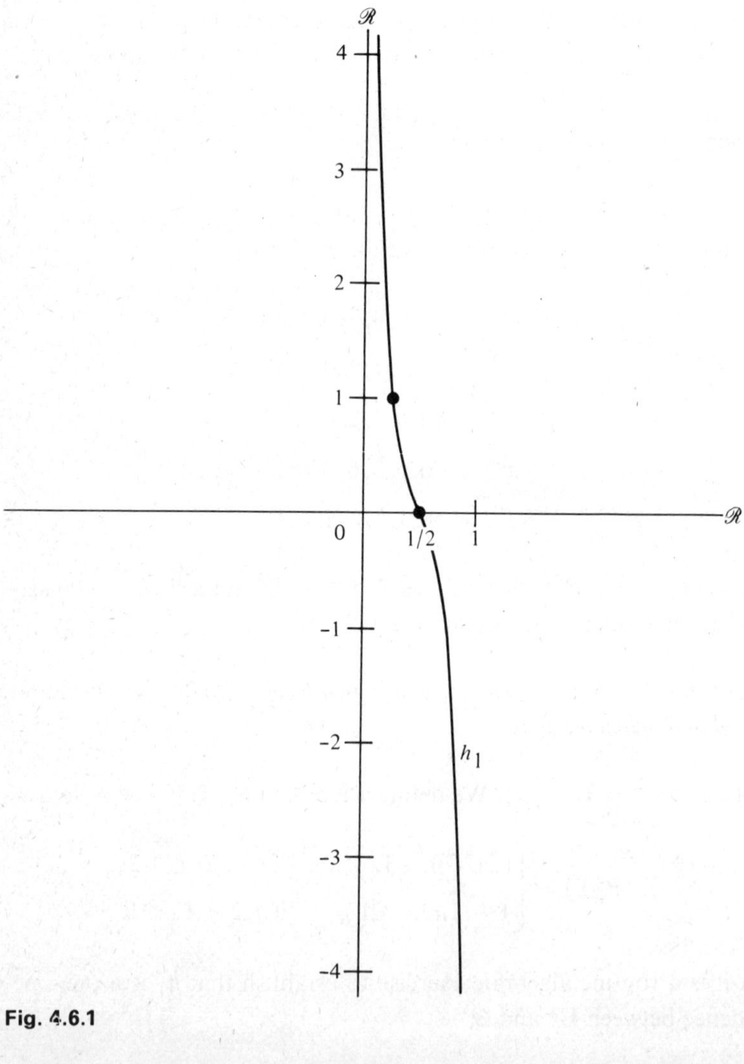

Fig. 4.6.1

We next show that there are essentially as many points on the line as on the (complex) plane.

Theorem 4.6.4. *$\mathscr{C}$ is in one-one correspondence with $\mathscr{R}$.*

PROOF: We establish this correspondence by considering one-one mappings of $\mathscr{C}$ onto $\mathscr{D}^{②}$, $\mathscr{D}^{②}$ onto $\mathscr{D}$, $\mathscr{D}$ onto $\mathscr{R}$, as in Figure 4.6.2.

$$\mathscr{C} \overset{G_1}{\to} \mathscr{D}^{②} \overset{G_2}{\to} \mathscr{D} \overset{G_3}{\to} \mathscr{R}.$$

Fig. 4.6.2

The second of these is the key mapping; we first consider the other two.

Let g_1 be a one-one mapping of $\mathscr{R}$ onto $\mathscr{D}$. (The existence of such a map is guaranteed by the preceding theorem.) Then g_1^{-1} $(=G_3)$ is a one-one mapping of $\mathscr{D}$ onto $\mathscr{R}$. Second, $\mathscr{C} = \{\langle\Gamma, \Delta\rangle : \Gamma, \Delta \in \mathscr{R}\}$; define $G_1 : \mathscr{C} \to \mathscr{D}^{②}$ by $G_1(\langle\Gamma, \Delta\rangle) = \langle g_1(\Gamma), g_1(\Delta)\rangle$. Then G_1 is a one-one mapping of $\mathscr{C}$ onto $\mathscr{D}^{②}$.

We now define a mapping of $\mathscr{D}^{②}$ into $\mathscr{D}$. Let d and e be arbitrary sequences of $\mathscr{D}$. We must assign a unique sequence $d^{\#}$ in $\mathscr{D}$ to each such $\langle d, e\rangle$ in such a way that each sequence in $\mathscr{D}$ gets assigned. Ostensibly, we could do this by interspersing the terms of d and e (one at a time), i.e., by assigning to $\langle d, e\rangle$ the sequence $\langle d_0, e_0, d_1, e_1, d_2, e_2, \ldots\rangle$. Clearly, this sequence is an element of $\mathscr{D}$. The difficulty lies in the fact that not every sequence of $\mathscr{D}$ is obtained in this way. In the first place we would only obtain sequences whose second term is 0. More crucially, sequences of $\mathscr{D}$ such as $\langle 0, 2, 7, 0, 1, 0, 1, 0, 1, \ldots\rangle$ could not be obtained because all odd terms from a certain index onward are 0. (The sequence e contributes the odd terms; if $e_N = 0$, there is some $M > N$ with $e_M > 0$.)

In order to avoid the first difficulty we simply define $d^{\#}{}_0$ to be 0 and begin the interspersion with $d^{\#}{}_1 = d_1$. Regarding the second difficulty, we agree to intersperse (finite) sequences of terms rather than individual terms. Following the use of any nonzero term d_K, when we next choose terms of d, we select all terms $d_{K+1}, d_{K+2}, \ldots, d_{K+L}$ until the first nonzero term has been selected. [Conceivably, $L = 1$ here; in any event, condition (4.6.3) guarantees that such an L exists.] The same criterion for the selection of blocks of terms applies to e.

We illustrate with an example.

Let $d = \langle 0, 4, 0, 0, 0, 3, 0, 1, 0, 0, 2, 7, \ldots\rangle$ and let $e = \langle 0, 3, 1, 4, 0, 4, 1, 1, 1, 0, 4, 8, \ldots\rangle$. Then $d^{\#}$ (corresponding to $\langle d, e\rangle$) will be a sequence beginning with $\langle 0, 4, 3, 0, 0, 0, 3, 1, 0, 1, 4, 0, 0, 2, 0, 4, 7, 1, \ldots\rangle$; i.e., $\langle 0, d_1, e_1, d_2, d_3, d_4, d_5, e_2, d_6, d_7, e_3, d_8, d_9, d_{10}, e_4, e_5, d_{11}, e_6, \ldots\rangle$.

More precisely, we define a mapping $G_2 : \mathscr{D}^{②} \to \mathscr{D}$ as follows: We first enumerate the nonzero terms of d and of e; thus, for $K = 1, 2, \ldots,$ we let

d_{N_K} and e_{M_K} be, respectively, the Kth nonzero terms of d and of e. We then let $G_2(\langle d, e\rangle)$ be the sequence $d^\#$ defined by

$$d^\#_{N_1} = d_{N_1},$$
$$d^\#_{N_K+M_K} = e_{M_K},$$
$$d^\#_{N_K{}^+ + M_K} = d_{N_K{}^+},$$
$$d^\#_L = 0,$$

if L is different from any of the above subscripts. This mapping G_2 is well-defined (by induction) and is one-one onto $\mathscr{D}$ (Exercise 4.6.6).

Finally, $G_3 \circ G_2 \circ G_1$ is a one-one mapping of $\mathscr{C}$ onto $\mathscr{R}$.

Corollary. *$\mathscr{C}$ is nondenumerable.*

EXAMPLE 4.6.5. Let $d = \langle 0, 4, 0, 0, 0, 3, 0, 1, 0, 0, 2, \ldots\rangle$ and $e = \langle 0, 3, 1, 4, 0, 4, 1, 1, 1, 0, 4, \ldots\rangle$ be as above. Then

$N_1 = 1,$	$d_{N_1} = 4,$	$M_1 = 1,$	$e_{M_1} = 3,$
$N_2 = 5,$	$d_{N_2} = 3,$	$M_2 = 2,$	$e_{M_2} = 1,$
$N_3 = 7,$	$d_{N_3} = 1,$	$M_3 = 3,$	$e_{M_3} = 4,$
$N_4 = 10,$	$d_{N_4} = 2,$	$M_4 = 5,$	$e_{M_4} = 4,$
		$M_5 = 6,$	$e_{M_5} = 1,$
		$M_6 = 7,$	$e_{M_6} = 1,$
		$M_7 = 8,$	$e_{M_7} = 1,$
		$M_8 = 10,$	$e_{M_8} = 4;$

$$d^\#_1 = d^\#_{N_1} = d_{N_1} = 4;$$
$$d^\#_2 = d^\#_{N_1+M_1} = e_{M_1} = 3;$$
$$d^\#_6 = d^\#_{N_2+M_1} = d_{N_2} = 3;$$
$$d^\#_7 = d^\#_{N_2+M_2} = e_{M_2} = 1;$$
$$d^\#_9 = d^\#_{N_3+M_2} = d_{N_3} = 1; \text{ etc.}$$

Note that $d^\#_0 = d^\#_3 = d^\#_4 = d^\#_5 = d^\#_8 = 0$. Thus, $d^\#$ (corresponding to $\langle d, e\rangle$) begins with

$$\langle 0, 4, 3, 0, 0, 0, 3, 1, 0, 1, \ldots\rangle.$$

Exercises

†4.6.1. Prove that the collection $\mathscr{D}$ of all infinite sequences $\langle d_0, d_1, d_2, \ldots, d_N, \ldots\rangle$—for which conditions (4.6.1), (4.6.2), and (4.6.3) hold—is a set.

†4.6.2. Show that $\leq_{\mathscr{D}}$ is a linear ordering on $\mathscr{D}$.

†4.6.3. Show that if $\rho < \sigma < 1$ (where ρ and σ are $\mathscr{R}$-rationals), then for $K = 1, 2, 3, \ldots,$ there are $\mathscr{R}$-rationals δ_K—of the form $\sum_{I=1}^{K} e_I/10^I$, where $e_I \in \mathscr{T}$ for $I = 1, 2, \ldots, K$—for which $\rho \leq \delta_K < \sigma$.

★4.6.4. Consider the proof of Theorem 4.6.1.
- **(a)** Show that for each d in $\mathscr{D}$, $f(d)$ is in **U**.
- **(b)** Use Definition by Induction to define α_N and β_N, $N = 1, 2, 3, \ldots.$
- **(c)** Let δ be as in the last paragraph of the proof. Use Definition by Induction to show that the sequence d, determined by δ, is well-defined.
- **(d)** Suppose that f is as in part (a), $\{L, U\}$ is any real number in **U**, and δ and d are sequences determined (as in the last paragraph of the proof) by $\{L, U\}$. Prove $f(d) = \{L, U\}$.

4.6.5. Let $\mathbf{U}^* = \mathbf{U} \sim \{1\}$. Determine a "trigonometric" function that establishes a one-one correspondence between $\mathbf{U}^*$ and $\mathscr{R}$.

4.6.6. Consider Theorem 4.6.4.
- **(a)** Show that the mapping G_2 can be defined by induction.
- **(b)** Show that G_2 is one-one onto $\mathscr{D}$.

4.6.7. Let $\mathscr{I}$ be the set of irrational numbers. Prove that $\mathscr{I}$ is in one-one correspondence with $\mathscr{R}$. (Useful for Exercises 6.3.9 and 6.5.5.)

4.7 Some Equivalents of the Axiom of Choice

We have been able to generalize the Cartesian product of sets so as to allow N factors for $N \geq 2$. We now extend the notion so as to allow infinitely many factors.

Consider the case of $N = 2$. The Cartesian product of the sets A and B, distinct or not, is the set $A \times B = \{\langle a, b\rangle : a \in A \text{ and } b \in B\}$. Let Z be the set of all functions f that are such that $f: 2 \to A \cup B$, $f(0) \in A$, and $f(1) \in B$. (Clearly, if at least one of the sets A or B is empty, $Z = \varnothing$.) There is a "natural" one–one correspondence, F, between Z and $A \times B$, defined as follows: For f in Z, let

$$F(f) = \langle a, b\rangle, \qquad \text{if } f(0) = a \text{ and } f(1) = b.$$

Thus, there would be no difficulty if we were to have defined Z, rather than $A \times B$, as the Cartesian product of A and B.

When we consider the case of "infinitely many" factors there are several reasons for preferring the latter approach. For well-ordered infinite sets it will be possible to extend the notion of "N-tuple" (in Chapter 5)—but (without assuming the Axiom of Choice—see Section 4.8) there may be nonwell-orderable sets. Moreover, the possibility of infinite repetition of a factor causes difficulties.

We say that $\langle I, f\rangle$ is an **indexing pair** for a set X if f is a function with domain I and range X. If $\langle I,f\rangle$ is an indexing pair for X, then for each x in X, $f^{-1}(x)$ serves as an "index for x". In other words, we can write x as "x_i" whenever $f(i) = x$. We first remark that *for every set X, $\langle X, I_X\rangle$ is always an indexing pair for X*; thus, *every set has an indexing pair.* (Recall that I_X is the identity function on X.) With respect to $\langle X, I_X\rangle$, each x in X indexes itself—i.e., $x_x = x$—because $I_X(x) = x$. Note that when we enumerated the elements of a countable set by means of the natural numbers to which they corresponded (under a fixed one–one correspondence, f) we were actually using $\langle N,f\rangle$ or $\langle \omega, f\rangle$ as an indexing pair.

If $\langle I,f\rangle$ is an indexing pair for X and if f is not one–one, then some x in X will have several indices; this is often useful, as our first example indicates.

†EXAMPLE 4.7.1. Let $X = \{3, 7\}$.

(1) Let $I_1 = \omega$. Let $f_1 : \omega \to X$ be such that

$$f_1(N) = \begin{cases} 3, & \text{if } N \text{ is even;} \\ 7, & \text{if } N \text{ is odd.} \end{cases}$$

Then $\langle I_1, f_1\rangle$ is an indexing pair for X. Each odd natural number is an index for 3 and each even natural number is an index for 7. If we want to indicate a denumerable alternation of 3's and 7's, $\langle I_1, f_1\rangle$ would be a reasonable indexing pair for X.

(2) Let $I_2 = 2$, let $f_2(0) = 3$, and let $f_2(1) = 7$. Then (with respect to I_2 and f_2) 0 is an index for 3 and 1 is an index for 7.

Let X be a set and let $\langle I,f\rangle$ be an indexing pair for X. Then each x in X is of the form x_i for at least one i in I. We define the (**generalized**) **Cartesian product of** X, **indexed by** $\langle I, f\rangle$, written "$\bigtimes_{I,f} X$" as

$$\text{"}\{g :\ \ g : I \to \cup X \text{ and } g(i) \in x_i \text{ for each } i \text{ in } I\}\text{"}.$$

EXAMPLE 4.7.2. Let X; I_1, f_1, I_2, f_2 be as in Example 4.7.1. Then $\cup X = 7$.

(1) $\bigtimes_{I_1,f_1} X = \{g :\ \ g : \omega \to 7 \text{ and } g(2K) < 3 \text{ for each } K \text{ in } \omega\}$;
if $g(0) = g(1) = \cdots = g(10) = 2, g(2K) = 1$ for $K \geq 6$ and $g(2K + 1) = 5$ for $K \geq 5$, then $g \in \bigtimes_{I_1,f_1} X$.

(2) $\bigtimes_{I_2,f_2} X = \{g: \ g:2 \to 7 \text{ and } g(0) < 3\}$. Clearly, $\bigtimes_{I_2,f_2} X$ has $3 \cdot 7$ elements.

†EXAMPLE 4.7.3. Suppose that we want to consider $A \times B$ in terms of the generalized Cartesian product. Let $X = \{A, B\}$ and $I = 2$; define

$$f:2 \to \cup X$$

(where $\cup X = A \cup B$), so that $f(0) = A\,(= x_0)$ and $f(1) = B\,(= x_1)$. Then $\bigtimes_{I,f} X = \{g: \ g:2 \to A \cup B, g(0) \in A, \text{ and } g(1) \in B\}$.

Observe that $\langle \varnothing, \varnothing \rangle$ is an indexing pair for $\varnothing$ and that $\bigtimes_{\varnothing,\varnothing} \varnothing = \{\varnothing\} = 1$. On the other hand, it is obvious that if $\varnothing \in X$, then for any indexing pair $\langle I, f \rangle$ for X, $\bigtimes_{I,f} X = \varnothing$. The definition of Cartesian product is reminiscent of that of the statement of the Axiom of Choice. Thus, it is not too surprising that the nonemptiness of $\bigtimes_{I,f} X$ is related to the Axiom of Choice.

There are many sentences of set theory that can be shown to be equivalent to the Axiom of Choice. Each such sentence, s, is true if we assume the Axiom of Choice (together with the other axioms of set theory). Moreover, if we take s to be an axiom of set theory instead of the Axiom of Choice, then we can prove the Axiom of Choice (by means of s together with the other axioms of set theory). Note that we cannot prove either s or its negation on the basis of the axioms introduced in Chapter 1, for, as we previously remarked, we cannot prove either the Axiom of Choice or its negation on the basis of these axioms.

We now consider some equivalents of the Axiom of Choice, each of which essentially asserts the existence of a function. (Thus, the existence of the function in question cannot be guaranteed on the basis of the previous axioms.) We first list the various sentences. Recall that *X is a partition of $\cup X$ iff X is a nonempty set of nonempty pairwise disjoint sets.*

Axiom of Choice (Choice): *Let X be a nonempty set of nonempty sets. Then there is a function F mapping X into $\cup X$ with the property that $F(Y) \in Y$ for each Y in X.*

Product Rule (Prod.): *Let X be a nonempty set of nonempty sets. Then with respect to any indexing pair $\langle I, f \rangle$ for X, $\bigtimes_{I,f} X \neq \varnothing$.*

Multiplicative Principle (Mult.): Let X be a partition of $\cup X$. *Then there exists a set A with the property that $A \cap X$ is a unit set for each Y in X.*

Disjoint Form of the Axiom of Choice (Disj. Choice): *Let X be a partition of $\cup X$. Then there is a function F mapping X into $\cup X$ with the property that $F(Y) \in Y$ for each Y in X.*

Baer's Axiom (Baer): *Let R be a binary relation. Then there is a function F that is such that $F \subseteq R$ and dom F = dom R.*

Theorem 4.7.1

(1) *The Product Rule,*

(2) *The Disjoint Form of the Axiom of Choice,*

(3) *The Multiplicative Principle, and*

(4) *Baer's Axiom*

are all equivalents of the Axiom of Choice.

PROOF: Figure 4.7.1 illustrates our procedure in proving the required equivalences.

$$\begin{array}{ccc} \text{(Choice)} \rightarrow \text{(Prod.)} \rightarrow \text{(Mult.)} \\ \uparrow \qquad\qquad\qquad\quad \downarrow \\ \text{(Baer)} \leftarrow \text{(Disj. Choice)} \end{array}$$

Fig. 4.7.1

(*Choice*) → (*Prod.*): Let X be a nonempty set of nonempty sets and let $\langle I, f \rangle$ be an indexing pair for X. By the Axiom of Choice, there is a function F that is such that $F : X \rightarrow \cup X$ and $F(Y) \in Y$ for each Y in X. But then $F \circ f \in \bigtimes_{I,f} X$.

(*Prod.*) → (*Mult.*): Let X be a partition of $\cup X$. By the Product Rule, $\bigtimes_{X,I_X} X \neq \varnothing$. Let $f \in \bigtimes_{X,I_X} X$ and let A = range f. Then for each Y in X, $f(Y) \in Y$. Moreover, if $Y' (\neq Y) \in X$, then $f(Y') \in Y'$; hence, $f(Y') \notin Y$ because f is a function and $Y \cap Y' = \varnothing$. Thus, $A \cap Y = \{f(Y)\}$.

(*Mult.*) → (*Disj. Choice*): Let X be a partition of $\cup X$ and let A be a set for which $A \cap Y$ is a unit set, say $\{a_Y\}$, for each Y in X. Let

$$F = \{u \in X \times \cup X : \text{for some } Y \text{ in } X, u = \langle Y, a_Y \rangle\}.$$

Clearly, F is a choice function on X.

(*Disj. Choice*) → (*Baer*): Let R be a binary relation. If $R = \varnothing$, we can take $F = R$. Otherwise, for each a in dom R, let $Y_a = \{\langle a, b \rangle \in R\}$. The collection of Y_a, for a in dom R is a set; let $X = \{Y_a : a \in \text{dom } R\}$. Then X is a partition of $\cup X$ (= R). Let $f : X \rightarrow \cup X$ be such that $f(Y_a) \in Y_a$ for each Y_a in X. Define F

as follows: For each a in dom R, let $F(a)$ be the second coordinate of $f(Y_a)$. Then F is a function; moreover, dom F = dom R. If $\langle a, F(a)\rangle \in F$, then $\langle a, F(a)\rangle = f(Y_a) \in Y_a \subseteq R$.

(Baer) $\rightarrow$ *(Choice)*: Let X be a nonempty set of nonempty sets. Let $R = \{\langle Y, y\rangle \in X \times \cup X : y \in Y\}$. By Baer's Axiom, there exists a function F for which dom F = dom $R = X$ and $F \subseteq R$. Thus, for each Y in X, $\langle Y, F(Y)\rangle \in R$; hence, $F(Y) \in Y$.

In the next section and in Chapters 5 and 6, we will consider other equivalents of the Axiom of Choice.

Exercises

4.7.1. Let a, b, c, d, e, and f be sets, let $A = \{a, b, c\}$, $B = \{d, e, f\}$, $C = \{a, c, e\}$, $D = \{d\}$, and let $X = \{A, B, C, D\}$. Let $f: 6 \rightarrow X$ be defined by $f(0) = f(4) = A$, $f(1) = f(2) = B$, $f(3) = C$, and $f(5) = D$.

- **(a)** Show that $\langle 6, f\rangle$ is an indexing pair for X.
- **(b)** Determine all indices for A, for B, for C, and for D (with respect to $\langle 6, f\rangle$).
- **(c)** Let $g: 6 \rightarrow \{a, b, c, d, e, f\}$ be defined by $g(0) = g(3) = g(4) = a$, $g(1) = g(2) = g(5) = d$. Show that $g \in \bigtimes_{6,f} X$.
- **(d)** How many elements are in $\bigtimes_{6,f} X$?

4.7.2. Let X be a unit set. Show that the collection of indexing pairs for X is not a set.

4.7.3. Prove, directly, that the Axiom of Choice is equivalent to the Disjoint Form of the Axiom of Choice.

4.7.4. In the proof of (Disj. Choice) $\rightarrow$ (Baer), show that the collection of Y_a, $a \in$ dom R, is a set.

4.7.5. Prove that the following sentence is equivalent to the Axiom of Choice: For every function f there is a function g that is such that dom g = dom f and if $x \in$ dom f and $f(x) \neq \varnothing$, then $g(x) \in f(x)$.

4.8 Well-ordering Theorem

The Well-ordering Theorem asserts that every set is well-orderable. This theorem, which utilizes the Axiom of Choice, was first proved by Zermelo in 1904; the proof we will give is due to Devidé (1965). Because the proof is somewhat lengthy, we will break it up into several parts. The Well-ordering Theorem is another equivalent of the Axiom of Choice.

Let S be a nonempty set and suppose that F is a choice function on $\mathscr{P}(S) \sim \{\varnothing\}$. For $Z\,(\neq \varnothing) \subseteq S$, let $Z' = Z \sim \{F(Z)\}$. A subset $\mathscr{A}$ of $\mathscr{P}(S)$ is said to be **distinguished (with respect to** F**)** if the following are true:

(4.8.1) $S \in \mathscr{A}$.
(4.8.2) If $Z\,(\neq \varnothing) \in \mathscr{A}$, then $Z' \in \mathscr{A}$.
(4.8.3) If $\mathscr{B}\,(\neq \varnothing) \subseteq \mathscr{A}$, then $\cap\mathscr{B} \in \mathscr{A}$.

It is easily established that if $\mathscr{A}$ is a distinguished subset of $\mathscr{P}(S)$, then $\varnothing \in \mathscr{A}$.

†EXAMPLE 4.8.1. For every nonempty set A of natural numbers, let N_A be the ($\leq$-) first element of A. Then $F = \{\langle A, N_A\rangle\} : A \in \mathscr{P}(\omega) \sim \{\varnothing\}\}$ is a choice function on $\mathscr{P}(\omega) \sim \{\varnothing\}$. Let $\mathscr{A} = \{\omega \sim N : N \in \omega\} \cup \{\varnothing\}$. We show that $\mathscr{A}$ satisfies conditions (4.8.1), (4.8.2), and (4.8.3) with respect to F.
(4.8.1) $\omega = \omega \sim 0 \in \mathscr{A}$.
(4.8.2) If $Z\,(\neq \varnothing) \in \mathscr{A}$, then for some N, $Z = \omega \sim N$; therefore, $Z' = Z \sim \{N_Z\} = \omega \sim N^+ \in \mathscr{A}$.
(4.8.3) Suppose that $\mathscr{B}\,(\neq \varnothing) \subseteq \mathscr{A}$. If $\varnothing \in \mathscr{B}$, then $\cap\mathscr{B} = \varnothing \in \mathscr{A}$. Otherwise, each B in $\mathscr{B}$ is of the form $\omega \sim N_B$. If $\{N_B : B \in \mathscr{B}\}$ is bounded, it must have a maximum, $N_{\mathscr{B}}$. Then $\cap\mathscr{B} = \omega \sim N_{\mathscr{B}}$. If $\{N_B : B \in \mathscr{B}\}$ is unbounded, then $\cap\mathscr{B} = \varnothing \in \mathscr{A}$.
Thus, $\mathscr{A}$ is distinguished.

Theorem 4.8.1. *Let S be a nonempty set for which there exists a choice function, F, on $\mathscr{P}(S) \sim \{\varnothing\}$. Then there is a smallest distinguished subset of $\mathscr{P}(S)$.*

PROOF: Clearly $\mathscr{P}(S)$ is distinguished; therefore,

$$\mathscr{B} = \{\mathscr{A} \in \mathscr{P}\mathscr{P}(S) : \mathscr{A} \text{ is distinguished}\}$$

is nonempty. Let $\mathscr{A}^* = \cap\mathscr{B}$. For each $\mathscr{A}$ in $\mathscr{B}$, $S \in \mathscr{A}$. Therefore, $S \in \mathscr{A}^*$. Similarly, $\mathscr{A}^*$ satisfies (4.8.2) and (4.8.3), above; hence, $\mathscr{A}^*$ is distinguished. If $\mathscr{A}$ is distinguished, then $\mathscr{A}^* \subseteq \mathscr{A}$; thus, $\mathscr{A}^*$ is the smallest distinguished subset of $\mathscr{P}(S)$.

Assume that S is a nonempty set for which there exists a choice function, F, on $\mathscr{P}(S) \sim \{\varnothing\}$. Let $\mathscr{A}^*$ be the set whose existence was just established in Theorem 4.8.1 [i.e., let $\mathscr{A}^*$ be the smallest distinguished subset of $\mathscr{P}(S)$].

We will call an element X of $\mathscr{A}^*$ **comparable** if for every Y in $\mathscr{A}^*$, $X \subseteq Y$ or $Y \subseteq X$; otherwise, we will say that X is **incomparable**. An element Z of $\mathscr{A}^*$ will be called **superincomparable** if Z includes every incomparable element of $\mathscr{A}^*$. Clearly, *a superincomparable set Z is comparable*; for otherwise, there would be a set Y in $\mathscr{A}^*$ such that $Y \nsubseteq Z$ and $Z \nsubseteq Y$. But this would say that Y is incomparable, and hence that $Y \subseteq Z$.

EXAMPLE 4.8.2. Let F be the choice function on $\omega \sim N$ discussed in Example 4.8.1. Then it is easily established, by induction on N, that each set $\omega \sim N$ is an element of every distinguished subset of $\mathscr{P}(\omega)$, as is $\varnothing$. Hence, the distinguished subset $\mathscr{A}$ $(= \{\omega \sim N : N \in \omega\} \cup \{\varnothing\})$ is, in fact, the smallest such distinguished subset, $\mathscr{A}^*$. Note that every element of $\mathscr{A}^*$ $(= \mathscr{A})$ is comparable. We will soon see that this is no coincidence.

Theorem 4.8.2. *Let S be a nonempty set and suppose that F is a choice function on $\mathscr{P}(S) \sim \{\varnothing\}$. Let $\mathscr{A}^*$ be the smallest distinguished subset of $\mathscr{P}(S)$.*
(1) *S is superincomparable.*
(2) $T = \cap\{Z \in \mathscr{A}^* : Z \textit{ is superincomparable}\} \in \mathscr{A}^*$.
(3) *T is superincomparable.*

PROOF: (1) S includes all elements of $\mathscr{A}^*$—hence, S includes all incomparable elements of $\mathscr{A}^*$.

(2) By (1), $\{Z \in \mathscr{A}^* : Z \text{ is superincomparable}\} \neq \varnothing$; thus, (2) follows from condition (4.8.3).

(3) If Y is incomparable and Z is superincomparable, then $Y \subseteq Z$; hence, $Y \subseteq T = \cap\{Z \in \mathscr{A}^* : Z \text{ is superincomparable}\}$. Therefore, T is superincomparable.

Theorem 4.8.3. *Let S be a nonempty set and suppose that F is a choice function on $\mathscr{P}(S) \sim \{\varnothing\}$. Then $\mathscr{A}^*$ is linearly ordered by inclusion ($\mathscr{I}_{\mathscr{A}^*}$).*

PROOF: Let T be as in Theorem 4.8.2, and let

(4.8.4) $$\mathscr{A}^\# = \{Z \in \mathscr{A}^* : \text{either } T \subseteq Z \text{ or both } T \neq \varnothing \text{ and } Z \subseteq T'\}.$$

We claim that $\mathscr{A}^\#$ is distinguished (from which it will follow that $\mathscr{A}^\# = \mathscr{A}^*$).

Clearly, $T \subseteq S$; thus, (4.8.1) is true of $\mathscr{A}^\#$. Suppose that $Z\,(\neq \varnothing) \in \mathscr{A}^\#$; we wish to show that $Z' \in \mathscr{A}^\#$. We consider three cases; in the first two, we assume that $T \neq \varnothing$.

(1) $Z \subseteq T'$. Then $Z' \subset Z \subseteq T'$.
(2) $Z = T$. Then $Z' = T'$.
(3) $T \subset Z$. If $F(Z) \notin T$, then $T \subset Z'$. Otherwise, $Z'\,(= Z \sim \{F(Z)\}) \nsubseteq T$ and $T \nsubseteq Z'$. Thus, Z' is incomparable; by Theorem 4.8.2(3), $Z' \subseteq T$. (Contradiction.)

Let $\mathscr{B}\ (\neq \varnothing) \subseteq \mathscr{A}^{\#}$. If for some B in $\mathscr{B}$, $B \subseteq T'$, then $\cap\mathscr{B} \subseteq B \subseteq T'$. Otherwise, $T \subseteq B$ for every B in $\mathscr{B}$. Thus, $T \subseteq \cap\mathscr{B}$.

Now that we have shown that $\mathscr{A}^{\#}$ is distinguished (and hence that $\mathscr{A}^{\#} = \mathscr{A}^*$), it follows from (4.8.4) that T' is comparable. Thus, suppose there exists an incomparable element, Y, in $\mathscr{A}^*$. Then because T is superincomparable (hence also comparable), $Y \subset T$. Because $T' = T \sim \{F(T)\}$ is comparable, $Y \subset T'$. It follows that T' is superincomparable and thus that $T \subseteq T'$. This contradiction establishes that every two elements of $\mathscr{A}^*$ are $\mathscr{I}_{\mathscr{A}^*}$-comparable. Because $\mathscr{I}_X$ is a partial ordering on $\mathscr{P}(X)$ for every set X, the theorem follows.

As a consequence of the theorem, note that every element of $\mathscr{A}^*$ is comparable. Thus the set T, of Theorems 4.8.2 and 4.8.3, is empty because $\varnothing$ is superincomparable.

Theorem 4.8.4. *Let S be a nonempty set and suppose that F is a choice function on $\mathscr{P}(S) \sim \{\varnothing\}$. Then $\mathscr{A}^*$ is well-ordered by $\mathscr{I}_{\mathscr{A}^*}{}^{-1}$.*

PROOF: The inverse of a linear ordering is a linear ordering; thus, we need only show that every nonempty subset, $\mathscr{V}$, of $\mathscr{A}^*$ has a first element with respect to "including".

Let $\mathscr{V}\ (\neq \varnothing) \subseteq \mathscr{A}^*$. If $\mathscr{V} = \{\varnothing\}$, then $\varnothing$ is the first element of $\mathscr{V}$ (with respect to $\mathscr{I}_{\mathscr{A}^*}{}^{-1}$). Otherwise, let

$$\mathscr{W} = \{Z \in \mathscr{A}^* : \text{if } V \in \mathscr{V}, \text{ then } V \subseteq Z\}.$$

Clearly, $S \in \mathscr{W}$; let $V_0 = \cap\mathscr{W}$. Then $V_0 \neq \varnothing$. (Why?) Also, $V_0 \in \mathscr{A}^*$, by condition (4.8.3); by condition (4.8.2), $V_0{}' \in \mathscr{A}^*$. Suppose that $V_0 \notin \mathscr{V}$. Then $V \in \mathscr{V}$ implies $V \subseteq V_0{}'$. (Verify.) Therefore, $V_0{}' \in \mathscr{W}$; hence, $V_0 \subseteq V_0{}'$. This contradiction guarantees that $V_0 \in \mathscr{V}$; surely V_0 is the first element of $\mathscr{V}$ with respect to $\mathscr{I}_{\mathscr{A}^*}{}^{-1}$.

It follows that under the hypotheses of the preceding theorem, every subset of $\mathscr{A}^*$ is well-orderable.

Theorem 4.8.5. The **Well-ordering Theorem** *(A.C.) Every set is well-orderable.*

PROOF: $\varnothing$ is well-orderable.

Let S be a nonempty set, let F be a choice function on $\mathscr{P}(S) \sim \{\varnothing\}$, and let $\mathscr{A}^*$ be the smallest distinguished subset of $\mathscr{P}(S)$. Define $G : S \to \mathscr{A}^*$ as follows: For s in S, let

$$\mathscr{A}_s = \{Z \in \mathscr{A}^* : s \in Z\}.$$

(Clearly, $S \in \mathscr{A}_s$ for all s in S.) Let $G(s) = \cap\mathscr{A}_s$; thus $s \in G(s)$. It suffices to show that G is one–one, for then S is in one–one correspondence with a well-orderable set. By Exercise 2.4.16, we need only show that $F \circ G(s) = s$ for all s in S. Otherwise, for some s in S, $s \in (G(s))'$. Therefore, $(G(s))' \in \mathscr{A}_s$; hence, $G(s) \subseteq (G(s))'$. (Contradiction.)

Theorem 4.8.6. *The Well-ordering Theorem implies the Axiom of Choice.*

(The proof will be left as an exercise.)

Exercises

4.8.1. Let S be a nonempty set and suppose that F is a choice function on $\mathscr{P}(S) \sim \{\varnothing\}$. Let $\mathscr{A}^*$ be the smallest distinguished subset of $\mathscr{P}(S)$. Prove that $\varnothing \in \mathscr{A}^*$.

4.8.2. In Example 4.8.1, determine a distinguished subset of $\mathscr{P}(\omega)$ other than $\mathscr{A}$ and $\mathscr{P}(\omega)$.

4.8.3. Let S be a nonempty finite set and let F be a choice function on $\mathscr{P}(S) \sim \{\varnothing\}$. Describe $\mathscr{A}^*$.

4.8.4. In the proof of Theorem 4.8.4, show the following:

(a) $V_0 \neq \varnothing$.

(b) If $V_0 \notin \mathscr{V}$ and $V \in \mathscr{V}$, then $V \subseteq V_0'$.

4.8.5. Prove Theorem 4.8.6; thus, prove that the Well-ordering Theorem implies the Axiom of Choice.

5 Ordinal Numbers

5.1 Similarity Mappings

Recall that a well-ordering on a set Y is a linear ordering on Y with respect to which every nonempty subset of Y has a first element.

Suppose that R is a linear ordering on a set X. We called a subset A of X an initial segment of X (with respect to R) or an initial segment of $\langle X, R\rangle$ if whenever $x_0 \in A$, all elements of X that are R-related to x_0 (i.e., that satisfy xRx_0) are also in A. Initial segments of X other than X itself were called proper initial segments. We defined a section of X (with respect to R)—or a section of $\langle X, R\rangle$—to be a subset A of X that is such that with respect to some x in X, $A = \{a \in X : aRx \text{ and } a \neq x\}$. We denoted such subsets A by "$\text{sect}_R(x)$". We observed that sections of X are always proper initial segments. The main theorem that we proved (in Section 2.7) on well-orderings stated that every proper initial segment of X is also a section of X iff R is a well-ordering.

We now define two linearly ordered systems $\langle X_1, R_1\rangle$ and $\langle X_2, R_2\rangle$ to be **similar** if there is a one–one mapping F of X_1 onto X_2 that is such that for all x, y in X_1, xR_1y iff $F(x)R_2F(y)$. In this case F is called a **similarity mapping**; we write "$\langle X_1, R_1\rangle \cong \langle X_2, R_2\rangle$". We will sometimes say that "F is a similarity mapping of $\langle X_1, R_1\rangle$ onto $\langle X_2, R_2\rangle$." (When the relations R_1 and R_2 are understood, we speak of a similarity mapping of X_1 onto X_2; we also say that X_1 and X_2 are similar, and we write "$X_1 \cong X_2$".)

EXAMPLE 5.1.1. With respect to the natural orderings, the (first three) embedding isomorphisms discussed in Chapter 3 are similarity mappings:

(1) Let $F : \omega \to \mathscr{Z}$ be defined by $F(N) = \overline{N}$, $N \in \omega$. Then F is a similarity mapping of the (set of) natural numbers onto the (set of) nonnegative integers.

(2) Let $G : \mathscr{Z} \to \mathscr{Q}$ be defined by $G(\mathscr{M}) = [\langle \mathscr{M}, \overline{1}\rangle]$, $\mathscr{M} \in \mathscr{Z}$. Then G is a similarity mapping of $\mathscr{Z}$ onto a subset of the rationals.

(3) Let $H : \mathscr{Q} \to \mathscr{R}$ be defined by $H(\alpha) = \{L_\alpha, U_\alpha\}$, where for each α in $\mathscr{Q}$, $L_\alpha = \{\beta \in \mathscr{Q} : \beta <_{\mathscr{Q}} \alpha\}$ and $U_\alpha = \mathscr{Q} \sim L_\alpha$. Then H is a similarity mapping of the rationals onto a subset of the reals.

(4) Furthermore, let $J : \mathscr{R} \to \mathscr{C}$ be defined by $J(\Gamma) = \langle \Gamma, \overset{\equiv}{0}\rangle$, $\Gamma \in \mathscr{R}$. Then J is a similarity mapping of the reals onto a subset of the complex numbers. Here, we define an order relation R on range J as follows:

$$\langle \Gamma_1, \overset{\equiv}{0}\rangle R \langle \Gamma_2, \overset{\equiv}{0}\rangle \text{ iff } \Gamma_1 \leq_{\mathscr{R}} \Gamma_2.$$

Part (4) illustrates the following method of "generating" a similarity map. If $\langle X, R_X\rangle$ is a linearly ordered system and if F is a one-one correspondence of X onto a set Y, define R_X on Y as follows: For y_1, y_2 in Y, let

$$y_1 R_Y y_2 \text{ iff } F^{-1}(y_1) R_X F^{-1}(y_2).$$

(See Exercise 5.1.1.)

Clearly, *every linearly ordered system is similar to itself.* Moreover, *let* $\langle X_I, R_I\rangle$, $I = 1, 2, 3$, *be linearly ordered systems. If* F_1 *is a similarity mapping of* $\langle X_1, R_1\rangle$ *onto* $\langle X_2, R_2\rangle$, then F_1^{-1} *is a similarity mapping of* $\langle X_2, R_2\rangle$ *onto* $\langle X_1, R_1\rangle$; *if* in addition, F_2 *is a similarity mapping of* $\langle X_2, R_2\rangle$ *onto* $\langle X_3, R_3\rangle$, *then* $F_2 \circ F_1$ *is a similarity mapping of* $\langle X_1, R_1\rangle$ *onto* $\langle X_3, R_3\rangle$ (Exercise 5.1.2). Thus, *if* $\mathscr{L}$ *is any set of linearly ordered systems, the relations* S *on* $\mathscr{L}$*—defined by* $\langle X_1, R_1\rangle S \langle X_2, R_2\rangle$ *iff* $\langle X_1, R_1\rangle$ *is similar to* $\langle X_2, X_2\rangle$*—is an equivalence relation on* $\mathscr{L}$.

If R_1 *and* R_2 *are distinct linear orderings* (*or even well-orderings*) *on the same set* X, $\langle X, R_1\rangle$ *and* $\langle X, R_2\rangle$ *may or may not be similar. They will always be similar if* X *is finite* (Exercise 5.1.5).

EXAMPLE 5.1.2. Let R_1 be the natural ordering on 3 and let R_2 be the well-ordering on 3 indicated by 2, 0, 1. Then $\langle 3, R_1 \rangle$ and $\langle 3, R_2 \rangle$ are similar; the similarity mapping F is defined by $F(0) = 2, F(1) = 0, F(2) = 1$; $\langle 3, R_1 \rangle$ is also similar to $\langle 3, R_1^{-1} \rangle$; here, the similarity mapping G is defined by $G(0) = 2$, $G(1) = 1$, $G(2) = 0$.

†EXAMPLE 5.1.3

(1) Let R be the well-ordering on ω indicated by

$$1, 2, 3, \ldots, N, N^+, \ldots; 0.$$

There is no similarity mapping of the well-ordered system $\langle \omega, R \rangle$ onto $\langle \omega, \leq \rangle$ because ω has a last element with respect to R—but no last element with respect to $\leq$.

(2) Let S be the well-ordering of ω indicated by

$$5, 4, 3, 2, 1, 0, 6, 7, 8, \ldots, N, N^+, \ldots.$$

Clearly, $\langle \omega, S \rangle \cong \langle \omega, \leq \rangle$.

We now concentrate on similarity mappings defined on well-ordered systems. The next example illustrates that a well-ordered system can be similar to a proper subsystem of itself.

EXAMPLE 5.1.4

(1) Let $F : \omega \to \omega_+$ be defined by $F(N) = N^+$, for N in ω. Then F is a similarity mapping of $\langle \omega, \leq \rangle$ onto $\langle \omega_+, \leq \rangle$.

(2) Let $G : \omega \to \omega \sim \{5\}$ be defined by

$$G(N) = \begin{cases} N, & N < 5; \\ N^+, & N \geq 5. \end{cases}$$

Then G is also a similarity mapping of $\langle \omega, \leq \rangle$ onto a proper subsystem of itself.

Note that in each of the mappings of the preceding example, no element "moves backward". We now show that this is characteristic of similarity mappings of well-ordered systems onto subsystems.

Theorem 5.1.1. *Let F be a similarity mapping of a well-ordered system $\langle X, R \rangle$ onto its subsystem $\langle Y, R \rangle$. Then for every x in X, $xRF(x)$.*

PROOF: Otherwise, $A = \{x \in X : x \not R F(x)\} = \{x \in X : F(x) R x \text{ and } F(x) \neq x\} \neq \varnothing$; let x_0 be its first element. Then

(5.1.1) $$F(x_0) R x_0 \text{ and } F(x_0) \neq x_0;$$

because F is a similarity mapping with domain X,

$$F(F(x_0)) R F(x_0) \text{ and } F(F(x_0)) \neq F(x_0).$$

Thus, $F(x_0) \in A$; according to (5.1.1), x_0 cannot be the first element of A.

Various limitations on similarity mappings result from this theorem. We first restrict the subsystems to which a well-ordered system can be similar.

Corollary 1. *A well-ordered system $\langle X, R\rangle$ is not similar to any of its sections, $\langle B, R\rangle$.*

PROOF: Let $B = \text{sect}_R(x)$ for some x in X. If F is a similarity mapping of $\langle X, R\rangle$ onto $\langle B, R\rangle$, then $F(x) \in B$. Thus, $F(x) R x$ and $F(x) \neq x$, in contradiction of the theorem.

Next, we exclude all but the obvious similarity mapping of a well-ordered system onto itself.

Corollary 2. *Let F be a similarity mapping of a well-ordered system $\langle X, R\rangle$ onto itself. Then $F = I_X$.*

PROOF: Otherwise, $C = \{x \in X : F(x) \neq x\} \neq \varnothing$; let $c \in C$. According to the theorem,

$$c R F(c) \text{ and } c \neq F(c).$$

But F^{-1} is also a similarity mapping of $\langle X, R\rangle$ onto itself; hence,

$$F^{-1}(c) R c \text{ and } F^{-1}(c) \neq c.$$

This contradicts the theorem.

Finally, we restrict the similarity mappings between two well-ordered systems.

Corollary 3. *Let $\langle X, R\rangle$ and $\langle Y, S\rangle$ be well-ordered systems. There is at most one similarity mapping of $\langle X, R\rangle$ onto $\langle Y, S\rangle$.*

PROOF: If F and G were distinct similarity mappings of $\langle X, R\rangle$ onto $\langle Y, S\rangle$, then $G^{-1} \circ F$ would be a similarity mapping of $\langle X, R\rangle$ onto itself. But $G^{-1} \circ F \neq I_X$.

Corollary 4. *A well-ordered system cannot be similar to two distinct initial segments of the same well-ordered system.*

(The proof is similar to that of Corollary 3; it is left as an exercise.)

Our next goal, the Comparability Theorem for well-ordered systems, is an affirmative statement; it asserts that either two well-ordered systems are similar or else one is similar to a subsystem of the other. We must first prove the following "lemma".

Recall that a set $\mathscr{F}$ of functions is said to be nested if $f_1, f_2 \in \mathscr{F}$ implies either $f_1 \subseteq f_2$ or $f_2 \subseteq f_1$.

Theorem 5.1.2. *Let $\langle X, R\rangle$ and $\langle Y, S\rangle$ be any two well-ordered systems. Let $\mathscr{F}$ be the set of similarity mappings from initial segments of $\langle X, R\rangle$ onto initial segments of $\langle Y, S\rangle$.*
(1) *If $f \in \mathscr{F}$ and if A is an initial segment of dom f, then $f \upharpoonright A \in \mathscr{F}$.*
(2) *$\mathscr{F}$ is nested.*
(3) *$\cup\mathscr{F}$ is the "largest" element of $\mathscr{F}$ (with respect to inclusion).*

PROOF: (1) Clearly, A (= dom $f \upharpoonright A$) is an initial segment of X. We must show that range $f \upharpoonright A$ is an initial segment of Y. Otherwise, there are y_1 in range $f \upharpoonright A$ and y_0 in $Y \sim$ (range $f \upharpoonright A$) for which $y_0 S y_1$. Let x_1 be the unique argument in A for which $f(x_1) = y_1$. Then for all x in (dom f) $\sim A$, $x_1 R x$; hence, $y_1 S f(x)$ and $y_1 \neq f(x)$. It follows that $y_0 \notin$ range f. This contradicts the hypothesis that $f \in \mathscr{F}$.

(2) Let $f_1, f_2 \in \mathscr{F}$. Dom f_1 and dom f_2 are each initial segments of X; thus, one must be a subset of the other. Suppose that dom $f_1 \subseteq$ dom f_2. Then $f_2 \upharpoonright$ dom $f_1 \in \mathscr{F}$, by part (1). Thus, f_1 and $f_2 \upharpoonright$ dom f_1 are each similarity mappings of $\langle$dom $f_1, R\rangle$ onto initial segments of $\langle Y, S\rangle$. By Corollary 4 of Theorem 5.1.1, $f_1 = f_2 \upharpoonright$ dom f_1; i.e., $f_1 \subseteq f_2$.

(3) By Exercise 2.4.15, $\cup\mathscr{F}$ is a one–one function, dom $\cup\mathscr{F} = \cup\{\text{dom } f : f \in \mathscr{F}\}$, and range $\cup\mathscr{F} = \cup\{\text{range } f : f \in \mathscr{F}\}$. One easily verifies the following:

(a) Dom $\cup\mathscr{F}$ is an initial segment of X.
(b) Range $\cup\mathscr{F}$ is an initial segment of Y.
(c) $\cup\mathscr{F}$ is a similarity mapping.

Hence, $\cup\mathscr{F} \in \mathscr{F}$. For g in $\mathscr{F}$, g $\subseteq \cup\mathscr{F}$. Therefore, $\cup\mathscr{F}$ is the largest element of $\mathscr{F}$.

Theorem 5.1.3. The Comparability Theorem for Well-ordered Systems. *Two well-ordered systems* $\langle X, R\rangle$ *and* $\langle Y, S\rangle$ *are similar or (excl.) exactly one of these systems is similar to a section of the other; in the latter case the section, in question, is unique.*

PROOF: Suppose, arbitrarily, that $\langle Y, S\rangle$ is not similar to any initial segment of $\langle X, R\rangle$. Let $\mathscr{F}$ be the set of similarity mappings from initial segments of $\langle X, R\rangle$ onto initial segments of $\langle Y, S\rangle$. By Theorem 5.1.2, $\cup\mathscr{F} \in \mathscr{F}$ and $\cup\mathscr{F}$ includes all elements of $\mathscr{F}$. Moreover, dom $\cup\mathscr{F}$ and range $\cup\mathscr{F}$ are initial segments of X and Y, respectively. Thus, range $\cup\mathscr{F}$ must be a proper initial segment (i.e., a section) of Y. Let y_0 be the first element of $Y \sim$ Range $\cup\mathscr{F}$ (with respect to S). Suppose that dom $\cup\mathscr{F} \neq X$; let x_0 be the first element of $X \sim$ dom $\cup\mathscr{F}$ (with respect to R). It then follows that $\cup\mathscr{F} \cup \{\langle x_0, y_0\rangle\}$ is an element of $\mathscr{F}$ that properly includes $\cup\mathscr{F}$. This cannot be; thus $\cup\mathscr{F}$ must be a similarity mapping of X onto a section of Y.

Corollary 4 of Theorem 5.1.1 guarantees the uniqueness of the section of Y; it also assures that if $\langle X, R\rangle$ and $\langle Y, S\rangle$ are similar, then neither is similar to a section of the other. If both $\langle X, R\rangle$ were similar to a section of $\langle Y, S\rangle$ and $\langle Y, S\rangle$ were similar to a section of $\langle X, R\rangle$, then $\langle X, R\rangle$ would be similar to its own section, in contradiction of Corollary 1 of Theorem 5.1.1.

Corollary. *One of two distinct initial segments of the same well-ordered system is properly included in the other.*

Exercises

†5.1.1. (a) Let $\langle X, R_X\rangle$ be a linearly ordered system, let F map X one-one onto a set Y, and define R_Y on Y by

$$y_1 R_Y y_2 \text{ iff } F^{-1}(y_1) R_X F^{-1}(y_2),\ y_1, y_2 \in Y.$$

Prove that R_Y linearly orders Y.

(b) Prove that if the linearly ordered systems $\langle X_1, R_1\rangle$ and $\langle X_2, R_2\rangle$ are similar, then X_1 is well-ordered by R_1 iff X_2 is well-ordered by R_2.

†5.1.2. Let $\langle X_1, R_1\rangle$, $\langle X_2, R_2\rangle$, and $\langle X_3, R_3\rangle$ be linearly ordered systems. Prove that if F_1 is a similarity mapping of $\langle X_1, R_1\rangle$ onto $\langle X_2, R_2\rangle$ and F_2 is a similarity mapping of $\langle X_2, R_2\rangle$ onto $\langle X_3, R_3\rangle$, then the following are true:

(a) F_1^{-1} is a similarity mapping of $\langle X_2, R_2\rangle$ onto $\langle X_1, R_1\rangle$.

(b) $F_2 \circ F_1$ is a similarity mapping of $\langle X_1, R_1\rangle$ onto $\langle X_3, R_3\rangle$.

5.1.3. In Theorem 5.1.1 and in each of its corollaries, replace each occurrence of "well-ordered" by "linearly ordered". Determine counterexamples.

5.1.4. Determine an example of a similarity mapping between a well-ordered system $\langle X, R\rangle$ and a subsystem $\langle Y, R\rangle$ in which both Y and $X \sim Y$ are infinite.

†5.1.5. **(a)** Prove that if R_1 and R_2 are any well-orderings on the same finite set X, then $\langle X, R_1\rangle$ and $\langle X, R_2\rangle$ are similar.

(b) Is (a) true if we replace "finite" by "denumerable"?

5.1.6. Prove Corollary 4 of Theorem 5.1.1.

5.1.7. In the proof of part (3) of Theorem 5.1.2, show the following:

(a) Dom $\cup\mathscr{F}$ is an initial segment of X.

(b) Range $\cup\mathscr{F}$ is an initial segment of Y.

(c) $\cup\mathscr{F}$ is a similarity mapping.

†5.1.8. Suppose the well-ordered systems $\langle X, R\rangle$ and $\langle Y, S\rangle$ are similar. Let A be a subset of a proper initial segment of Y (with respect to S). Prove that $\langle X, R\rangle \not\cong \langle A, S\rangle$.

5.2 Ordinal Numbers

The magnitude of finite sets can be expressed in terms of natural numbers. Each natural number N is well-ordered by $\leq \cap N^{②}$, which we will denote by "$\leq_N$". However, as we mentioned in the preceding section, the well-ordering is not of primary significance in the sense that for every N-element set A and linear ordering R on A, (R is a well-ordering) and $\langle A, R\rangle \cong \langle N, \leq_N\rangle$.

We now extend the concept of "natural number" into the realm of the transfinite. In this chapter we will study ordinal numbers, which express both the magnitude and (up to similarity) a specific well-ordering on a well-ordered set. We will introduce a new axiom (Substitution) by means of which we will be able to show that every well-ordered system corresponds to a unique ordinal number. Furthermore, two well-ordered systems $\langle A, R\rangle$ and $\langle B, S\rangle$ will correspond to the same ordinal number if and only if they are similar. In Chapter 6 we will consider other "transfinite" generalizations of these concepts. Much of the basic work in transfinite arithmetic is due to Cantor, Sierpiński, and Tarski.

Cantor formulated the concept of ordinal numbers in connection with his studies of trigonometric series. In Section 5.12 we will consider some applications of ordinal numbers in analysis.

In preaxiomatic set theory an ordinal number is usually introduced as an "equivalence class consisting of all sets that are similar to a given well-ordered set X". (Of course, the similarity is understood to be with respect to a specific well-ordering on X.) The difficulty with this approach is that such an equivalence class is not a set if $X \neq \varnothing$. (See Exercise 5.2.1.) In our "axiomatic" approach we introduce ordinal number systems as prototype well-ordered systems that have the property that every well-ordered system is similar to one such prototype. Our definition of "ordinal number" will be such that the finite ordinals are the natural numbers. Natural numbers X are well-ordered by the relation R_X defined by xR_Xy, for x, y in X, iff $x \in y$ or $x = y$; such a relation, R_X, will also serve to well-order ordinal numbers X.

For every set X, let $\in_X$ be the relation,

$$\{\langle x, y\rangle \in X^{②} : x \in y \text{ or } x = y\}.$$

Thus, for arbitrary sets x, y, and X, $x \in_X y$ iff $x, y \in X$, and either $x \in y$ or $x = y$. We always have $I_X \subseteq \in_X$; thus, $\in_X$ is always *reflexive*.

†EXAMPLE 5.2.1

(1) $\in_3 = \{\langle 0, 0\rangle, \langle 0, 1\rangle, \langle 0, 2\rangle, \langle 1, 1\rangle, \langle 1, 2\rangle, \langle 2, 2\rangle\}$
$= \{\langle 0, 1\rangle, \langle 0, 2\rangle, \langle 1, 2\rangle\} \cup I_3$.

(2) Let $X = \{1, 4, 8, 9\}$. Then
$\in_X = \{\langle 1, 4\rangle, \langle 1, 8\rangle, \langle 1, 9\rangle, \langle 4, 8\rangle, \langle 4, 9\rangle, \langle 8, 9\rangle\} \cup I_X$.

(3) Let $Y = \{1, \{1\}, \omega\}$. Then
$\in_Y = \{\langle 1, \{1\}\rangle, \langle 1, \omega\rangle\} \cup I_Y$.

(4) Let $Z = \{\{1\}, \{2\}\}$. Then $\in_Z = I_Z$.

A set X is called an **ordinal (number)** if $\in_X$ well-orders X and

(5.2.1) for each x in X, $x = \text{sect}_{\in_X}(x)$.

If α is an ordinal number, we will refer to $\langle \alpha, \in_\alpha\rangle$ as an **ordinal system**; hereafter, we will write this as "$\langle \alpha, \in\rangle$".

EXAMPLE 5.2.2

(1) 0 is an ordinal number.

(2) $3\,(=\{0, 1, 2\})$ is an ordinal number. In fact, $\in_3$ well-orders 3 in such a way that $0 = \text{sect}_{\in_3}(0)$, $1 = \{0\} = \text{sect}_{\in_3}(1)$, and $2 = \{0, 1\} = \text{sect}_{\in_3}(2)$. Thus property (5.2.1) is true for $X = 3$.

(3) ω is well-ordered by $\in_\omega$ (which is the same as $\leq$). Moreover, $0 = \text{sect}_{\in_\omega}(0)$; if $N\,(\neq 0) \in \omega$, then for some unique K in ω, $N = K^+ = \{0, 1, 2, \ldots, K\} = \text{sect}_{\in_\omega}(K^+)$. Thus, ω is an ordinal.

(4) $X = \{0, 1, 3\}$, although well-ordered by $\in_X$, is not an ordinal because $3 = \{0, 1, 2\} \neq \{0, 1\} = \text{sect}_{\in_X}(3)$.

(5) Let $Y = \{0, 1, \{1\}\}$. Here, $\{1\} = \text{sect}_{\in_Y}(\{1\})$; hence, for each y in Y, $y = \text{sect}_{\in_Y}(y)$. However, $\in_Y$ is not even a linear ordering on Y because $0 \notin \{1\}$ and $\{1\} \notin 0$. Thus, Y is not an ordinal.

Our first theorem shows that we need not have specified the well-ordering in the definition of an ordinal number.

Theorem 5.2.1. *A set X is an ordinal number if and only if there is a well-ordering R on X that is such that*

(5.2.2) *for each x in X, $x = \text{sect}_R(x)$.*

PROOF: By definition, if X is an ordinal, then $\in_X$ is such a well-ordering on X.

Let R be an arbitrary well-ordering on a set X with the stipulated property (5.2.2). Let $x, y\,(\neq x) \in X$. Then xRy iff $x \in \text{sect}_R(y)$. But $\text{sect}_R(y) = y$; thus xRy iff $x \in y$. It follows that $R = \in_X$ and that X is an ordinal.

We have shown that *the only well-ordering R on a set X that can have property* (5.5.2) *is* $\in_X$. Recall that for an arbitrary set X,

$$\mathscr{I}_X = \{\langle x, y\rangle \in X^{②} : x \subseteq y\}.$$

We now show that *whenever X is an ordinal then*, in fact, $\in_X$ *is the same as the inclusion relation on* $X, \mathscr{I}_X$. This is true of natural numbers and of ω; it is not true of arbitrary sets.

†EXAMPLE 5.2.3. If Example 5.2.1, $\mathscr{I}_3 = \in_3$, $\mathscr{I}_X = \in_X$, and $\mathscr{I}_Z = \in_Z$. However, in part (3), $Y = \{1, \{1\}, \omega\}$, and $\mathscr{I}_Y = \{\langle 1, \omega\rangle, \langle\{1\}, \omega\rangle\} \cup I_Y \neq \in_Y = \{\langle 1, \omega\rangle, \langle 1, \{1\}\rangle\} \cup I_Y$.

Theorem 5.2.2. *A set X is an ordinal number if and only if $\mathscr{I}_X$ well-orders X in such a way that*

(5.2.3) *for each x in X, $x = \text{sect}_{\mathscr{I}_X}(x)$.*

PROOF: If $\mathscr{I}_X$ well-orders X in accordance with property (5.2.3), then by the preceding theorem, X is an ordinal (and $\mathscr{I}_X = \in_X$).

Now suppose X is an ordinal number. According to the definition, in order to show that $\mathscr{I}_X$ well-orders X and has property (5.2.3), it suffices to show that $\mathscr{I}_X = \in_X$. Thus, let x and y be distinct elements of X; we must show that $x \in y$ iff $x \subset y$.

Suppose that $x \in y$. Then $x \in \text{sect}_{\in_X}(y)\,(= y)$. Because $x \notin \text{sect}_{\in_X}(x)$, either $\text{sect}_{\in_X}(x) \subset \text{sect}_{\in_X}(y)$ or (excl.) $\text{sect}_{\in_X}(y) \subset \text{sect}_{\in_X}(x)$. The first alternative must be the case; thus, $x = \text{sect}_{\in_X}(x) \subset \text{sect}_{\in_X}(y) = y$.

Suppose that $x \notin y$. Then because $\in_X$ is a linear ordering, $y \in x$. The above argument, with the roles of x and y interchanged, yields $y \subset x$. Hence, $x \not\subset y$.

Corollary. *If X is an ordinal number, then* $\mathscr{I}_X = \in_X$.

Note that the converse of this corollary is false; see Examples 5.2.1(4) and 5.2.3.

Many statements about natural numbers are true with "ordinal" substituted for "natural".

Theorem 5.2.3. *Every ordinal number is ordinary.*

PROOF: $\alpha \notin \text{sect}_{\in_\alpha}(\alpha) = \alpha$.

Theorem 5.2.4. *Every ordinal number is transitive.*

PROOF: We must show that if α is an ordinal number and if $x \in \alpha$, then $x \subseteq \alpha$. Let $x \in \alpha$. For any element y in x, $y \in x = \text{sect}_{\in_\alpha}(x) \subset \alpha$. Thus, $y \in \alpha$, and hence $x \subseteq \alpha$, as was to be shown.

Corollary 1. *Let α be an ordinal number. If $x \in \alpha$, then $x \subset \alpha$.*

PROOF: By Theorem 5.2.4, if $x \in \alpha$, then $x \subseteq \alpha$; by Theorem 5.2.3, if $x \in \alpha$, then $x \neq \alpha$.

Corollary 2. *Every element of an ordinal number is transitive.*

PROOF: Let α be an ordinal number; let $x \in y$, $y \in z$, $z \in \alpha$. By the above, $y \in \alpha$ and $x \in \alpha$. Thus, $x \in_\alpha y \in_\alpha z$; because $\in_\alpha$ is a well-ordering (and is, consequently, a transitive relation), it follows that $x \in_\alpha z$. Thus, $x \in z$, and z is a transitive set.

Theorem 5.2.5. *For distinct ordinal numbers α and β, either $\alpha \in \beta$ or (excl.) $\beta \in \alpha$.*

PROOF: Ordinals are transitive and ordinary; thus, not both $\alpha \in \beta$ and $\beta \in \alpha$.

One of these ordinal systems is similar to an initial segment of the other; we can assume that F is a similarity mapping of $\langle \alpha, \in \rangle$ onto an initial segment of $\langle \beta, \in \rangle$. Let $A = \{a \in \alpha : F(a) = a\}$. Suppose that for x in α, $\text{sect}_{\in_\alpha}(x)\ (= x) \subseteq A$. By Theorem 5.1.2(1), $F \upharpoonright x\ (= I_x)$ is a similarity mapping onto an initial segment—hence onto a section—of β. Thus, $x\ (= \text{range } F \upharpoonright x) \in \beta$. If $F(x) \neq x$, then because $F(x) \notin \text{sect}_{\in_\beta}(x) = x$, it follows that $x \in F(x)$. Because F is order-preserving, $x \notin \text{range } F$. But then range F is not a section of β. Thus, $F(x) = x$ and $x \in A$. By the Principle of Transfinite Induction, $A = \alpha$. Thus, $F = I_\alpha$ and α $(= \text{range } F)$ is a section of β. Therefore, $\alpha \in \beta$.

Corollary 1. The Law of Trichotomy for Ordinal Numbers. *For every two ordinal numbers α and β exactly one of the following holds:*

$$\alpha \in \beta;\ \alpha = \beta;\ \beta \in \alpha.$$

Corollary 2. *For distinct ordinal numbers α and β, either $\alpha \subset \beta$ or (excl.) $\beta \subset \alpha$.*

Theorem 5.2.6. *Every element of an ordinal number is an ordinal number.*

PROOF: Let α be an ordinal number and let $a \in \alpha$. Then $a \subset \alpha$; consequently, $\in_a = \in_\alpha \cap a^{②}$. Thus, $\in_a$ well-orders a.

Let $x \in a$; then $\text{sect}_{\in_a}(x) \subseteq \text{sect}_{\in_\alpha}(x) = x$. If $y \in x \in a$, then by the transitivity of a, $y \in a$. Therefore, $y \in_a x$, and $y \in \text{sect}_{\in_a}(x)$. Thus, $x = \text{sect}_{\in_a}(x)$. Hence, a is an ordinal.

For further reference we note a fact that was used in the proof of the preceding theorem.

Corollary 1. *If α is an ordinal number and if $\beta \in \alpha$, then $\in_\beta = \in_\alpha \cap \beta^{②}$.*

Corollary 2. *An ordinal number is a transitive set of ordinal numbers.*

We next consider the "converse" of Corollary 2.

Theorem 5.2.7. *A transitive set of ordinal numbers is an ordinal number.*

PROOF: Let A be a transitive set of ordinals. By definition, $\in_A$ is reflexive; it is antisymmetric, by Theorem 5.2.5; it is transitive, by Theorem 5.2.4. Thus, $\in_A$ is a partial ordering. Let $B\ (\neq \varnothing) \subseteq A$ and let $\beta \in B$. If $\beta \cap B = \varnothing$, then, by Theorem 5.2.5, β is the first element of B; otherwise, the first element of $\beta \cap B$ is the first element of B. Thus, $\in_A$ is a well-ordering. Furthermore, the transitivity of A implies that for α in A, $\text{sect}_{\in_A}(\alpha) = \alpha$. Thus, A is an ordinal.

Corollary 1. *If α is an ordinal number, then α^+ is an ordinal number.*

Corollary 2. *Every natural number is an ordinal number.*

Corollary 3. *If A is a set of ordinal numbers, then $\cup A$ is an ordinal number; moreover, $(\cup A)^+$ is an ordinal number that includes A. In particular, if α is an ordinal number, then $\cup \alpha$ is an ordinal number.*

Corollary 4. *ω is an ordinal number.*

Corollary 5. *Each "iterated successor of ω"—*

$$\omega^+, \omega^{++}, \omega^{+++}, \ldots$$

—is an ordinal number.

Corollary 6. *The collection $\mathcal{O}$ consisting of all the ordinal numbers is not a set.*

(The proof of these corollaries is Exercise 5.2.10.)

We would like to apply Corollary 3 to the collection of the "iterated successors of ω" of Corollary 5; unfortunately, we cannot as yet guarantee that this is a set. We consider this problem in the next section.

We note that *the first four Peano Postulates have their analogues for ordinal numbers.* Thus, *let α and β be sets.*

(1) 0 *is an ordinal number.*
(2) *If α is an ordinal number, then α^+ is an ordinal number.*
(3) *For no ordinal number α does $\alpha^+ = 0$.*
(4) *If α and β are ordinal numbers and if $\alpha^+ = \beta^+$, then $\alpha = \beta$.*

The counterpart of Peano Postulate (5) (the Principle of Induction) for ordinal numbers would require a "set containing all ordinals". Corollary 6 of Theorem 5.2.7 implies that no such set can exist.

We can define inequalities for ordinal numbers, much as we did for natural numbers. Let "$\alpha <_{\mathcal{O}} \beta$" mean "$\alpha \in \beta$" and let "$\alpha \leq_{\mathcal{O}} \beta$" mean "$\alpha <_{\mathcal{O}} \beta$ or $\alpha = \beta$". Note that $\alpha \leq_{\mathcal{O}} \beta$ *iff* $\alpha \in_{\beta^+} \beta$. *The Law of Trichotomy for Ordinal Numbers can be reformulated to say that for any two ordinals α and β exactly one of the following holds:*

$$\alpha <_{\mathcal{O}} \beta; \qquad \alpha = \beta; \qquad \beta <_{\mathcal{O}} \alpha.$$

As usual, we define the symbols "$>_{\mathcal{O}}$" and "$\geq_{\mathcal{O}}$" in terms of "$<_{\mathcal{O}}$" and "$\leq_{\mathcal{O}}$".

If α and β are natural numbers, then $\alpha \leq_{\mathcal{O}} \beta$ iff $\alpha \leq \beta$. Although $\leq$ is a well-ordering on the (set of) natural numbers, we cannot even say that $\leq_{\mathcal{O}}$ is a relation on the ordinals. *The ordinals do not constitute a set*, and relations have only been defined on sets. However, *$\leq_{\mathcal{O}}$ determines a well-ordering on any set of ordinals.*

We agree to drop the subscript "$\mathcal{O}$" in this notation.

Theorem 5.2.8. *Let α and β be ordinal numbers. Let $\langle A, R_A \rangle$ and $\langle B, R_B \rangle$ be any relational systems that are such that $\langle A, R_A \rangle \cong \langle \alpha, \in \rangle$ and $\langle B, R_B \rangle \cong \langle \beta, \in \rangle$. Then $\alpha < \beta$ iff $\langle A, R_A \rangle$ is similar to a section of $\langle B, R_B \rangle$.*

(The proof is an exercise.)

Theorem 5.2.8 affords us a means of verifying ordinal inequalities. For if we can find "representative systems" $\langle A, R_A \rangle$ and $\langle B, R_B \rangle$ that are similar to $\langle \alpha, \in \rangle$ and $\langle \beta, \in \rangle$, respectively, and that are such that $\langle A, R_A \rangle$ is similar to a section of $\langle B, R_\beta \rangle$, then $\alpha < \beta$.

Our definition of an ordinal number is due to von Neumann; we conclude this section with another definition, due to R. Robinson.

A set A is said to be **well-founded** if every nonempty subset B of A contains an element b for which $b \cap B = \varnothing$.

†EXAMPLE 5.2.4

(1) Every natural number N is well-founded. Each nonempty subset B of N has a first element K (with respect to $\in_N$); if any element of K were in B, this would contradict the assertion that K is the first element of B.

(2) Similarly, every set of natural numbers is well-founded.

(3) $X = \{1, \{2\}\}$ is well-founded because the nonempty subsets of X are 1, $\{\{2\}\}$, and X.

(4) If a set Y is not ordinary, then Y is not well-founded. Here $Y \in Y$; hence, $\{Y\}$ is a nonempty subset of Y. However, Y is the unique element of $\{Y\}$ and $Y \in Y \cap \{Y\}$.

(*Note:* On the basis of our axioms we do not know whether or not any such nonordinary set Y exists. In fact, in Chapter 6 we will introduce an axiom that excludes such nonordinary sets.)

It follows from part (4) of Example 5.2.4 that *every well-founded set is ordinary. The elements of well-founded sets must also be ordinary* (see Exercise 5.2.13a).

Clearly, *every set X that is well-ordered by $\in_X$ is well-founded* because if B is a nonempty subset of X and if b is the first element of B, then $b \cap B = \varnothing$. Thus, *every ordinal number as well as every set of ordinal numbers is well-founded.* On the other hand, Example 5.2.4(3) shows that not every well-founded set X is well-ordered by $\in_X$. Thus, well-foundedness is the more general notion, and it is natural to seek the properties that when combined with well-foundedness assure that a set X is an ordinal.

Theorem 5.2.9. *A set X is an ordinal number if and only if it has the following properties:*

(1) *X is transitive.*

(2) *X is well-founded.*

(3) *Every two elements of X are $\in_X$-comparable.*

PROOF: We already know that every ordinal number has these properties.

Suppose that X is an arbitrary set with these three properties. By definition, $\in_X$ is reflexive. Suppose that x, y, and z are distinct elements of X. If $x \in y$ and $y \in x$, then $y \in \{x, y\} \cap x$ and $x \in \{x, y\} \cap y$; this would contradict the assumption that X is well-founded. Thus, $\in_X$ is antisymmetric. If $x \in y \in z$, then by property (3), either $x \in z$ or $z \in x$. Suppose that $z \in x$. Then $z \in \{x, y, z\} \cap x$, $x \in \{x, y, z\} \cap y$, $y \in \{x, y, z\} \cap z$. Again, we would contradict the well-foundedness of X. Hence, $x \in z$. It follows that $\in_X$ is a transitive relation and, consequently, that it is a partial ordering of X. Let B be any nonempty subset of X. By (2), there is some b_0 in B for which $b_0 \cap B = \varnothing$. Then b_0 is the first element of B with respect to $\in_X$, and X is well-ordered by $\in_X$. Finally, as in the proof of Theorem 5.2.7, the transitivity of X implies that for each x in X, $x = \text{sect}_{\in_X}(x)$. Consequently, X is an ordinal.

Other alternative definitions of an ordinal will be considered in the exercises.

Exercises

5.2.1. Let X be a unit set. Prove that the collection of sets similar to X (ordered by I_X) is not a set.

5.2.2. Let $X = \{1, 2, \{1, 2\}, 3, \omega\}$.

- **(a)** Determine $\in_X$.
- **(b)** Determine $\mathscr{I}_X$.
- **(c)** Is X well-ordered by $\in_X$? By $\mathscr{I}_X$?
- **(d)** Is X well-founded?
- **(e)** Is X an ordinal number?

5.2.3. Let $X = \{1, 2, 4, \{2\}, \{2, 4\}\}$. Determine the following sections:

- **(a)** $\text{sect}_{\in_X}(2)$.
- **(b)** $\text{sect}_{\mathscr{I}_X}(2)$.
- **(c)** $\text{sect}_{\in_X}(4)$.
- **(d)** $\text{sect}_{\mathscr{I}_X}(4)$.
- **(e)** $\text{sect}_{\in_X}(\{2\})$.
- **(f)** $\text{sect}_{\mathscr{I}_X}(\{2\})$.
- **(g)** $\text{sect}_{\in_X}(\{2, 4\})$.
- **(h)** $\text{sect}_{\mathscr{I}_X}(\{2, 4\})$.

5.2.4. Which of the following sets X are (1) well-ordered by $\in_X$, (2) well-ordered by $\mathscr{I}_X$, and (3) such that (5.2.1) is true?

- **(a)** $\{2, 4, 7\}$.
- **(b)** $\{1, \{1\}, \{\{1\}\}\}$.
- **(c)** $\{0, 1, \{1\}, 2, \{2\}\}$.
- **(d)** $\{1, \omega, \omega^+\}$.
- **(e)** $\{\omega^+, \omega^{++}, \{\omega\}\}$.

†5.2.5. Prove that for any set x and ordinal α, if $x \in \alpha$, then $x^+ \in \alpha^+$.

†5.2.6. Prove that every finite ordinal is a natural number.

5.2.7. Prove that $\mathscr{P}(\omega)$ is not an ordinal.

5.2.8. Let A be a set of ordinals. Prove that the following sentences are equivalent:

(a) A is transitive.

(b) For all ordinals α and β, if $\alpha \in A$ and $\beta \notin A$, then $\alpha < \beta$.

5.2.9. Determine a set A of ordinals that satisfies the following:

(1) Whenever $a \in A$, then $\alpha^+ \in A$.

(2) A is not an ordinal.

5.2.10. Prove the six corollaries of Theorem 5.2.7.

†5.2.11. Prove that every set of ordinals is properly included in an ordinal.

5.2.12. Prove Theorem 5.2.8.

5.2.13. Suppose that X is well-founded. Prove (a), (b), and (c).

(a) Every element of X is ordinary.

(b) X has no N-element subset $\{x_1, x_2, \ldots, x_N\}$ such that

$$x_1 \in x_2 \in x_3 \in \cdots \in x_N \in x_1.$$

(c) X has no denumerable subset $\{x_1, x_2, \ldots, x_N, \ldots\}$ such that

$$\ldots x_{N+} \in x_N \in \cdots \in x_3 \in x_2 \in x_1.$$

(d) Prove that there is no denumerable set of ordinals, $\alpha_1, \alpha_2, \ldots, a_N, \ldots$ satisfying $\cdots \alpha_{N+} < \alpha_N < \cdots < \alpha_3 < \alpha_2 < \alpha_1$.

(Used in Theorems 5.7.5 and 5.11.1 and in Exercise 6.1.12.)

5.2.14. Give an example of a set X that is transitive and well-founded and that has the following properties:

(a) Not every two elements of X are $\in_X$-comparable.

(b) Not every element of X is transitive.

★5.2.15. **(a)** Prove that a set X is an ordinal number[1] iff the following all hold:

(1) $0 \in X$ or $0 = X$.

(2) For every element x in X, either $x^+ \in X$ or $x^+ = X$.

(3) For every subset Y of X either $\cup Y \in X$ or $\cup Y = X$.

(b) Prove that a set X is an ordinal number iff (1′), (2′), (3′) hold, where each of these is obtained from (1), (2), (3), respectively, by replacing "or" by "or (excl.)".

5.2.16. **★(a)** Prove that a set X is an ordinal number iff every transitive proper subset of X is an element of X.

(b) Determine a proper subset of 3 that is not an element of 3.

(c) Determine a transitive subset of 3 that is not an element of 3.

★5.2.17. **(a)** Prove that a well-founded set X obeys the following **Generalized Induction Principle**: Let $A \subseteq X$ be such that for each x in X, $x \cap X \subseteq A$ implies $x \in A$. Then $A = X$.

(b) Prove that a set X is an ordinal number[2] iff the following all hold:

(1) X is transitive.

(2) X is well-founded.

(3) Every element of X is transitive.

---

[1] Zermelo's definition of an ordinal number.

[2] Gödel's definition of an ordinal number.

5.3 Axiom of Substitution

We have already commented upon the necessity of adding a new axiom in order to guarantee the existence of more ordinals. This axiom will also provide us with the means of demonstrating that every well-ordered set is similar to a (unique) ordinal.

Before introducing the Axiom of Substitution we pause to consider another collection that, intuitively, one would expect to be a set.

†EXAMPLE 5.3.1. We know that $\varnothing$ is a set; by the Pairing Axiom each of the following is also a set:

$$\varnothing, \{\varnothing\}, \{\{\varnothing\}\}, \{\{\{\varnothing\}\}\}, \{\{\{\{\varnothing\}\}\}\}, \text{etc.}$$

The collection, $\mathscr{J}$, consisting of all these sets is (intuitively speaking) in "one–one correspondence" with ω. Yet on the basis of our previous axioms we cannot guarantee that $\mathscr{J}$ is a set. For we know of no set X whose union is $\mathscr{J}$. Each element of $\mathscr{J}$ is an element of $\mathscr{P}^K(\omega)$, by which we mean $\underbrace{\mathscr{P}(\mathscr{P}(\cdots \mathscr{P}(\mathscr{P}}_{K \text{ such}}(\omega))\cdots))$. But again we cannot find a set that includes *all* these power sets.

Similarly, the Axioms of Subsets and Choice cannot readily be utilized, for each of these asserts the sethood of certain subcollections of an already known set.

We will speak of "variables" and "constants" instead of "set variables" and "set constants".

Let s be a formula of set theory that has exactly two free variables, x and y. Then s is said to be **bimonadic** if it has the following property: For all constants c_1, c_2, c_3, if "$s\begin{pmatrix} x & y \\ c_1 & c_2 \end{pmatrix}$" and "$s\begin{pmatrix} x & y \\ c_1 & c_3 \end{pmatrix}$" are both true, then c_2 and c_3 denote the same set. The formula s is called **strictly bimonadic** if s is bimonadic and if for every constant c_1 there is some constant c_2 for which "$s\begin{pmatrix} x & y \\ c_1 & c_2 \end{pmatrix}$" is true.

†EXAMPLE 5.3.2

(1) Let s_1 be "y is a natural number and $y = x^+$." Then s_1 is bimonadic. For if c_1 denotes a natural number, then $c_1{}^+$ denotes the unique set for which "$s_1\begin{pmatrix} x & y \\ c_1 & c_2 \end{pmatrix}$" is true. If c_1 denotes a set that is not a natural number, then "$s_1\begin{pmatrix} x & y \\ c_1 & c_2 \end{pmatrix}$" is false for every c_2.

(2) Let s_2 be "$y = x^+$". Then s_2 is strictly bimonadic.

(3) Let s_3 be "$x \in y$". Then s_3 is not bimonadic; "$s_3\begin{pmatrix} x & y \\ 0 & 1 \end{pmatrix}$" and "$s_3\begin{pmatrix} x & y \\ 0 & 2 \end{pmatrix}$" are both true.

Frequently, when no confusion seems likely, we will not distinguish between sets and the constants that denote them. (We, of course, assume that distinct sets are denoted by distinct constants. On the other hand, a set can be denoted by more than one constant.)

Axiom of Substitution:[3] *Let s be a bimonadic formula. Then for every set A there is a set B consisting of all sets b for which "$s\begin{pmatrix} x & y \\ a & b \end{pmatrix}$" is true for some a in A.*

With respect to a set A, the bimonadic formula s, in effect, defines a collection $\mathscr{F}$ of ordered pairs; $\mathscr{F}$ acts like a function in the sense that if $\langle c_1, c_2 \rangle$ and $\langle c_1, c_3 \rangle$ are both "in $\mathscr{F}$", then (the set denoted by) c_2 = (the set denoted by) c_3. Observe that the "domain" of this pseudofunction is a subset X of A. The Axiom of Substitution asserts that $\mathscr{F}''(A)$ $[=\mathscr{F}''(X)]$ is a set.

Let A be a set and let s be a *strictly* bimonadic formula. The Axiom of Substitution guarantees that the collection of c_2 corresponding to a in A—is a set, B. For a in A, we let b_a be the unique set corresponding to a (by means of the formula s), and we let $B = \{b_a : a \in A\}$. It then follows, by our previous axioms, that $\{\langle a, b_a \rangle : a \in A\}$ is a set and, hence, a function with domain A. But if we are given that F is a function (hence, a set) with domain A, then we do not need to use the Axiom of Substitution to show that range F is a set. (Why not?)

The Axiom of Substitution is used when the collection under consideration is not a subcollection of any known set. Thus, the Axiom of Substitution is used in cases where the Axiom of Subsets does not apply. In fact, it can be proved that the Axiom of Substitution implies the Axiom of Subsets (Exercise 5.3.7).

We now show how the Axiom of Substitution guarantees that the collection, $\mathscr{J}$, of Example 5.3.1, is a set. Observe that the elements of $\mathscr{J}$ are obtained from 0 by successively applying the "unit set operation".

Let x, y, and z be sets. We say that y is x-**linked to** z if there exists a set A satisfying the following:

(5.3.1) $z \in A$.

(5.3.2) If $a\,(\neq z) \in A$, then $a = \{b\}$ for some b in A.

(5.3.3) $y \in A$ and $\{y\} \notin A$.

(5.3.4) $x \in \omega$ and A is an x^+-element set.

[3] For a formulation of the Axiom of Substitution in a formal treatment see the Axiom of Replacement in (Cohen, pp. 52–53); see also (Levy).

†EXAMPLE 5.3.3

(1) Every set y is 0-linked to itself. Let $A = \{y\}$; A clearly satisfies conditions (5.3.1), (5.3.2), (5.3.3), and (5.3.4) with "0" in the role of "x" and "y" in lieu of "z".

(2) $\{\{0\}\}$ is 2-linked to 0: here, let $A = \{0, \{0\}, \{\{0\}\}\}$.

(3) 2 is not x-linked to 0 for any set x: If it were, there would have to be a set A containing 2 and containing some element b for which $2 = \{b\}$. But 2 is a 2-element set, whereas for every set b, $\{b\}$ is a unit set.

†EXAMPLE 5.3.4. Let s be the formula, "y is x-linked to 0." We leave as an exercise the verification that s is bimonadic. Then by the Axiom of Substitution, there is a set C consisting of all sets c_2 for which there is an N in ω such that "$s\begin{pmatrix} x & y \\ N & c_2 \end{pmatrix}$" is true. This set C contains 0, by Example 5.3.3(1); C contains $\{0\}$ because with $N = 1$ and $A = \{0, \{0\}\}$, "$s\begin{pmatrix} x & y \\ 1 & \{0\} \end{pmatrix}$" is true. Furthermore, C contains $\{\{0\}\}$, by Example 5.3.3(2). By continuing this process, it should be clear that the set C expresses (in set theory) the collection that we denoted by "$\mathscr{J}$" in Example 5.3.1.

Consider the collection consisting of those ordinals

$$0, 1, 2, \ldots, N, N^+, \ldots, \omega, \omega^+, \omega^{++}, \ldots$$

that (up to this point) we know exist. In anticipation of its role in ordinal arithmetic, we denote this collection by "$\omega 2$".

Theorem 5.3.1. *$\omega 2$ is a set.*

PROOF: Let X be the collection consisting of

$$\omega, \omega^+, \omega^{++}, \ldots;$$

it suffices to show that X is a set, for then $\omega 2 = \omega \cup X$ is a set. In order to utilize the Axiom of Substitution, we must express the "one-one correspondence" between X and ω by means of a formula of set theory. We proceed along the same lines as in Examples 5.3.3 and 5.3.4.

For sets x, y, and z we say that y is an **x-successor of** z if there exists a set A satisfying the following:

(1) $z \in A$.

(2) If $a\,(\neq z) \in A$, then $a = b^+$ for some b in A.

(3) $y \in A$ and $y^+ \notin A$.

(4) $x \in \omega$ and A is an x^+-element set.

Note that ω is a 0-successor of itself (with $A = \{\omega\}$); ω^+ is a 1-successor of ω (with $A = \{\omega, \omega^+\}$), etc. Let s be the formula "y is an x-successor of ω." Then, by the Axion of Substitution, X is a set; in fact,

$$X = \{c_2 : \text{for some } N \text{ in } \omega, \text{“}s\begin{pmatrix} x & y \\ N & c_2 \end{pmatrix}\text{” is true}\}.$$

Here, we have used the easily verifiable fact that s is bimonadic.

Corollary. *$\omega 2$ is an ordinal number.*

PROOF: Clearly, $\omega 2$ is a transitive set of ordinals.

The ordinal $\omega 2$ serves as the "gateway to higher ordinals". We had to introduce a new axiom in order to assure its existence (as a set). Now that we know of $\omega 2$, the existence of still "higher" ordinals will follow.

For every ordinal α and every natural number N, let $S_N(\alpha)$ be the "N-fold successor of α"; i.e., $a^{++\cdots+}$ (N times). (We will say that α is the 0-fold successor of itself.) According to Corollary 1 of Theorem 5.2.7 (and induction), each such $S_N(\alpha)$ is an ordinal.

EXAMPLE 5.3.5

(1) $N = S_N(0), N \in \omega$.

(2) $N + K = S_N(K) = S_K(N), N, K \in \omega$.

(3) $S_2(\omega) = \omega^{++}$.

(4) $\omega 2 = \omega \cup \{\alpha : \alpha \text{ is an } N\text{-fold successor of } \omega\}$.

For each N in ω, $S_N(\omega 2)$ is an ordinal that includes $\omega 2$ and contains every ordinal $S_K(\omega 2)$, for $K < N$. The proofs of Theorem 5.3.1 and its corollary can be modified to show the existence of the ordinal

"$\omega 3$", which we define to be "$\omega 2 \cup \{S_N(\omega 2) : N \in \omega\}$".

Clearly, we could continue in this fashion and define the ordinals

$$\omega 4, \omega 5, \omega 6, \ldots.$$

In order to proceed with our listing of ordinals, we must introduce the following definition. If $\alpha = \beta^+$ for some ordinal number β, then α (is an ordinal that) is called a **successor ordinal**. If α is a nonzero ordinal and is not a successor ordinal, then α is called a **limit ordinal**. Clearly, *an ordinal α is a successor ordinal iff it contains a last element with respect to $\in_\alpha$*.

EXAMPLE 5.3.6

(1) For each N in ω_+, N is a successor ordinal because $N = K^+$ for some (unique) natural number K.

(2) ω is a limit ordinal.

(3) $S_K(\omega)$ is a successor ordinal, $K = 1, 2, \ldots$.

(4) $\omega 2$ and $\omega 3$ are each limit ordinals.

Without the Axiom of Substitution the only limit ordinal whose existence we could confirm was ω. We will presently see that the Axiom of Substitution guarantees the existence of so many limit ordinals that the totality of these no longer constitutes a set!

Theorem 5.3.2. *For every ordinal number α, there is a smallest limit ordinal containing α.*

We denote this ordinal by $\mathscr{L}(\alpha)$. Thus, $\mathscr{L}(\alpha)$ is a limit ordinal; if $\beta > \alpha$ and β is a limit ordinal, then $\mathscr{L}(\alpha) \leq \beta$.

PROOF: By means of the Axiom of Substitution, it is easily verified that the collection of all N-fold successors of α constitutes a (nonempty) set, $\mathscr{S}_\alpha$. By Corollary 3 of Theorem 5.2.7, $\cup \mathscr{S}_\alpha$ is an ordinal number.

Suppose that $\cup \mathscr{S}_\alpha$ is a successor ordinal; then $\cup \mathscr{S}_\alpha = \gamma \cup \{\gamma\}$ for some ordinal γ. Moreover, $\gamma \in \cup \mathscr{S}_\alpha$; hence $\gamma \in S_K(\alpha)$ for some K in ω. It follows from Exercise 5.2.5 that $\gamma^+ \in S_{K^+}(\alpha) \subseteq \cup \mathscr{S}_\alpha = \gamma^+$. Therefore, γ^+ is not ordinary, in contradiction of Theorem 5.2.3; thus, $\cup \mathscr{S}_\alpha$ is a limit ordinal.

If β is any limit ordinal for which $\alpha < \beta$, then $\alpha \in \beta$; by induction, each $S_K(\alpha)$ is in β. Hence, for each K in ω, $S_K(\alpha) \subseteq \beta$. Therefore, $\cup \mathscr{S}_\alpha \subseteq \beta$; i.e., $\mathscr{L}(\alpha) \leq \beta$.

Corollary 1. *For every set A of ordinal numbers there is a smallest limit ordinal that properly includes A.*

We denote this ordinal by "$\mathscr{L}(A)$". (Note that because ordinal numbers are themselves sets of ordinals, this simply extends our previous notation.)

PROOF: We know that $\cup A$ is an ordinal. If $\cup A = 0$, then clearly $\mathscr{L}(A) = \omega\,(= \mathscr{L}(0))$. If $\cup A$ is a limit ordinal that properly includes A, then $\cup A = \mathscr{L}(A)$. Otherwise, $\cup A\ (= A)$ is a limit ordinal or (excl.) else $\cup A$ is a successor ordinal; if we call this ordinal α, then we seek $\mathscr{L}(\alpha)$.

Corollary 2. *There is no set consisting of all limit ordinals.*

EXAMPLE 5.3.7

(1) $\mathscr{L}(1) = \cup\{S_N(1)\} = \omega$.
(2) $\mathscr{L}(\{1, 4, 13\}) = \mathscr{L}(\cup\{1, 4, 13\}) = \mathscr{L}(13) = \omega$.
(3) $\mathscr{L}(\{2, 4, 6, \ldots, 2N, \ldots\}) = \cup\{2, 4, 6, \ldots, 2N, \ldots\} = \omega$.
(4) $\mathscr{L}(\omega) = \omega 2$.

For sets x, y, and z we say that y is an x-**limit of** z if there exists a set A of ordinals satisfying the following:

(1) $z \in A$.
(2) If $\alpha\,(\neq z) \in A$, then $\alpha = \mathscr{L}(\beta)$ for some ordinal β in A.
(3) $y \in A$ and $\mathscr{L}(y) \notin A$.
(4) $x \in \omega$ and A is an x^+-element set.

According to this definition, if the set z is not an ordinal, then no set is an x-limit of z (because the set A must consist (entirely) of ordinals one of which is z).

EXAMPLE 5.3.8

(1) $\omega 2$ is a 2-limit of 5; here, let $A = \{5, \omega, \omega 2\}$.
(2) $\omega 5$ is a 4-limit of ω; here, let $A = \{\omega, \omega 2, \omega 3, \omega 4, \omega 5\}$.

Let s be the formula, "y is an x-limit of 0." By the Axiom of Substitution, the collection of c_2—such that for some N in ω, "$s\begin{pmatrix} x & y \\ N & c_2 \end{pmatrix}$" is true—is a set, A. Moreover, A consists of the limit ordinals

$$\omega, \omega 2, \omega 3, \omega 4, \ldots.$$

Note that A is not itself an ordinal number; in fact, A is not transitive. But A is a set of ordinal numbers, and thus $\cup A$ is an ordinal number. Clearly, $\cup A$ is the smallest limit ordinal that includes A. In anticipation of ordinal exponentiation, we denote "$\cup A$" by "ω^2".

It should now be fairly obvious that by means of the Axiom of Substitution we can obtain more and more ordinals. First (in anticipation of ordinal addition) we write "$\mathscr{L}(\omega^2)$" as "$\omega^2 + \omega$"; continuing in this manner we obtained the limit ordinals

(5.3.5) $$\omega^2 + \omega, \omega^2 + \omega 2, \omega^2 + \omega 3, \ldots.$$

By modifying the technique used to guarantee the existence of ω^2, we obtain the smallest (limit) ordinal greater than those listed in (5.3.5); we write this as "$\omega^2 2$". Continuing in this fashion, we obtain

(5.3.6) $$\omega^2 2 + \omega, \omega^2 2 + \omega 2, \omega^2 2 + \omega 3, \ldots;$$

the smallest (limit) ordinal greater than any of those in (5.3.6) is denoted by "$\omega^2 3$". Similarly, we obtain

(5.3.7) $$\omega^2 4, \omega^2 5, \omega^2 6, \ldots.$$

The smallest limit ordinal greater than any of those in (5.3.7) is denoted by "ω^3". We eventually obtain

(5.3.8) $$\omega^4, \omega^5, \omega^6, \ldots.$$

We can again find a formula of set theory that will guarantee the existence of a set containing all of these—and hence the existence of a smallest (limit) ordinal greater than any of those indicated in (5.3.8); we denote this ordinal by "ω^ω". It is fairly clear that this method of obtaining "higher" ordinals can be continued much further, but this is a rather tedious (and inefficient) method of procedure. In the next section we will prove the transfinite analogue of Definition by Induction; by means of this we will be able to define the arithmetical operations on the ordinals. Sums, products, and powers of ordinals will be ordinals, and all the above ordinals will thereby be obtained in a simpler manner.

Theorem 5.3.3. The Counting Theorem. *For every well-ordered system $\langle X, R\rangle$ there is a unique ordinal number α such that $\langle X, R\rangle$ is similar to $\langle \alpha, \in\rangle$.*

PROOF: Let $\langle X, R\rangle$ be a well-ordered system. Let s be the formula, "x is an initial segment of X (with respect to R), y is an ordinal, and $\langle x, R\rangle \cong \langle y, \in\rangle$". Corollary 4 of Theorem 5.1.1 guarantees that s is bimonadic. We utilize the Axiom of Substitution. Let

$$A = \left\{c_2 : \text{for some } c_1 \text{ in } \mathscr{P}(X), \text{``}s\begin{pmatrix} x & y \\ c_1 & c_2 \end{pmatrix}\text{'' is true}\right\}.$$

Each c_2 in A is an ordinal; according to Theorem 5.2.7, in order to show that A is an ordinal, it suffices to show that A is transitive.

Suppose that β and γ are ordinals for which $\beta \in \gamma$, $\beta \notin A$, and $\gamma \in A$. By the Comparability Theorem, $\langle X, R\rangle$ is similar to a section of $\langle \beta, \in\rangle$. Consequently, $\langle \gamma, \in\rangle$ must also be similar to a section of $\langle \beta, \in\rangle$. But this cannot be, because $\langle \beta, \in\rangle$ is a section of $\langle \gamma, \in\rangle$.

Clearly, A serves as the (unique) ordinal α of the statement of the theorem.

The Counting Theorem provides us with a means of enumerating the elements of a well-ordered set X (with respect to a fixed well-ordering R on X). For let α be the unique ordinal such that $\langle X, R\rangle \cong \langle \alpha, \in\rangle$, and let F be the similarity mapping. We write "x_0" for the unique element x such that $F(x) = 0$, "x_1" for the unique element x such that $F(x) = 1$, etc., and we can speak of the 0th, 1st, ..., ωth, ... element of X.

EXAMPLE 5.3.9. Let $X = \omega$, let R_1 be $\leq$; let R_2 be the well-ordering indicated by

$$2, 3, 4, \ldots, N, N^+, \ldots, 0, 1;$$

let R_3 be the well-ordering indicated by

$$0, 2, 4, \ldots, 2N, 2N + 2, \ldots, 1, 3, 5, \ldots, 2N + 1, 2N + 3, \ldots.$$

(1) For each N in ω, the Nth element of X, with respect to R_1, is N.
(2) $\langle X, R_2 \rangle \cong \langle \omega^{++}, \in \rangle$; for each N in ω, the Nth element of X is $N + 2$; the ωth element is 0 and the ω^+th element is 1.
(3) $\langle X, R_3 \rangle \cong \langle \omega 2, \in \rangle$; for each N in ω, the Nth element of X is $2N$; the ωth element is 1, the ω^+th element is 3, the ω^{++}th element is 5, etc.

Assume the Axiom of Choice. Then by Theorem 4.1.6 and the Well-ordering Theorem (4.8.5), $\mathscr{P}(\omega)$ is a nondenumerable well-orderable set. Let R well-order $\mathscr{P}(\omega)$. Without the Axiom of Substitution we cannot find an ordinal α for which $\langle \mathscr{P}(\omega), R \rangle \cong \langle \alpha, \in \rangle$. (For if such an ordinal could be shown to exist, then so could $\omega 2$.)

The Counting Theorem, which utilizes the Axiom of Substitution, guarantees the existence of this ordinal α. Thus, the Axioms of Choice and Substitution together imply the existence of a nondenumerable ordinal. In Section 5.12 we will show that the Axiom of Choice is not necessary for this, i.e., that Substitution (with the Axioms of Chapter 1) yields this result.

Hereafter, we will assume Substitution among our axioms.

Exercises

†5.3.1. **(a)** Show that the formula s of Example 5.3.4 is bimonadic.
(b) Which of the following are bimonadic? strictly bimonadic?

(1) $y \subseteq x$. (2) $y = \mathscr{P}(x)$.
(3) $y = \cup x$. (4) $y = \cap x$.
(5) For some $z, y \in \{x, z\}$. (6) For every $z, y \in \{x, z\}$.

5.3.2. Consider the set C of Example 5.3.4.
(a) Show that $\in_C$ is not a partial ordering.
(b) Define a well-ordering on C.

5.3.3. **†(a)** Use the Axiom of Substitution to show that the collection of $\mathscr{P}^K(\omega)$, $K \in \omega$, is a set.
(b) Now use part (a) (without any further use of Substitution) to show that the collection $\mathscr{J}$, of Example 5.3.1, is a set.

5.3.4. Suppose that in formulating the Axiom of Substitution we had replaced "bimonadic" by "strictly bimonadic". Show that in the presence of the other axioms of set theory, the existence of a set can be proved using the Axiom of Substitution iff it can be proved using this alternative formulation.

†5.3.5. Let s be a bimonadic formula. Use the Axiom of Substitution (and other axioms) to show that for every set A,

$$\{\langle c_1, c_2\rangle : \text{for some (set) } c_2 \text{ and for some } c_1 \text{ in } A, \text{“}s\begin{pmatrix} x & y \\ c_1 & c_2 \end{pmatrix}\text{” is true}\}$$

is a function.

5.3.6. Consider the sentence

s : There is an inductive set containing ω.

(a) Show that the Axiom of Substitution implies s.
(b) Which new ordinals are guaranteed by s (without Substitution)?

Now assume the Axiom of Substitution.

(c) Prove that an ordinal α is a limit ordinal iff α is inductive.
(d) Determine an inductive set that is not an ordinal.

†5.3.7. Prove that the Axiom of Substitution implies the Axiom of Subsets.

5.3.8. What is wrong with the following "proof" that the Axiom of Substitution implies the Axiom of Union?

For each set A let s_A be the formula "$y \in x$ and $x \in A$".

Then $\cup A = \{c_2 : \text{for some } a \text{ in } A, \text{“}s\begin{pmatrix} x & y \\ a & c_2 \end{pmatrix}\text{” is true}\}$.

5.3.9. In the proof of Theorem 5.3.2, show that for each ordinal α, the collection of all N-fold successors of α is a set.

†5.3.10. **(a)** Prove that the union of a nonempty set of limit ordinals is a limit ordinal.
(b) Show, by means of examples, that the union of a nonempty set of successor ordinals can be either a successor ordinal or a limit ordinal.
(c) Determine a set A of ordinals that contains both limit ordinals and successor ordinals and whose union is (1) a limit ordinal, (2) a successor ordinal.

†5.3.11. **(a)** Show that for every limit ordinal α, $\cup\alpha = \alpha$.
(b) Show that for every successor ordinal β, $(\cup\beta)^+ = \beta$.
(c) Show that a nonzero ordinal α is a successor ordinal iff it has a last element with respect to $\in_\alpha$.
(d) Show that if A is a nonempty set of ordinals, then $\cup A$ is a limit ordinal iff $\alpha \notin \cup A$ for every α in A.

†5.3.12. Prove that if α is a limit ordinal, then $x < \alpha$ implies $x^+ < \alpha$.

5.3.13. **(a)** Prove Corollary 2 of Theorem 5.3.2—i.e., show that there is no set consisting of all limit ordinals.
(b) Show that there is no set consisting of all successor ordinals.

5.3.14. Suppose that we know of the ordinals $\omega, \omega^2, \omega^3, \ldots, \omega^N, \ldots$. Determine a bimonadic formula that, because of the Axiom of Substitution, guarantees the existence of a set containing all of these.

5.3.15. In the proof of the Counting Theorem (5.3.3), prove the final assertion that A serves as the unique ordinal α of the statement of the theorem.

★5.3.16. Prove that every set is contained in a transitive set. (Used in Exercise 6.1.12.)

5.4 Definition by Transfinite Induction

The fact that the collection $\mathcal{O}$ of ordinals does not constitute a set presents us with various difficulties. We know that every set A of ordinals is well-ordered (by $\in_{(\cup A)^+} \cap A^{②}$). However, if $\mathcal{O}$ is a collection, but not a set, of ordinals, we cannot say that $\mathcal{O}$ is well-ordered because relations, and in particular, well-orderings, have only been defined on sets.

Suppose that $\mathcal{O}$ is the collection of all ordinals and s is a formula—with one free variable, x—pertaining to ordinals. We will sometimes be concerned with sentences such as, "For every α, $s\begin{pmatrix} x \\ \alpha \end{pmatrix}$" (which means "For every set α, if α is an ordinal, then $s\begin{pmatrix} x \\ \alpha \end{pmatrix}$"). In order to show that such a sentence is true, we may assume it to be false and attempt to derive a contradiction. Thus, we may assume that for some α, $s\begin{pmatrix} x \\ \alpha \end{pmatrix}$ is false. Let α_0 be an ordinal for which $s\begin{pmatrix} x \\ \alpha_0 \end{pmatrix}$ is false; $\alpha_0{}^+$ is a nonempty well-ordered set. At this point we can utilize properties of well-ordered sets, such as the Principle of Transfinite Induction (Theorem 2.7.3). Thus, let A be the subset of $\alpha_0{}^+$ consisting of ordinals γ for which $s\begin{pmatrix} x \\ \gamma \end{pmatrix}$ is true. If we can show that for each γ in A, whenever $\operatorname{sect}_{\in_{\alpha_0{}^+}}(\gamma)\,(= \gamma)$ is *included* in A, then $\gamma \in A$, the Principle of Transfinite Induction assures us that $A = \alpha_0{}^+$. In particular, $\alpha_0 \in A$. But this contradicts the previous statement that $s\begin{pmatrix} x \\ \alpha_0 \end{pmatrix}$ is false.

In Chapter 3, we defined addition, multiplication, and exponentiation of natural numbers. These were binary operations on ω, and could be defined by means of Definition by Induction (Theorem 2.8.1). In order to define these arithmetical processes for ordinals, we would like to extend Definition by Induction so as to apply to the ordinals (which do not constitute a set). We repeat the statement of Theorem 2.8.1 so that we may compare this with the new theorem on Definition by Transfinite Induction.

Definition by Induction: *Let X be a nonempty set, let $x_0 \in X$, and let $G : X \to X$. Then there is a unique function $F : \omega \to X$ that satisfies the following:*
(1) $F(0) = x_0$.
(2) *For all N in ω,* $F(N^+) = G(F(N))$.

Thus, for each nonempty set X, Definition by Induction, essentially, asserts the existence of a function whose domain is the set of ordered pairs $\langle G, x_0 \rangle$, where G is a function from X into itself and $x_0 \in X$—and whose range is the set of functions from ω to X. Definition by Transfinite Induction, in effect, will say that there exists a "correspondence" between the collection of ordered triples, $\langle s, c_0, \alpha \rangle$—where s is a strictly bimonadic formula, c_0 is an arbitrary set, and α is an ordinal—and functions f defined on α. In the statement of the Theorem on Definition by Transfinite Induction, we no longer specify (in advance) the set X that played so central a role in Definition by Induction; in the proof of the present theorem the role of X is taken by a set B whose existence follows from the Axiom of Substitution.

For an ordinal α, an α-**sequence** is defined to be a function with domain α. An α-sequence whose range is included in the set A is called an α-**sequence in** A.

Theorem 5.4.1. Definition by Transfinite Induction. *Let s be a (fixed) strictly bimonadic formula. Then for every set c_0 and every nonzero ordinal number α there is a unique α-sequence f that satisfies the following:*

(5.4.1) $f(0) = c_0$.
(5.4.2) *For every successor ordinal $\beta^+ < \alpha$, $f(\beta^+) = c_2$, where c_2 is the unique set for which "$s\begin{pmatrix} x & y \\ f(\beta) & c_2 \end{pmatrix}$" is true.*
(5.4.3) *For every limit ordinal $\beta < \alpha$, the collection of $f(\gamma)$ for γ in β is a set and*

$$f(\beta) = \cup\{f(\gamma) : \gamma \in \beta\}.$$

PROOF: Let s be a strictly bimonadic formula, let c_0 be a set, and let α be a nonzero ordinal.

For $\beta < \alpha$, we say that a β-sequence F is s-**suitable** provided $\beta = 0$—or $\beta > 0$ and the following hold:

(a) $F(0) = c_0$.
(b) For every successor ordinal $\gamma^+ < \beta$, $F(\gamma^+)$ is the unique set c_2 for which $s\begin{pmatrix} x & y \\ F(\gamma) & c_2 \end{pmatrix}$ is true.
(c) For every limit ordinal $\delta < \alpha$, $F(\delta) = \cup\{F(\gamma) : \gamma \in \delta\}$.

Let S be the strictly bimonadic formula, "$x \in \alpha$ and y is an s-suitable x-sequence—or $x \notin \alpha$ and $y = 0$." According to the Axiom of Substitution there is a set B consisting of all sets for which $S\begin{pmatrix} x & y \\ \gamma & b \end{pmatrix}$ is true for some γ in α. Let $f = \cup B$. Then f is an α-sequence that satisfies properties (5.4.1), (5.4.2), and (5.4.3).

If g is a function with domain α and with properties (5.4.1), (5.4.2), and (5.4.3), let $D = \{\beta < \alpha : f(\beta) = g(\beta)\}$. We apply the Principle of Transfinite Induction (Theorem 2.7.3) to show that $D = \alpha$. Clearly, $0 \in D$ because $f(0) = c_0 = g(0)$. For $\beta\ (\neq 0) < \alpha$, assume that $\beta \subseteq D$. If $\beta = \gamma^+$, then $\gamma < \beta$ and $f(\gamma) = g(\gamma)$. It follows that $f(\beta) = c_2 = g(\beta)$, where c_2 is the unique set for which "$s\begin{pmatrix} x & y \\ f(\gamma) & c_2 \end{pmatrix}$", and hence "$s\begin{pmatrix} x & y \\ g(\gamma) & c_2 \end{pmatrix}$", are true. If β is a limit ordinal, then $E = \{f(\gamma) : \gamma < \beta\}$ is a set; because $f(\gamma) = g(\gamma)$ for all $\gamma < \beta$, it follows that $E = \{g(\gamma) : \gamma < \beta\}$. Hence, $f(\beta) = \cup E = g(\beta)$.

EXAMPLE 5.4.1. Let s be the strictly bimonadic formula, "$y = x^+$".
(1) If $c_0 = 0$ and if $\alpha = \omega 2$, then $f = I_{\omega 2}$.
(2) If $c_0 = 5$ and if $\alpha = \omega 2$, then

$$f = \{\langle N, N + 5\rangle : N < \omega\} \cup \{\langle S_N(\omega), S_N(\omega)\rangle : N < \omega\}.$$

(3) If $c_0 = \omega$ and if $\alpha = \omega 2$, then

$$f = \{\langle N, S_N(\omega)\rangle : N < \omega\} \cup \{\langle S_N(\omega), S_N(\omega 2)\rangle : N < \omega\}.$$

There is another form of Definition by Transfinite Induction that we will frequently utilize in lieu of Theorem 5.4.1.

Theorem 5.4.2. Alternative Form of Definition by Transfinite Induction. *Let A be a nonempty set and let α be a nonzero ordinal number. For $\beta < \alpha$, let B_β be the set of β-sequences in A; let $B = \cup\{B_\beta : \beta < \alpha\}$. Let $G : B \to A$. Then there is a unique α-sequence, F, in A that satisfies*

$$F(\beta) = G(F \upharpoonright \beta), \textit{ for all } \beta < \alpha.$$

PROOF: Let A, α, and B be as indicated, and let $G : B \to A$. Let s_G be the strictly bimonadic formula, "For some β in α, x is a β-sequence in A and $y = x \cup \{\langle \beta, G(x)\rangle\}$—or x is not a β-sequence in A for any $\beta < \alpha$ and $y = 0$." Then, by conditions (5.4.1), (5.4.2), and (5.4.3) of the original form of Definition by Transfinite Induction, there is a unique α-sequence, f, that satisfies the following:

(1) $f(0) = \varnothing$.

(2) For every successor ordinal β^+ in α, $f(\beta^+) = c_2$, where c_2 is the unique set such that "$s_G\begin{pmatrix} x & y \\ f(\beta) & c_2 \end{pmatrix}$" is true.

(3) For every limit ordinal $\beta < \alpha$, the collection of $f(\gamma)$, for $\gamma < \beta$, is a set and $f(\beta) = \cup\{f(\gamma) : \gamma < \beta\}$.

It can be shown that (2) can be restated as follows:

(2′) For every successor ordinal $\beta^+ < \alpha$, $f(\beta^+) = f(\beta) \cup \{\langle \beta, G(f \upharpoonright \beta)\rangle\}$.

Define F on α by

$F(\beta)$ is the second coordinate of the (unique) ordered pair of

$$f(\beta^+) \sim f(\beta),\ \beta < \alpha.$$

Then F is the unique α-sequence that is such that for all $\beta < \alpha$,

$$F(\beta) = G(F \upharpoonright \beta).$$

†EXAMPLE 5.4.2. Let $A = \alpha = \omega 2$. Then B consists of all β-sequences in $\omega 2$ for $\beta < \omega 2$. Let $G : B \to A$ be defined as follows: For a β-sequence b in B, let $G(b) = \beta^+$. Then the unique function F from $\omega 2$ into $\omega 2$ is defined as follows:

$F \upharpoonright 0 = 0$ and $G(0) = 0^+ = 1$; thus, $F(0) = G(F \upharpoonright 0) = 1$.
$F \upharpoonright 1$ $(= \{\langle 0, 1\rangle\})$ is a 1-sequence in $\omega 2$; thus, $F(1) = G(F \upharpoonright 1) = 2$.

By induction on N, $F(N) = G(F \upharpoonright N) = N^+$, $N < \omega$.

Furthermore, $F \upharpoonright \omega = \{\langle N, N^+\rangle : N < \omega\}$; thus, $F \upharpoonright \omega$ is an ω-sequence in $\omega 2$. It follows that $F(\omega) = G(F \upharpoonright \omega) = \omega^+$; by induction, $F(S_N(\omega)) = S_{N^+}(\omega)$ for all $N < \omega$. Thus, $F = \{\langle \gamma, \gamma^+\rangle : \gamma < \omega 2\}$.

Exercises

5.4.1. Let s be the formula, "x is an ordinal and $y = \mathscr{L}(x)$ (the smallest limit ordinal containing x)—or x is not an ordinal and $y = 0$." Let $c_0 = 6$ and let $\alpha = \omega 2$. Determine the function f with domain α whose existence (and uniqueness) is guaranteed by Definition by Transfinite Induction.

5.4.2. Answer Exercise 5.4.1 with $c_0 = \omega$ (instead of $c_0 = 6$).

5.4.3. Let s be the sentence, "x is an ordinal and y is the smallest ordinal that contains denumerably many limit ordinals containing x—or x is not an ordinal and $y = 0$." Let $c_0 = 0$.

(a) Determine the function f with domain ω whose existence is guaranteed by Definition by Transfinite Induction.

(b) Determine the function g with domain $\omega 2$ whose existence is similarly guaranteed.

5.4.4. Consider the proof of Theorem 5.4.1.

(a) Verify that S is strictly bimonadic.

(b) Verify that f is an α-sequence that satisfies properties (5.4.1), (5.4.2), and (5.4.3).

5.4.5. Consider Theorem 5.4.2 and its proof.

(a) How do we know that for each $\beta < \alpha$, B_β is a set?

(b) How do we know that B is a set?

(c) Verify that for every successor ordinal $\beta^+ < \alpha, f(\beta^+) = f(\beta) \cup \{\langle \beta, G(f \upharpoonright \beta)\rangle\}$.

(d) Why isn't f itself the function we seek?

(e) Verify that $F(\beta) = G(F \upharpoonright \beta)$ for all $\beta < \alpha$.

(f) Show the uniqueness of F.

5.4.6. In Example 5.4.2, is F an element of B?

5.4.7. Let $\alpha = \omega 2$, let $A = \omega^3$, and let B be the set of β-sequences in A for $\beta < \omega 2$. Let $G : B \to A$ be defined as follows. For b in B, let $G(b) = \mathscr{L}(\beta)$, if b is a β-sequence. Determine the (unique) $\omega 2$-sequence F satisfying $F(\beta) = G(F \upharpoonright \beta)$, $\beta < \omega 2$.

5.5 Ordinal Addition

We now utilize Definition by Transfinite Induction to define ordinal addition, just as we used Definition by Induction to define addition of natural numbers.

Let s be the strictly bimonadic formula, "$y = x^+$". Then for all ordinal numbers α and β with $\beta \neq 0$, there is a (unique) β-sequence $f_{\alpha,\beta}$ that satisfies the following:

(1) $f_{\alpha,\beta}(0) = \alpha$.
(2) For every successor ordinal $\gamma^+ < \beta, f_{\alpha,\beta}(\gamma^+) = (f_{\alpha,\beta}(\gamma))^+$.
(3) For every limit ordinal $\gamma < \beta, f_{\alpha,\beta}(\gamma) = \cup\{f_{\alpha,\beta}(\delta) : \delta < \gamma\}$.

In (2), $(f_{\alpha,\beta}(\gamma))^+$ is the unique set c_2 such that "$s\begin{pmatrix} x & y \\ f_{\alpha,\beta}(\gamma) & c_2 \end{pmatrix}$" is true.

A routine proof by Transfinite Induction shows that for all ordinals α and γ, and all ordinals β_1 and β_2, both of which contain γ,

$$f_{\alpha,\beta_1}(\gamma) = f_{\alpha,\beta_2}(\gamma).$$

Thus, we can define

$$\text{“}\alpha +_{\mathcal{O}} \beta\text{” to be “}f_{\alpha,\beta^+}(\beta)\text{”}$$

with the knowledge that this definition is independent of the second subscript (provided it is an ordinal containing β).

We note that *if* α *and* β *are both natural numbers, then* $\alpha +_{\mathcal{O}} \beta = \alpha + \beta$. Therefore, *ordinal addition extends natural number addition.* Henceforth, we will simply write "$\alpha + \beta$" in lieu of "$\alpha +_{\mathcal{O}} \beta$" for all ordinals α and β. This notation is consistent with our usage in Section 5.3.

†EXAMPLE 5.5.1

(1) $\omega + 3 = f_{\omega,4}(3) = (f_{\omega,4}(2))^+ = (f_{\omega,4}(1))^{++}$
$= (f_{\omega,4}(0))^{+++} = \omega^{+++} = S_3(\omega).$

(2) $3 + \omega = f_{3,\omega^+}(\omega) = \cup\{f_{3,\omega^+}(N) : N < \omega\}$
$= \cup\{f_{3,\omega^+}(0), f_{3,\omega^+}(1), f_{3,\omega^+}(2), \ldots\}$
$= \cup\{3, 4, 5, \ldots\} = \omega.$

(3) $\omega + \omega = f_{\omega,\omega^+}(\omega) = \cup\{f_{\omega,\omega^+}(N) : N < \omega\}$
$= \cup\{f_{\omega,\omega^+}(0), f_{\omega,\omega^+}(1), f_{\omega,\omega^+}(2), \ldots\}$
$= \cup\{\omega, \omega^+, \omega^{++}, \ldots\} = \omega 2.$

We will usually write "$\alpha + N$" instead of "$S_N(\alpha)$", for N in ω.

Parts (1) and (2) show that, in general, ordinal addition is *not* commutative. Thus, unlike the addition operation on other sets of numbers that we have considered, here we must be careful about the order in which these summands are taken.

†EXAMPLE 5.5.2

(1) $\omega + \omega^2 = \cup\{f_{\omega,(\omega^2)^+}(\delta) : \delta < \omega^2\}$
$= \cup\{\omega, \omega + 1, \omega + 2, \ldots, \omega 2, \omega 2 + 1, \omega 2 + 2, \ldots,$
$\omega N, \omega N + 1, \omega N + 2, \ldots\} = \omega^2.$

(2) $(\omega + 5) + \omega^2 = \cup\{f_{\omega+5,(\omega^2)^+}(\delta) : \delta < \omega^2\}$
$= \cup\{\omega + 5, \omega + 6, \ldots, \omega 2, \omega 2 + 1, \omega 2 + 2, \ldots,$
$\omega N, \omega N + 1, \omega N + 2, \ldots\} = \omega^2.$

(3) $\omega 2 + \omega^2 = \cup\{\omega 2, \omega 2 + 1, \omega 2 + 2, \ldots, \omega 3, \omega 3 + 1,$

$$\omega 3 + 2, \ldots, \omega N, \omega N + 1, \omega N + 2, \ldots\} = \omega^2.$$

(4) By transfinite induction it can be shown that if $\alpha < \omega^2$, then

$$\alpha + \omega^2 = \omega^2.$$

(5) $\omega^2 + 1 = (\omega^2)^+ \neq 1 + \omega^2 = \omega^2.$

(6) $$\begin{aligned}\omega^2 + \omega &= \cup\{f_{\omega^2,\omega^+}(\delta) : \delta < \omega\} \\ &= \cup\{\omega^2, \omega^2 + 1, \omega^2 + 2, \ldots, \omega^2 + N, \ldots\} \\ &= \mathscr{L}(\omega^2).\end{aligned}$$

(7) $$\begin{aligned}\omega^2 + \omega^2 &= \cup\{\omega^2 + \alpha : \alpha < \omega^2\} \\ &= \mathscr{L}(\{\omega^2 + \alpha : \alpha < \omega^2\}) = \omega^2 2.\end{aligned}$$

Part (6) illustrates the following "law": For every ordinal α, $\alpha + \omega = \mathscr{L}(\alpha)$.

Theorem 5.5.1. *For every two ordinal numbers, α and β, $\alpha + \beta$ is an ordinal number.*

PROOF: Let α and β be arbitrary ordinals; we must show that $\alpha + \beta\,(= f_{\alpha,\beta^+}(\beta))$ is an ordinal.

Let A be the subset of β^+ consisting of those ordinals γ for which $\alpha + \gamma$ is an ordinal. We apply the Principle of Transfinite Induction to show that $A = \beta^+$.

First, $0 \in A$ because $\alpha + 0 = f_{\alpha,1}(0) = \alpha$.

Suppose that for an ordinal $\gamma\,(\neq 0)$ in β^+, we know that $\gamma\,(= \text{seg}_{\in\beta^+}(\gamma)) \subseteq A$. If γ is a successor ordinal, let $\gamma = \delta^+$. Then

$$\alpha + \gamma = f_{\alpha,\gamma^+}(\gamma) = (f_{\alpha,\gamma^+}(\delta))^+ = (f_{\alpha,\delta^+}(\delta))^+ = (\alpha + \delta)^+.$$

Because $\delta < \gamma$, $\alpha + \delta$ is an ordinal. By Corollary 1 of Theorem 5.2.7, the successor of an ordinal is an ordinal; hence, $\gamma \in A$. If γ is a limit ordinal, then

$$\begin{aligned}\alpha + \gamma &= f_{\alpha,\gamma^+}(\gamma) = \cup\{f_{\alpha,\gamma^+}(\rho) : \rho < \gamma\} \\ &= \cup\{f_{\alpha,\rho^+}(\rho) : \rho < \gamma\} = \cup\{\alpha + \rho : \rho < \gamma\}.\end{aligned}$$

Also, $\{\alpha + \rho : \rho < \gamma\}$ is a set of ordinals. Consequently, $\alpha + \gamma$, as the union of a set of ordinals, is an ordinal (by Corollary 3 of Theorem 5.2.7).

We have shown that $\gamma \subseteq A$ implies $\gamma \in A$ for each $\gamma < \beta^+$. By transfinite induction, $A = \beta^+$; hence, $\beta \in A$ and $\alpha + \beta$ is an ordinal.

Let R_A and R_B be linear orderings on the disjoint sets A and B, respectively. Let

$$R_{A,B} = R_A \cup R_B \cup (A \times B).$$

Then $R_{A,B}$ *is a linear ordering on* $A \cup B$; in fact, *for* $x, y \in A \cup B$, $xR_{A,B}y$ *iff (exactly) one of the following holds:*

(1) $x, y \in A$ *and* xR_Ay.
(2) $x, y \in B$ *and* xR_By.
(3) $x \in A$ *and* $y \in B$.

We call $R_{A,B}$ the **natural ordering of** $A \cup B$ (with respect to $\langle R_A, R_B \rangle$). Note that *if* A *and* B *are each nonempty, then* $R_{A,B} \neq R_{B,A}$.

Clearly, if R_A *and* R_B *are both well-orderings, then so are* $R_{A,B}$ *and* $R_{B,A}$.

EXAMPLE 5.5.3. Let $A = \{a, b\}$ and let $B = \{c, d, e\}$, where a, b, c, d, and e are distinct. Let R_A and R_B be represented by the chain diagrams of Fig. 5.5.1.

(a) (b)

Fig. 5.5.1

Then $R_{A,B}$, the natural ordering on $A \cup B$, is represented by the chain diagrams of Figure 5.5.2(a) and $R_{B,A}$ by the chain diagram of Figure 5.5.2(b).

(a) (b)

Fig. 5.5.2

Theorem 5.5.2. *Let* α *and* β *be any two ordinal numbers. Let* $\langle A, R_A \rangle$ *and* $\langle B, R_B \rangle$ *be any two well-ordered systems for which* $A \cap B = \varnothing$, $\langle A, R_A \rangle \cong \langle \alpha, \in \rangle$, *and* $\langle B, R_B \rangle \cong \langle \beta, \in \rangle$. *Let* $R_{A,B}$ *be the natural (well-) ordering on* $A \cup B$. *Then* $\langle A \cup B, R_{A,B} \rangle \cong \langle \alpha + \beta, \in \rangle$.

PROOF: Let F_1 be the similarity map of $\langle A, R_A \rangle$ onto $\langle \alpha, \in \rangle$ and let F_2 be the similarity mapping of $\langle B, R_B \rangle$ onto $\langle \beta, \in \rangle$. Let $\mathscr{B} = \{\gamma : \alpha < \gamma < \alpha + \beta\}$. Then one readily verifies that $\langle \mathscr{B}, \in_{\alpha+\beta} \rangle$ is similar to $\langle \beta, \in \rangle$; let F_3 be this similarity mapping. Clearly, $F_1 \cup (F_3{}^{-1} \circ F_2)$ is the required similarity mapping of $\langle A \cup B, R_{A,B} \rangle$ onto $\langle \alpha + \beta, \in \rangle$.

The proof of Theorem 5.5.2 indicates that for any two ordinals α and β, we can always determine well-ordered systems—$\langle A, R_A \rangle$, similar to $\langle \alpha, \in \rangle$, and $\langle B, R_B \rangle$, similar to $\langle \beta, \in \rangle$—that are such that $A \cap B = \varnothing$. For take $\langle A, R_A \rangle = \langle \alpha, \in \rangle$ and take $\langle B, R_B \rangle = \langle \mathscr{B}, \in_{\alpha+\beta} \rangle$, as defined above. A second example is given by $\langle \alpha \times \{1\}, R_1 \rangle$ and $\langle \beta \times \{2\}, R_2 \rangle$, where $\langle \gamma, 1 \rangle R_1 \langle \delta, 1 \rangle$ iff $\gamma \leq \delta < \alpha$, and $\langle \rho, 2 \rangle R_2 \langle \sigma, 2 \rangle$ iff $\rho \leq \sigma < \beta$.

In Section 5.9 we will indicate a canonical method of selecting the representative well-ordered systems.

Theorem 5.5.2 affords us a convenient method of verifying an ordinal equation of the form

$$\alpha + \beta = \gamma. \tag{5.5.1}$$

For let $\langle A, R_A \rangle$, $\langle B, R_B \rangle$, and $\langle C, R_C \rangle$ be any well-ordered systems that are such that $\langle A, R_A \rangle \cong \langle \alpha, \in \rangle$, $\langle B, R_B \rangle \cong \langle \beta, \in \rangle$, $\langle C, R_C \rangle \cong \langle \gamma, \in \rangle$, and $A \cap B = \varnothing$. Let $R_{A,B}$ be the natural well-ordering on $A \cup B$. Then if $\langle A \cup B, R_{A,B} \rangle \cong \langle C, R_C \rangle$, according to Theorems 5.3.3 and 5.5.2, equation (5.5.1) must follow. Similarly, according to Theorem 5.2.8, we can also use this method to verify an inequality of the form

$$\alpha + \beta < \gamma$$

by showing that $\langle A \cup B, R_{A,B} \rangle$ is similar to a section of $\langle C, R_C \rangle$. We will utilize this "representative-systems" method in several of the remaining proofs in this section.

Example 5.5.1 showed, by means of a counterexample, that ordinal addition is noncommutative. The next two theorems yield properties of ordinal addition that have their analogues in the addition of natural numbers.

Theorem 5.5.3. *For every ordinal number* α,

$$\alpha + 0 = 0 + \alpha = \alpha.$$

PROOF: $\alpha \cap \varnothing = \varnothing \cap \alpha = \varnothing$ and $\varnothing$ is a well-ordering on itself. Clearly, $\in_\alpha$ is the natural well-ordering on α ($= \alpha \cup \varnothing = \varnothing \cup \alpha$) with respect to $\in_\alpha$ and $\varnothing$ (respectively, $\varnothing$ and $\in_\alpha$).

We will presently see that *if* α *is any infinite ordinal, then* 0 *is the only natural number* N *that "commutes with* α*", i.e., that satisfies* $\alpha + N = N + \alpha$.

Theorem 5.5.4. *Let α and β be any two ordinal numbers. Then*

$$\alpha + \beta^+ = (\alpha + \beta)^+.$$

(The proof is straightforward and is left as an exercise.)

Corollary 1. *For every ordinal α, $\alpha + 1 = \alpha^+$.*

Corollary 2. *Let α and β $(\neq 0)$ be any two ordinal numbers. Then $\alpha + \beta$ is a successor ordinal if and only if β is a successor ordinal.*

PROOF: Suppose that β is a successor ordinal; i.e., suppose that $\beta = \gamma^+$ for some ordinal γ. Then $\alpha + \beta = \alpha + \gamma^+ = (\alpha + \gamma)^+$, and $\alpha + \beta$ is a successor ordinal.

Suppose that β is a limit ordinal. Let $\langle A, R_1 \rangle$ be similar to $\langle \alpha, \in \rangle$ and $\langle B, R_2 \rangle$ be similar to $\langle \beta, \in \rangle$, where $A \cap B = \varnothing$. Then B has no last element with respect to R_2. Consequently, $A \cup B$ has no last element with respect to the associated natural ordering, R_3. Because $\langle \alpha + \beta, \in \rangle$ is similar to $\langle A \cup B, R_3 \rangle, \alpha + \beta$ has no last element. Clearly $\alpha + \beta \neq 0$; thus, $\alpha + \beta$ is a limit ordinal.

Corollary 3. *Let α and β $(\neq 0)$ be any two ordinal numbers. Then $\alpha + \beta$ is a limit ordinal if and only if β is a limit ordinal.*

In general, $\alpha^+ + \beta \neq (\alpha + \beta)^+$; for example, let $\alpha = 0$ and $\beta = \omega$. Then $\alpha^+ + \omega = 1 + \omega = \omega \neq \omega + 1$.

Theorem 5.5.5. *Let β be a limit ordinal. Then for every ordinal number α,*

$$\alpha + \beta = \cup\{\alpha + \gamma : \gamma < \beta\}.$$

PROOF:

$$\begin{aligned} \alpha + \beta = f_{\alpha,\beta^+}(\beta) &= \cup\{f_{\alpha,\beta^+}(\gamma) : \gamma < \beta\} \\ &= \cup\{f_{\alpha,\gamma^+}(\gamma) : \gamma < \beta\} = \cup\{\alpha + \gamma : \gamma < \beta\}. \end{aligned}$$

(Although we could have proved this theorem using the method of representative well-ordered systems, here the direct method is clearly the easier one.)

Theorems 5.5.3, 5.5.4, and 5.5.5 describe the sum $\alpha + \beta$ according to whether the *second* summand β is (1) zero, (2) a successor ordinal, or (3) a limit ordinal. Thus the following are true:

(1) *If* $\beta = 0$, $\alpha + \beta = \alpha$.
(2) *If* $\beta = \gamma^+$, $\alpha + \beta = (\alpha + \gamma)^+$.
(3) *If* β *is a limit ordinal,* $\alpha + \beta = \cup\{\alpha + \gamma : \gamma < \beta\}$.

In contrast to the noncommutativity of ordinal addition, we next show that associativity holds.

Theorem 5.5.6. *Let* α, β, *and* γ *be any ordinal numbers. Then*

(5.5.2) $$\alpha + (\beta + \gamma) = (\alpha + \beta) + \gamma.$$

PROOF: We can choose well-ordered systems $\langle A, R_A\rangle$, similar to $\langle\alpha, \in\rangle$, $\langle B, R_B\rangle$, similar to $\langle\beta, \in\rangle$, and $\langle C, R_C\rangle$, similar to $\langle\gamma, \in\rangle$ such that $\{A, B, C\}$ is pairwise disjoint. Recall that $A \cup (B \cup C) = (A \cup B) \cup C$; we have agreed to omit parentheses. Let $R_{B,C}$ be the natural ordering of $B \cup C$ with respect to $\langle R_B, R_C\rangle$ and let $R_{A,B\cup C}$ be the natural ordering of $A \cup B \cup C$ with respect to $\langle R_A, R_{B,C}\rangle$. Similarly, let $R_{A,B}$ be the natural ordering of $A \cup B$ with respect to $\langle R_A, R_B\rangle$ and let $R_{A\cup B,C}$ be the natural ordering of $A \cup B \cup C$ with respect to $\langle R_{A,B}, R_C\rangle$. We know that $\langle A \cup B \cup C, R_{A,B\cup C}\rangle$ is similar to $\langle\alpha + (\beta + \gamma), \in\rangle$ and that $\langle A \cup B \cup C, R_{A\cup B,C}\rangle$ is similar to $\langle(\alpha + \beta) + \gamma, \in\rangle$. In order to show the equality of the ordinals in equation (5.5.2), it suffices to show that $R_{A,B\cup C} = R_{A\cup B,C}$. But this is the case because for each of these well-orderings R of $A \cup B \cup C$, xRy iff one of the following holds:

$$x, y \in A \text{ and } xR_Ay,$$

$$x, y \in B \text{ and } xR_By,$$

$$x, y \in C \text{ and } xR_Cy,$$

$$x \in A \text{ and } y \in B,$$

$$x \in A \text{ and } y \in C,$$

$$x \in B \text{ and } y \in C.$$

As usual, we generally omit the parentheses in writing finite sums.

EXAMPLE 5.5.4. Let $\alpha = \omega$, $\beta = \omega + 5$, $\gamma = \omega^2$.
(1) $\alpha + \beta + \gamma = \beta + \alpha + \gamma = \omega^2$.
(2) $\alpha + \gamma + \beta = \omega^2 + \omega + 5$.
(3) $\gamma + \alpha + \beta = \omega^2 + \omega 2 + 5$.
(4) $\beta + \gamma + \alpha = \omega^2 + \omega$.
(5) $\gamma + \beta + \alpha = \omega^2 + \omega 2$.

Theorem 5.5.7. *Let α, β, and γ be any ordinal numbers. If $\alpha + \beta = \alpha + \gamma$, then $\beta = \gamma$.*

PROOF: Consider well-ordered systems $\langle B, R_B \rangle$, similar to $\langle \beta, \in \rangle$, and $\langle C, R_C \rangle$, similar to $\langle \gamma, \in \rangle$, satisfying $\alpha \cap B = \alpha \cap C = \varnothing$. Let F_1 and F_2 be the respective similarity mappings (from $\langle B, R_B \rangle$ onto $\langle \beta, \in \rangle$ and from $\langle C, R_C \rangle$ onto $\langle \gamma, \in \rangle$). If $R_{\alpha,B}$ and $R_{\alpha,C}$ are the respective natural orderings on $\alpha \cup B$ and $\alpha \cup C$, if $G_1 = I_\alpha \cup F_1$ and $G_2 = I_\alpha \cup F_2$, then G_1 is a similarity mapping from $\langle \alpha \cup B, R_{\alpha,B} \rangle$ onto $\langle \alpha + \beta, \in \rangle$ and G_2 is a similarity mapping from $\langle \alpha \cup C, R_{\alpha,C} \rangle$ onto $\langle \alpha + \gamma, \in \rangle$. Clearly, $G_2^{-1} \circ G_1$ is a similarity mapping of $\langle \alpha \cup B, R_{\alpha,B} \rangle$ onto $\langle \alpha \cup C, R_{\alpha,C} \rangle$ in which every element of α goes into itself. It follows that $\langle B, R_B \rangle \cong \langle C, R_C \rangle$ and, consequently, that $\langle \beta, \in \rangle \cong \langle \gamma, \in \rangle$. But then $\beta = \gamma$.

†EXAMPLE 5.5.5. In general, for ordinals α, β, and γ, $\beta + \alpha = \gamma + \alpha$ does not imply $\beta = \gamma$. Take $\alpha = \omega$, $\beta = 2$, $\gamma = 1$. Then $2 + \omega = \omega = 1 + \omega$. Suppose that we applied the method of the preceding proof here (with respect to $R_{B,\alpha}$ and $R_{C,\alpha}$). If b were the counterpart of 1 in B, then

$$G_2^{-1} \circ G_1(b_1)$$

would be 0 (in α).

Theorem 5.5.7 implies some inequalities involving ordinal addition; we leave the verification of these as exercises.

Corollary 1. *For all ordinal numbers α, β, and γ, if $\beta < \gamma$, then $\alpha + \beta < \alpha + \gamma$.*

Corollary 2. *Let α and β ($\neq 0$) be any ordinal numbers. Then $\alpha < \alpha + \beta$.*

Exercises

†5.5.1. Prove that for all ordinals α and γ, and for all ordinals β_1 and β_2 both of which contain γ,

$$f_{\alpha,\beta_1}(\gamma) = f_{\alpha,\beta_2}(\gamma).$$

5.5.2. Use the definition of ordinal addition to determine each of these ordinals in the form $S_N(\omega)$, $N = 0, 1, 2, \ldots$.

(a) $\omega + 4$. **(b)** $4 + \omega$.
(c) $(\omega + 2) + 3$. **(d)** $(2 + \omega) + 3$.
(e) $3 + (\omega + 2)$. **(f)** $3 + (2 + \omega)$.

5.5.3. Simplify, wherever possible. [Do not use "$\mathscr{L}(A)$", etc.]

(a) $\omega^2 + \omega^3$. **(b)** $\omega^3 + \omega^2$.

(c) $\omega^2 + (\omega + 1) + \omega^2$. **(d)** $(\omega^6 + \omega^2 3 + 1) + \omega^6$.

(e) $\omega^8 + \omega^6 + \omega^2 + 5 + \omega^\omega$.

†5.5.4. Prove that for all natural numbers M and N,

$$M +_{\mathscr{O}} N = M + N.$$

†5.5.5. Let $A \cap B = \varnothing$, let R_A be a linear ordering on A and let R_B be a linear ordering on B.

(a) Prove that $R_{A,B}$, the natural ordering on $A \cup B$, is a linear ordering.

(b) Prove that if A and B are each nonempty, then $R_{A,B} \neq R_{B,A}$.

(c) Prove that if R_A and R_B are each well-orderings, then so is $R_{A,B}$.

5.5.6. In the proof of Theorem 5.5.2, verify that $\langle \mathscr{B}, \in_{\alpha+\beta} \rangle \cong \langle \beta, \in \rangle$.

5.5.7. Prove Theorem 5.5.4.

5.5.8. **(a)** Determine all possible ordinals $\alpha_1 + \alpha_2 + \alpha_3$, where $\{\alpha_1, \alpha_2, \alpha_3\} = \{1, 2, \omega\}$.

(b) Show that there are only two distinct ordinal sums $\alpha_1 + \alpha_2 + \alpha_3$, where $\{\alpha_1, \alpha_2, \alpha_3\} = \{1, \omega, \omega 2\}$.

(Used in Exercise 5.11.3.)

5.5.9. Determine a 3-element set of ordinals $\{\alpha_1, \alpha_2, \alpha_3\}$ such that the number of distinct sums, $\alpha_1 + \alpha_2 + \alpha_3$, is as indicated.

(a) 1.

(b) 3.

(c) 5.

Note: This, together with the preceding exercise, show that three ordinals, in all possible arrangements, can yield K sums for $1 \leq K \leq 5$. Exercise 5.11.3 will show that this is impossible for $K = 6 (= 3!)$.

5.5.10. Prove by transfinite induction that for every ordinal α, $0 + \alpha = \alpha$.

5.5.11. Find examples illustrating each.

(a) α and $\alpha + \beta$ can both be limit ordinals.

(b) α can be a limit ordinal and $\alpha + \beta$ a successor ordinal.

(c) α and $\alpha + \beta$ can both be successor ordinals.

(d) α can be a successor ordinal and $\alpha + \beta$ a limit ordinal.

5.5.12. Prove Theorem 5.5.5 by the representative-well-ordered-systems method.

†5.5.13. Prove that for all ordinal numbers α and β $(\neq 0)$, $\alpha + \beta$ is the smallest ordinal that includes $\{\alpha + \gamma : \gamma < \beta\}$.

5.5.14. **(a)** Prove Corollary 1 of Theorem 5.5.7; i.e., prove that for all ordinals α, β, and γ, $\beta < \gamma$ implies $\alpha + \beta < \alpha + \gamma$.

(b) Find a counterexample to show that $\beta < \gamma$ does not imply $\beta + \alpha < \gamma + \alpha$.

5.5.15. **(a)** Prove Corollary 2 of Theorem 5.5.7. Thus, show that for all ordinals α and $\beta (\neq 0)$, $\alpha < \alpha + \beta$.

(b) Find ordinals α and β for which $\beta = \alpha + \beta$.

†5.5.16. Prove that for all ordinals α, β, and γ, the following hold:

(a) If $\alpha \leq \beta$, then $\alpha + \gamma \leq \beta + \gamma$.

(b) $\beta \leq \alpha + \beta$.

†5.5.17. Show that if α is an infinite ordinal, then for every nonzero natural number N, $N + \alpha = \alpha < \alpha + N$.

5.6 Ordinal Multiplication

Now we apply Definition by Transfinite Induction to define ordinal multiplication. Let α be a (constant) ordinal number and let s_α be the strictly bimonadic formula, "x is an ordinal number and $y = x + \alpha$—or x is not an ordinal number and $y = 0$." Then for any ordinal number $\beta > 0$, there is a (unique) function $g_{\alpha,\beta}$ with domain β that satisfies the following:

(1) $g_{\alpha,\beta}(0) = 0$.
(2) For every successor ordinal $\gamma^+ < \beta$, $g_{\alpha,\beta}(\gamma^+) = g_{\alpha,\beta}(\gamma) + \alpha$.
(3) For every limit ordinal $\gamma < \beta$, $g_{\alpha,\beta}(\gamma) = \cup\{g_{\alpha,\beta}(\delta) : \delta < \gamma\}$.

In (2), $g_{\alpha,\beta}(\gamma^+)$ is the unique set c_2 for which "$s_\alpha\begin{pmatrix} x & y \\ g_{\alpha,\beta}(\gamma) & c_2 \end{pmatrix}$" is true.

For all ordinals α and γ and for all ordinals β_1 and β_2 both of which contain γ, $g_{\alpha,\beta_1}(\gamma) = g_{\alpha,\beta_2}(\gamma)$; the proof is again by transfinite induction. We define "$a \cdot_{\mathcal{O}} \beta$" to be "$g_{\alpha,\beta^+}(\beta)$". As was the case for ordinal addition, *if* α *and* β *are natural numbers, then* $\alpha \cdot_{\mathcal{O}} \beta = \alpha \cdot \beta$. Again, we agree to write "$\alpha \cdot \beta$" in lieu of "$\alpha \cdot_{\mathcal{O}} \beta$" (for all ordinals α and β); we frequently omit even the dot and write "$\alpha\beta$". The ordinals denoted (in Section 5.3) by "$\omega 2$", "$\omega 3$", etc., are the same in the new sense.

†EXAMPLE 5.6.1

(1) $\omega 3 = g_{\omega,4}(3) = g_{\omega,4}(2) + \omega = g_{\omega,4}(1) + \omega + \omega =$
$g_{\omega,4}(0) + \omega + \omega + \omega = 0 + \omega + \omega + \omega = \omega + \omega + \omega$.
(2) $3 \cdot \omega = g_{3,\omega^+}(\omega) = \cup\{g_{3,\omega^+}(N) : N \in \omega\} =$
$\cup\{g_{3,\omega^+}(0), g_{3,\omega^+}(1), g_{3,\omega^+}(2), \ldots\} = \cup\{0, 3, 6, \ldots\} = \omega$.
(3) $\omega \cdot \omega = g_{\omega,\omega^+}(\omega) = \cup\{g_{\omega,\omega^+}(N) : N \in \omega\} =$
$\cup\{g_{\omega,\omega^+}(0), g_{\omega,\omega^+}(1), g_{\omega,\omega^+}(2), \ldots\} = \cup\{0, \omega, \omega 2, \ldots\} = \omega^2$.
(4) $(\omega + 3)\omega = \cup\{\omega + 3, \omega 2 + 3, \omega 3 + 3, \ldots\} = \omega^2$.
(5) $\omega(\omega + 3) = \cup\{\omega, \omega 2, \omega 3, \ldots\} + \omega + \omega + \omega = \omega^2 + \omega 3$.

(6) $(\omega + 3)(\omega + 2) = \omega^2 + (\omega + 3) + (\omega + 3) = \omega^2 + \omega 2 + 3.$

(7) $$\begin{aligned}\omega \cdot \omega^2 &= \cup\{\omega\alpha : \alpha \in \omega^2\}\\ &= \cup\{\omega, \omega 2, \omega 3, \ldots, \omega\omega, \omega(\omega + 1), \omega(\omega + 2), \ldots,\\ &\qquad \omega(\omega 2), \omega(\omega 2 + 1), \omega(\omega 2 + 2), \ldots\}\\ &= \cup\{\omega, \omega 2, \omega 3, \ldots, \omega^2, \omega^2 + \omega, \omega^2 + \omega 2, \ldots,\\ &\qquad \omega^2 2, \omega^2 2 + \omega, \omega^2 2 + \omega 2, \ldots\}\\ &= \omega^3.\end{aligned}$$

We see that ordinal multiplication, like ordinal addition, is noncommutative.

Example 5.6.1(1) suggests that *for every ordinal number* α *and every natural number* $N > 1$, $\alpha N = \alpha + \alpha + \cdots + \alpha$ (N times). The proof of this is by a routine induction on N. In fact, in Section 5.9, when we extend addition so as to allow for "transfinite sums", we will see that for all ordinals α and β, $\alpha\beta = \alpha + \alpha + \cdots + \alpha + \cdots$ ("β times").

In analogy with Theorem 5.5.1, we have the following:

Theorem 5.6.1. *For all ordinal numbers* α *and* β, $\alpha \cdot \beta$ *is an ordinal number.*

The proof, which is left as an exercise, utilizes the fact that $\alpha + \beta$ is an ordinal, and proceeds along the lines of that proof.

It follows almost immediately from the definition that *for all ordinals* α *and* β, the following are true:

(5.6.1) $\alpha 0 = 0.$

(5.6.2) $\alpha\beta^+ = \alpha\beta + \alpha.$

If β *is a limit ordinal,* $\alpha\beta = \cup\{\alpha\gamma : \gamma \in \beta\}$.

From (5.6.1) and (5.6.2) we conclude that *for all ordinals* α, $\alpha 1 = \alpha$.

Theorem 5.5.2 suggests defining a well-ordering on $A \times B$ that is such that for appropriate well-ordering R_A on A and R_B on B, $\langle A, R_A\rangle \cong \langle \alpha, \in\rangle$ and $\langle B, R_B\rangle \cong \langle \beta, \in\rangle$. Unfortunately, the definition of ordinal multiplication forces us to define this ordering in a peculiar way. Had we defined $\alpha\beta$ so that the roles of α and β were reversed, a more "natural" ordering on $A \times B$ could have been used; however, this would have been contrary to the traditional definition.

Let R_A and R_B be linear orderings on the sets A and B, respectively. (Here, we need not require that A and B be disjoint.) We define $R_{A \times B}$ on $A \times B$ by $\langle a_1, b_1\rangle R_{A \times B} \langle a_2, b_2\rangle$ iff either of the following holds:

$$b_1 \neq b_2 \text{ and } b_1 R_B b_2,$$

$$b_1 = b_2 \text{ and } a_1 R_A a_2.$$

(The reader should verify that $R_{A \times B}$ is a linear ordering.) $R_{A \times B}$ is called the **antilexicographical ordering on** $A \times B$ **(with respect to** $\langle R_A, R_B \rangle$**).** *If A and B are distinct nonempty sets,* $R_{A \times B} \neq R_{B \times A}$*; and if* R_A *and* R_B *are well-orderings, so are* $R_{A \times B}$ *and* $R_{B \times A}$.

The ordered pairs $\langle a, b \rangle$ are arranged as are two-letter words in a dictionary (lexicon) except that the ordering of the second letters takes precedence over that of the first letters.

EXAMPLE 5.6.2. Let $A = \{a, b\}$ and let $B = \{c, d, e\}$, where a, b, c, d, and e are distinct. Let R_A and R_B be represented by the chain diagrams of Figure 5.6.1(a) and $R_{A \times B}$ by the chain diagram of Figure 5.6.1(b).

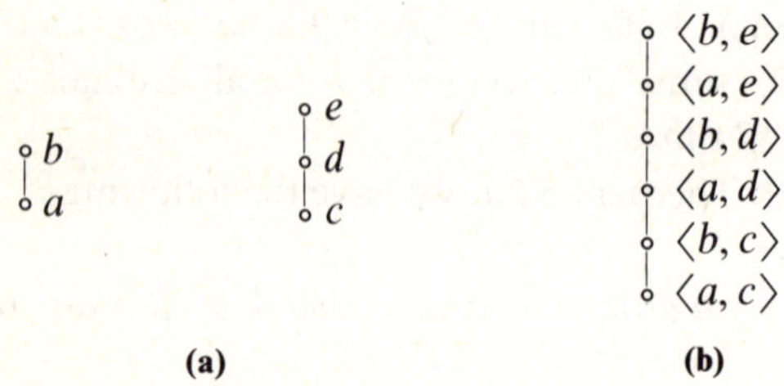

Fig. 5.6.1

In the graphic diagram for $A \times B$, if two points appear on distinct horizontal lines we take the lower one first; if two points lie on the same horizontal line, we take the left one first. (See Figure 5.6.2.)

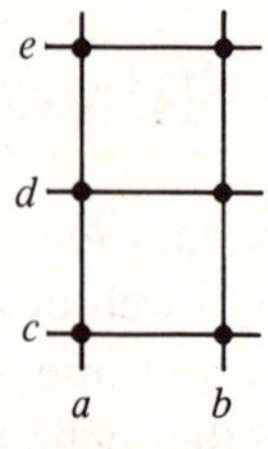

Fig. 5.6.2

EXAMPLE 5.6.3. Let $A = \omega$, let $B = 2$, and consider the natural ordering on each. Then $A \times B$ is represented by the graphic diagram of Figure 5.6.3(a) and $B \times A$ by the diagram of Figure 5.6.3(b). We see that $\langle \omega \times 2, R_{\omega \times 2} \rangle \cong \langle \omega 2, \in \rangle$, whereas $\langle 2 \times \omega, R_{2 \times \omega} \rangle \cong \langle 2\omega, \in \rangle$ $(= \langle \omega, \in \rangle)$.

Theorem 5.6.2. *Let* α *and* β *be any two ordinal numbers. Let* $\langle A, R_A \rangle$ *and* $\langle B, R_B \rangle$ *be any well-ordered systems that are such that* $\langle A, R_A \rangle \cong \langle \alpha, \in \rangle$ *and*

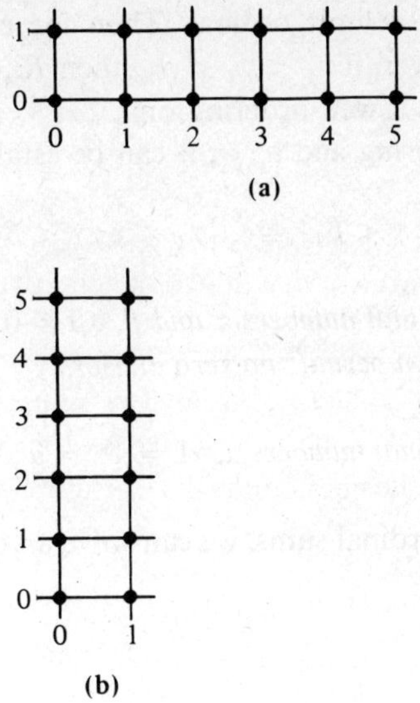

Fig. 5.6.3

$\langle B, R_B\rangle \cong \langle \beta, \in\rangle$. Let $R_{A\times B}$ be the antilexicographical ordering on $A \times B$. Then $\langle A \times B, R_{A\times B}\rangle \cong \langle \alpha\beta, \in\rangle$.

PROOF: Clearly, $\langle A \times B, R_{A\times B}\rangle \cong \langle \alpha \times \beta, R_{\alpha\times\beta}\rangle$; thus, it suffices to consider the latter system. The well-ordered system $\langle \alpha \times \beta, R_{\alpha\times\beta}\rangle$ is similar to a unique ordinal system, $\langle \gamma, \in\rangle$. We claim that $\gamma = \alpha\beta$. Otherwise, let β_1 be the first ordinal β that is such that for some ordinal α, $\langle \alpha \times \beta, R_{\alpha\times\beta}\rangle \not\cong \langle \alpha\beta, \in\rangle$; let α_1 be any such ordinal α. Clearly $\beta_1 \neq 0$, because for all ordinals α, $\alpha \times 0 = \alpha 0 = 0$.

Suppose that $\beta_1 = {\beta_0}^+$. Then

$$\alpha_1\beta_1 = \alpha_1{\beta_0}^+ = \alpha_1\beta_0 + \alpha_1.$$

Also,

$$\langle \alpha_1\beta_0, \in\rangle \cong \langle \alpha_1 \times \beta_0, R_{\alpha_1\times\beta_0}\rangle$$

and

$$\langle \alpha_1, \in\rangle \cong \langle \alpha_1 \times \{\beta_0\}, R_{\alpha_1\times\{\beta_0\}}\rangle.$$

Clearly, $(\alpha_1 \times \beta_0) \cap (\alpha_1 \times \{\beta_0\}) = \varnothing$. Thus,

$$\begin{aligned}\langle \alpha_1\beta_0 + \alpha_1, \in\rangle &\cong \langle (\alpha_1 \times \beta_0) \cup (\alpha_1 \times \{\beta_0\}), R_{\alpha_1\times\beta_0,\alpha_1\times\{\beta_0\}}\rangle \\ &= \langle \alpha_1 \times {\beta_0}^+, R_{\alpha_1\times{\beta_0}^+}\rangle \\ &= \langle \alpha_1 \times \beta_1, R_{\alpha_1\times\beta_1}\rangle.\end{aligned}$$

Suppose that β_1 is a limit ordinal. Then for each $\gamma < \beta_1$, $\langle \alpha_1\gamma, \in \rangle \cong \langle \alpha_1 \times \gamma : R_{\alpha_1 \times \gamma} \rangle$. Moreover, if $\gamma_1 < \gamma_2 < \beta_1$, then $R_{\alpha_1 \times \gamma_1} \subseteq R_{\alpha_1 \times \gamma_2}$. It follows that $\cup\{R_{\alpha_1 \times \gamma} : \gamma < \beta_1\}$ is a well-ordering on $\cup\{\alpha_1 \times \gamma : \gamma < \beta_1\}$. (Verify this.) With respect to this ordering and $\in_{\alpha_1\beta_1}$, it can be established that

$$\alpha_1\beta_1 = \cup\{\alpha_1\gamma : \gamma < \beta_1\} \cong \cup\{\alpha_1 \times \gamma : \gamma < \beta_1\} = \alpha_1 \times \beta_1.$$

Corollary 1. *For all ordinal numbers α and β, $\alpha\beta = 0$ if and only if $0 \in \{\alpha, \beta\}$. Thus, ordinal multiplication permits no zero divisors.*

Corollary 2. *For all ordinal numbers α, $\alpha 1 = 1\alpha = \alpha$.*

As was the case for ordinal sums, we can solve an ordinal equation

$$\alpha\beta = \gamma \tag{5.6.3}$$

or an ordinal inequality

$$\alpha\beta < \gamma \tag{5.6.4}$$

by the representative-systems method. Thus, if $\langle A, R_A \rangle \cong \langle \alpha, \in \rangle$, $\langle B, R_B \rangle \cong \langle \beta, \in \rangle$, and $\langle C, R_C \rangle \cong \langle \gamma, \in \rangle$, then equation (5.6.3) is true iff $\langle A \times B, R_{A \times B} \rangle \cong \langle C, R_C \rangle$, and inequality (5.6.4) is true iff $\langle A \times B, R_{A \times B} \rangle$ is similar to a section of $\langle C, R_C \rangle$.

We can also utilize the representative-systems method to prove multiplicative associativity and right distributivity for ordinals (Exercise 5.6.8).

Theorem 5.6.3. *Let α, β, and γ be any ordinal numbers.*

(1) $\alpha(\beta\gamma) = (\alpha\beta)\gamma$.

(2) $\alpha(\beta + \gamma) = \alpha\beta + \alpha\gamma$.

†EXAMPLE 5.6.4. Left distributivity does not hold: For example,

$$(1 + 2)\omega = \omega \neq \omega 2 = 1\omega + 2\omega.$$

Theorem 5.6.4. *Let $\alpha\,(\neq 0)$, β, and γ be any ordinal numbers and let $\beta < \gamma$. Then $\alpha\beta < \alpha\gamma$.*

PROOF: Suppose that $\beta < \gamma$. Then for any ordinal $\delta < \alpha$, $\langle \{\delta\} \times \beta, R_{\{\delta\} \times \beta} \rangle$ is similar to a section of $\langle \{\delta\} \times \gamma, R_{\{\delta\} \times \gamma} \rangle$. We can use this to show, by transfinite induction, that $\langle \alpha \times \beta, R_{\alpha \times \beta} \rangle$ is similar to a section of $\langle \alpha \times \gamma, R_{\alpha \times \gamma} \rangle$. Therefore, $\alpha\beta < \alpha\gamma$.

Corollary. *Let $\alpha\,(\neq 0)$, β, and γ be any ordinal numbers. Then $\alpha\beta = \alpha\gamma$ implies $\beta = \gamma$.*

†EXAMPLE 5.6.5. For ordinals $\alpha\,(\neq 0)$, β, and γ, $\beta < \gamma$ does *not* imply $\beta\alpha < \gamma\alpha$. Let $\alpha = \omega$, $\beta = 1$, $\gamma = 2$; then $1\omega = 2\omega$.

The following weaker inequality can be proved along the lines of the preceding proof.

Theorem 5.6.5. *Let α, β, γ, and δ be any ordinal numbers. If $\alpha \leq \gamma$ and $\beta \leq \delta$, then $\alpha\beta \leq \gamma\delta$.*

Other inequalities involving ordinal multiplication will be given in the exercises.

Theorem 5.6.6. *Let α and β be any ordinal numbers. Then $\alpha\beta$ is a limit ordinal if and only if the following are true:*

(1) *α and β are nonzero.*

(2) *(At least) one of these is a limit ordinal.*

PROOF: We first show that (1) and (2) are both necessary for $\alpha\beta$ to be a limit ordinal. If either α or β is zero, then by Corollary 1 of Theorem 5.6.2, $\alpha\beta = 0$. If both are successor ordinals, let $\alpha = \gamma^+$ and $\beta = \delta^+$. Then

$$\alpha\beta = \gamma^+\delta^+ = \gamma^+\delta + \gamma^+ = (\gamma^+\delta + \gamma)^+.$$

Next, suppose that (1) holds. First, let β be a limit ordinal. Clearly, $\alpha\beta \neq 0$; thus (according to Exercise 5.3.11), in order to show that $\alpha\beta$ is a limit ordinal, it suffices to show that $\alpha\beta \subseteq \cup(\alpha\beta)$. But $\alpha\beta = \cup\{\alpha\gamma : \gamma < \beta\}$. Thus, if $x \in \alpha\beta$, then $x \in \delta \in \alpha\gamma$ for some $\gamma < \beta$ and some $\delta < \alpha\gamma$. By Theorem 5.6.4,

$$x \in \delta \in \alpha\gamma \in \alpha\beta.$$

By the transitivity of $\alpha\beta$, $x \in \delta \in \alpha\beta$. Consequently, $x \in \cup(\alpha\beta)$. Finally, suppose that α is a limit ordinal and $\beta\,(= \beta_0{}^+)$ is a successor ordinal. Then $\alpha\beta = \alpha\beta_0 + \alpha$. By Corollary 3 of Theorem 5.5.4, $\alpha\beta$ is a limit ordinal.

Exercises

†5.6.1. Show that for all ordinals α and γ and for all ordinals β_1 and β_2 both of which contain γ, $g_{\alpha,\beta_1}(\gamma) = g_{\alpha,\beta_2}(\gamma)$.

5.6.2. Use the definition of ordinal multiplication to determine each of these ordinals in the form ωN, $N = 1, 2, 3, \ldots$.

(a) 5ω. (b) $2(\omega 5)$.
(c) $5(\omega 2)$. (d) $5(2\omega)$.
(e) $(\omega 5)2$.

5.6.3. Simplify each of the indicated ordinals.

(a) $(\omega + 1)3$. (b) $3(\omega + 1)$.
(c) $3\omega + 1$. (d) $(\omega + 2)\omega$.
(e) $\omega(\omega + 2)$. (f) $(\omega + 2)(\omega + 1)$.
(g) $(\omega + 1)(\omega + 2)$.

†5.6.4. (a) Show that for every ordinal number α and every natural number $N > 1$, $aN = \alpha + \alpha + \cdots + \alpha$ (N times).

(b) Show that for every infinite ordinal α and every natural number $N > 1$, $\alpha N \neq N\alpha$.

(c) Let α be a nonzero ordinal. Prove that a necessary and sufficient condition that α be a limit ordinal is that $N\alpha = \alpha$ for every nonzero natural number N.

5.6.5. Prove Theorem 5.6.1. Thus, show that for any two ordinals α and β, $\alpha\beta$ is an ordinal.

†5.6.6. Prove that for all ordinals α and β, (a), (b), and (c) are true:

(a) $\alpha 0 = 0$.

(b) $\alpha\beta^+ = \alpha\beta + \alpha$.

(c) If β is a limit ordinal, $\alpha\beta = \cup\{\alpha\gamma : \gamma < \beta\}$.

(d) If β is a successor ordinal, state a necessary and sufficient condition (on α) in order that $\alpha\beta = \cup\{\alpha\gamma : \gamma < \beta\}$.

5.6.7. Consider the proof of Theorem 5.6.2.

(a) Show that $\langle A \times B, R_{A \times B}\rangle \cong \langle \alpha \times \beta, R_{\alpha \times \beta}\rangle$.

(b) Verify that $\cup\{R_{\alpha_1 \times \gamma} : \gamma < \beta_1\}$ is a well-ordering on $\cup\{\alpha_1 \times \gamma : \gamma < \beta_1\}$.

(c) Verify that $\cup\{\alpha_1\gamma : \gamma < \beta_1\} \cong \cup\{\alpha_1 \times \gamma : \gamma < \beta_1\}$.

(d) Prove the two corollaries of Theorem 5.6.2.

5.6.8. Prove Theorem 5.6.3.

5.6.9. Determine all possible products $\alpha_1 \cdot \alpha_2 \cdot \alpha_3$, where $\{\alpha_1, \alpha_2, \alpha_3\} = \{3, \omega, \omega 2\}$.

★5.6.10. For each $N \geq 1$, determine an N-element set of ordinals $\{\alpha_1, \alpha_2, \ldots, \alpha_N\}$ such that $\{\alpha_1 \cdot \alpha_2 \cdot \ldots \cdot \alpha_N\}$ is an $N!$-element set. [$N! = N \cdot (N \mathbin{\dot{-}} 1) \cdot \ldots \cdot 3 \cdot 2 \cdot 1$.]

5.6.11. Write out the transfinite inductive argument used in the proof of Theorem 5.6.4.

5.6.12. Prove Theorem 5.6.5.

5.6.13. Prove that for all nonzero ordinal numbers α and β, $\alpha\beta \geq \max\{\alpha, \beta\}$.

5.6.14. Let α and β be ordinals and let $\alpha < \beta$.

†(a) Prove that if γ is a successor ordinal, then $\alpha\gamma < \beta\gamma$.

(b) Determine infinite ordinals α and β, as above, and a limit ordinal γ such that $\alpha\gamma = \beta\gamma$.

†5.6.15. Let α, β, and γ be any ordinal numbers. Prove (a) and (b).

(a) $\alpha\beta < \alpha\gamma$ implies $\beta < \gamma$.

(b) $\beta\alpha < \gamma\alpha$ implies $\beta < \gamma$.

(c) Suppose $\alpha \neq 0$. Does $\alpha\beta \leq \alpha\gamma$ imply $\beta \leq \gamma$?

5.7 Ordinal Subtraction and Division

We define subtraction of ordinal numbers in terms of ordinal addition. The few theorems we present follow almost immediately from previous considerations. We will have somewhat more to say about division of ordinal numbers, which can be defined in terms of ordinal multiplication.

Theorem 5.7.1. *Let α and β be any ordinal numbers with $\beta \leq \alpha$. Then there is a unique ordinal number γ such that $\alpha = \beta + \gamma$.*

PROOF: If $\beta = \alpha$, then $\alpha + 0 = \alpha$, whereas if $\gamma > 0$, $\alpha + \gamma > \alpha$.

If $\beta < \alpha$, then $\beta \in \alpha$. Let $C = \{\zeta < \alpha : \beta < \zeta\}$. Then $\langle C, \in_\alpha \rangle$ is well-ordered; let γ be the unique ordinal such that $\langle C, \in_\alpha \rangle \cong \langle \gamma, \in \rangle$. By Theorem 5.5.2, $\alpha = \beta + \gamma$.

If δ is any ordinal such that $\alpha = \beta + \delta$, then $\beta + \delta = \beta + \gamma$. By Theorem 5.5.7, $\delta = \gamma$.

Corollary 1. *If α, β, and γ are any ordinal numbers that satisfy $\alpha \leq \beta \leq \alpha + \gamma$, then $\beta = \alpha + \delta$ for $\delta \leq \gamma$.*

(The proof of this corollary is left as an exercise.)

Corollary 2. *If α is an infinite ordinal number, then there is a unique ordinal number γ such that $\alpha = \omega + \gamma$.*

For $\alpha \geq \beta$, we define, "$\alpha -_{\mathcal{O}} \beta$" to be the unique ordinal γ such that $\alpha = \beta + \gamma$. Thus, $\alpha = \beta + (\alpha -_{\mathcal{O}} \beta)$. The ordinal "$\alpha -_{\mathcal{O}} \beta$" is called the "**difference between α and β**".

For natural numbers M and N with $M \leq N$, $N \dot{-} M = N -_{\mathcal{O}} M$; henceforth, we write "$\alpha \dot{-} \beta$" instead of "$\alpha -_{\mathcal{O}} \beta$".

EXAMPLE 5.7.1

(1) $\omega \dot{-} 2 = \omega$ because $2 + \omega = \omega$.

(2) $(\omega + 4) \dot{-} 2 = \omega + 4$ because

$$2 + (\omega + 4) = (2 + \omega) + 4 = \omega + 4.$$

On the other hand, $\omega + (4 \dot{-} 2) = \omega + 2$.

(3) $(\omega + 1) \dot{-} \omega = 1$.

(4) $\omega 2 \dot{-} \omega = \omega$ because $\omega + \omega = \omega 2$.

†EXAMPLE 5.7.2. Let α, β, γ, and γ' be any ordinals with $\beta \geq \gamma$ and $\beta \geq \gamma'$. In general, observe the following:
(1) $\alpha + (\beta \dot{-} \gamma) \neq (\alpha + \beta) \dot{-} \gamma$. [Consider Example 5.7.1(2).]
(2) $\beta \dot{-} \gamma = \beta \dot{-} \gamma'$ does not imply $\gamma = \gamma'$. (Take $\beta = \omega, \gamma = 0, \gamma' = 1$.)
(3) $(\beta \dot{-} \gamma) + \gamma \neq \beta$. (Take $\beta = \omega, \gamma = 1$.)

Theorem 5.7.2. *If α, β, and γ are ordinal numbers and if $\beta \leq \alpha$, then $\gamma(\alpha \dot{-} \beta) = \gamma\alpha \dot{-} \gamma\beta$.*

PROOF: By Theorem 5.6.5, $\gamma\beta \leq \gamma\alpha$; thus, $\gamma\alpha \dot{-} \gamma\beta$ is defined. Let $\delta = \alpha \dot{-} \beta$; then $\alpha = \beta + \delta$. Also, $\gamma\alpha = \gamma(\beta + \delta) = \gamma\beta + \gamma\delta$. Thus, $\gamma(\alpha \dot{-} \beta) = \gamma\delta = \gamma\alpha \dot{-} \gamma\beta$.

†EXAMPLE 5.7.3. In general, for $\beta \leq \alpha$,

$$(\alpha \dot{-} \beta)\gamma \neq \alpha\gamma \dot{-} \beta\gamma.$$

Take $\alpha = 2$, $\beta = 1$, $\gamma = \omega$. Then

$$(\alpha \dot{-} \beta)\gamma = (2 \dot{-} 1)\omega = 1\omega = \omega \neq 0 = \omega \dot{-} \omega = 2\omega - 1\omega = \alpha\gamma \dot{-} \beta\gamma.$$

The following inequalities involving ordinal subtraction also hold.

Theorem 5.7.3. *Let α_1, α_2, and β be any ordinal numbers.*
(1) *If $\beta \leq \alpha_1 < \alpha_2$, then $\alpha_1 \dot{-} \beta < \alpha_2 \dot{-} \beta$.*
(2) *If $\alpha_1 < \alpha_2 \leq \beta$, then $\beta \dot{-} \alpha_2 \leq \beta \dot{-} \alpha_1$.*

PROOF: We prove (1) and leave (2) as an exercise. (1) Let $\delta_1 = \alpha_1 \dot{-} \beta$ and let $\delta_2 = \alpha_2 \dot{-} \beta$. Then $\beta + \delta_1 = \alpha_1 < \alpha_2 = \beta + \delta_2$.
By trichotomy and Corollary 1 of Theorem 5.5.7, $\delta_1 < \delta_2$.

The Division Algorithm (for ω) (Exercise 3.3.12) states that for natural numbers M (> 0) and N there are unique natural numbers K and R that satisfy

$$N = K \cdot M + R \quad \text{and} \quad 0 \leq R < M.$$

Here, K can be considered the quotient and R the remainder. The Division Algorithm can be stated for integers (Exercise 3.4.14); we now show that it also applies to ordinal numbers.

Theorem 5.7.4. The Division Algorithm for Ordinal Numbers. *Let α and β be any ordinal numbers with $\beta > 0$. Then there are unique ordinal numbers γ and ρ that satisfy*

(5.7.1) $$\alpha = \beta\gamma + \rho \quad \textit{and} \quad 0 \leq \rho < \beta.$$

PROOF: We first show the existence of ordinals γ and ρ satisfying (5.7.1). Clearly, $\alpha \leq \beta\alpha$. Hence, $\{\delta: \delta \leq \alpha \leq \beta\delta\} \neq \varnothing$; let δ_1 be the first element of this set. If $\alpha = \beta\delta_1$, let $\gamma = \delta_1$ and $\rho = 0$. If $\alpha < \beta\delta_1$, δ_1 cannot be 0. Nor can δ_1 be a limit ordinal because $\alpha < \beta\delta_1 = \cup\{\beta\delta' : \delta' < \delta_1\}$ would imply $\alpha < \beta\delta_0$ for some $\delta_0 < \delta_1$. Thus, δ_1 is a successor ordinal; let $\delta_1 = \delta_2{}^+$. It follows that

$$\beta\delta_2 < \alpha < \beta\delta_1 = \beta(\delta_2 + 1) = \beta\delta_2 + \beta.$$

By Corollary 1 of Theorem 5.7.1, $\alpha = \beta\delta_2 + \rho$, for $\rho < \beta$.

We now show the uniqueness of γ and ρ. Suppose that $\alpha = \beta\gamma_1 + \rho_1 = \beta\gamma_2 + \rho_2$ are two representations of the form (5.7.1). If $\gamma_1 \neq \gamma_2$, suppose, arbitrarily, that $\gamma_1 < \gamma_2$. Then $\gamma_1{}^+ \leq \gamma_2$; hence,

$$\alpha = \beta\gamma_1 + \rho_1 < \beta\gamma_1 + \beta = \beta(\gamma_1 + 1) \leq \beta\gamma_2 \leq \beta\gamma_2 + \rho_2 = \alpha.$$

This is a contradiction. Thus, $\gamma_1 = \gamma_2$. If $\rho_1 \neq \rho_2$, say $\rho_1 < \rho_2$, then

$$\alpha = \beta\gamma_1 + \rho_1 < \beta\gamma_1 + \rho_2 = \beta\gamma_2 + \rho_2 = \alpha.$$

This, too, is a contradiction. Thus, $\rho_1 = \rho_2$.

EXAMPLE 5.7.4

(1) Let $\alpha = 5$ and $\beta = \omega$. Then $5 = \omega 0 + 5$.
(2) Let $\alpha = \omega$ and $\beta = 5$. Then $\omega = 5\omega + 0$.
(3) Let $\alpha = \omega^2$ and $\beta = \omega 2$. Then $\omega^2 = (\omega 2)\omega + 0$.
(4) Let $\alpha = \omega 2 + 5$ and $\beta = \omega + 1$. Then $\omega 2 + 5 = (\omega + 1)2 + 4$.

If $\alpha = \beta\gamma$, we say that β is a **left divisor of** α, that γ is a **right divisor of** α, and that β and γ are each **divisors of** α. *Every ordinal α is both a left and a right divisor of itself*; *1 is also both a left and right divisor of every ordinal α*. Clearly, *every divisor β of a nonzero ordinal α satisfies $\beta \leq \alpha$.*

EXAMPLE 5.7.5

(1) Every nonzero natural number is a left divisor of ω, as is ω; these constitute all the left divisors of ω. The only right divisors of ω are ω and 1.
(2) The only divisors of $\omega + 1$ are itself and 1. Note that for $M, N < \omega$, it follows that $MN < \omega$, $M\omega = \omega$, and $\omega M > \omega$ for $M \geq 2$.
(3) The left divisors of $\omega 2$ are the nonzero natural numbers, ω, and $\omega 2$; the right divisors of $\omega 2$ are 1, 2, and $\omega 2$.

Theorem 5.7.5. *Every nonzero ordinal number α has a finite number of distinct right divisors.*

PROOF: Suppose that α (> 0) has infinitely many right divisors. Then there is a denumerable set of these; let D be such a set. We can enumerate the elements δ_I of D in such a way that $\delta_0 < \delta_1 < \delta_2 < \cdots < \delta_N < \cdots$. To each δ_I there corresponds at least one left divisor of α; let γ_I be the smallest such left divisor. Thus, $\gamma_I \delta_I = \alpha$ for each I in ω. For $I < J$, if $\gamma_I \leq \gamma_J$, then, by Theorems 5.6.4 and 5.6.5, $\alpha = \gamma_I \delta_I < \gamma_I \delta_J \leq \gamma_J \delta_J = \alpha$. Thus, for all I, J in ω, with $I < J$, $\gamma_I > \gamma_J$. This means that there is a denumerable set of ordinals satisfying

$$\cdots \gamma_{N^+} < \gamma_N < \cdots < \gamma_2 < \gamma_1 < \gamma_0,$$

in contradiction of Exercise 5.2.13.

Theorem 5.7.6. *Let α be a nonzero ordinal number. Then α is a successor ordinal if and only if it has a finite number of left divisors.*

PROOF: Suppose that α is a successor ordinal. Let β be a left divisor of α. Then by the Corollary of Theorem 5.6.4, there is a unique right divisor, γ, of α corresponding to β (i.e., there is a unique ordinal γ for which $\beta\gamma = \alpha$). Moreover, by Theorem 5.6.6, γ is a successor ordinal. Each such right divisor γ corresponds to a unique left divisor, for, if $\beta_1 < \beta_2$, then by Exercise 5.6.14(1), $\beta_1\gamma < \beta_2\gamma$. Thus, there is a one-one correspondence between the left and right divisors of a successor ordinal. Hence, there are only finitely many left divisors.

Suppose that α is a limit ordinal. Then by Exercise 5.6.4(3), $N\alpha = \alpha$ for every nonzero N in ω.

An ordinal number α is said to be (a) **prime** if $\alpha > 1$ and if $\alpha = \beta\gamma$ implies either $\beta = \alpha$ or $\gamma = \alpha$. Ordinals $\alpha > 1$ that are not primes are called **composites**. Clearly, *a natural number is a prime* (in the sense of Section 3.3) *iff it is a prime ordinal.*

†EXAMPLE 5.7.6

(1) ω is a prime because for $M, N < \omega$, $MN < \omega$; for $\alpha > \omega$ and for every nonzero ordinal β, $\alpha\beta > \omega$ and $\beta\alpha > \omega$.
(2) $\omega + 1$ is a prime, according to Example 5.7.5(2).
(3) $\omega 2$ is a composite; for each $N \geq 2$ (in ω), ωN and ω^N are each composites.
(4) ω^ω is a prime.

We say that $\alpha_1\alpha_2 \ldots \alpha_N$ is a **prime factorization of the ordinal** α if the following are true.

(5.7.2) $\alpha = \alpha_1\alpha_1 \cdots \alpha_N$.
(5.7.3) $1 \leq N < \omega$.
(5.7.4) α_I is a prime for $1 \leq I \leq N$.

Observe that the order of the factors is, generally, crucial. Recall that every natural number has a unique prime factorization. We say that α has a **unique prime factorization** if whenever $\alpha_1\alpha_2\cdots\alpha_N$ and $\beta_1\beta_2\cdots\beta_M$ are each prime factorizations of α, ($M = N$ and) for each I, $1 \le I \le N$, and for each J, $1 \le J \le M$, α_I and β_J each appear the same number of times among the ordinals $\alpha_1, \alpha_2, \ldots, \alpha_N$ as among $\beta_1, \beta_2, \ldots, \beta_M$. [Strictly speaking, we should have defined a prime factorization as an ordered N-tuple $\langle\alpha_1, \alpha_2, \ldots, \alpha_N\rangle$ for which (5.7.2), (5.7.3), and (5.7.4) hold. But then a natural number with two distinct prime factors would not have a unique prime factorization.] An (infinite) ordinal number need not have a unique prime factorization, even in our intended sense.

EXAMPLE 5.7.7

(1) $N\omega = \omega$ for each N in ω_+; hence, if P and Q are distinct (natural number) primes, then $P\omega$ and $Q\omega$ are distinct prime factorizations of ω. Clearly, ω, even though it is a prime, has infinitely many distinct prime factorizations.

(2) $$\begin{aligned}(\omega + 1)\omega &= g_{\omega+1,\omega+1}(\omega)\\ &= \cup\{0, \omega + 1, (\omega + 1) + (\omega + 1), (\omega + 1) + (\omega + 1)\\ &\quad + (\omega + 1), \ldots\}\\ &= \cup\{0, \omega + 1, \omega 2 + 1, \omega 3 + 1, \ldots\}\\ &= \omega^2.\end{aligned}$$

But $\omega \cdot \omega = \omega^2$ [Example 5.6.1(3)]. The ordinals ω and $\omega + 1$ are each primes. Hence, the prime factorization of ω^2 is not unique. Clearly, part (1) indicates that there are, in fact, infinitely many prime factorizations of ω^2. [Begin by considering $(N\omega)\omega$.]

(3) $\omega + 1$ has a unique prime factorization—itself.

(4) $\omega + 2$ has (exactly) two prime factorizations—itself and $2(\omega + 1)$.

Theorem 5.7.7. *Every ordinal number $\alpha > 1$ has a prime factorization;* i.e., *for some N in ω_+, there are primes $\alpha_1, \alpha_2, \ldots, \alpha_N$ for which $\alpha = \alpha_1 \cdot \alpha_2 \cdot \ldots \cdot \alpha_N$.*

PROOF: Let γ_1 (>1) be the smallest right divisor of α. Let β_1 be the smallest ordinal (possibly 1) such that $\alpha = \beta_1\gamma_1$. Observe that γ_1 must be prime, for otherwise $\gamma_1 = \delta_1\delta_2$ with $1 < \delta_2 < \gamma_1$; but then $\alpha = (\beta\delta_1)\delta_2$, contradicting the assertion that γ_1 is the smallest such right divisor of α.

If $\beta_1 = 1$, we are finished because then α is prime. If $\beta_1 > 1$, then $\beta_1 < \alpha$, as otherwise, $\beta_1\gamma_1 > \alpha$. We can write $\beta_1 = \beta_2\gamma_2$, where γ_2 (> 1) (possibly $= \beta_1$) is the smallest right divisor of β_1, and β_2 is the smallest corresponding left divisor. Again, γ_2 is prime; moreover, either $\beta_2 = 1$, in which case $\alpha = \beta_1\gamma_1 = \gamma_2\gamma_1$ is a prime factorization of α, or else $\beta_2 < \beta_1$.

If this process does not terminate in a finite number of steps, there is a denumerable set $\{\alpha, \beta_1, \beta_2, \ldots, \beta_N, \beta_{N^+}, \ldots\}$ with

$$\cdots \beta_{N^+} < \beta_N < \cdots < \beta_2 < \beta_1 < \alpha.$$

Because this is not the case, for some N, $\alpha = \gamma_N \cdot \ldots \cdot \gamma_2 \cdot \gamma_1$ is a factorization of α into primes.

In Section 5.10 we will define "infinite products"—i.e., products in which there are infinitely many factors. In the light of this, Theorem 5.7.7 takes on an added significance. If α is such an infinite product, α must also be a finite product of prime factors.

Exercises

5.7.1. Prove Corollary 1 of Theorem 5.7.1.

5.7.2. Determine the following ordinals:

(a) $(\omega + 7) \dot{-} \omega$.

(b) $(\omega + 8) \dot{-} (\omega + 5)$.

(c) $(\omega + 8) \dot{-} 5$.

(d) $(\omega 3 + 2) \dot{-} \omega$.

(e) $(\omega 3 + 2) \dot{-} (\omega + 5)$.

(f) $\omega^2 \dot{-} (\omega 7 + \omega 3 + 4)$.

5.7.3. Show that for all N in ω, if $\beta \le \alpha$ and $\alpha \dot{-} \beta = \omega + N$, then $\alpha = \omega + N$.

5.7.4. State an analogue of Theorem 5.5.2 for subtraction; i.e., state a necessary and sufficient condition for the equality $\alpha \dot{-} \beta = \gamma$ in terms of well-ordered systems.

5.7.5. Determine ordinals $\alpha\,(\ge \omega)$ and $\beta\,(\le \alpha)$ for which the equation $\alpha = x + \beta$ is

(a) Solvable (for x) but not uniquely solvable.

(b) Not solvable.

(c) Uniquely solvable.

5.7.6. Prove Theorem 5.7.3(2).

5.7.7. Let α, β, and γ be any ordinals satisfying $\gamma \le \beta \le \alpha$. Which of the following are true?

(a) $(\alpha \dot{-} \beta) + \gamma = \alpha \dot{-} (\beta \dot{-} \gamma)$.

(b) $(\alpha \dot{-} \gamma) \dot{-} (\beta \dot{-} \gamma) = \alpha \dot{-} \beta$.

5.7.8. Let α and β be as indicated. Determine the unique γ and ρ for which $\alpha = \beta\gamma + \rho$ and $\rho < \beta$.

(a) $\alpha = \omega + 3$, $\beta = 3$.

(b) $\alpha = \omega 2 + 3$, $\beta = \omega + 1$.

(c) $\alpha = \omega^\omega$, $\beta = \omega$.

(d) $\alpha = \omega^2 + \omega 4 + 5$, $\beta = \omega^2 + 1$.

(e) $\alpha = \omega^\omega + \omega^2 + \omega 2 + 1$, $\beta = \omega^4$.

(f) $\alpha = \omega^4$, $\beta = \omega^\omega + \omega^2 + \omega 2 + 1$.

5.7.9. **(a)** Determine all the right divisors of ω^2.

(b) Determine all the left divisors of ω^2.

***5.7.10.** Determine the first three infinite primes.

5.7.11. Determine all prime factorizations of each of the following:

(a) $\omega 6$.

(b) $\omega 8 + 4$.

(c) $\omega^2 + \omega$.

(d) $\omega + P$, where P is a prime (natural number).

5.7.12. Prove that every two ordinals have a greatest common right divisor.

5.7.13. Is the Axiom of Choice necessary for the proof of Theorem 5.7.5?

5.8 Ordinal Exponentiation

For any given ordinal number α, let s_α be the strictly bimonadic formula,

"x is an ordinal number and $y = x \cdot \alpha$—

or x is not an ordinal number and $y = 0$."

Then for every ordinal $\beta > 0$ there is a function $h_{\alpha.\beta}$ with domain β that has the properties:

(1) $h_{0,\beta}(0) = 0$, $h_{\alpha,\beta}(0) = 1$ for $\alpha > 0$.

(2) For every successor ordinal γ^+ in β, $h_{\alpha,\beta}(\gamma^+) = h_{\alpha,\beta}(\gamma) \cdot \alpha$.

(3) For every limit ordinal γ in β, $h_{\alpha,\beta}(\gamma) = \cup\{h_{\alpha,\beta}(\delta) : \delta < \gamma\}$.

Here, in (2), $h_{\alpha,\beta}(\gamma^+)$ is the unique set c_2 such that "$s_\alpha\begin{pmatrix} x & y \\ h_{\alpha,\beta}(\gamma) & c_2 \end{pmatrix}$" is true.

Again, a transfinite induction argument shows that for all ordinals α, β_1, β_2, and γ, with $\gamma \in \beta_1 \cap \beta_2$,

$$h_{\alpha,\beta_1}(\gamma) = h_{\alpha,\beta_2}(\gamma).$$

We define

"α^β" to be "$h_{\alpha,\beta^+}(\beta)$"

for α and β not both 0.

Again, *if α and β are natural numbers, α^β is the same in the ordinal sense as in the natural number sense.*

The ordinals previously denoted (in Section 5.3) *by* "ω^2", "ω^3", "ω^ω", *etc., are the same in the new sense.*

EXAMPLE 5.8.1

(1) $\omega^3 = h_{\omega,4}(3) = h_{\omega,4}(2) \cdot \omega = h_{\omega,4}(1) \cdot \omega \cdot \omega$
$= h_{\omega,4}(0) \cdot \omega \cdot \omega \cdot \omega = 1 \cdot \omega \cdot \omega \cdot \omega = \omega \cdot \omega \cdot \omega.$

(2) $3^\omega = h_{3,\omega^+}(\omega) = \cup\{h_{3,\omega^+}(N) : N < \omega\}$
$= \cup\{h_{3,\omega^+}(0), h_{3,\omega^+}(1), h_{3,\omega^+}(2), h_{3,\omega^+}(3), \ldots\}$
$= \cup\{1, 3, 9, 27, \ldots\}$
$= \omega.$

(3) $\omega^\omega = h_{\omega,\omega^+}(\omega) = \cup\{h_{\omega,\omega^+}(N) : N < \omega\}$
$= \cup\{h_{\omega,\omega^+}(0), h_{\omega,\omega^+}(1), h_{\omega,\omega^+}(2), h_{\omega,\omega^+}(3), \ldots\}$
$= \cup\{1, \omega, \omega^2, \omega^3, \ldots\}.$

Theorem 5.8.1. *For all ordinal numbers α and β, not both 0, α^β is an ordinal number.*

The proof is left as an exercise; here, we utilize the fact that $\alpha \cdot \beta$ is an ordinal.
Note that *a necessary and sufficient condition for α^β to be a successor ordinal is for (at least) one of the following conditions to hold:*

(1) $\alpha = 1$ *and β is arbitrary.*
(2) $\alpha \neq 0$ *and* $\beta = 0$.
(3) *α is a successor ordinal and β is a natural number.*

For all ordinals α and β:

(5.6.1) $\alpha^0 = 1$ *for* $\alpha > 0$.

(5.6.2) $\alpha^{\beta^+} = \alpha^\beta\alpha$, *$\alpha$ and β not both* 0; $0^1 = 0$.

If β is a limit ordinal,

$$\alpha^\beta = \cup\{\alpha^\gamma : \gamma < \beta \text{ (and } \alpha^\gamma \text{ is defined)}\}$$

From (5.6.1) and (5.6.2) we conclude that *for every ordinal* α,

$$\alpha^1 = \alpha.$$

For ordinals α and β, let $\langle A, R_A\rangle$ and $\langle B, R_B\rangle$ be arbitrary well-ordered systems that are similar to $\langle \alpha, \in\rangle$ and $\langle \beta, \in\rangle$, respectively. For $A = \varnothing$, let $Z(A, B) = \mathscr{F}(\varnothing, B) = \{f : f \text{ is a function from } B \text{ into } \varnothing\}$. Thus, $Z(\varnothing, \varnothing) = \{\varnothing\}$ and $Z(\varnothing, B) = \varnothing$ for $B \neq \varnothing$. Let $W_{Z(\varnothing,\varnothing)} = \{\langle\varnothing, \varnothing\rangle\}$ and let $W_{Z(\varnothing,B)} = \varnothing$ for $B \neq \varnothing$. Suppose that $A \neq \varnothing$. Let a_0 be the first element of A. Let $Z(A, B)$ be the set of functions f mapping B into A for which $f(b) = a_0$ for all but a finite number of b in B. If f and g are distinct elements of $Z(A, B)$, clearly, there are only finitely many arguments on which they differ; let b^* $(= b^*_{f,g})$ be the last element b of B such that $f(b) \neq g(b)$. Define the relation $W(= W_{Z(A,B)})$ on $Z(A, B)$ as follows: For f, g in $Z(A, B)$,

$$fWg \text{ if either } f = g \text{ or if } f(b^*)R_A g(b^*).$$

Observe that $Z(A, \varnothing) = \{\varnothing\}$ (whether or not $A = \varnothing$). We will generally write "$\langle Z(A, B), W\rangle$" in lieu of "$\langle Z(A, B), W_{Z(A,B)}\rangle$".

EXAMPLE 5.8.2. Let $\langle A, R_A\rangle = \langle \omega^+, \in\rangle$ and $\langle B, R_B\rangle = \langle \omega, \in\rangle$. Consider $Z(\omega^+, \omega)$; this consists of those ω-sequences Γ in ω^+, for which there is some M ($< \omega$) such that $\Gamma_N = 0$ for all $N \geq M$. Each of the following sequences are in $Z(\omega^+, \omega)$:

$$\Gamma = \langle \omega, 5, 1, 0, 0, 0, \ldots, 0, 0, 0, \ldots\rangle,$$

$$\Delta = \langle 1, 1, 2, 0, 0, 0, \ldots, 0, 0, 0, \ldots\rangle,$$

$$\Phi = \langle 1, 0, 1, 0, 9, 0, 0, 0, \ldots, 0, 0, 0, \ldots\rangle,$$

$$\Psi = \langle 0, 0, 0, 0, 12, 0, 0, 0, \ldots, 0, 0, 0, \ldots\rangle.$$

Furthermore, W is a transitive relation and $\Gamma W \Delta W \Phi W \Psi$.

Theorem 5.8.2. *For all ordinal numbers α and β not both 0 and for all well-ordered systems $\langle A, R_A\rangle$ and $\langle B, R_B\rangle$ that are similar to $\langle \alpha, \in\rangle$ and $\langle \beta, \in\rangle$, respectively, $\langle \alpha^\beta, \in\rangle$ is similar to $\langle Z(A, B), W\rangle$.*

PROOF: The proof is by induction (with respect to β). Thus, we will assume that the indicated similarity holds for $\alpha^{\beta'}$ with $\beta' < \beta$, α arbitrary (but $\alpha \neq 0$ or $\beta' \neq 0$); we must show that this implies the indicated similarity for α^β where α is arbitrary (but $\alpha \neq 0$ or $\beta \neq 0$). Assume $\langle A, R_A\rangle \cong \langle \alpha, \in\rangle$ and $\langle B, R_B\rangle \cong \langle \beta, \in\rangle$. We can use the second similarity map to index the elements b of B. For $\gamma < \beta$, let $B_\gamma = \{b_\delta \in B : \delta < \gamma\}$.

If $\alpha = 0$ and $\beta \neq 0$, $\langle \alpha^\beta, \in\rangle = \langle 0, 0\rangle = \langle Z(\varnothing, B), W\rangle$. Hereafter, we will assume that $\alpha > 0$. If $\beta = 0$, $\langle \alpha^\beta, \in\rangle = \langle 1, \in\rangle = \langle Z(A, \varnothing), W\rangle$.

If $\beta = 1$, $\langle Z(A, B), W\rangle \cong \langle A, R_A\rangle \cong \langle \alpha, \in\rangle = \langle \alpha^1, \in\rangle$.

If $\beta = \gamma^+ > 1$, let α be an arbitrary nonzero ordinal and let g be the similarity mapping of $\langle B, R_B\rangle$ onto $\langle \beta, \in\rangle$. Clearly, $B_\gamma = B \sim \{b_\gamma\} \cong \langle \gamma, \in\rangle$; in fact, $g \upharpoonright B_\gamma$ is the similarity map.

We first show that

$$\langle Z(A, B), W\rangle \cong \langle Z(A, B_\gamma) \times Z(A, \{b_\gamma\}),\ R_{Z(A,B_\gamma)\times Z(A,\{b_\gamma\})}\rangle,$$

where $R_{X\times Y}$ is the antilexicographical ordering on $X \times Y$ of Section 5.6. To this end, we define

$$H : Z(A \times B) \to Z(A, B_\gamma) \times Z(A, \{b_\gamma\})$$

as follows: For f in $Z(A, B)$, let $f_\gamma : \{b_\gamma\} \to A$ be such that $f_\gamma(b_\gamma) = f(b_\gamma)$. Let $H(f) = \langle f \restriction B_\gamma, f_\gamma \rangle$. (The verification that H is the required similarity mapping is left as an exercise.)

By our inductive hypothesis, $Z(A, B_\gamma) \cong \langle \alpha^\gamma, \in \rangle$; clearly, $\langle Z(A, \{b_\gamma\}), W \rangle \cong \langle \alpha, \in \rangle$. By Theorem 5.6.2 and the above,

$$\begin{aligned}\langle Z(A, B), W \rangle &\cong \langle Z(A, B_\gamma) \times Z(A, \{b_\gamma\}), R_{Z(A,B_\gamma) \times Z(A,\{b_\gamma\})} \rangle \\ &\cong \langle \alpha^\gamma \cdot \alpha, \in \rangle \\ &= \langle \alpha^\beta, \in \rangle.\end{aligned}$$

Finally, let β be a limit ordinal. For $\gamma < \beta$, let G_γ be the similarity map of $\langle Z(A, B_\gamma), W \rangle$ onto $\langle \alpha^\gamma, \in \rangle$. Define $G : Z(A, B) \to \alpha^\beta$ as follows: For f in $Z(A, B)$, let $\gamma(f)$ be the smallest nonzero ordinal γ with the property that $f(b_\delta) = a_0$ for all δ satisfying $\gamma \leq \delta < \beta$. (Thus $\gamma(f)$ is a successor ordinal.) Let

$$G(f) = G_{\gamma(f)}(f \restriction B_{\gamma(f)}).$$

G is the required similarity mapping of $\langle Z(A, B), W \rangle$ onto $\langle \alpha^\beta, \in \rangle$ [Exercise 5.8.5(2)].

Corollary. *For all ordinal numbers α and β not both 0, and for all well-ordered systems $\langle A, R_A \rangle$ and $\langle B, R_B \rangle$ that are similar to $\langle \alpha, \in \rangle$ and $\langle \beta, \in \rangle$ respectively, the following are true:*

(1) *$\langle Z(A, B), W \rangle$ is a well-ordered system.*

(2) *For each $\gamma < \beta$, $\langle Z_\gamma(A, B), W_\gamma \rangle$ is a well-ordered system.*

Theorem 5.8.3. *Let α, β, and γ be ordinal numbers that are such that $\alpha \neq 0$ or $\beta\gamma \neq 0$. Then the following are true:*

(1) $\alpha^{\beta+\gamma} = \alpha^\beta \alpha^\gamma$.

(2) $(\alpha^\beta)^\gamma = \alpha^{\beta\gamma}$.

PROOF: We prove (1) by induction (with respect to γ) and leave the proof of (2) as an exercise.

For $\gamma = 0$ and for all ordinals β and nonzero ordinals α,

$$\alpha^{\beta+\gamma} = \alpha^{\beta+0} = \alpha^\beta = \alpha^\beta 1 = \alpha^\beta \alpha^0 = \alpha^\beta \alpha^\gamma.$$

Suppose that (1) holds for all triples $\langle \alpha, \beta, \gamma' \rangle$ with $\gamma' < \gamma$ and (appropriate) α and β. First suppose $\gamma = \delta + 1$. Then

$$\begin{aligned}\alpha^{\beta+\gamma} &= \alpha^{\beta+(\delta+1)} = \alpha^{(\beta+\delta)+1} \\ &= \alpha^{\beta+\delta}\alpha = (\alpha^\beta\alpha^\delta)\alpha = \alpha^\beta(\alpha^\delta\alpha) \\ &= \alpha^\beta\alpha^{\delta+1} = \alpha^\beta\alpha^\gamma.\end{aligned}$$

Now suppose that γ is a limit ordinal. If $\alpha = 1$, then $\alpha^{\beta+\gamma} = 1 = \alpha^{\beta}\alpha^{\gamma}$. If $\alpha > 1$, then $\beta + \gamma$ is a limit ordinal, as is α^{γ}. Thus,

$$\begin{aligned}\alpha^{\beta+\gamma} &= \cup\{\alpha^{\delta} : \delta < \beta + \gamma\}\\ &= \cup\{\alpha^{\beta+\delta} : \delta < \gamma\}\\ &= \cup\{\alpha^{\beta}\alpha^{\delta} : \delta < \gamma\}\\ &= \cup\{\alpha^{\beta}\rho : \rho < \alpha^{\gamma}\}\\ &= \alpha^{\beta}\alpha^{\gamma}.\end{aligned}$$

Other laws of ordinal exponentiation will be presented in the exercises.

Exercises

5.8.1. Prove that for all natural numbers $N \geq 2$, $N^{\omega} = \omega$.

5.8.2. Determine each.

(a) $5^{\omega+1}$. **(b)** $(\omega + 1)^3$. **(c)** $(\omega + 2)^{\omega}$.

(d) $(\omega + 2)^{\omega+1}$. **(e)** $(\omega^2 + \omega)^{\omega}$. **(f)** $3^{\omega 3+5}$.

5.8.3. Prove Theorem 5.8.1.

†5.8.4. Prove that for all ordinals α and β, α^{β} is a successor ordinal iff at least one of the following conditions holds:

(1) $\alpha = 1$ and β is arbitrary.

(2) $\alpha \neq 0$ and $\beta = 0$.

(3) α is a successor ordinal and β is a natural number.

5.8.5. In the proof of Theorem 5.8.2, show the following:

(a) H is a similarity mapping.

(b) G is a similarity mapping.

5.8.6. Prove Theorem 5.8.3(2).

†5.8.7. Let α, β, γ, and δ be ordinals satisfying $1 < \alpha$, $1 < \delta$, $\beta < \gamma$. Prove the following:

(a) $\alpha^{\beta} < \alpha^{\gamma}$.

(b) $\alpha < \alpha^{\delta}$.

(c) $\beta \leq \alpha^{\beta}$. Find an example with equality.

5.8.8. Let $\alpha (> 1)$ and $\gamma (> 0)$ be ordinals. Prove that there is a unique ordinal β for which $\alpha^{\beta} \leq \gamma < \alpha^{\beta^+}$. (Used in Section 5.11.)

5.9 Transfinite Addition

We now extend the definition of ordinal addition so as to permit infinitely many summands.

Let α be an ordinal number. Recall that an α-sequence is a function whose domain is α, and that an α-sequence is *in* A if its range is included in A. An α-**ordinal-sequence** is defined to be an α-sequence whose range is a set of ordinals. For an α-ordinal-sequence Γ, "Γ_γ" stands for "$\Gamma(\gamma)$" ($\gamma < \alpha$); we frequently write "$[\Gamma_\gamma]_{\gamma<\alpha}$" for "$\Gamma$". $[\Gamma_\gamma]_{\gamma<\alpha}$ is called an **increasing α-ordinal-sequence** if whenever $\gamma_1 < \gamma_2 < \alpha$, it follows that $\Gamma_{\gamma_1} < \Gamma_{\gamma_2}$. If Γ is an α-ordinal-sequence and if $\delta < \gamma$, the δ-sequence $\Gamma \upharpoonright \delta \, (= [\Gamma_\gamma]_{\gamma<\delta})$ is called an **initial segment of** Γ.

†EXAMPLE 5.9.1

(1) Let $\Gamma : \omega 2 + 5 \to \{7\}$. Then Γ is an $\omega 2 + 5$-ordinal-sequence and $\Gamma_\gamma = 7$ for all $\delta < \omega 2 + 5$. We can indicate Γ by writing

$$[7, 7, 7, \ldots; 7, 7, 7, \ldots; 7, 7, 7, 7, 7].$$

(2) Let $\Phi : \omega 4 \to 4$ be defined by

$$\Phi_\gamma = N \qquad \text{for } \gamma = \omega N + M \text{ with } N < 4 \text{ and } M < \omega.$$

This $\omega 4$-sequence is indicated by

$$[0, 0, 0, \ldots; 1, 1, 1, \ldots; 2, 2, 2, \ldots; 3, 3, 3, \ldots].$$

(3) Let $\Lambda : \omega + 4 \to \omega^+$, where $\Lambda_\gamma = 0$ if γ is an even natural number, $\Gamma_\gamma = 1$ if γ is an odd natural number, and $\Lambda_\gamma = \omega$ if $\omega \le \gamma < \omega + 4$. This $\omega + 4$-sequence is indicated by

$$[0, 1, 0, 1, 0, 1, \ldots; \omega, \omega, \omega, \omega].$$

We apply (the second version of) Definition by Transfinite Induction (Theorem 5.4.2) to define transfinite addition.

Let α be a nonzero ordinal number and let

$$[\Gamma_\gamma]_{\gamma<\alpha} \tag{5.9.1}$$

be an α-ordinal-sequence. Clearly, R, the range of Γ, is a set of ordinals; thus, $(\cup R)^+$ is an ordinal number that is such that $\Gamma_\gamma < (\cup R)^+$ for all $\gamma < \alpha$. Let $\sigma = (\cup R)^+ \cdot \alpha$. For $\gamma < \alpha$, let B_γ be the set of all γ-sequences in σ, and let $B = \cup\{B_\gamma : \gamma < \alpha\}$. Define $G : B \to \sigma$ as follows: For each γ-sequence Δ in B, the following are true:

(1) If $\gamma = 0$, let $G(\Delta) = 0$.
(2) If $\gamma = \rho^+$, let $G(\Delta) = \Delta_\rho + \Gamma_\rho$.
(3) If α is a limit ordinal, let $G(\Delta) = \cup$ range Δ.

Then by Theorem 5.4.2, there is a unique α-sequence (in fact, an α-ordinal-sequence) Σ, in σ, defined by

$$\Sigma_\gamma = G(\Sigma \upharpoonright \gamma), \qquad \gamma < \alpha.$$

Let $\hat{\Sigma}$ be the following α^+-ordinal-sequence:

$$\hat{\Sigma} = \begin{cases} \Sigma \cup \{\langle \alpha, \Sigma_\delta + \Gamma_\delta \rangle\}, & \text{if } \alpha = \delta^+; \\ \Sigma \cup \{\langle \alpha, \cup \text{ range } \Sigma \rangle\}, & \text{if } \alpha \text{ is a limit ordinal.} \end{cases}$$

Hence, $\hat{\Sigma}$ is determined by the sequence (5.9.1); we sometimes indicate this dependency by writing

$$\text{“}\hat{\Sigma}_{[\Gamma_\gamma]_{\gamma<\alpha}}\text{”}.$$

The sequence $\hat{\Sigma}$ is known as the **sequence of partial sums corresponding to** (5.9.1) or, simply, the **associated sequence of partial sums**. Finally, we define

$$\text{“}\sum_{\gamma<\alpha} \Gamma_\gamma\text{” to be “}\hat{\Sigma}_\alpha\text{”};$$

$\sum_{\gamma<\alpha} \Gamma_\gamma$ is called the **transfinite sum associated with** (5.9.1). For convenience, we define

$$\text{“}\sum_{\gamma<0} \Gamma_\gamma\text{” to be “0”.}$$

When $\alpha = \omega$, the transfinite sum associated with $[\Gamma_\gamma]_{\gamma<\omega}$ is usually called the **sum of an infinite series**.

It is easily established, by transfinite induction, that *for any $\delta < \alpha$, the transfinite sum associated with the initial segment $[\Gamma_\gamma]_{\gamma<\delta}$ obeys the following laws:*

(5.9.2) *If $\delta = \rho^+$, then* $\displaystyle \sum_{\gamma<\delta} \Gamma_\gamma = \left(\sum_{\gamma<\rho} \Gamma_\gamma\right) + \Gamma_\rho.$

(5.9.3) *If δ is a limit ordinal, then* $\displaystyle \sum_{\gamma<\delta} \Gamma_\gamma = \cup \{\sum_{\gamma<\mu} \Gamma_\gamma : \mu < \delta\}.$

From (5.9.3) it follows that *if δ is a limit ordinal $\leq \alpha$ and if $\eta < \sum_{\gamma<\delta} \Gamma_\gamma$, then $\eta < \sum_{\gamma<\mu} \Gamma_\gamma$ for some $\mu < \delta$.*

†EXAMPLE 5.9.2

(1) We consider the ordinal-sequence $[\Gamma_\gamma]_{\gamma<\omega 2+5}$ of Example 5.9.1(1),

$$[7, 7, 7, \ldots; 7, 7, 7, \ldots; 7, 7, 7, 7, 7].$$

Here, $\Gamma_\gamma = 7$ for all $\gamma < \omega 2 + 5$; $R = \text{range}([\Gamma_\gamma]_{\gamma < \omega 2 + 5}) = \{7\}$; $(\cup R)^+ = 8$. Thus, $\sigma = 8(\omega 2 + 5) = \omega 2 + 40$. For $\gamma < \omega 2 + 5$, B_γ is the set of all γ-sequences $[\Delta_\delta]_{\delta<\gamma}$, where each $\Delta_\delta < \omega 2 + 40$; $B = \cup\{B_\gamma : \gamma < \omega 2 + 5\}$. Let $\Delta = [\Delta_\delta]_{\delta<\gamma} \in B$.

(a) $G(0) = 0$.
(b) If $\gamma = \rho^+$, $G(\Delta) = \Delta_\rho + \Gamma_\rho = \Delta_\rho + 7$.
(c) If γ is a limit ordinal, $G(\Delta) = \cup$ range Δ.

The $(\omega 2 + 5)$-sequence Σ is defined as follows:

$$\Sigma_0 = G(\Sigma \restriction 0) = G(0) = 0;$$

$$\Sigma_1 = G(\Sigma \restriction 1).$$

Clearly, $\Sigma \restriction 1$ is a 1-sequence; hence, by clause (2) of the definition of G, p. 230,

$$\Sigma_1 = \Sigma_0 + \Gamma_0 = 0 + 7 = 7;$$

$$\Sigma_2 = G(\Sigma \restriction 2) = \Sigma_1 + \Gamma_1 = 7 + 7 = 14.$$

By induction on N,

$$\Sigma_N = 7N, \qquad \text{for all } N < \omega;$$

$$\Sigma_\omega = \cup\{\Sigma_N : N < \omega\} = \cup\{7N : N < \omega\} = \omega.$$

By induction on N,

$$\Sigma_{\omega + N} = \omega + 7N, \qquad N < \omega.$$

$$\begin{aligned} \Sigma_{\omega 2} &= \cup\{\Sigma_\delta : \delta < \omega 2\} \\ &= \cup\{\Sigma_\delta : \delta < \omega\} \cup \cup\{\Sigma_\delta : \omega \leq \delta < \omega 2\} \\ &= \omega \cup \cup\{\omega + 7N : N < \omega\} \\ &= \omega \cup \omega 2 \\ &= \omega 2; \end{aligned}$$

$$\Sigma_{\omega 2 + N} = \omega 2 + 7N, \qquad \text{for } N < 5.$$

The associated sequence of partial sums $\hat{\Sigma}$ is defined to be

$$\Sigma \cup \{\langle \omega 2 + 5, \Sigma_{\omega 2 + 4} + \Gamma_{\omega 2 + 4} \rangle\}.$$

Thus,

$$\sum_{\gamma < \omega 2 + 5} \Gamma_\gamma = \hat{\Sigma}_{\omega 2 + 5} = (\omega 2 + 28) + 7 = \omega 2 + 35.$$

Note that $\hat{\Sigma}$ is indicated by

$$[0, 7, 14, \ldots, \omega, \omega + 7, \omega + 14, \ldots, \omega 2, \omega 2 + 7, \omega 2 + 14, \omega 2 + 21, \omega 2 + 28, \omega 2 + 35].$$

(2) $[\Delta_\gamma]_{\gamma < \omega 4}$ is indicated by

$$[0, 0, 0, \ldots; 1, 1, 1, \ldots; 2, 2, 2, \ldots; 3, 3, 3, \ldots].$$

Here,

$$\Sigma_N = 0, \qquad \text{for } N < \omega.$$

Hence,

$$\Sigma_\omega = \cup\{0\} = 0.$$

By induction,

$$\Sigma_{\omega + N} = N, \qquad N < \omega.$$

Therefore,

$$\Sigma_{\omega 2} = 0 \cup \cup\{N : N < \omega\} = \omega.$$

Furthermore,

$$\Sigma_{\omega 2 + N} = \omega + 2N, \qquad N < \omega;$$

$$\Sigma_{\omega 3} = \omega \cup \cup\{\omega + 2N : N < \omega\} = \omega \cup \omega 2 = \omega 2;$$

$$\Sigma_{\omega 3 + N} = \omega 2 + 3N, \qquad N < \omega.$$

Thus,

$$\sum_{\gamma < \omega 4} \Delta_\gamma = \hat{\Sigma}_{\omega 4} = \cup \text{ range } \Sigma = \omega 2 \cup \cup\{\omega 2 + 3N : N < \omega\}$$

$$= \omega 2 \cup \omega 3 = \omega 3.$$

The associated sequence of partial sums is given by

$$[0, 0, 0, 0, \ldots, 0, 1, 2, 3, \ldots; \omega, \omega + 2, \omega + 4, \omega + 6, \ldots; \omega 2, \omega 2 + 3, \omega 2 + 6, \omega 2 + 9, \ldots; \omega 3].$$

(3) $[\Lambda_\gamma]_{\gamma < \omega + 4}$ is indicated by

$$[0, 1, 0, 1, 0, 1, \ldots; \omega, \omega, \omega, \omega].$$

By induction,

$$\Sigma_{2K} = \Sigma_{2K+1} = K, \qquad \text{for } K < \omega;$$

$$\Sigma_\omega = \cup\{\Sigma_{2K} : K < \omega\} \cup \cup\{\Sigma_{2K+1} : K < \omega\}$$

$$= \cup\{K : K < \omega\} \cup \cup\{K : K < \omega\} = \omega \cup \omega = \omega;$$

$$\Sigma_{\omega+N} = \omega N^+, \qquad N < 4;$$

$$\sum_{\gamma<\omega+4} = \hat{\Sigma}_{\omega+4} = \omega 5.$$

Thus, the associated sequence of partial sums is given by

$$[0, 1, 1, 2, 2, 3, 3, \ldots, \omega, \omega 2, \omega 3, \omega 4, \omega 5].$$

Clearly, *transfinite addition extends (finite) ordinal addition*; in fact, *for* $1 \leq N < \omega$,

$$\sum_{\gamma<N^+} \Gamma_\gamma = \Gamma_0 + \Gamma_1 + \cdots + \Gamma_N.$$

Transfinite addition is "noncommutative" in the sense that a rearrangement of the order of the summands, Γ_γ, may affect the sum.

EXAMPLE 5.9.3

(1) Let $\Gamma_0 = \omega^3, \Gamma_\omega = \omega^2$, and for $2 \leq N < \omega, \Gamma_N = \omega$. Then $\sum_{N<\omega^+} \Gamma_N = \omega^3 + \omega^2 2$.

(2) Let $\Delta_0 = \omega^2, \Delta_\omega = \omega^3, \Delta_N = \Gamma_N = \omega$ for $2 \leq N < \omega$. Then $\sum_{N<\omega^+} \Delta_N = \omega^3$.

Let α be an ordinal number and let $\delta < \alpha$. For every α-sequence $[\Gamma_\gamma]_{\gamma<\alpha}$, surely

$$\sum_{\gamma<\delta} \Gamma_\gamma \leq \sum_{\gamma<\alpha} \Gamma_\gamma.$$

Theorem 5.9.1. *Let α be a limit ordinal. A sufficient condition for $\sum_{\gamma<\alpha} \Gamma_\gamma$ to be a limit ordinal is that*

$$\sum_{\gamma<\mu} \Gamma_\gamma < \sum_{\gamma<\alpha} \Gamma_\gamma, \qquad \textit{for every } \mu < \alpha.$$

PROOF: According to (5.9.3),

$$\sum_{\gamma<\alpha} \Gamma_\gamma = \cup\{\sum_{\gamma<\mu} \Gamma_\gamma : \mu < \alpha\};$$

thus, $\sum_{\gamma<\alpha} \Gamma_\gamma$ is the union of a nonempty set of ordinals. Exercise 5.3.11(d) asserts that the indicated inequality is sufficient for $\sum_{\gamma<\alpha} \Gamma_\gamma$ to be a limit ordinal.

Corollary 1. *If α is a limit ordinal and if $\Gamma_\gamma > 0$ for all $\gamma < \alpha$, then $\sum_{\gamma<\alpha} \Gamma_\gamma$ is a limit ordinal.*

Corollary 2. *If α is a limit ordinal and if $\sum_{\gamma<\alpha} \Gamma_\gamma$ is a successor ordinal, then there exists an ordinal number μ such that $\mu < \alpha$ and such that if ρ satisfies $\mu \leq \rho < \alpha$, then*

$$\Gamma_\rho = 0 \qquad \textit{and} \qquad \sum_{\gamma<\alpha} \Gamma_\gamma = \sum_{\gamma<\rho} \Gamma_\gamma.$$

The next theorem shows that ordinal multiplication can be regarded as a special case of transfinite addition.

Theorem 5.9.2. *Let α and β be ordinal numbers. Let $\Gamma_\gamma = \beta$ for all $\gamma < \alpha$. Then*

$$\sum_{\gamma<\alpha} \Gamma_\gamma = \beta\alpha.$$

[The proof involves a routine transfinite induction on α (with fixed β); it is left as an exercise.]

Corollary. *Let α be an ordinal number. Let $\Gamma_\gamma = 1$ for all $\gamma < \alpha$. Then*

$$\sum_{\gamma<\alpha} \Gamma_\gamma = \alpha.$$

EXAMPLE 5.9.4

(1) $\omega 2 = \omega + \omega = \sum_{\gamma<2} \Gamma_\gamma$, where $\Gamma_0 = \Gamma_1 = \omega$.

(2) Let $\Delta_\gamma = 2$ for all $\gamma < \omega$. Then

$$\sum_{\gamma<\omega} \Delta_\gamma = \cup\{2N : N < \omega\} = \omega = 2\omega.$$

Let α be a nonzero ordinal number and let $[\Gamma_\gamma]_{\gamma<\alpha}$ be an α-ordinal-sequence. For $\alpha_1 \leq \alpha_2 < \alpha$, let

$$\text{“}\sum_{\alpha_1\leq\gamma<\alpha_2} \Gamma_\gamma\text{” stand for “}\sum_{\gamma<\alpha_2} \Gamma_\gamma \dot{-} \sum_{\gamma<\alpha_1} \Gamma_\gamma\text{”.}$$

Observe that

$$\sum_{\alpha_1 \leq \gamma < \alpha_1} \Gamma_\gamma = 0.$$

This extends our previous notation in the sense that

$$\sum_{\gamma < \alpha} \Gamma_\gamma = \sum_{0 \leq \gamma < \alpha} \Gamma_\gamma$$

because

$$\begin{aligned} \sum_{\gamma < \alpha} \Gamma_\gamma &= \sum_{\gamma < \alpha} \Gamma_\gamma \dot{-} 0 \\ &= \sum_{\gamma < \alpha} \Gamma_\gamma \dot{-} \sum_{\gamma < 0} \Gamma_\gamma \\ &= \sum_{0 \leq \gamma < \alpha} \Gamma_\gamma. \end{aligned}$$

Note that for all ordinals α_1 and α_2 and all $(\alpha_1 + \alpha_2)$-sequences $[\Gamma_\gamma]_{\gamma < \alpha_1 + \alpha_2}$,

(5.9.4) $$\sum_{\gamma < \alpha_1 + \alpha_2} \Gamma_\gamma = \sum_{\gamma < \alpha_1} \Gamma_\gamma + \sum_{\alpha_1 \leq \gamma < \alpha_1 + \alpha_2} \Gamma_\gamma.$$

We now formulate an "associative law" for transfinite addition. To this end, suppose that for some nonzero ordinal δ_0, $[\Lambda_\mu]_{\mu < \delta_0}$ is an increasing sequence of nonzero ordinals. Let

"Λ_{0^-}" be "0";

for $\mu < \delta_0$, let

"$\Lambda_{(\mu^+)^-}$" be "Λ_μ",

and let

"Λ_{μ^-}" be "$\cup\{\Lambda_\gamma : \gamma < \mu\}$" if μ is a limit ordinal.

Let $\alpha > 0$ and let $[\Gamma_\gamma]_{\gamma < \alpha}$ be an α-ordinal-sequence. Let $\delta \leq \alpha$ and let $[\Lambda_\mu]_{\mu < \delta^+}$ be an increasing sequence of ordinals in α^+ that is such that $\Lambda_\delta = \alpha$. Thus,

$$\Lambda_0 < \Lambda_1 < \Lambda_2 < \cdots < \Lambda_\mu < \cdots < \Lambda_\delta = \alpha, \mu \leq \delta.$$

For $\mu < \delta^+$, define

$$\Xi_\mu = \sum_{\Lambda_{\mu^-} \leq \gamma < \Lambda_\mu} \Gamma_\gamma;$$

Ξ_μ is called the μth **interval sum**.

EXAMPLE 5.9.5. Consider $[\Gamma_\gamma]_{\gamma<\omega 3+1}$, where $\Gamma_\gamma = 1$ for all $\gamma < \omega 3 + 1$. Let $\delta = \omega 2$ and let

$$\Lambda_\mu = \begin{cases} \mu^+, & \text{if } \mu < \omega; \\ \mu, & \text{if } \omega \leq \mu < \omega 2; \\ \omega 3 + 1, & \text{if } \mu = \omega 2. \end{cases}$$

Thus, $[\Lambda_\mu]_{\mu<\omega 2}$ is indicated by

$$[1, 2, 3, \ldots; \omega, \omega + 1, \omega + 2, \ldots; \omega 3 + 1].$$

Observe that

$$\Lambda_{\omega^-} = \omega = \Lambda_\omega,$$

but that

$$\Lambda_{(\omega 2)^-} = \omega 2 \neq \omega 3 + 1 = \Lambda_{\omega 2}.$$

Thus

$$\Xi_\mu = \sum_{\Lambda_{\mu^-} \leq \gamma < \Lambda_\mu} \Gamma_\gamma = \Gamma_{\Lambda_{\mu^-}} = 1, \qquad \text{for } \mu < \omega;$$

$$\Xi_\omega = 0 \qquad (\text{because } \Lambda_{\omega^-} = \Lambda_\omega);$$

$$\Xi_\mu = \sum_{\Lambda_{\mu^-} \leq \gamma < \Lambda_\mu} \Gamma_\gamma = \Gamma_{\Lambda_{\mu^-}} = 1, \qquad \text{for } \omega < \mu < \omega 2;$$

$$\begin{aligned} \Xi_{\omega 2} &= \sum_{\Lambda_{(\omega 2)^-} \leq \gamma < \Lambda_{\omega 2}} \Gamma_\gamma = \sum_{\omega 2 \leq \gamma < \omega 3 + 1} \Gamma_\gamma \\ &= (\omega 3 + 1) \dot{-} \omega 2 = \omega + 1. \end{aligned}$$

Observe that by (5.9.4),

$$\begin{aligned} \sum_{\mu<\delta^+} \Xi_\mu &= \sum_{\mu<(\omega 2)^+} \Xi_\mu \\ &= \left(\sum_{\mu<\omega} \Xi_\mu\right) + \Xi_\omega + \left(\sum_{\omega<\mu<\omega 2} \Xi_\mu\right) + \Xi_{\omega 2} \\ &= 1\omega + 0 + 1\omega + (\omega + 1) \\ &= \omega 3 + 1 \\ &= \sum_{\gamma<\omega 3+1} \Gamma_\gamma \\ &= \sum_{\gamma<\alpha} \Gamma_\gamma. \end{aligned}$$

Theorem 5.9.3. *Let α be a nonzero ordinal, let $[\Gamma]_{\gamma<\alpha}$ be an α-ordinal-sequence and let $\delta \leq \alpha$. Let $[\Lambda_\mu]_{\mu<\delta^+}$ be any increasing sequence of ordinals in α^+ for which $\Lambda_\delta = \alpha$; let $[\Xi_\mu]_{\mu<\delta^+}$ be the associated sequence of μth interval sums. Then*

$$\sum_{\gamma<\alpha} \Gamma_\gamma = \sum_{\mu<\delta^+} \Xi_\mu.$$

PROOF: The proof is by transfinite induction (with respect to δ, rather than α).

Suppose δ is the smallest ordinal for which the following exist:

(1) An ordinal $\alpha \geq \delta$.
(2) An α-sequence $[\Gamma_\gamma]_{\gamma<\alpha}$.
(3) An increasing sequence of ordinals in α^+, $[\Lambda_\mu]_{\mu<\delta^+}$, for which $\Lambda_\delta = \alpha$ and for which

$$\sum_{\gamma<\alpha} \Gamma_\gamma \neq \sum_{\mu<\delta^+} \Xi_\mu.$$

Thus, $\delta \neq 0$; for let α be an arbitrary ordinal, let Γ be as in (2) and let Λ be as in (3). Then $[\Lambda_\mu]_{\mu<0^+} = \{\langle 0, \alpha\rangle\}$. Furthermore,

$$\sum_{\mu<\delta^+} \Xi_\mu = \Xi_0 = \sum_{\Lambda_{0^-} \leq \gamma < \Lambda_0} \Gamma_\gamma = \sum_{0\leq\gamma<\alpha} \Gamma_\gamma = \sum_{\gamma<\alpha} \Gamma_\gamma.$$

Clearly, δ cannot be a successor ordinal, ρ^+. For let Γ be as in (2) and let Λ be as in (3). Then

$$\begin{aligned}\sum_{\gamma<\alpha} \Gamma_\gamma &= \sum_{\gamma<\Lambda_\delta} \Gamma_\gamma + \sum_{\Lambda_\delta \leq \gamma < \Lambda_{\delta^+}} \Gamma_\gamma \\ &= \left(\sum_{\mu<\delta} \Xi_\mu\right) + \Xi_\delta \\ &= \sum_{\mu<\delta^+} \Xi_\mu, \qquad \text{by (5.9.2).}\end{aligned}$$

Finally, δ cannot be a limit ordinal. For let Γ be as in (2) and let Λ be as in (3). We consider two cases.

First, if $\Lambda_{\delta^-} (= \cup\{\Lambda_\gamma : \gamma < \delta\}) = \Lambda_\delta$, then $\Xi_\delta = 0$; moreover,

$$\begin{aligned}\sum_{\gamma<\alpha} \Gamma_\gamma &= \cup\left\{\sum_{\gamma<\rho} \Gamma_\gamma : \rho < \alpha\right\} \\ &= \cup\left\{\sum_{\gamma<\Lambda_\eta} \Gamma_\gamma : \eta < \delta\right\} \\ &= \cup\left\{\sum_{\mu<\eta^+} \Xi_\mu : \eta^+ < \delta\right\} \\ &= \sum_{\mu<\delta} \Xi_\mu \\ &= \sum_{\mu<\delta^+} \Xi_\mu.\end{aligned}$$

Otherwise,

$$\begin{aligned}
\sum_{\gamma<\alpha} \Gamma_\gamma &= \sum_{\gamma<\Lambda_{\delta^-}} \Gamma_\gamma + \sum_{\Lambda_{\delta^-}\leq\gamma<\Lambda_\delta} \Gamma_\gamma \\
&= \cup\left\{\sum_{\gamma<\Lambda_\eta} \Gamma_\gamma : \eta < \delta\right\} + \Xi_\delta \\
&= \cup\left\{\sum_{\mu<\eta^+} \Xi_\mu : \eta^+ < \delta\right\} + \Xi_\delta \\
&= \left(\sum_{\mu<\delta} \Xi_\mu\right) + \Xi_\delta \\
&= \sum_{\mu<\delta^+} \Xi_\mu .
\end{aligned}$$

Theorem 5.9.4. Right Distributivity for Transfinite Sums. *Let α and μ be ordinal numbers and let $[\Gamma_\gamma]_{\gamma<\alpha}$ be an α-ordinal-sequence. Let $\Delta_\gamma = \mu\Gamma_\gamma$ for $\gamma < \alpha$. Then*

$$\mu \cdot \sum_{\gamma<\alpha} \Gamma_\gamma = \sum_{\gamma<\alpha} \Delta_\gamma . \tag{5.9.5}$$

PROOF: The proof is by transfinite induction (with respect to α).

If μ or α is 0, both sides of (5.9.5) are 0; assume otherwise. Suppose that right distributivity holds for β-sequences with $\beta < \alpha$.

If $\alpha = \rho^+$, then

$$\begin{aligned}
\mu \cdot \sum_{\gamma<\alpha} \Gamma_\gamma &= \mu\left(\left(\sum_{\gamma<\rho} \Gamma_\gamma\right) + \Gamma_\rho\right) \\
&= \left(\mu \cdot \sum_{\gamma<\rho} \Gamma_\gamma\right) + \mu\Gamma_\rho \\
&= \sum_{\gamma<\rho} \Delta_\gamma + \Delta_\rho \\
&= \sum_{\gamma<\alpha} \Delta_\gamma .
\end{aligned}$$

Suppose that α is a limit ordinal. If $\sum_{\gamma<\alpha} \Gamma_\gamma = 0$, then $\Gamma_\gamma = 0$ for all $\gamma < \alpha$; hence, each Δ_γ is 0 and

$$\mu \cdot \sum_{\gamma<\alpha} \Gamma_\gamma = 0 = \sum_{\gamma<\alpha} \Delta_\gamma .$$

If $\sum_{\gamma<\alpha} \Gamma_\gamma$ is a successor ordinal, then by Corollary 2 of Theorem 5.9.1, for some $\alpha_0 < \alpha, \Gamma_\gamma = 0$ and hence $\Delta_\gamma = 0$ for $\alpha_0 \leq \gamma < \alpha$. Thus,

$$\mu \cdot \sum_{\gamma<\alpha} \Gamma_\gamma = \mu \cdot \sum_{\gamma<\alpha_0} \Gamma_\gamma = \sum_{\gamma<\alpha_0} \Delta_\gamma = \sum_{\gamma<\alpha} \Delta_\gamma.$$

If $\sum_{\gamma<\alpha} \Gamma_\gamma$ is a limit ordinal, then $\mu \cdot \sum_{\gamma<\alpha} \Gamma_\gamma =$

$$\cup\left\{\mu\eta : \eta < \sum_{\gamma<\alpha} \Gamma_\gamma\right\}. \tag{5.9.6}$$

Also, $\sum_{\gamma<\alpha} \Delta_\gamma =$

$$\cup\left\{\sum_{\gamma<\delta} \Delta_\gamma : \delta < \alpha\right\}. \tag{5.9.7}$$

We leave as an exercise the verification that the ordinal numbers (5.9.6) and (5.9.7) are the same.

The sum of two ordinals can be considered in terms of the union of disjoint representative well-ordered systems. Rather than consider arbitrary well-ordered systems for infinite sums, we introduce a canonical representation.

Let α be an ordinal number and let $[\Gamma_\gamma]_{\gamma<\alpha}$ be any α-sequence. We can well-order the set $B = \cup\{\Gamma_\gamma \times \{\gamma\}, \gamma < \alpha\}$ as follows:

$$\langle\delta_1, \gamma_1\rangle R \langle\delta_2, \gamma_2\rangle \qquad \text{if either } \gamma_1 < \gamma_2 \text{ or both } \gamma_1 = \gamma_2 \text{ and } \delta_1 \leq \delta_2.$$

(The verification that R is a well-ordering is left as an exercise.)

Theorem 5.9.5. *Let α be an ordinal number and let $[\Gamma_\gamma]_{\gamma<\alpha}$ be an α-ordinal-sequence. Let $\langle B, R\rangle$ be as above. Then*

$$\langle B, R\rangle \cong \left\langle \sum_{\gamma<\alpha} \Gamma_\gamma, \in \right\rangle.$$

PROOF: The proof is by transfinite induction with respect to α.

If $\alpha = 0$, then

$$B = R = \sum_{\gamma<\alpha} \Gamma_\gamma = \in_{\sum_{\gamma<\alpha} \Gamma_\gamma} = 0.$$

Let $\alpha > 0$ and suppose the result holds for all $\beta < \alpha$. Let $[\Gamma_\gamma]_{\gamma<\alpha}$ and $\langle B, R\rangle$ be as indicated.

Let $\alpha = \rho^+$. Then $[\Gamma_\gamma]_{\gamma<\rho}$ is an initial segment of $[\Gamma_\gamma]_{\gamma<\alpha}$. Let

$$B_\rho = \cup\{\Gamma_\gamma \times \{\gamma\} : \gamma < \rho\};$$

$\langle B_\rho, R\rangle$ is an initial segment of $\langle B, R\rangle$. Moreover, by the inductive hypothesis, $\langle B_\rho, R\rangle \cong \langle \sum_{\gamma<\rho} \Gamma_\gamma, \in\rangle$; let F be the similarity map. We can define a similarity mapping G of B onto $\sum_{\gamma<\alpha} \Gamma_\gamma$ by

$$G(\langle\delta,\gamma\rangle) = \begin{cases} F(\langle\delta,\gamma\rangle, & \text{if } \gamma < \rho; \\ \left(\sum_{\mu<\rho} \Gamma_\mu\right) + \delta, & \text{if } \gamma = \rho. \end{cases}$$

Let α be a limit ordinal. For each $\beta < \alpha$, let $B_\beta = \cup\{\Gamma_\gamma \times \{\gamma\} : \gamma < \beta\}$. Then $\langle B_\beta, R\rangle$ is an initial segment of $\langle B, R\rangle$ and $\langle B_\beta, R\rangle$ is similar to the initial segment $\langle \sum_{\gamma<\beta} \Gamma_\gamma, \in\rangle$ of $\langle \sum_{\gamma<\alpha} \Gamma_\gamma, \in\rangle$; let F_β be the similarity map. Let $\mathscr{F} = \{F_\beta : \beta < \alpha\}$; $\mathscr{F}$ is a nested set of similarity maps. It is not difficult to verify that $\cup\mathscr{F}$ is a similarity map of B onto $\sum_{\gamma<\alpha} \Gamma_\gamma$. (See the proof of Theorem 5.1.2.)

Corollary. *Let α be an ordinal number and let $[\Gamma_\gamma]_{\gamma<\alpha}$ be an α-ordinal-sequence. Let $\langle B, R\rangle$ be as above. Then*

$$\sum_{\gamma<\alpha} \Gamma_\gamma = \sigma,$$

where σ is the unique ordinal number such that $\langle B, R\rangle \cong \langle\sigma, \in\rangle$.

Exercises

5.9.1. Determine the sums of the infinite series associated with the indicated ω-sequences:

(a) $[1, 2, 3, \ldots]$. **(b)** $[1^2, 2^2, 3^2, \ldots]$.
(c) $[\omega, \omega, \omega, \ldots]$. **(d)** $[3, 2, 1, 0, 3, 2, 1, 0, \ldots]$.
(e) $[\omega + 1, \omega + 2, \omega + 3, \ldots]$. **(f)** $[\omega, \omega 2, \omega 3, \ldots]$.
(g) $[\omega, \omega^2, \omega^3, \ldots]$.

5.9.2. Determine the transfinite sums associated with the indicated sequences:

(a) $\beta_\gamma = \begin{cases} \omega, \text{ for } \gamma < \omega; \\ 1, \text{ for } \omega \le \gamma < \omega + 3. \end{cases}$

(b) $\beta_\gamma = \begin{cases} \omega, \text{ for } \gamma < \omega; \\ 1, \text{ for } \omega \le \gamma < \omega 2. \end{cases}$

(c) $\beta_\gamma = \begin{cases} 1, \text{ for } \gamma < \omega; \\ \omega, \text{ for } \omega \le \gamma < \omega 2. \end{cases}$

5.9.3. Determine $\sum_{\gamma<\alpha} \Gamma_\gamma$ in each case.

(a) $\alpha = \omega 2$, $\Gamma_\gamma = 2$ for all $\gamma < \omega 2$.

(b) $\alpha = \omega 3$, $\Gamma_\gamma = 2$ for all $\gamma < \omega 2$, $\Gamma_\gamma = \omega$ for $\omega 2 \leq \gamma < \omega 3$.

(c) $\alpha = \omega^2$, $\Gamma_\gamma = 2$ for all $\gamma < \omega^2$.

5.9.4. Let α be an ordinal and let $\Gamma_\gamma = \gamma$ for all $\gamma < \alpha$. Determine $\sum_{\gamma<\alpha} \Gamma_\gamma$ in each case.

(a) $\alpha = \omega 2$. **(b)** $\alpha = \omega 2 + 3$. **(c)** $\alpha = \omega 3$.

(d) $\alpha = \omega^2$. **(e)** $\alpha = \omega^3$. **(f)** $\alpha = \omega^\omega$.

†5.9.5. Verify (5.9.2) and (5.9.3).

5.9.6. **(a)** Prove that the sufficient condition of Theorem 5.9.1 is not necessary. In fact, determine a limit ordinal α and an α-sequence $[\Gamma_\gamma]_{\gamma<\alpha}$ for which $\sum_{\gamma<\alpha} \Gamma_\gamma$ is a limit ordinal, but $\sum_{\gamma<\delta} \Gamma_\gamma = \sum_{\gamma<\alpha} \Gamma_\gamma$ for some $\delta < \alpha$.

(b) Let α be a limit ordinal. State a necessary and sufficient condition for $\sum_{\gamma<\alpha} \Gamma_\gamma$ to equal $\sum_{\gamma<\delta} \Gamma_\gamma$ for some $\delta < \alpha$.

5.9.7. **(a)** Prove Theorem 5.9.2.

(b) Prove the corollary of Theorem 5.9.2.

5.9.8. In the proof of Theorem 5.9.4 for the case in which α is a limit ordinal, show that the ordinal numbers (5.9.6) and (5.9.7) are the same.

5.9.9. Let α be an ordinal number and let $[\Gamma_\gamma]_{\gamma<\alpha}$ be an α-ordinal-sequence. Prove that the relation R on $\cup\{\Gamma_\gamma \times \{\gamma\} : \gamma < \alpha\}$—defined by $\langle \delta_1, \gamma_1 \rangle R \langle \delta_2, \gamma_2 \rangle$ iff either $\gamma_1 < \gamma_2$ or both $\gamma_1 = \gamma_2$ and $\delta_1 \leq \delta_2$—is a well-ordering. (Useful for Section 6.10.)

5.9.10. In the limit ordinal case of the proof of Theorem 5.9.5, show that $\cup\mathscr{F}$ is a similarity map of B onto $\sum_{\gamma<\alpha} \Gamma_\gamma$.

5.9.11. **(a)** Prove that for every ordinal α and for all α-ordinal-sequences $[\Gamma_\gamma]_{\gamma<\alpha}$ and $[\Delta_\gamma]_{\gamma<\alpha}$, if $\Gamma_\gamma \leq \Delta_\gamma$ for all $\gamma < \alpha$, then $\sum_{\gamma<\alpha} \Gamma_\gamma \leq \sum_{\gamma<\alpha} \Delta_\gamma$.

(b) Find a counterexample if both instances of "$\leq$" are replaced by "$<$".

5.10 Transfinite Multiplication

We apply Theorem 5.4.2 to extend the definition of multiplication to the case of infinitely many factors.

Let α be a nonzero ordinal number and let

$$[\Gamma_\gamma]_{\gamma<\alpha} \tag{5.10.1}$$

be an α-ordinal-sequence. As in Section 5.9, if R = range Γ, then $(\cup R)^+$ is an ordinal greater than each $\Gamma_\gamma, \gamma < \alpha$. Let $\pi = ((\cup R)^+)^\alpha$. For $\gamma < \alpha$, let C_γ be the set of all γ-sequences in π, and let $C = \cup\{C_\gamma : \gamma < \alpha\}$. Define $H : C \to \pi$ as follows. For each γ-sequence Δ in C,

(1) If $\gamma = 0$, let $H(\Delta) = 1$.
(2) If $\gamma = \rho^+$, let $H(\Delta) = \Delta_\rho \cdot \Gamma_\rho$.
(3) If γ is a limit ordinal, let $H(\Delta) = \cup$ range Δ.

There is a unique α-ordinal-sequence, Π, in π defined by

$$\Pi_\gamma = H(\Pi \upharpoonright \gamma), \qquad \gamma < \alpha.$$

Define an intermediary α-sequence $\overline{\Pi}$ as follows. For $\gamma < \alpha$,

$$\overline{\Pi}_\gamma = \begin{cases} \Pi_\gamma, & \text{if for all } \rho < \gamma, \Pi_\rho > 0; \\ 0, & \text{otherwise.} \end{cases}$$

Let $\hat{\Pi}$ be the following α^+-ordinal-sequence:

$$\hat{\Pi} = \overline{\Pi} \cup \{\langle \alpha, 0 \rangle\}, \qquad \text{if for some } \rho < \alpha, \pi_\rho = 0;$$

otherwise,

$$\hat{\Pi} = \begin{cases} \Pi \cup \{\langle \alpha, \Pi_\delta \cdot \Gamma_\delta \rangle\}, & \text{if } \alpha = \delta^+; \\ \Pi \cup \{\alpha, \cup \text{ range } \Pi\}, & \text{if } \alpha \text{ is a limit ordinal.} \end{cases}$$

We sometimes indicate the dependency of $\hat{\Pi}$ upon the sequence (5.10.1) by writing "$\hat{\Pi}_{[\Gamma_\gamma]_{\gamma<\alpha}}$". $\hat{\Pi}$ is called the **sequence of partial products corresponding to** (5.10.1), or simply the **associated sequence of partial products**. We define

$$\text{“}\prod_{\gamma<\alpha} \Gamma_\gamma\text{” to be “}\hat{\Pi}_\alpha\text{”.}$$

$\prod_{\gamma<\alpha} \Gamma_\gamma$ is known as the **transfinite product associated with** (5.10.1). For convenience, we define

$$\text{“}\prod_{\gamma<0} \Gamma_\gamma\text{” to be “1”.}$$

For every ordinal α and α-ordinal-sequence $[\Gamma_\gamma]_{\gamma<\alpha}$, the following are true: *If $\rho^+ < \alpha$, then*

$$\prod_{\gamma<\rho^+} \Gamma_\gamma = \left(\prod_{\gamma<\rho} \Gamma_\gamma\right) \cdot \Gamma_\rho;$$

(5.10.2) *If ρ $(<\alpha)$ is a limit ordinal and if $\Gamma_\gamma > 0$ for all $\gamma < \rho$, then*

$$\prod_{\gamma<\rho} \Gamma_\gamma = \cup\left\{\prod_{\gamma<\mu} \Gamma_\gamma : \mu < \rho\right\}.$$

It follows from (5.10.2) that *if* $\rho\,(\leq\alpha)$ *is a limit ordinal and if* $\eta < \prod_{\gamma<\rho} \Gamma_\gamma$, *then* $\eta < \prod_{\gamma<\mu} \Gamma_\gamma$ *for some* $\mu < \rho$.

If γ_0 *is the smallest index for which* $\Gamma_\gamma = 0$, *then*

$$\prod_{\gamma<\delta} \Gamma_\gamma \begin{cases} > 0, & \textit{for } \delta \leq \gamma_0; \\ = 0, & \textit{for } \delta > \gamma_0. \end{cases}$$

EXAMPLE 5.10.1. We reconsider the $(\omega + 5)$-sequence of Examples 5.9.1(1) and 5.9.2(1):

$$[7, 7, 7, \ldots; 7, 7, 7, \ldots; 7, 7, 7, 7, 7].$$

As in Example 5.9.2(1), $(\cup R)^+ = 8$; thus, $\pi = 8^{\omega 2+5} = \omega^2 \cdot 8^5$. For $\gamma < \omega 2 + 5$, C_γ is the set of all γ-sequences $[\Delta_\delta]_{\delta<\gamma}$, where each $\Delta_\delta < \omega^2 \cdot 8^5$. Let $C = \cup\{C_\gamma : \gamma < \omega 2 + 5\}$; let $\Delta = [\Delta_\delta]_{\delta<\gamma} \in C$.

(1) $H(0) = 1$.
(2) If $\gamma = \rho^+$, $H(\Delta) = \Delta_\rho \cdot \Gamma_\rho = \Delta_\rho \cdot 7$.
(3) If γ is a limit ordinal, $H(\Delta) = \cup$ range Δ.

The sequence Π is defined as follows:

$$\Pi_0 = H(\Pi \restriction 0) = H(0) = 1;$$
$$\Pi_1 = H(\Pi \restriction 1) = \Pi_0 \cdot \Gamma_0 = 1 \cdot 7 = 7.$$

By induction on N,

$$\Pi_N = 7^N, \qquad N < \omega;$$
$$\Pi_\omega = \cup\{\Pi_N : N < \omega\} = \omega.$$

By induction on N,

$$\Pi_{\omega+N} = \omega \cdot 7^N, \qquad N < \omega;$$
$$\Pi_{\omega 2} = \omega^2;$$
$$\Pi_{\omega 2+N} = \omega^2 \cdot 7^N, \qquad N < 4.$$

Because $\Pi_\gamma > 0$ for all $\gamma < \omega 2 + 5$, $\overline{\Pi} = \Pi$. Thus,

$$\prod_{\gamma<\omega 2+5} \Gamma_\gamma = \hat{\Pi}_{\omega 2+5} = \Pi_{\omega 2+4} \cdot 7 = \omega^2 \cdot 7^5.$$

EXAMPLE 5.10.2

(1) Consider the $\omega 4$-sequence $[\Delta_\gamma]_{\gamma<\omega 4}$ in which

$$\Delta_{\omega N+M} = \begin{cases} 1, & \text{for } N = 0 \text{ and } M < \omega; \\ N + M, & \text{for } 1 \leq N < 4 \text{ and } M < \omega. \end{cases}$$

This is indicated by

$$[1, 1, 1, \ldots; 1, 2, 3, \ldots; 2, 3, 4, \ldots; 3, 4, 5, \ldots].$$

Here,

$$\Pi_M = 1, \qquad \text{for all } M < \omega;$$

$$\Pi_\omega = \Pi_{\omega+1} = 1;$$

$$\Pi_{\omega+M} = M!, \qquad \text{for all } M < \omega;$$

$$\Pi_{\omega 2} = \omega;$$

$$\Pi_{\omega 2+M} = \omega \cdot (M+1)!, \qquad \text{for all } M < \omega;$$

$$\Pi_{\omega 3} = \omega^2;$$

$$\Pi_{\omega 3+M} = \omega^2 \cdot (M+2)!, \qquad \text{for } 1 \leq M < \omega;$$

$$\prod_{\gamma<\omega 4} \Delta_\gamma = \Pi_{\omega 4} = \omega^3.$$

(2) Let $\Lambda : \omega + 4 \to \omega^+$ be indicated by

$$[\omega, \omega, \omega, \ldots; 1, 0, 1, 0].$$

Then,

$$\Pi_N = \omega^N, \qquad N < \omega;$$

$$\Pi_\omega = \Pi_{\omega+1} = \omega^\omega;$$

$$\Pi_{\omega+N} = 0, \qquad \text{for } N = 2, 3;$$

$$\prod_{\gamma<\omega+4} \Lambda_\gamma = \hat{\Pi}_{\omega+4} = 0.$$

Transfinite multiplication generalizes (finite) ordinal multiplication in the sense that

$$\prod_{\gamma<N^+} \Gamma_\gamma = \Gamma_0 \cdot \Gamma_1 \cdot \ldots \cdot \Gamma_N$$

for $1 \leq N < \omega$. Transfinite multiplication is "noncommutative" in that the order of the Γ_γ affects the product.

Clearly, the analogue of Theorem 5.9.1 goes through with "$\prod$" replacing each occurrence of "$\sum$". The analogues of its corollaries are true if "$\prod$" replaces each occurrence of "$\sum$" and "1" replaces "0". However, the remark preceding Theorem 5.9.1 is not true in the case of products; a partial product (for $\beta < \alpha$) can be greater than the product over α factors. This occurs if (and only if) there is at least one zero factor [as in Example 5.10.2(2)].

Similarly, a generalized associativity law holds; we can replace each occurrence of "$\sum$" by "$\prod$" and "sums" by "products" in Theorem 5.9.3. (Of course, $\prod_{\alpha_1 \le \gamma < \alpha_2} \Gamma_\gamma$ must be properly defined.)

The characterization of transfinite products in terms of "infinite Cartesian products" fails (Exercise 5.10.8).

Finally, just as ordinal multiplication could be regarded as a special case of transfinite addition (Theorem 5.9.2), ordinal exponentiation can be thought of as a particular case of transfinite multiplication.

Theorem 5.10.1. *If α and β are any ordinal numbers not both* 0, *then*

$$\alpha^\beta = \prod_{\gamma<\beta} \Gamma_\gamma, \text{ where for each } \gamma < \beta, \Gamma_\gamma = \alpha.$$

(The proof is Exercise 5.10.9.)

Exercises

5.10.1. Determine the transfinite product associated with the sequences $[\beta_\gamma]_{\gamma<\omega}$.

(a) $[0, 1, 2, 3, \ldots]$. **(b)** $[1, 2, 3, 4, \ldots]$.

(c) $[4, 3, 2, 1, 4, 3, 2, 1, \ldots]$. **(d)** $[\omega + 1, \omega + 2, \omega + 3, \ldots]$.

(e) $[\omega, \omega^2, \omega^3, \ldots]$. **(f)** $[\omega, \omega^2 + \omega, \omega^3 + \omega^2, \omega^4 + \omega^3, \ldots]$.

5.10.2. Determine the transfinite product associated with the indicated sequences:

(a) $\beta_\gamma = \begin{cases} \omega, \text{ for } \gamma < \omega; \\ 2, \omega \le \gamma < \omega + 3. \end{cases}$

(b) $\beta_\gamma = \begin{cases} \omega, \text{ for } \gamma < \omega; \\ 2, \text{ for } \omega \le \gamma < \omega 2. \end{cases}$

(c) $\beta_\gamma = \begin{cases} 2, \text{ for } \gamma < \omega; \\ \omega, \text{ for } \omega \le \gamma < \omega 2. \end{cases}$

5.10.3. Determine $\prod_{\gamma<\alpha} \Gamma_\gamma$ in each case.

(a) $\alpha = \omega 2, \Gamma_\gamma = 2$ for all $\gamma < \omega 2$.

(b) $\alpha = \omega 3, \Gamma_\gamma = 2$ for all $\gamma < \omega 2, \Gamma_\gamma = \omega^2$ for $\omega 2 \le \gamma < \omega 3$.

(c) $\alpha = \omega^2, \Gamma_\gamma = 2$ for all $\gamma < \omega^2$.

5.10.4. Let $\Gamma_\gamma = \gamma^+$ for all $\gamma < \alpha$. Determine $\prod_{\gamma<\alpha} \Gamma_\gamma$ in each case:

(a) $\alpha = \omega 2$. **(b)** $\alpha = \omega 2 + 3$. **(c)** $\alpha = \omega 3$.
(d) $\alpha = \omega^2$. **(e)** $\alpha = \omega^3$. **(f)** $\alpha = \omega^\omega$.

†5.10.5. Let α be an ordinal number and let $[\Gamma_\gamma]_{\gamma<\alpha}$ be an α-ordinal-sequence. Show the following:

(a) If $\rho^+ < \alpha$, then $\prod_{\gamma<\rho^+} \Gamma_\gamma = \left(\prod_{\gamma<\rho} \Gamma_\gamma\right) \cdot \Gamma_\rho$.

(b) If ρ is a limit ordinal $< \alpha$ and if $\Gamma_\gamma > 0$ for all $\gamma < \rho$, then

$$\prod_{\gamma<\rho} \Gamma_\gamma = \cup\left\{\prod_{\gamma<\mu} \Gamma_\gamma : \mu < \rho\right\}.$$

†5.10.6. State and prove the multiplicative analogue of Theorem 5.9.1.

5.10.7. State and prove the multiplicative analogue of Theorem 5.9.3.

★,†5.10.8. Show that the characterization of finite ordinal multiplication in terms of Cartesian products and antilexicographical orderings does not extend to transfinite ordinal multiplication. (*Hint*: Determine a set X and an indexing pair $\langle I, f\rangle$ for X for which $\bigtimes_{I,f} X$ cannot be antilexicographically ordered.)

5.10.9. Prove Theorem 5.10.1.

5.10.10. **(a)** Prove that for every ordinal α and all α-sequences $[\Gamma_\gamma]_{\gamma<\alpha}$ and $[\Delta_\gamma]_{\gamma<\alpha}$, if $\Gamma_\gamma \le \Delta_\gamma$ for all $\gamma < \alpha$, then $\prod_{\gamma<\alpha} \Gamma_\gamma \le \prod_{\gamma<\alpha} \Delta_\gamma$.

(b) Find a counterexample to part (a) if both instances of "$\le$" are replaced by "$<$".

5.11 Cantor Normal Form

It is well known that every nonzero natural number N can be uniquely represented in terms of powers of 10 as follows:

(5.11.1)
$$N = 10^{E_1}M_1 + 10^{E_2}M_2 + \cdots + 10^{E_K}M_K,$$
where $K, E_1, E_2, \ldots, E_K, M_1, M_2, \ldots, M_K$ are natural numbers satisfying $K > 0$, $N > E_1 > E_2 > \cdots > E_K \ge 0$, and $1 \le M_I < 10$ for $1 \le I \le K$.

In fact, (5.11.1) is the basis for our usual decimal system notation.

Example 5.11.1

(1) $627 = 10^2 \cdot 6 + 10^1 \cdot 2 + 10^0 \cdot 7$; here, $K = 3, E_1 = 2, E_2 = 1, E_3 = 0, M_1 = 6, M_2 = 2, M_3 = 7$.

(2) $4{,}000{,}607 = 10^6 \cdot 4 + 10^2 \cdot 6 + 10^0 \cdot 7$.

(3) $7 = 10^0 \cdot 7$.

Moreover, it is common practice to utilize the binary system in computer programming; here, "2" replaces "10" everywhere in the representation (5.11.1). More generally, for every natural number $R > 1$, we can let "R" replace "10" everywhere in (5.11.1).

EXAMPLE 5.11.2
(1) $19 = 2^4 \cdot 1 + 2^1 \cdot 1 + 2^0 \cdot 1$.
(2) $57 = 2^5 \cdot 1 + 2^4 \cdot 1 + 2^3 \cdot 1 + 2^0 \cdot 1$.
(3) $35 = 3^3 \cdot 1 + 3^1 \cdot 2 + 3^0 \cdot 2$.

We now consider the unique representation of ordinal numbers in terms of powers of ω. This representation is the basis for our usual notation for ordinal numbers. A more general expansion will be considered in Exercise 5.11.2; there we will show that ordinal numbers can be represented in terms of the powers of μ, for every ordinal $\mu > 1$.

Theorem 5.11.1. Cantor Normal Form. *Every ordinal number $\alpha > 0$ has a unique representation of the form*

$$(5.11.2)\quad \begin{cases} \alpha = \omega^{\beta_1}M_1 + \omega^{\beta_2}M_2 + \cdots + \omega^{\beta_K}M_K, \\ \text{where } K, M_1, M_2, \ldots, M_K \text{ are nonzero natural numbers and } \beta_1, \beta_2, \ldots, \\ \beta_K \text{ are ordinal numbers that satisfy } \alpha \geq \beta_1 > \beta_2 > \cdots > \beta_K \geq 0. \end{cases}$$

PROOF: The smallest ordinal δ such that $\alpha < \omega^\delta$ is a successor ordinal. (See Exercise 5.8.8.) Let $\delta = \beta_1{}^+$. Then

$$(5.11.3)\qquad \omega^{\beta_1}1 \leq \alpha < \omega^{\beta_1{}^+} = \omega^{\beta_1}\omega.$$

Thus, for some nonzero natural number M_1,

$$(5.11.4)\qquad \omega^{\beta_1}M_1 \leq \alpha < \omega^{\beta_1}M_1{}^+;$$

in fact, for some ordinal α_1,

$$(5.11.5)\qquad \alpha = \omega^{\beta_1}M_1 + \alpha_1.$$

If $\alpha_1 = 0$, we have a representation of the form (5.11.2); in any event, $\alpha_1 < \omega^{\beta_1}$, as otherwise (5.11.5) would contradict (5.11.4). If $\alpha_1 > 0$, we can repeat the previous argument (with "α_1" replacing "α") and obtain

$$\alpha_1 = \omega^{\beta_2}M_2 + \alpha_2,$$

where M_2 is a nonzero natural number and $0 \leq \alpha_2 < \omega^{\beta_2}$. Again, if $\alpha_2 = 0$, we have a representation

$$\alpha = \omega^{\beta_1}M_1 + \omega^{\beta_2}M_2$$

of the form (5.11.2); otherwise, we repeat the argument for α_2. This process must terminate after a finite number of steps; otherwise, there would be a sequence of ordinals satisfying

$$\ldots \alpha_{N^+} < \omega^{\beta_{N^+}} \leq \alpha_N < \omega^{\beta_N} \leq \cdots \leq \alpha_2 < \omega^{\beta_2} \leq \alpha_1 < \omega^{\beta_1} \leq \alpha.$$

This would contradict Exercise 5.2.13(d).

Thus, for some natural number K, there is a representation

$$\alpha = \omega^{\beta_1}M_1 + \omega^{\beta_2}M_2 + \cdots + \omega^{\beta_K}M_K \text{ of the form (5.11.2).}$$

We now consider the uniqueness of the representation (5.11.2). Suppose that

$$\begin{aligned}\alpha &= \omega^{\beta_1}M_1 + \omega^{\beta_2}M_2 + \cdots + \omega^{\beta_K}M_K \\ &= \omega^{\gamma_1}N_1 + \omega^{\gamma_2}N_2 + \cdots + \omega^{\gamma_L}N_L\end{aligned}$$

are two representations of α that satisfy the conditions for the Cantor normal form.

Suppose that $\beta_1 \neq \gamma_1$; arbitrarily, say $\beta_1 < \gamma_1$. Then $\omega^{\beta_1^+} \leq \omega^{\gamma_1} \leq \alpha$; this contradicts (5.11.3). Thus, $\beta_1 = \gamma_1$.

Suppose that $M_1 \neq N_1$; arbitrarily, suppose that $M_1 < N_1$. Then

$$\omega^{\beta_2}M_2 + \cdots + \omega^{\beta_K}M_K = \omega^{\gamma_1}(N_1 \dot{-} M_1) + \omega^{\gamma_2}N_2 + \cdots + \omega^{\gamma_L}N_L.$$

However, it is easily established, by induction on K, that

$$\begin{aligned}\omega^{\beta_2}M_2 + \cdots + \omega^{\beta_K}M_K &< \omega^{\beta_1} \\ &= \omega^{\gamma_1} \\ &\leq \omega^{\gamma_1}(N_1 \dot{-} M_1) + \omega^{\gamma_2}N_2 + \cdots + \omega^{\gamma_L}N_L.\end{aligned}$$

Clearly, we can repeat these arguments R times, where $R = \min\{K, L\}$, to show that $\beta_I = \gamma_I, M_I = N_I, I = 1, 2, \ldots, R$. Suppose, for example, that $K < L$. Then

$$\begin{aligned}\alpha &= \omega^{\beta_1}M_1 + \omega^{\beta_2}M_2 + \cdots + \omega^{\beta_K}M_K \\ &= \omega^{\beta_1}M_1 + \omega^{\beta_2}M_2 + \cdots + \omega^{\beta_K}M_K + \omega^{\gamma_{K+1}}N_{K+1} + \omega^{\gamma_{K+2}}N_{K+2} \\ &\quad + \cdots + \omega^{\gamma_L}N_L.\end{aligned}$$

Thus,

$$0 = \omega^{\gamma_{K+1}}N_{K+1} + \omega^{\gamma_{K+2}}N_{K+2} + \cdots + \omega^{\gamma_L}N_L.$$

Clearly,

$$N_{K+1} = N_{K+2} = \cdots = N_L = 0.$$

The uniqueness of the representation (5.11.2) follows.

In the Cantor normal form we will write "ω" instead of "ω^1" and "N" instead of "$\omega^0 N$". Also, we will write "ω^β" in lieu of "$\omega^\beta 1$".

EXAMPLE 5.11.3
(1) Each of the following is already in Cantor normal form:
  (a) K, where K is a nonzero natural number.
  (b) $\omega + 4$.
  (c) $\omega^2 + \omega 5 + 3$.
  (d) $\omega^{\omega+2} + \omega^\omega$.
(2) The Cantor normal form of $(\omega + 1)^2$ is $\omega^2 + \omega + 1$.

We utilize the Cantor-normal-form representation in order to define commutative arithmetic "operations" on the ordinals.[4]

Let α and β be nonzero ordinals whose Cantor normal forms are

$$\alpha = \omega^{\gamma_1} M_1 + \omega^{\gamma_2} M_2 + \cdots + \omega^{\gamma_K} M_K,$$
$$\beta = \omega^{\delta_1} N_1 + \omega^{\delta_2} N_2 + \cdots + \omega^{\delta_L} N_L.$$

We can arrange the distinct exponents among $\gamma_1, \gamma_2, \ldots, \gamma_K; \delta_1, \delta_2, \ldots, \delta_L$ in a decreasing sequence. Thus, let

$$\{\rho_1, \rho_2, \ldots, \rho_R\} = \{\gamma_1, \gamma_2, \ldots, \gamma_K; \delta_1, \delta_2, \ldots, \delta_L\},$$

where $\rho_1 > \rho_2 > \cdots > \rho_R$. Then

$$(5.11.6)\quad \begin{cases} \alpha = \omega^{\rho_1} P_1 + \omega^{\rho_2} P_2 + \cdots + \omega^{\rho_R} P_R, \\ \beta = \omega^{\rho_1} Q_1 + \omega^{\rho_2} Q_2 + \cdots + \omega^{\rho_R} Q_R, \\ \text{where } P_1, P_2, \ldots, P_R; Q_1, Q_2, \ldots, Q_R \text{ are natural numbers (possibly 0),} \\ \text{and for } 1 \le H \le R, \\ P_H = \begin{cases} M_I, & \text{if } \rho_H = \gamma_I \text{ for some (unique) } I, 1 \le I \le H; \\ 0, & \text{otherwise;} \end{cases} \\ Q_H = \begin{cases} N_J, & \text{if } \rho_H = \delta_J \text{ for some (unique) } J, 1 \le J \le H; \\ 0, & \text{otherwise.} \end{cases} \end{cases}$$

We define the **(Hessenberg) natural sum of α and β**, $\alpha +_{\mathscr{H}} \beta$, to be α, if $\beta = 0$, β, if $\alpha = 0$; if neither is 0, then with reference to (5.11.6),

$$\alpha +_{\mathscr{H}} \beta = \omega^{\rho_1}(P_1 + Q_1) + \omega^{\rho_2}(P_2 + Q_2) + \cdots + \omega^{\rho_R}(P_R + Q_R).$$

[4] The remainder of this section can be omitted, if desired; this material will not be utilized in subsequent development.

In general, $\alpha +_{\mathscr{H}} \beta \neq \alpha + \beta$. The Hessenberg sum is "natural" in the sense that we add ordinals according to their representation in terms of powers of ω—just as we add natural numbers according to their representation in terms of powers of 10.

EXAMPLE 5.11.4

(1) If $M, N < \omega$, $M +_{\mathscr{H}} N = N +_{\mathscr{H}} M = M + N$.

(2) $\omega +_{\mathscr{H}} \omega = \omega + \omega = \omega 2$.

(3) If $N < \omega$, $\omega +_{\mathscr{H}} N = \omega + N$, but $N +_{\mathscr{H}} \omega \neq N + \omega$; in fact,

$$N +_{\mathscr{H}} \omega = \omega +_{\mathscr{H}} N = \omega + N \neq \omega = N + \omega.$$

(4)
$$\begin{aligned}(\omega + 1) +_{\mathscr{H}} (\omega + 1) &= \omega 2 + 2 \neq \omega 2 + 1 \\ &= (\omega + 1) + (\omega + 1).\end{aligned}$$

(5)
$$\begin{aligned}&(\omega^\omega + \omega^4 3 + \omega^2 7 + 1) +_{\mathscr{H}} (\omega^5 + \omega^4 3 + 1) \\ &\quad = (\omega^5 + \omega^4 3 + 1) +_{\mathscr{H}} (\omega^\omega + \omega^4 3 + \omega^2 7 + 1) \\ &\quad = \omega^\omega + \omega^5 + \omega^4 6 + \omega^2 7 + 2; \\ &(\omega^\omega + \omega^4 3 + \omega^2 7 + 1) + (\omega^5 + \omega^4 3 + 1) \\ &\quad = \omega^\omega + \omega^5 + \omega^4 3 + 1; \\ &(\omega^5 + \omega^4 3 + 1) + (\omega^\omega + \omega^4 + \omega^2 7 + 1) = \omega^\omega + \omega^4 + \omega^2 7 + 1.\end{aligned}$$

Clearly, *the natural sum* $+_{\mathscr{H}}$, unlike the sum $+$, *is commutative* (*as well as associative*). *Thus, for all ordinals* α, β, *and* γ,

$$\alpha +_{\mathscr{H}} \beta = \beta +_{\mathscr{H}} \alpha;$$

$$(\alpha +_{\mathscr{H}} \beta) +_{\mathscr{H}} \gamma = \alpha +_{\mathscr{H}} (\beta +_{\mathscr{H}} \gamma)$$

(Exercise 5.11.7). Because of the associative law, we agree to drop parentheses in writing such sums. Moreover, *for all ordinals* α *and* β,

$$\alpha +_{\mathscr{H}} \beta \geq \max\{\alpha + \beta, \beta + \alpha\}$$

(Exercise 5.11.10). In parts (4) and (5) of Example 5.11.4, there is strict inequality. It follows, by induction, that *for all N ordinals*, $\alpha_1, \alpha_2, \ldots, \alpha_N$,

$$\alpha_1 +_{\mathscr{H}} \alpha_2 +_{\mathscr{H}} \cdots +_{\mathscr{H}} \alpha_N \geq \max \alpha_{\psi(1)} + \alpha_{\psi(2)} + \cdots + \alpha_{\psi(N)},$$

where ψ *ranges over all "permutations" of* $\{1, 2, \ldots, N\}$—*i.e., over all one-one correspondences of* $\{1, 2, \ldots, N\}$ *onto itself.*

We now define the **(Hessenberg) natural product of two ordinals**, $\alpha \cdot_{\mathscr{H}} \beta$. It is to be 0 if either α or β is zero. Otherwise, let

$$\alpha = \omega^{\gamma_1} M_1 + \omega^{\gamma_2} M_2 + \cdots + \omega^{\gamma_K} M_K$$

and

$$\beta = \omega^{\delta_1}N_1 + \omega^{\delta_2}N_2 + \cdots + \omega^{\delta_L}N_L$$

be the Cantor normal forms of α and β, respectively. Let

$$\{\sigma_1, \sigma_2, \ldots, \sigma_S\} = \{\gamma_I +_{\mathscr{H}} \delta_J : 1 \le I \le K, 1 \le J \le L\},$$

where $1 \le S \le KL$ and

$$\gamma_1 +_{\mathscr{H}} \delta_1 = \sigma_1 > \sigma_2 > \cdots > \sigma_S = \gamma_K +_{\mathscr{H}} \delta_L.$$

For $1 \le U \le S$, let $T_U = \Sigma\, M_I N_J$, where the summation is taken over all indices $I, J, 1 \le I \le K, 1 \le J \le L$, for which $\gamma_I +_{\mathscr{H}} \delta_J = \sigma_U$. Then

$$\alpha \cdot_{\mathscr{H}} \beta = \omega^{\sigma_1}T_1 + \omega^{\sigma_2}T_2 + \cdots + \omega^{\sigma_S}T_S.$$

In general, $\alpha \cdot_{\mathscr{H}} \beta \ne \alpha \cdot \beta$.

EXAMPLE 5.11.5. Let $\alpha = \omega^3 4 + \omega 5$ and let $\beta = \omega^2 2 + \omega 4$. These ordinals are already in Cantor normal form. Here,

$$K = 2, \qquad L = 2;$$

$$\gamma_1 = 3, \qquad \gamma_2 = 1;$$

$$\delta_1 = 2, \qquad \delta_2 = 1;$$

$$\sigma_1 = \gamma_1 +_{\mathscr{H}} \delta_1 = 5, \qquad \sigma_2 = \gamma_1 +_{\mathscr{H}} \delta_2 = 4,$$

$$\sigma_3 = \gamma_2 +_{\mathscr{H}} \delta_1 = 3, \qquad \sigma_4 = \gamma_2 +_{\mathscr{H}} \delta_2 = 2.$$

In this example, $S = KL$; thus, for each $U, 1 \le U \le S$, there is only one pair of indices $\langle I, J\rangle$ for which $\gamma_1 +_{\mathscr{H}} \delta_J = \sigma_U$.

For $U = 1, \gamma_1 +_{\mathscr{H}} \delta_1 = \sigma_1; T_1 = M_1N_1 = 4 \cdot 2 = 8.$
For $U = 2, \gamma_1 +_{\mathscr{H}} \delta_2 = \sigma_2; T_2 = M_1N_2 = 4 \cdot 4 = 16.$
For $U = 3, \gamma_2 +_{\mathscr{H}} \delta_1 = \sigma_3; T_3 = M_2N_1 = 5 \cdot 2 = 10.$
For $U = 4, \gamma_2 +_{\mathscr{H}} \delta_2 = \sigma_4; T_4 = M_2N_2 = 5 \cdot 4 = 20.$

Thus,

$$\alpha \cdot_{\mathscr{H}} \beta = \beta \cdot_{\mathscr{H}} \alpha = \omega^5 8 + \omega^4 16 + \omega^3 10 + \omega^2 20.$$

Observe that

$$\begin{aligned}\alpha \cdot \beta &= (\omega^3 4 + \omega 5) \cdot (\omega^2 2 + \omega 4)\\ &= (\omega^3 4 + \omega 5)(\omega^2 2) + (\omega^3 4 + \omega 5)(\omega 4)\\ &= \omega^5 2 + \omega^4 4;\\ \beta \cdot \alpha &= (\omega^2 2 + \omega 4) \cdot (\omega^3 4 + \omega 5)\\ &= (\omega^2 2 + \omega 4)(\omega^3 4) + (\omega^2 2 + \omega 4)(\omega 5)\\ &= \omega^5 4 + \omega^3 5.\end{aligned}$$

EXAMPLE 5.11.6. Let

$$\alpha = \omega^{\omega+1} + \omega^{\omega}2 + \omega 3 + 2$$

and let

$$\beta = \omega^3 4 + \omega^2.$$

Thus,

$$K = 4, \qquad L = 2;$$

$$\gamma_1 = \omega + 1, \qquad \gamma_2 = \omega, \qquad \gamma_3 = 1, \qquad \gamma_4 = 0;$$

$$\delta_1 = 3, \qquad \delta_2 = 2;$$

$$\sigma_1 = \gamma_1 +_{\mathscr{H}} \delta_1 = \omega + 4,$$

$$\sigma_2 = \gamma_1 +_{\mathscr{H}} \delta_2 = \gamma_2 +_{\mathscr{H}} \delta_1 = \omega + 3,$$

$$\sigma_3 = \gamma_2 +_{\mathscr{H}} \delta_2 = \omega + 2,$$

$$\sigma_4 = \gamma_3 +_{\mathscr{H}} \delta_1 = 4,$$

$$\sigma_5 = \gamma_3 +_{\mathscr{H}} \delta_2 = \gamma_4 +_{\mathscr{H}} \delta_1 = 3,$$

$$\sigma_6 = \gamma_4 +_{\mathscr{H}} \delta_2 = 2.$$

Here, $S < KL$; the pairs $\langle I, J \rangle$ for which $\gamma_I +_{\mathscr{H}} \delta_J = \sigma_U$ are indicated above. Thus,

$$T_1 = M_1 N_1 = 1 \cdot 4 = 4;$$

$$T_2 = M_1 N_2 +_{\mathscr{H}} M_2 N_1 = 1 \cdot 1 + 2 \cdot 4 = 9;$$

$$T_3 = M_2 N_2 = 2 \cdot 1 = 2;$$

$$T_4 = M_3 N_1 = 3 \cdot 4 = 12;$$

$$T_5 = M_3 N_2 + M_4 N_1 = 3 \cdot 1 + 2 \cdot 4 = 11;$$

$$T_6 = M_4 N_2 = 2 \cdot 1 = 2.$$

Thus,

$$\alpha \cdot_{\mathscr{H}} \beta = \omega^{\omega+4}4 + \omega^{\omega+3}9 + \omega^{\omega+2}2 + \omega^4 12 + \omega^3 11 + \omega^2 2.$$

The natural product is commutative as well as associative; i.e., for all ordinals α, β, *and* γ,

$$\alpha \cdot_{\mathscr{H}} \beta = \beta \cdot_{\mathscr{H}} \alpha;$$

$$(\alpha \cdot_{\mathscr{H}} \beta) \cdot_{\mathscr{H}} \gamma = \alpha \cdot_{\mathscr{H}} (\beta \cdot_{\mathscr{H}} \gamma).$$

Clearly, $\alpha \cdot_{\mathscr{H}} \beta \geq \max\{\alpha \cdot \beta, \beta \cdot \alpha\}$, but, as Example 5.11.5 shows, *this can be a strict inequality*.

Exercises

5.11.1. Determine the Cantor normal form for each ordinal.

(a) $(\omega + 2)3$.

(b) $(\omega + 4)^2 + 4$.

(c) $(\omega^2 + 5)(\omega + 3)$.

(d) $(\omega + 1)\omega^2 + (\omega + 2)5$.

(e) $(\omega^\omega + \omega 5)(\omega^3 2 + \omega^2 3 + 4)$.

(f) $(\omega^3 2 + \omega^2 3 + 4)(\omega^\omega + \omega 5)$.

***5.11.2.** **(A)** Let $\mu > 1$ be an ordinal number. Show that every ordinal number $\alpha > 0$ has a unique representation of the form

$$\alpha = \mu^{\beta_1}\gamma_1 + \mu^{\beta_2}\gamma_2 + \cdots + \mu^{\beta_K}\gamma_K,$$

where K is a nonzero natural number; $\gamma_1, \gamma_2, \ldots, \gamma_K, \beta_1, \beta_2, \ldots, \beta_K$ are ordinals satisfying $\gamma_I < \mu$, $1 \le I \le K$, and $\alpha \ge \beta_1 > \beta_2 > \cdots > \beta_K \ge 0$. (Note that this proves the assertions made at the beginning of this section.)

(B) "μ" is called the **base of the representation**. Express the ordinals (a), (b), (c), (d), (e), and (f) of Exercise 5.11.1 to the base (1) 10, (2) $\omega + 1$, (3) $\omega + 2$, and (4) ω^2.

***5.11.3.** We previously showed that by considering the $3!(= 6)$ possible arrangements of 3 ordinals we could have 1, 2, 3, 4, or 5 distinct sums. (See Exercises 5.5.8 and 5.5.9.) Use the Cantor normal form to show that 3 ordinals can never yield 6 distinct sums.

5.11.4. Let

$$\alpha = \omega^7 2 + \omega^4 3 + \omega^2 5 + 1,$$

$$\beta = \omega^4 3 + \omega^2 2 + \omega + 2,$$

$$\gamma = \omega^3 2 + \omega^2 4 + 6.$$

Determine the following ordinals:

(a) $\alpha +_{\mathscr{H}} \beta$. **(b)** $\alpha +_{\mathscr{H}} \gamma$.

(c) $\beta +_{\mathscr{H}} \gamma$. **(d)** $\alpha +_{\mathscr{H}} \beta +_{\mathscr{H}} \gamma$.

5.11.5. Let α, β, and γ be as in Exercise 5.11.4. Determine the following ordinals:

(a) $\alpha \cdot_{\mathscr{H}} \beta$. **(b)** $\alpha \cdot_{\mathscr{H}} \gamma$.

(c) $\beta \cdot_{\mathscr{H}} \gamma$. **(d)** $\alpha \cdot_{\mathscr{H}} \beta \cdot_{\mathscr{H}} \gamma$.

5.11.6. Let

$$\alpha = \omega^{\omega+2}3 + \omega^{\omega+1}2 + \omega^4 + 3,$$

$$\beta = \omega^\omega + \omega 2 + 1,$$

$$\gamma = \omega^{\omega 2}3 + \omega^{\omega+1} + \omega^\omega.$$

Determine the following ordinals:

(a) $\alpha \cdot_{\mathscr{H}} \beta$. **(b)** $\alpha \cdot_{\mathscr{H}} \gamma$.

(c) $\beta \cdot_{\mathscr{H}} \gamma$. **(d)** $\alpha \cdot_{\mathscr{H}} \beta \cdot_{\mathscr{H}} \gamma$.

5.11.7. Prove the commutativity and associativity of natural addition. Thus, prove that for all ordinal numbers α, β, and γ, the following are true:

(a) $\alpha +_{\mathscr{H}} \beta = \beta +_{\mathscr{H}} \alpha$.

(b) $(\alpha +_{\mathscr{H}} \beta) +_{\mathscr{H}} \gamma = \alpha +_{\mathscr{H}} (\beta +_{\mathscr{H}} \gamma)$.

5.11.8. Prove the commutativity and associativity of natural multiplication. Thus, prove that for all ordinal numbers α, β, and γ, the following are true:

(a) $\alpha \cdot_{\mathscr{H}} \beta = \beta \cdot_{\mathscr{H}} \alpha$.

(b) $(\alpha \cdot_{\mathscr{H}} \beta) \cdot_{\mathscr{H}} \gamma = \alpha \cdot_{\mathscr{H}} (\beta \cdot_{\mathscr{H}} \gamma)$.

5.11.9. Do either of the distributive laws hold for natural multiplication over natural addition? Thus, let α, β, and γ be arbitrary ordinals.

(a) Does $\alpha \cdot_{\mathscr{H}} (\beta +_{\mathscr{H}} \gamma) = \alpha \cdot_{\mathscr{H}} \beta +_{\mathscr{H}} \alpha \cdot_{\mathscr{H}} \gamma$?

(b) Does $(\alpha +_{\mathscr{H}} \beta) \cdot_{\mathscr{H}} \gamma = \alpha \cdot_{\mathscr{H}} \gamma +_{\mathscr{H}} \beta \cdot_{\mathscr{H}} \gamma$?

5.11.10. Prove that for all ordinals α and β, the following are true:

(a) $\alpha +_{\mathscr{H}} \beta \geq \max\{\alpha + \beta, \beta + \alpha\}$.

(b) $\alpha \cdot_{\mathscr{H}} \beta \geq \max\{\alpha \cdot \beta, \beta \cdot \alpha\}$.

5.12 Denumerable Ordinal Numbers

The set of finite ordinals, i.e., the set of natural numbers, is often called the **first number class**; the collection of denumerable ordinals is called the **second number class**. In this section we consider these first two number classes.

Theorem 5.12.1. *The second number class is a set.*

PROOF: Let

$$A = \{x \in \{\omega\} \times \omega^{②} : x = \langle \omega, R \rangle, \text{ where } R \text{ is a well-ordering on } \omega\}.$$

By the Counting Theorem, each such $\langle \omega, R \rangle$ in A is similar to a unique ordinal system $\langle \alpha, \in \rangle$, in which α is clearly denumerable. Moreover, every ordinal system $\langle \beta, \in \rangle$, in which β is denumerable, corresponds to at least one $\langle \omega, R_\beta \rangle$ in A. For if f is a one-one mapping of ω onto β, let $M R_\beta N$ iff $f(M) \in f(N)$. Let s be "$x \in A$ and y is the first coordinate of an ordinal system that is similar to x." Then s is bimonadic. By the Axiom of Substitution, there exists a set

$$S = \left\{c_2 : \text{for some } c_1 \text{ in } A, \text{ ``} s\begin{pmatrix} x & y \\ c_1 & c_2 \end{pmatrix}\text{'' is true}\right\}.$$

By the above, S is the second number class.

Theorem 5.12.2. *Let $\omega_1 = \omega \cup S$, where S is the second number class. Then ω_1 is an uncountable ordinal number.*

PROOF: ω_1 is a transitive set of ordinals; hence ω_1 is an ordinal number. Clearly, ω_1 is infinite; if it were denumerable, then we would have $\omega_1 \in \omega_1$.

Corollary. *ω_1 is the smallest uncountable ordinal number.*

PROOF: If α is an uncountable ordinal, distinct from ω_1, then either $\alpha \in \omega_1$ or (excl.) $\omega_1 \in \alpha$. If $\alpha \in \omega_1$, then α would be countable.

Thus, ω_1, the smallest uncountable ordinal, consists of all the countable ordinals. Recall that ω is the smallest infinite ordinal and consists of all the finite ordinals. In Chapter 6 we will generalize this situation.

Clearly, we can repeat the considerations of Section 5.3 to show the existence of such other uncountable ordinals as

$$\omega_1{}^+, \omega_1{}^{++}, \ldots, \omega_1 2, \ldots, \omega_1{}^2, \ldots, \omega_1{}^\omega, \text{etc.}$$

In Chapter 6 we will consider still other "large" ordinals. We return now to further considerations of denumerable ordinals.

We know that if α and β are both in the first number class, then so are $\alpha + \beta$, $\alpha \cdot \beta$, and α^β.

Theorem 5.12.3. *If α and β are in the second number class, then so are the following:*
(1) $\alpha + \beta$.
(2) $\alpha \cdot \beta$.
(3) α^β.

PROOF: Let $\langle A, R_A\rangle \cong \langle \alpha, \in\rangle$ and let $\langle B, R_B\rangle \cong \langle \beta, \in\rangle$, where, in part (1), $A \cap B = \varnothing$.

(1) $\langle A \cup B, R_{A\cup B}\rangle \cong \langle \alpha + \beta, \in\rangle$. The sets A and B are each denumerable; hence by Theorem 4.3.1, so is $A \cup B$. The denumerability of $\alpha + \beta$ follows.

(2) $$\langle A \times B, R_{A\times B}\rangle \cong \langle \alpha \cdot \beta, \in\rangle.$$

By Theorem 4.3.2, $A \times B$ is denumerable; hence, so is $\alpha \cdot \beta$.

(3) $$\langle Z(A, B), R_{Z(A,B)}\rangle \cong \langle \alpha^\beta, \in\rangle.$$

Recall that $Z(A, B)$ is the set of functions f that are such that $f : B \to A$ and $f(b)$ differs from a_0, the first element of A, for only finitely many b in B. With respect to a specific one-one correspondence between ω_+ and $A \sim \{a_0\}$, we can write

$$A \sim \{a_0\} = \{a_1, a_2, \ldots, a_M, \ldots\};$$

similarly, we can write

$$B = \{b_0, b_1, b_2, \ldots, b_N, \ldots\}.$$

For f in $Z(A, B)$, let $B_f = \{b \in B : f(b) \neq a_0\}$ and let $A_f = f''(B_f)$. For each $K \geq 1$, let Z_K be the set of functions f in $Z(A, B)$ such that the sum of the subscripts of elements of $A_f \cup B_f$ is K; let Z_0 be the unit set consisting of the function that is equal to a_0 everywhere.

Clearly, each Z_K is finite; we can define a specific enumeration R_K of Z_K for $K \geq 1$ as follows: f in Z_K corresponds to a sequence

$$\langle b_{M_1}, a_{N_1}, b_{M_2}, a_{N_2}, \ldots, b_{M_R}, a_{N_R}\rangle,$$

where

(1) $B_f = \{b_{M_1}, b_{M_2}, \ldots, b_{M_R}\}$.
(2) $M_1 < M_2 < \cdots < M_R$.
(3) $a_{N_I} = f(B_{M_I}), I = 1, 2, \ldots, R$.
(4) $M_1 + N_1 + M_2 + N_2 + \cdots + M_R + N_R = K$.

These sequences can be ordered according to the lexicographical ordering of their subscripts; we thereby induce an ordering R_K on Z_K. (See Example 5.12.1.)

Observe that $Z(A, B)$ is the disjoint union of $\{Z_K : K < \omega\}$. For $f, g \in Z(A, B)$, let

$$f R g \text{ if } f \in Z_K, g \in Z_L, \qquad K < L,$$

or, if for some K, we have $f, g \in Z_K$ and $f R_K g$. Clearly, R determines an enumeration of $Z(A, B)$. Thus, $Z(A, B)$ and, consequently, α^β are denumerable.

Note that we did not employ the Axiom of Choice in this proof. In part (3), enumerations of the elements of $A \sim \{a_0\}$ and of B exist by virtue of the assumed denumerability of these two sets. All other enumerations in the proof could be defined in terms of these two by means of formulas of set theory. [Another proof of (3) is considered in Exercise 5.12.4.]

†EXAMPLE 5.12.1. In the proof of part (3), the first few cells, Z_0, Z_1, Z_2, Z_3, together with the first few functions—$f_0, f_1, \ldots, f_7$—of $Z(A, B)$ in the enumeration R are (in terms of the corresponding sequences)

Z_0 : the function that takes the constant value a_0 everywhere;

$Z_1 : \langle b_0, a_1\rangle$;

$Z_2 : \langle b_0, a_2\rangle, \langle b_1, a_1\rangle$;

$Z_3 : \langle b_0, a_1, b_1, a_1\rangle, \langle b_0, a_3\rangle, \langle b_1, a_2\rangle, \langle b_2, a_1\rangle$.

Thus,

$$f_4(b) = \begin{cases} a_1, & \text{if } b = b_0 \text{ or if } b = b_1; \\ a_0, & \text{otherwise.} \end{cases}$$

EXAMPLE 5.12.2. According to Theorem 5.12.3, each of the following ordinals is denumerable:

$$\omega^2, \omega^2 + \omega, \omega^\omega, (\omega^\omega)^\omega (= \omega^{(\omega^2)}), \omega^{(\omega^\omega)} \cdot (\omega^2 + \omega 3).$$

Theorem 5.12.4. *Let α be a denumerable limit ordinal. Then there exists an increasing ω-ordinal-sequence in α, $[\Gamma_N]_{N<\omega}$, such that*

$$\sum_{N<\omega} \Gamma_N = \alpha.$$

PROOF: By hypothesis, α is denumerable; thus, there is a one-one correspondence between ω and α. Enumerate the ordinals less than α by means of this correspondence; for $N < \omega$, let β_N be the Nth such ordinal. (Clearly, if $\alpha > \omega$, the enumeration fails to preserve magnitude.) We use Definition by Induction (Theorem 2.8.1) to define an increasing ω-ordinal-sequence in α, $[\Gamma_N]_{N<\omega}$, as follows:

$\Gamma_0 = \beta_0$;

Γ_{N^+} is the ordinal—with smallest subscript larger than N—that is larger than Γ_N.

It is easily verified that

$$\sum_{N<\omega} \Gamma_N = \cup\{\Gamma_N : N < \omega\} = \cup\{\beta_N : N < \omega\} = \alpha.$$

Note that Theorem 5.12.4 is an existence theorem; it does not yield an effective procedure for selecting such an ω-sequence.

Theorem 5.12.5. *Let A be a set that satisfies the following:*

(1) $0 \in A$.

(2) *If $a \in A$, then $a^+ \in A$.*

(3) *Whenever $\{a_N : N < \omega\}$ is a subset of A that is such that $a_N \subset a_{N^+}$ for all N in ω, it follows that $\cup\{a_N : N < \omega\} \in A$.*

Then A includes the first two number classes.

The proof of this theorem is left as an exercise. From the theorem we easily deduce the following.

Corollary. Transfinite Induction for Countable Ordinals. *Let s be a formula with one free variable, x. Suppose the following:*

(1) $s\binom{x}{0}$ *is true.*

(2) *If* $s\binom{x}{a}$ *is true, then* $s\binom{x}{a^+}$ *is true.*

(3) *Whenever* $a = \{a_N : N < \omega \text{ and } a_N \subset a_{N^+} \text{ for all } N \text{ in } \omega\}$, *then the truth of*

$$\text{For all } a_N \text{ in } a,\ s\binom{x}{a_N}$$

implies the truth of

$$s\binom{x}{\cup a}.$$

It follows that

$$s\binom{x}{\alpha} \text{ is true for every countable ordinal } \alpha.$$

Theorem 5.12.6. *(A.C.) Let α be in the second number class. Let $[\Gamma_\gamma]_{\gamma<\alpha}$ be an α-ordinal-sequence in $\omega_1 \sim \{0\}$. Then $\sum_{\gamma<\alpha} \Gamma_\gamma$ is in the second number class.*

(The proof is left as an exercise.)

Corollary 1. *(A.C.) For $\alpha < \omega_1$, ω_1 is not the sum of an α-ordinal-sequence in ω_1.*

As a result of Theorem 5.12.6, we can now demonstrate how extensive the second number class is. By induction, we can define the following sequence of denumerable ordinals:

$$\alpha_0 = 1, \alpha_1 = \omega, \alpha_2 = \omega^\omega, \alpha_3 = \omega^{(\omega^\omega)},$$

and, in general,

$$\alpha_{N^+} = \omega^{\left(\omega^{\cdot^{\cdot^{\cdot^{(\omega^{(\omega^\omega)})}}}}\right)}$$

(with N such "exponential levels"). Let

$$\varepsilon_0 = \cup\{\alpha_N : N < \omega\} = \sum_{N<\omega} \alpha_N.$$

Corollary 2. *(A.C.)* ε_0 *is denumerable.*

PROOF: $\{\alpha_N\}$ *is a denumerable "increasing" sequence of denumerable ordinals.*

Corollary 3. *(A.C.)* ${\varepsilon_0}^{\varepsilon_0}$ *is denumerable.*

The first two number classes have important applications in analysis.

An **open interval** is a set of real numbers of the form $\{\Gamma \in \mathscr{R}: \Gamma_1 < \Gamma < \Gamma_2\}$, for $\Gamma_1, \Gamma_2 \in \mathscr{R}$. A set of real numbers is **open** if it is a union of (a set of) open intervals. Unions of open sets are open; but intersections of open sets need not be open.

Let $S \subseteq \mathscr{R}$. A real number Γ_0 is called an **accumulation point of** S if every open interval containing Γ_0 contains points of $S \sim \{\Gamma_0\}$. A **closed** set is a subset of $\mathscr{R}$ that contains all its accumulation points. Intersections of closed sets are closed; but unions of closed sets need not be closed. It is easily established that S is closed iff $\mathscr{R} \sim S$ is open.

Every ordinal α has a unique expression of the form $\delta + N$, where δ is either 0 or a limit ordinal, and where N is a natural number. We will say that α is **odd** or **even** according to whether N is odd or even.

We apply Definition by Transfinite Induction to define sets $F_\alpha, G_\alpha, H_\alpha, \alpha < \omega_1$ (F_α will be the set of F_α-sets, etc.).

Let the F_0-sets be the closed sets. For each ordinal α, $0 < \alpha < \omega_1$, let the F_α-sets be denumerable unions of $F_{\beta N}$-sets for $\beta_N < \alpha$, $N = 1, 2, \ldots$, if α is odd; let the F_α-sets be denumerable intersections of $F_{\beta N}$-sets for $\beta_N < \alpha$, $N = 1, 2, \ldots$, if α is even. For $\beta < \alpha < \omega_1$, $F_\beta \subset F_\alpha$. However, were we to continue this process beyond ω_1, no new sets would be obtained. Let

$$\mathscr{B}_0 = \{B: B \text{ is an } F_\alpha\text{-set for some } \alpha < \omega_1\}.$$

The set $\mathscr{B}_0$ is known as the set of **Borel sets** (of $\mathscr{R}$).

The set $\mathscr{B}_0$ can be generated in other ways. Let the G_0-sets be the open sets. For each ordinal α, $0 < \alpha < \omega_1$, let the G_α-sets be denumerable intersections of G_{β_N}-sets for $\beta_N < \alpha$, $N = 1, 2, \ldots$, if α is odd; let the G_α-sets be denumerable unions of G_{β_N}-sets for $\beta_N < \alpha$, $N = 1, 2, \ldots$, if α is even. For $\beta < \alpha < \omega_1$, $G_\beta \subset G_\alpha$. Again, the process of obtaining new sets terminates at ω_1. Moreover, $\mathscr{B}_0 = \{S: S \in G_\alpha \text{ for some } \alpha < \omega_1\}$.

Let $F: \mathscr{R} \to \mathscr{R}$. Then F is said to be **continuous** if $F^{-1''}(O)$ is open for every open subset O of $\mathscr{R}$. By an **ω-function-sequence** we will mean an ω-sequence whose range consists of functions f from $\mathscr{R}$ into $\mathscr{R}$. Let $g: \mathscr{R} \to \mathscr{R}$. Then g is said to be the **limit of an ω-function-sequence** $[f_N]_{N<\omega}$ if for every $x \in \mathscr{R}$ and for every open set O containing $g(x)$ there is an M $(= M_{l,x})$ such that $f_N(x) \in O$ for all $N \geq M$. Let the H_0-functions be the continuous functions. For ordinals α, $0 < \alpha < \omega_1$, let the H_α-functions be the limits of ω-function-sequences $[f_N]_{N<\omega}$, where each $f_N \in H_{\beta_N}$ for some $\beta_N < \alpha$. Then for $\beta < \alpha < \omega_1$, $H_\beta \subset H_\alpha$; again, the process terminates at ω_1. Let $\mathscr{B}_1 = \{f: f \in H_\alpha \text{ for some } \alpha < \omega_1\}$. The set $\mathscr{B}_1$ is known as the set of **Baire functions**. A necessary and sufficient condition for a subset of $\mathscr{R}$ to be a Borel set is that its characteristic function be a Baire function. [For further discussion of Borel sets and Baire functions, see (Goffman, Chap. 11) and (Halmos, 1950, Secs. 15 and 51)].

Let $S \subseteq \mathscr{R}$. Let S_0 be the set of accumulation points of S. For each ordinal α, $0 < \alpha < \omega_1$, let S_α be the set consisting of all real numbers that are accumulation points of S_β for every $\beta < \alpha$. Then for every $\delta < \omega_1$ there are sets S such that for $\beta < \alpha \leq \delta$, $S_\alpha \subset S_\beta$. Moreover, for every set S of real numbers there is an ordinal $\delta < \omega_1$ such that for $\delta < \varepsilon < \omega_1$, $S_\delta = S_\varepsilon$.

Exercises

5.12.1. Let α and β be in the second number class and let $\alpha < \beta$. Give an example in each case.

(a) $\beta \dot{-} \alpha$ is in the first number class.

(b) $\beta \dot{-} \alpha$ is in the second number class.

5.12.2. Where was the Axiom of Substitution used in the proof of Theorem 5.12.1?

5.12.3. In the proof of Theorem 5.12.3, determine Z_4 and R_4. (See Example 5.12.1.)

5.12.4. †**(a)** Use Theorem 4.3.1 to prove that if α and β are each in the second number class, then so is α^β.

(b) Show that the Axiom of Choice is not necessary even in this proof.

5.12.5. (A.C.) Let A be a denumerable set of denumerable ordinals. Show that there is a denumerable ordinal β that is such that $\alpha < \beta$ for all α in A.

***5.12.6.** An ordinal number α is called an **ε-number** if $\omega^\alpha = \alpha$.

(a) Prove that ε_0 is the smallest ε-number.

(b) What is the Cantor normal form of an ε-number?

(Used in Exercise 6.11.7.)

5.12.7. Can there be an increasing ω-sequence $[\Gamma_N]_{N<\omega}$ in α, for which $\sum_{N<\omega} \Gamma_N = \alpha$, where α is a denumerable successor ordinal?

5.12.8. **(a)** Prove Theorem 5.12.5.

(b) Prove the Corollary of Theorem 5.12.5.

5.12.9. **(a)** Prove Theorem 5.12.6.

(b) State a multiplicative version of Theorem 5.12.6.

5.12.10. (A.C.) Show that there are nondenumerably many denumerable limit ordinals.

5.13 Maximal Principles

Now that we have the Axiom of Substitution at our disposal, we are in a position to consider several other equivalents of the Axiom of Choice (Choice) [or of the Well-ordering Theorem (WO). See Section 4.8.] The first of these is simply a reformulation of the Well-ordering Theorem in terms of ordinals.

Ordinal Formulation of Well-ordering Theorem (Ord): *Every set is in one-one correspondence with an ordinal number.*

We say that a linearly ordered subsystem $\langle B, R\rangle$ of a partially ordered system $\langle A, R\rangle$ is a **maximal linearly ordered subsystem** provided that $B \subset C \subseteq A$ implies R does not linearly order C.

EXAMPLE 5.13.1. Let $A = 4 \cup \{\{1\}, \{1, 2\}\}$. Then $\langle A, \in_A\rangle$ is a partially ordered system. $\langle 4, \in\rangle$ is a maximal linearly ordered system as are $\langle B, \in_B\rangle$, where $B = \{0, 1, \{1\}, \{1, 2\}\}$, and $\langle C, \in_C\rangle$, where $C = 3 \cup \{\{1, 2\}\}$. (See Figure 5.13.1.)

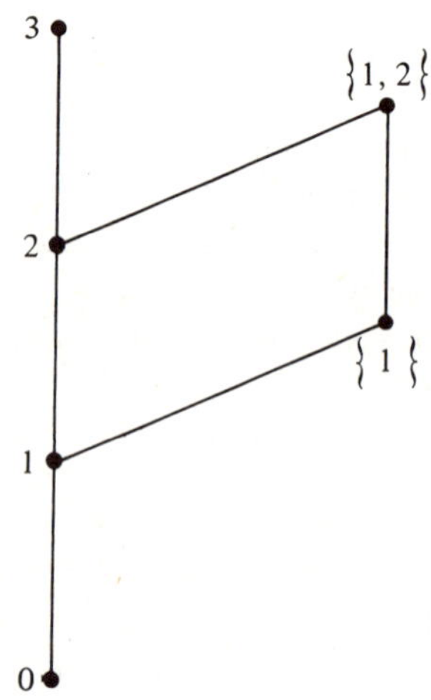

Fig. 5.13.1

A set x is said to be **of finite character** provided $y \in x$ iff every finite subset of y is in x.

EXAMPLE 5.13.2. For every set X, $\mathscr{P}(X)$ is of finite character. For if $y \in \mathscr{P}(X)$ and z is a finite subset of X, then $z \subseteq y \subseteq X$; hence, $z \in \mathscr{P}(X)$. Moreover, if u is an arbitrary set for which every finite subset of u is in $\mathscr{P}(X)$, then for each v in u, $\{v\} \in \mathscr{P}(X)$. Thus, $\{v\} \subseteq X$. In other words, $v \in u$ implies $v \in X$. It follows that $u \in \mathscr{P}(X)$.

EXAMPLE 5.13.3. 3 is not of finite character because $2 \in 3$, $\{1\} \subseteq 2$, but $\{1\} \notin 3$.

Maximal Principles

Hausdorff Maximal Principle (Hausdorff). *Every partially ordered system has a maximal linearly ordered subsystem.*

Zorn's Lemma (Zorn). *If $\langle A, R\rangle$ is a partially ordered system in which every linearly ordered subsystem, $\langle B, R\rangle$, has an upper bound, then $\langle A, R\rangle$ has a maximal element.*

Teichmüller–Tukey Lemma (T.T.L.). *Every set of finite character has a maximal element with respect to inclusion.*

Theorem 5.13.1

(a) *The Ordinal Formulation of the Well-ordering Theorem,*
(b) *The Hausdorff Maximal Principle,*
(c) *Zorn's Lemma, and*
(d) *The Teichmüller–Tukey Lemma*

are all equivalents of the Axiom of Choice.

PROOF: Figure 5.13.2 illustrates our procedure:

$$\begin{array}{ccccc} & \text{(Choice)} \rightarrow & \text{(WO)} & \rightarrow \text{(Ord)} & \\ & \uparrow & & \downarrow & \\ \text{(T.T.L.)} & \leftarrow & \text{(Zorn)} & \leftarrow & \text{(Hausdorff)} \end{array}$$

Fig. 5.13.2

(Choice) $\rightarrow$ *(WO)*: See Section 4.8.

(WO) $\rightarrow$ *(Ord)*: By the Well-ordering Theorem, for every set X, there is a relation R on X that well-orders X. By the Counting theorem, $\langle X, R\rangle$ is similar to a unique ordinal system, $\langle \alpha, \in\rangle$. Thus, X is in one–one correspondence with α.

(Ord) $\rightarrow$ *(Hausdorff)*: Let $\langle X, R\rangle$ be a partially ordered system and suppose that F is a one–one mapping of the ordinal number α onto X. Define $G : \alpha \rightarrow X$ as follows:

$$G(\beta) = \begin{cases} F(\beta), & \text{if for all } \gamma < \beta, \text{ either } G(\gamma)RF(\beta) \text{ or } F(\beta)RG(\gamma); \\ F(0), & \text{otherwise.} \end{cases}$$

Then range G is linearly ordered by R because every two elements of range G are R-comparable. If $x \in X \sim$ range G, then $x = F(\beta)$ for some $\beta < \alpha$. Hence, for some $\gamma < \beta$, $G(\gamma)\not R F(\beta)$ and $F(\beta)\not R G(\gamma)$—i.e., $G(\gamma)\not R x$ and $x\not R G(\gamma)$. Thus, range $G \cup \{x\}$ is not linearly ordered by R; consequently, $\langle \text{range}\, G, R\rangle$ is a maximal linearly ordered subsystem of $\langle A, R\rangle$.

(*Hausdorff*) → (*Zorn*): Let $\langle A, R\rangle$ be a partially ordered system in which every linearly ordered subsystem has an upper bound. According to the Hausdorff Maximal Principle, there is a maximal linearly ordered subsystem, $\langle B, R\rangle$. Let b_0 be an upper bound for $\langle B, R\rangle$. Thus, for all b in B, bRb_0. If $b_0 \notin B$, then $B \cup \{b_0\}$ would be linearly ordered by R, contradicting the assertion that $\langle B, R\rangle$ is a maximal linearly ordered subsystem of $\langle A, R\rangle$. It follows that $b_0 \in B$. If $a\,(\neq b_0) \in A$ and b_0Ra, then $a \notin B$ and $B \cup \{a\}$ is linearly ordered by R. This contradiction establishes b_0 as an R-maximal element of A.

(*Zorn*) → (*T.T.L.*): Let A be of finite character. For every set A, $\mathscr{I}_A$ is a partial ordering on A. In order to invoke Zorn's Lemma, we must show that every linearly ordered subsystem $\langle B, \mathscr{I}_B\rangle$ has an upper bound. Consider $\cup B$; to show that $\cup B \in A$ (and hence $\cup B$ is automatically an upper bound for B with respect to inclusion), it suffices to show that an arbitrary finite subset, $C = \{c_1, c_2, \ldots, c_N\} \subseteq \cup B$, is in A. There are $b_1, b_2, \ldots, b_N$ in B such that $c_I \in b_I$, $1 \leq I \leq N$. Because B is linearly ordered by inclusion, there is b^* in $B\,(\subseteq A)$ such that $b_I \subseteq b^*$ for all I, $1 \leq I \leq N$. (Verify.) Hence, $C \subseteq b^*$. Because $b^* \in A$ and C is a finite subset of b^*, $C \in A$. Thus, every finite subset of $\cup B$ is in A.

(*T.T.L.*) → (*Choice*): Let X be a nonempty set of nonempty sets. Let $\mathscr{F} = \{f : f \text{ is a function, } \operatorname{dom} f \subseteq X, \text{ and for each } A \text{ in } \operatorname{dom} f, f(A) \in A\}$.

We claim that $\mathscr{F}$ is of finite character. If $f \in \mathscr{F}$ and if g is a (finite) subset of f, then g is a set of ordered pairs, $\langle A, a\rangle$, where $a \in A \in X$. Because f is a function, g is one, too—hence $g \in \mathscr{F}$. Now suppose that h is a set all of whose finite subsets are in $\mathscr{F}$. Then if $x \in h$, $\{x\} \in \mathscr{F}$; hence, x must be an ordered pair, $\langle A, a\rangle$, with $a \in A \in X$. If $\langle A, a\rangle, \langle A, b\rangle \in h$, then $h_0 = \{\langle A, a\rangle, \langle A, b\rangle\} \in \mathscr{F}$; therefore, h_0 is a function. Consequently, $a = b$ and h is a function. It follows from the preceding remarks that $h \in \mathscr{F}$.

Because $\mathscr{F}$ is of finite character, it has a maximal element, F, with respect to inclusion. If $\operatorname{dom} F \neq X$, then for any A in $X \sim \operatorname{dom} F$ and for any a in A, $F \cup \{\langle A, a\rangle\}$ would be an element of $\mathscr{F}$ that properly includes F. Hence, $\operatorname{dom} F = X$, and F is a choice function on X.

The Axiom of Choice has numerous applications in other fields of mathematics. We consider one such application in algebra and one in analysis.

The definitions of a ring, of an identity, and of a commutative group were given in Exercise 2.8.5. Let $\langle X, \langle F_1, F_2\rangle\rangle$ be a ring. Let $I \subseteq X$. (We write "F_K" for "$F_K \restriction I^{②}$", $K = 1, 2$.) Then $\langle I, \langle F_1, F_2\rangle\rangle$ is an **ideal** (of the ring) if $\langle I, F_1\rangle$ is a commutative group and if for every i in I and r in R, both iF_2r and rF_2i are in I. An **ideal** is said to be **maximal** if $I \subset X$ and if whenever $\langle J, \langle F_1, F_2\rangle\rangle$ is an ideal (of the ring) for which $I \subset J$, then $J = X$. The Axiom of Choice is sufficient to guarantee that in every ring that has an identity, every ideal (other than the ring itself) is "included" in a maximal ideal; without the Axiom of Choice we cannot prove this. (See Herstein, p. 99.)

Let $S \subseteq \mathscr{R}$. Let $m^*(S)$ be the greatest lower bound of the set of infinite sums

$$\sum_{I<\omega} (b_I - a_I),$$

where $S \subseteq \bigcup \{(a_I, b_I) : a_I < b_I, I < \omega\}$. Here, the $b_I - a_I$ can be any positive real numbers. A set S is said to be **measurable** if for every subset $A \subseteq \mathscr{R}$,

$$m^*(A) = m^*(A \cap S) + m^*(A \sim S).$$

Every Borel set (Section 5.12) is measurable, and it is difficult to conceive of a nonmeasurable set. However, the Axiom of Choice guarantees the existence of nonmeasurable sets (Royden, pp. 63–64). In fact, the axiom [2] (Exercise 4.2.7) implies that nonmeasurable sets exist [see (Sierpinski, 1927)].

Exercises

5.13.1. Show that the following sets are of finite character:
- **(a)** 0. **(b)** 1.
- **(c)** 2. **(d)** $3 \cup \{\{1\}\}$.

5.13.2. Let X and Y be arbitrary sets and let $f : X \to Y$. Let $\mathscr{F}$ be the set of all restrictions of f to subsets of X. Show that $\mathscr{F}$ is of finite character.

5.13.3. **(a)** Let X have at least 4 elements. Is the set of all 4-element subsets of X of finite character?
(b) Let X be an infinite set and let Y be the set of finite subsets of X. Is Y of finite character?

5.13.4. Is the set of all countable subsets of X of finite character if X is
(a) A denumerable set?
(b) An uncountable set?

5.13.5. Determine a partially ordered system, $\langle A, R\rangle$, in which some linearly ordered subsystem, $\langle B, R\rangle$, has no upper bound.

★5.13.6. **(a)** Let $\langle X, R\rangle$ be a partially ordered system. Let

$$\mathscr{L} = \{A \in \mathscr{P}(X) : \langle A, R\rangle \text{ is linearly ordered}\}.$$

Prove that $\mathscr{L}$ is of finite character.
(b) Prove directly that the Teichmüller–Tukey Lemma implies the Hausdorff Maximal Principle.

5.13.7. In the proof that Zorn's Lemma implies the Teichmüller–Tukey Lemma, show that there is some b^* in B such that $b_I \subseteq b^*$ for all I, $1 \leq I \leq N$.

★5.13.8. **Kuratowski's Lemma** states that every linearly ordered subsystem of a partially ordered system is "included" in a maximal linearly ordered subsystem. Show (directly) that Kuratowski's Lemma is equivalent to the Hausdorff Maximal Principle.

6

Order Types and Cardinal Numbers

6.1 Rank

We use Definition by Transfinite Induction to define a "nested" collection of sets that will play a key role in the introduction of two new kinds of transfinite numbers.

Let s be the strictly bimonadic formula, "$y = \mathscr{P}(x)$". Let $c_0 = 0$ and let α be a fixed nonzero ordinal. Then there is a unique function r_α with domain α that satisfies the following:

(1) $r_\alpha(0) = 0$.
(2) $r_\alpha(\gamma^+) = \mathscr{P}(r_\alpha(\gamma))$, if $\gamma^+ < \alpha$.
(3) $r_\alpha(\gamma) = \cup\{r_\alpha(\delta) : \delta < \gamma\}$, if γ is a limit ordinal in α.

In (2), $\mathscr{P}(r_\alpha(\gamma))$ is the unique set c_2 such that "$s\begin{pmatrix} x & y \\ r_\alpha(\gamma) & c_2 \end{pmatrix}$" is true. Again, it is easy to show that for all ordinals γ, α_1, and α_2 that are such that $\gamma \in \alpha_1 \cap \alpha_2$,

$$r_{\alpha_1}(\gamma) = r_{\alpha_2}(\gamma).$$

We define

$$\text{“}\mathbf{R}_\alpha\text{” to be “}r_{\alpha^+}(\alpha)\text{”}.$$

†EXAMPLE 6.1.1

(1) $\mathbf{R}_0 = r_1(0) = 0.$
(2) $\mathbf{R}_1 = r_2(1) = \mathscr{P}(r_1(0)) = \mathscr{P}(0) = \{0\} = 1.$
(3) $\mathbf{R}_2 = r_3(2) = \mathscr{P}(r_3(1)) = \mathscr{P}(r_2(1)) = \mathscr{P}(\mathbf{R}_1) = \mathscr{P}(1) = \{0, 1\} = 2.$
(4) $\mathbf{R}_3 = r_4(3) = \mathscr{P}(r_4(2)) = \mathscr{P}(r_3(2)) = \mathscr{P}(\mathbf{R}_2) = \mathscr{P}(2)$
$= \{0, 1, \{1\}, 2\} = 3 \cup \{\{1\}\}.$
(5) $\mathbf{R}_\omega = r_{\omega^+}(\omega) = \cup\{r_{\omega^+}(N) : N < \omega\} = \cup\{r_{N^+}(N) : N < \omega\}$
$= \cup\{\mathbf{R}_N : N < \omega\} = \cup\{0, 1, 2, 3 \cup \{\{1\}\}, \mathbf{R}_4, \mathbf{R}_5, \ldots\}.$

Theorem 6.1.1. *For every ordinal number α, the following are true:*
(1) $\mathbf{R}_{\alpha^+} = \mathscr{P}(\mathbf{R}_\alpha).$
(2) $\mathbf{R}_\alpha \in \mathbf{R}_{\alpha^+}.$
(3) If α is a limit ordinal,

$$\mathbf{R}_\alpha = \cup\{\mathbf{R}_\delta : \delta < \alpha\}.$$

(4) $\mathbf{R}_\alpha \subseteq \mathbf{R}_{\alpha^+}.$

PROOF: (1) and (2): $\mathbf{R}_\alpha \subseteq \mathbf{R}_\alpha$; hence

$$\mathbf{R}_\alpha \in \mathscr{P}(\mathbf{R}_\alpha) = \mathscr{P}(r_{\alpha^+}(\alpha)) = \mathscr{P}(r_{\alpha^{++}}(\alpha))$$
$$= r_{\alpha^{++}}(\alpha^+) = \mathbf{R}_{\alpha^+}.$$

(3): Suppose α is a limit ordinal. Then

$$\mathbf{R}_\alpha = r_{\alpha^+}(\alpha) = \cup\{r_\alpha(\delta) : \delta < \alpha\}$$
$$= \cup\{r_{\delta^+}(\delta) : \delta < \alpha\}$$
$$= \cup\{\mathbf{R}_\delta : \delta < \alpha\}.$$

(4): $\mathbf{R}_0 = \varnothing \subseteq \mathbf{R}_1.$

Assume that for all $\alpha < \alpha_0$, $\mathbf{R}_\alpha \subseteq \mathbf{R}_{\alpha^+}.$

Suppose that $\alpha_0 = \beta^+$. If $x \in \mathbf{R}_{\alpha_0}$, then by part (1) and our inductive hypothesis, $x \subseteq \mathbf{R}_\beta \subseteq \mathbf{R}_{\beta^+} = \mathbf{R}_{\alpha_0}$. It follows that $x \in \mathscr{P}(\mathbf{R}_{\alpha_0}) = \mathbf{R}_{\alpha_0^+}$; thus, $\mathbf{R}_{\alpha_0} \subseteq \mathbf{R}_{\alpha_0^+}.$

Suppose that α_0 is a limit ordinal. Then $\mathbf{R}_{\alpha_0} = \cup\{\mathbf{R}_\beta : \beta < \alpha_0\}$. Hence, if $x \in \mathbf{R}_{\alpha_0}$, then $x \in \mathbf{R}_\beta$ for some $\beta < \alpha$. By the inductive hypothesis, $x \in \mathbf{R}_{\beta^+}$; consequently, $x \subseteq \mathbf{R}_\beta \subseteq \mathbf{R}_{\alpha_0}$. Thus, $x \in \mathbf{R}_{\alpha_0^+}$. Again, $\mathbf{R}_{\alpha_0} \subseteq \mathbf{R}_{\alpha_0^+}.$

In the following corollaries, assume that α and β are ordinal numbers.

Corollary 1. *If* $\alpha \le \beta$, *then* $\mathbf{R}_\alpha \subseteq \mathbf{R}_\beta$.

PROOF: Let α be fixed.

If $\beta = \alpha$, then $\mathbf{R}_\alpha = \mathbf{R}_\beta$; hence, $\mathbf{R}_\alpha \subseteq \mathbf{R}_\beta$.

If β is a limit ordinal $> \alpha$, then

$$\mathbf{R}_\alpha \subseteq \cup\{\mathbf{R}_\gamma : \gamma < \beta\} = \mathbf{R}_\beta.$$

If $\beta = {\gamma_1}^+ > \alpha$ for some ordinal γ_1 and if $\alpha \le \gamma < \beta$ implies $\mathbf{R}_\alpha \subseteq \mathbf{R}_\gamma$, then by the inductive hypothesis together with the theorem,

$$\mathbf{R}_\alpha \subseteq \mathbf{R}_{\gamma_1} \subseteq \mathbf{R}_{\gamma_1^+} = \mathbf{R}_\beta.$$

Corollary 2. *If* $\alpha < \beta$, *then* $\mathbf{R}_\alpha \in \mathbf{R}_\beta$.

PROOF: $\alpha < \beta$ implies $\alpha^+ \le \beta$. Thus, $\mathbf{R}_\alpha \in \mathbf{R}_{\alpha^+} \subseteq \mathbf{R}_\beta$.

Corollary 3. *Every* $\mathbf{R}_\alpha$ *is ordinary.*

PROOF: Let α_0 be the smallest ordinal α such that $\mathbf{R}_\beta \in \mathbf{R}_\alpha$ for some $\beta \ge \alpha$. Surely, $\alpha_0 \ne 0$ because $\mathbf{R}_0 = \varnothing$.

If $\alpha_0 = {\alpha_1}^+$, then for some $\beta \ge \alpha_0$,

$$\mathbf{R}_\beta \in \mathbf{R}_{\alpha_0} = \mathscr{P}(\mathbf{R}_{\alpha_1}).$$

Therefore, according to Corollary 2,

$$\mathbf{R}_{\alpha_1} \in \mathbf{R}_\beta \subseteq \mathbf{R}_{\alpha_1};$$

this contradicts the definition of α_0.

If α_0 is a limit ordinal, then for $\beta \ge \alpha_0$,

$$\mathbf{R}_\beta \in \mathbf{R}_{\alpha_0} = \cup\{\mathbf{R}_\gamma : \gamma < \alpha_0\}.$$

Thus, $\mathbf{R}_\beta \in \mathbf{R}_\gamma$ for $\gamma < \alpha_0$, again contradicting our assumption about α_0.

Corollary 4. *If* $\alpha < \beta$, *then* $\mathbf{R}_\alpha \subset \mathbf{R}_\beta$.

PROOF: By Corollary 1, $\mathbf{R}_\alpha \subseteq \mathbf{R}_\beta$; by Corollary 2, $\mathbf{R}_\alpha \in \mathbf{R}_\beta$; by Corollary 3, $\mathbf{R}_\alpha \notin \mathbf{R}_\alpha$.

Theorem 6.1.2. *For every ordinal number* α, $\mathbf{R}_\alpha$ *is transitive.*

PROOF: First, $\mathbf{R}_0 = 0$, and 0 is transitive. Suppose that for all $\beta < \alpha$, $\mathbf{R}_\beta$ is transitive. If α is a successor ordinal, γ^+, and if $y \in x \in \mathbf{R}_\alpha$, then $y \in x \subseteq \mathbf{R}_\gamma$. Thus, by the hypothesized transitivity of $\mathbf{R}_\gamma$, $y \in \mathbf{R}_\gamma \subseteq \mathbf{R}_{\gamma^+} = \mathbf{R}_\alpha$. If α is a limit ordinal, then $y \in x \in \mathbf{R}_\alpha$ implies $y \in x \in \mathbf{R}_\beta$ for some $\beta < \alpha$. Thus, by the inductive hypothesis, $y \in \mathbf{R}_\beta \subseteq \mathbf{R}_\alpha$; hence, $y \in \mathbf{R}_\alpha$.

Let α be an ordinal number. A set x is said to have **rank** α if $x \in \mathbf{R}_{\alpha^+} \sim \mathbf{R}_\alpha$; x **has rank** if it has rank α for some ordinal α.

It follows almost immediately that x *has rank* α *iff* α *is the smallest ordinal* β *for which* $x \in \mathbf{R}_{\beta^+}$, and that *if* x *has rank* α, *then* α *is unique* (Exercise 6.1.6). Consequently, *if* β_0 *is the smallest ordinal* β *for which* $x \in \mathbf{R}_\beta$, *then* β_0 *must be a successor ordinal.*

(6.1.1) *If* x *has rank* α *and* $y \in x$, *then* y *has rank* β *for some* $\beta < \alpha$.

For if $x \in \mathbf{R}_{\alpha^+}$, then $x \subseteq \mathbf{R}_\alpha$; hence, $y \in \mathbf{R}_\alpha$.

For an arbitrary set x, let "$\rho(x)$" denote the rank of x, if x has rank; otherwise, let "$\rho(x)$" denote "$\{1\}$". Thus, $\rho(x)$ *is an ordinal number iff* x *has rank.*

†EXAMPLE 6.1.2

(1) $\rho(0) = 0$ because $0 \in \mathbf{R}_1 = \{0\}$ and $0 \notin 0 = \mathbf{R}_0$.
(2) $\rho(1) = 1$ because $1 \in \mathbf{R}_2 = \{0, 1\}$ and $1 \notin \mathbf{R}_1$.
(3) $\rho(2) = \rho(\{1\}) = 2$; see Example 6.1.1(3) and (4).
(4) $\rho(\mathbf{R}_\alpha) = \alpha$, for every ordinal α.

Theorem 6.1.3. *Every ordinal number* α *has rank* α.

PROOF: First, $\rho(0) = 0$, by Example 6.1.2(1). Suppose that for all $\beta < \alpha$, $\rho(\beta) = \beta$. If α is of the form γ^+, then $\gamma \in \mathbf{R}_{\gamma^+} = \mathbf{R}_\alpha$; by the transitivity of $\mathbf{R}_\alpha$, $\gamma \subseteq \mathbf{R}_\alpha$. Therefore, $\alpha = \gamma^+ = \gamma \cup \{\gamma\} \subseteq \mathbf{R}_\alpha$. Hence, $\alpha \in \mathbf{R}_{\alpha^+}$, and α has rank at most α. But $\gamma \in \alpha$ and, by (6.1.1), $\rho(\gamma) = \gamma$ implies that α has rank $> \gamma$; i.e., α has rank at least α. Thus, $\rho(\alpha) = \alpha$.

If α is a limit ordinal, then

$$\alpha = \{\beta : \beta < \alpha\} \subseteq \cup\{\mathbf{R}_{\beta^+} : \beta^+ < \alpha\} = \cup\{\mathbf{R}_\beta : \beta < \alpha\} = \mathbf{R}_\alpha.$$

Therefore, $\alpha \in \mathbf{R}_{\alpha^+}$ and $\rho(\alpha) \leq \alpha$. For each $\beta < \alpha$ there is an element, β, of α of rank β; according to (6.1.1), $\rho(\alpha) \geq \alpha$. Hence, $\rho(\alpha) = \alpha$.

EXAMPLE 6.1.3. Let $x = \{0, 2, 7\}$. We claim that $\rho(x) = 8$. Observe that $7 \in x$ and $\rho(7) = 7$; hence, $\rho(x) \geq 8$. For each y in x, $\rho(y) \leq 7$; thus, $x \subseteq \mathbf{R}_8$. Therefore, $x \in \mathbf{R}_9$ and $\rho(x) \leq 8$.

Theorem 6.1.4. *If a set has rank, it is well-founded.*

PROOF: Suppose that x has rank and let B be a nonempty subset of x. Every element of B has rank; hence, $\mathscr{B} = \{\beta : \beta = \rho(b)$ for some b in $B\} \neq \varnothing$. Let β_0 be the smallest ordinal in $\mathscr{B}$, and let b_0 be any element of B with $\rho(b_0) = \beta_0$. Every element of b_0 has rank $< \beta_0$; hence, $b_0 \cap B = \varnothing$.

Note that the Axiom of Choice is not used in the proof of this theorem.

We now introduce an axiom that asserts that every set has rank. Unlike previous axioms that guaranteed the existence of certain sets, our new axiom eliminates certain "irregular" sets.

Axiom of rank. *Every set has rank.*

We assume this axiom hereafter; Theorem 6.1.4 then guarantees that the following sentence, known as the **Axiom of Regularity**, is true: "*Every set is well-founded.*" In Exercise 6.1.12, we will show that, *assuming the Axiom of Substitution, the Axiom of Rank is equivalent to the Axiom of Regularity.*

Scott showed that the Axiom of Rank enables us to introduce order types and cardinal numbers without resorting to the Axiom of Choice; it guarantees that every ordered set corresponds to a unique order type and that every (arbitrary) set corresponds to a unique cardinal number. Much of the arithmetic of cardinal numbers depends upon the Axiom of Choice; in fact, many arithmetic properties are actually equivalent to the Axiom of Choice. Also, there are theorems of cardinal arithmetic that do not require the Axiom of Choice but whose proofs can be simplified by utilizing this axiom. In order to emphasize the relationship between the Axiom of Choice and cardinal arithmetic, we first study cardinal arithmetic without the Axiom of Choice. We then add this axiom and reconsider the subject.

We conclude this section with an important characterization of certain of the $\mathbf{R}_\alpha$.

A set X is called a **Zermelo Universe** if it has the following properties:

(1) If $A, B \in X$, then $\{A, B\} \in X$.
(2) If $A \in X$, then $\cup A \in X$.
(3) If $A \in X$, then $\mathscr{P}(A) \in X$.
(4) $\omega \in X$.
(5) X is transitive.

A Zermelo Universe is a set that reflects our basic assumptions about the collection of all sets. Thus, *each axiom introduced in Chapter 1 is true of a*

Zermelo Universe X when the phrase "element(s) of X" replaces "set(s)". For the Axioms of Pairing, Union, Power Set, and Infinity this is immediate. For the Null Set, this follows from clauses (4) and (5).

Consider, next, the modified Axiom of Extensionality. Certainly, if A and B are elements of (the Zermelo Universe) X, then $A = B$ iff both $A \subseteq B$ and $B \subseteq A$ because of the (usual) Axiom of Extensionality. A more interesting modification is the "relativization of the Axiom of Extensionality to X":

Let X be a Zermelo Universe and let $A, B \in X$.

Then $A = B$ iff both $A \cap X \subseteq B \cap X$ and $B \cap X \subseteq A \cap X$.

Surely, if $A = B$, then (for every set X), $A \cap X = B \cap X$; hence (by Extensionality) $A \cap X \subseteq B \cap X$ and $B \cap X \subseteq A \cap X$.

Suppose that $A \cap X \subseteq B \cap X$ and $B \cap X \subseteq A \cap X$. By the transitivity of X, $A \subseteq X$ and $B \subseteq X$; hence, $A \cap X = A$ and $B \cap X = B$. Thus, $A \subseteq B$ and $B \subseteq A$. By Extensionality, $A = B$.

We next consider the Axiom of Subsets. Let X be a Zermelo Universe and let $A \in X$. Let s be a formula that is such that x is its one free variable and each of its constants denotes an element of X (hence a set). Then the collection of a in A for which "$s\begin{pmatrix} x \\ a \end{pmatrix}$" is true is a set, according to the (usual) Axiom of Subsets. Call this B; clearly, $B \subseteq A$. Hence, by clause (3), $B \in \mathscr{P}(A) \in X$; by clause (5), $B \in X$.

Theorem 6.1.5. *Let α be an ordinal number. Then $\mathbf{R}_\alpha$ is a Zermelo Universe if and only if α is a limit ordinal and $\alpha > \omega$.*

PROOF: Let α be a limit ordinal $> \omega$.

(1) If $A, B \in \mathbf{R}_\alpha$, then $A \in \mathbf{R}_{\beta_1}$, $B \in \mathbf{R}_{\beta_2}$ for $\beta_1 < \alpha$, $\beta_2 < \alpha$. Let $\beta = \max\{\beta_1, \beta_2\}$; then $\{A, B\} \subseteq \mathbf{R}_\beta$. Furthermore, $\beta < \alpha$ and $\beta^+ < \alpha$; hence, $\{A, B\} \in \mathbf{R}_{\beta^+} \subset \mathbf{R}_\alpha$.

(2) Let $A \in \mathbf{R}_\alpha$ and let $x \in \cup A$. Then for some y in A, $x \in y \in A$. Because A has rank β, $\beta < \alpha$, y has rank $\gamma_1 < \beta$, and x has rank $\gamma_2 < \gamma_1 < \beta$. Consequently, each such x is in $\mathbf{R}_\beta$, and $\cup A \subseteq \mathbf{R}_\beta$. Therefore, $\cup A \in \mathbf{R}_{\beta^+} \subset \mathbf{R}_\alpha$.

(3) Let $A \in \mathbf{R}_\alpha$. Then $A \in \mathbf{R}_\gamma$ for some $\gamma < \alpha$. If $B \subseteq A$, then $B \in \mathbf{R}_{\gamma^+}$ and, hence, $\mathscr{P}(A) \subseteq \mathbf{R}_{\gamma^+}$. Also $\gamma < \alpha$ implies $\gamma^{++} < \alpha$. Thus, $\mathscr{P}(A) \in \mathbf{R}_{\gamma^{++}} \subset \mathbf{R}_\alpha$.

(4) $\omega \in \mathbf{R}_{\omega^+} \subset \mathbf{R}_\alpha$.

(5) $\mathbf{R}_\alpha$ is transitive, according to Theorem 6.1.2.

Suppose that α is an ordinal, but not a limit ordinal $> \omega$. If $\alpha = 0$ or $\alpha = \omega$, then $\omega \notin \mathbf{R}_\alpha$. If $\alpha = \beta^+$, then $\mathbf{R}_\beta \in \mathbf{R}_\alpha$, but $\mathscr{P}(\mathbf{R}_\beta) = \mathbf{R}_\alpha \notin \mathbf{R}_\alpha$.

Exercises

†6.1.1. Prove that for all ordinals γ, α_1, and α_2, with $\gamma \in \alpha_1 \cap \alpha_2$, $r_{\alpha_1}(\gamma) = r_{\alpha_2}(\gamma)$.

6.1.2. Determine $\mathbf{R}_4$.

6.1.3. Prove that for every natural number $N \geq 3$, the following are true:
(a) $N \subset \mathbf{R}_N$. **(b)** $\mathscr{P}(N) \subset \mathbf{R}_{N^+}$.

6.1.4. For each N in ω, how many elements are there in $\mathbf{R}_N$?

†6.1.5. Prove that for every ordinal α, if $x \subseteq y \in \mathbf{R}_\alpha$, then $x \in \mathbf{R}_\alpha$.

†6.1.6. **(a)** Prove that x has rank α iff α is the smallest ordinal β for which $x \in \mathbf{R}_{\beta^+}$.
(b) Prove that if x has rank α, then α is unique.
(c) Let β_0 be the smallest ordinal β for which $x \in \mathbf{R}_\beta$. Prove that β_0 is a successor ordinal.

†6.1.7. **(a)** Prove that if every element of a set x has rank, then x has rank.
(b) Describe the rank of x in terms of the ranks of its elements.

6.1.8. For each $N \leq 4$, determine all sets of rank N.

†6.1.9. **(a)** Show that every finite subset of ω has rank N for some N in ω.
(b) Show that every denumerable subset of ω has rank ω.

6.1.10. Determine the rank of each of the following sets:
(a) $\{3, 13, \omega\}$. **(b)** $\{\{\{\omega\}\}\}$.
(c) $\{\{\{0, 3, 4\}\}, \{2, \{\{4\}\}, \{5\}\}\}$. **(d)** $\{\mathbf{R}_{\omega+2}, \{\mathbf{R}_\omega, \mathbf{R}_{\omega+1}\}\}$.
(e) $\langle 4, 2\rangle$. **(f)** $+$ (i.e., addition on ω).
(g) $\mathscr{Z}$. **(h)** $\mathscr{Q}$.
(i) $\mathscr{R}$. **(j)** $\mathscr{C}$.

6.1.11. Let X have rank α. Determine the rank of each of the following sets:
(a) $\mathscr{P}(X)$. **(b)** $X^{②}$. **(c)** $\mathscr{F}(X, X)$.

★6.1.12. Theorem 6.1.4 shows that the Axiom of Rank implies the Axiom of Regularity. Exercise 5.2.13(c) shows that the Axiom of Regularity implies the following "**descending chain condition**": "There is no denumerable set $\{x_1, x_2, \ldots, x_N, \ldots\}$ for which $\cdots x_{N^+} \in x_N \in \cdots \in x_3 \in x_2 \in x_1$."
(a) Use the Axioms of Substitution and Dependent Choices (Exercise 4.2.8) to show that the descending chain condition implies the Axiom of Rank (and hence that the Axiom of Regularity implies the Axiom of Rank).
(b) Now use just Substitution to show this. (*Hint*: Use Exercise 5.3.16.)

6.1.13. Which properties of a Zermelo Universe pertain to each of the following?
(a) $\mathbf{R}_N$, $N < \omega$. **(b)** $\mathbf{R}_\omega$. **(c)** $\mathbf{R}_{\beta^+}$, β an infinite ordinal.

★6.1.14. Determine the smallest Zermelo Universe.

6.2 Order Types

According to the Counting Theorem, every well-ordered system $\langle A, R\rangle$ is similar to a unique ordinal system $\langle \alpha, \in\rangle$. Thus, ordinal systems serve to classify well-ordered systems. We now seek the counterpart of the ordinal concept for linearly ordered systems; the sets in question will be called "order types". Two linearly ordered systems $\langle A, R_A \rangle$ and $\langle B, R_B \rangle$ will be represented by the same order type iff $\langle A, R_A \rangle \cong \langle B, R_B \rangle$. Again, the collection of all linearly ordered systems that are similar to a given one will not be a set, in general. Moreover, it is by no means obvious how to find a representative linearly ordered system similar to a given one, $\langle A, R_A \rangle$. The Axiom of Rank will enable us to find a *set* consisting of *some* of the linearly ordered systems that are similar to $\langle A, R_A \rangle$.

Let $\langle A, R \rangle$ be a linearly ordered system. The **order type of** $\langle A, R \rangle$, written "$\mathscr{T}_{A,R}$", is defined to be the collection of all linearly ordered systems $\langle B, S \rangle$ that satisfy the following:

(1) $\langle A, R \rangle \cong \langle B, S \rangle$.
(2) If $\langle A, R \rangle \cong \langle C, T \rangle$, then $\rho(B) \leq \rho(C)$.

In other words, $\mathscr{T}_{A,R}$ is the collection of all linearly ordered systems $\langle B, S \rangle$ similar to $\langle A, R \rangle$, for which B is of lowest rank. We emphasize that we are not concerned with the collections $\mathscr{T}_{A,R}$ per se; rather, they are introduced in order to represent the intuitive notion of "order types" within set theory. Note that the linearly ordered system $\langle A, R \rangle$ is not necessarily an element of $\mathscr{T}_{A,R}$; this will be illustrated in our first two examples. We must show that each order type is a set (Theorem 6.2.1).

We write "$\mathscr{T}_\alpha$" for the order type of an ordinal system $\langle \alpha, \in \rangle$.

†EXAMPLE 6.2.1. We compute the order types $\mathscr{T}_{A,R}$, where A is an N-element set and $N \leq 4$. Recall that $\mathbf{R}_0 = 0$, $\mathbf{R}_1 = 1$, $\mathbf{R}_2 = 2$, $\mathbf{R}_3 = 3 \cup \{\{1\}\}$, and $\mathbf{R}_4 = \mathscr{P}(\mathbf{R}_3)$.

(1) $\mathscr{T}_0 = \{\langle 0, 0 \rangle\}$ because $\langle 0, 0 \rangle$ is not similar to any linearly ordered system other than itself.

(2) $\mathscr{T}_1 = \{\langle 1, \in \rangle\}$. For 1 is the only 1-element set of rank ≤ 1, and $\in_1$ is the unique linear ordering on 1.

(3) $\mathscr{T}_2 = \{\langle 2, \in \rangle, \langle 2, \in_2{}^{-1} \rangle\}$. Here, 2 is the only 2-element set of rank ≤ 2. Furthermore, $\in_2$ and $\in_2{}^{-1}$ are the only linear orderings on 2.

(4) $\mathscr{T}_3$: No 3-element set has rank < 3; there are four 3-element sets of rank 3. Thus, $\mathscr{T}_3 = \{\langle A, R \rangle : \rho(A) = 3$, A is a 3-element set, and R linearly orders $A\}$. There are 6 distinct linear orderings on every 3-element, thus $\mathscr{T}_3$ is a 24-element set.

(5) $\mathscr{T}_4$: $\rho(\mathbf{R}_3) = 3$ and $\mathbf{R}_3$ is the only 4-element set of rank ≤ 3. Thus, $\mathscr{T}_4 = \{\langle \mathbf{R}_3, R \rangle : R$ linearly orders $\mathbf{R}_3\}$. $\mathscr{T}_4$ is also a 24-element set. Note that $\langle 4, \in \rangle \notin \mathscr{T}_4$, whereas $\langle N, \in \rangle \in \mathscr{T}_N$ for $N < 4$.

(6) Recall that for every natural number N, if A is an N-element set and if R is any linear ordering on A, then $\langle A, R\rangle \cong \langle N, \in\rangle$. Thus, for $N \leq 4$, $\mathscr{T}_{A,R}$ is determined above. For example, if a and b are distinct and if $R = \{\langle a, a\rangle, \langle b, a\rangle, \langle b, b\rangle\}$, then

$$\mathscr{T}_{\{a,b\},R} = \{\langle 2, \in\rangle, \langle 2, \in_2{}^{-1}\rangle\}.$$

EXAMPLE 6.2.2

(1) $\mathscr{T}_\omega = \{\langle A, R\rangle : A \in \mathbf{R}_{\omega^+}, R$ linearly orders A, and $\langle A, R\rangle \cong \langle \omega, \in\rangle\}$. Note that for each N in ω, the elements of $\mathbf{R}_N$ are all finite; hence, the elements of $\mathbf{R}_\omega$ are all finite. Also, $\omega \in \mathbf{R}_{\omega^+}$.

(2) Let α be a denumerable ordinal. Then

$$\mathscr{T}_\alpha = \{\langle A, R\rangle : A \in \mathbf{R}_{\omega^+}, R \text{ linearly orders } A, \text{ and } \langle A, R\rangle \cong \langle \alpha, \in\rangle\}.$$

Theorem 6.2.1. *For every linearly ordered system, $\langle A, R\rangle$, $\mathscr{T}_{A,R}$, the order type of $\langle A, R\rangle$, is a set.*

PROOF: By the Axiom of Rank, A has rank α, for some ordinal α. Thus, $A \in \mathbf{R}_{\alpha+1}$; hence, $A \subseteq \mathbf{R}_\alpha$. Also,

$$\begin{aligned} A^{②} &= \{\langle \alpha_1, \alpha_2\rangle : \alpha_1, \alpha_2 \in A\} \\ &= \{\{\{\alpha_1\}, \{\alpha_1, \alpha_2\}\} : \alpha_1, \alpha_2 \in A\} \subseteq \mathscr{P}\mathscr{P}(A). \end{aligned}$$

Thus,

$$R \subseteq A^{②} \subseteq \mathscr{P}\mathscr{P}(A) \subseteq \mathscr{P}\mathscr{P}(\mathbf{R}_\alpha) = \mathbf{R}_{\alpha+2}.$$

Consequently, $R \in \mathbf{R}_{\alpha+3}$ and

$$\langle A, R\rangle = \{\{A\}, \{A, R\}\} \in \mathscr{P}\mathscr{P}(\mathbf{R}_{\alpha+3}) = R_{\alpha+5}.$$

Thus,

$$\begin{aligned} \mathscr{T}_{A,R} = \{x \in \mathbf{R}_{\alpha+5} : \; & x \text{ is a linearly ordered system, } \langle B, S\rangle, \langle B, S\rangle \cong \langle A, R\rangle, \\ & \text{and if } \langle C, T\rangle \text{ is a linearly ordered system that is similar to} \\ & \langle A, R\rangle, \text{ then } \rho(B) \leq \rho(C)\}. \end{aligned}$$

We now consider Cantor's characterizations of the order types of $\mathscr{Q}$, the set of rational numbers, and of $\mathscr{R}$, the set of real numbers. Recall that R densely orders A if whenever $a, b\ (\neq a) \in A$ and aRb, there is some c in $A \sim \{a, b\}$ for which aRc and cRb. (See Section 3.8.)

We denote "$\mathscr{T}_{\mathscr{Q}, \leq_{\mathscr{Q}}}$" by "$\eta$".

Theorem 6.2.2. *Let $\langle A, R\rangle$ be a linearly ordered system. Then $\mathcal{T}_{A,R} = \eta$ if and only if the following all hold:*

(6.2.1) *A has neither a first nor a last element (with respect to R).*

(6.2.2) *A is denumerable.*

(6.2.3) *R densely orders A.*

Thus, $\langle A, R\rangle$ satisfies (6.2.1), (6.2.2), and (6.2.3) if and only if $\langle A, R\rangle \cong \langle \mathcal{Q}, \leq_{\mathcal{Q}}\rangle$.

PROOF: We already know that $\langle \mathcal{Q}, \leq\rangle$ has properties (6.2.1), (6.2.2), and (6.2.3). (See Sections 3.8 and 4.5.) It is easily verified that if $\langle A, R\rangle \cong \langle \mathcal{Q}, \leq\rangle$, then $\langle A, R\rangle$ must have properties (6.2.1), (6.2.2), and (6.2.3).

Suppose that $\langle A, R\rangle$ is a linearly ordered system that has properties (6.2.1), (6.2.2), and (6.2.3). We must show that $\langle A, R\rangle \cong \langle \mathcal{Q}, \leq\rangle$. The sets $\mathcal{Q}$ and A are each denumerable; with respect to specific one-one correspondences with ω, we can enumerate $\mathcal{Q}$ as

$$q_0, q_1, q_2, \ldots, q_N, \ldots$$

and A as

$$a_0, a_1, a_2, \ldots, a_N, \ldots.$$

We utilize Definition by Induction in order to define a similarity mapping of $\mathcal{Q}$ onto A.

Let $\mathcal{P}^*(A)$ be the set of nonempty finite subsets of A. If $B \in \mathcal{P}^*(A)$, then for some $K < \omega$, $B = \{a_{I_0}, a_{I_1}, \ldots, a_{I_K}\}$, where $I_0 < I_1 < \cdots < I_K$; let $i_B = \{I_0, I_1, \ldots, I_K\}$, the set of indices of elements of B. Let $Q_K = \{q_0, q_1, \ldots, q_K, q_{K^+}\}$; for some J, $1 \leq J \leq K^{++}$, q_{K^+} is the Jth element of Q_K in the natural ordering, $\leq_2$. Let $G(B) = B \cup \{a_{M_0}\}$, where M_0 is the smallest index M in $\omega \sim i_B$ for which a_M is the Jth element (with respect to R) of $\{a_{I_0}, a_{I_1}, \ldots, a_{I_K}, a_M\}$. We note that properties (6.2.1), (6.2.2), and (6.2.3) of $\langle A, R\rangle$ assure the existence of such an M. Then (according to Definition by Induction) there is a unique function $F : \omega \to \mathcal{P}^*(A)$ that is such that $F(0) = \{a_0\}$ and for all $N < \omega$, $F(N^+) = G(F(N))$. Define $H : \mathcal{Q} \to A$ by letting $H(q_0) = a_0$, the unique element of $F(0)$, and by letting $H(q_{N^+})$ be the unique element of $F(N^+) \sim F(N)$. (See Example 6.2.3.) We claim that H is the required similarity map.

First, H is one-one because for $N > M$, $H(q_N) \notin \{H(q_0), H(q_1), \ldots, H(q_M)\}$. By induction (on N), we can show that for each $N \geq 0$ and for all $M \leq N$, q_M is the Ith element of $\{q_0, q_1, \ldots, q_N\}$ (with respect to $\leq_{\mathcal{Q}}$) iff $H(q_M)$ is the Ith element of $\{H(q_0), H(q_1), \ldots, H(q_N)\}$ (with respect to R), $1 \leq I \leq N^+$. It follows that for all $M, N < \omega$, $q_M \leq_{\mathcal{Q}} q_N$ iff $H(q_M) R H(q_N)$. Thus, H is order-preserving, and we need only show that it is onto A.

We know that $a_0 \in$ range H because $a_0 = H(q_0)$. Let $N > 0$ and suppose that for all $I < N$, $a_I \in$ range H. We want to show that $a_N \in$ range H. Let L be the highest index of $H^{-1}(a_I)$, $I < N$; then

$$\{a_0, a_1, a_2, \ldots, a_{N \dot{-} 1}\} \subseteq \{H(q_0), H(q_1), \ldots, H(q_L)\}.$$

If $a_N = H(q_I)$ for $I < L$, we are finished. Otherwise, consider

$$\{H(q_0), H(q_1), \ldots, H(q_L), a_N\}.$$

Because of properties (6.2.1), (6.2.2), and (6.2.3) of $\mathcal{Q}$, there exists some rational q_S, $S > L$, that is such that for every $I \leq L$, $q_S \leq_{\mathcal{Q}} q_I$ iff $a_N R H(q_I)$. Let S_0 be the first such S and suppose that q_{S_0} is the Ith element of $\{q_0, q_1, \ldots, q_{S_0}\}$. By the definition of S_0, for $M < S_0$, $H(q_M) \neq a_N$. We claim that $H(q_{S_0}) = a_N$; it suffices to show that

$$\text{for every } M < S_0, q_M <_{\mathcal{Q}} q_{S_0} \text{ iff } H(q_M) R a_N.$$

For then, a_N must be the element of $A \sim \{H(q_0), \ldots, H(q_{S_0 \dot{-} 1})\}$ with smallest index N^* that is the Ith element of $\{H(q_0), \ldots, H(q_{S_0 \dot{-} 1}), a_{N^*}\}$. [Recall that if $M < N$, then $a_M \in \{H(q_0), \ldots, H(q_{S_0 \dot{-} 1})\}$.] We already know that if $M \leq L$, then (by the definition of S_0) $q_M \leq_{\mathcal{Q}} q_{S_0}$ iff $H(q_M) R a_N$.

Let $L < M < S_0$. Again by the definition of S_0, we know that there is some $I \leq L$ such that either

(6.2.4) $$q_M <_{\mathcal{Q}} q_I \text{ and } H(q_I) R a_N$$

or else

(6.2.5) $$q_I <_{\mathcal{Q}} q_M \text{ and } a_N R H(q_I).$$

In case (6.2.4) we cannot have $q_{S_0} <_{\mathcal{Q}} q_M$ because, by the transitivity of $<_{\mathcal{Q}}$, we would have $q_{S_0} <_{\mathcal{Q}} q_I$. But $I \leq L$ and $H(q_I) R a_N$. Thus, $q_M <_{\mathcal{Q}} q_{S_0}$. Also, $q_M <_{\mathcal{Q}} q_I$ implies $H(q_M) R H(q_I)$ and $H(q_I) R a_N$, by (6.2.4). By the transitivity of R, $H(q_M) R a_N$.

In case (6.2.5) we cannot have $q_M <_{\mathcal{Q}} q_{S_0}$ because then for $I \leq L$, $q_I <_{\mathcal{Q}} q_{S_0}$, whereas $a_N R H(q_I)$. Thus, $q_{S_0} <_{\mathcal{Q}} q_M$. Here, $q_I <_{\mathcal{Q}} q_M$ implies $H(q_I) R H(q_M)$. But then $a_N R H(q_I)$ implies $a_N R H(q_M)$.

Corollary. *If $q_1, q_2 \in \mathcal{Q}$ and $q_1 <_{\mathcal{Q}} q_2$, let $\mathcal{Q}_{(q_1, q_2)} = \{q \in \mathcal{Q} : q_1 <_{\mathcal{Q}} q <_{\mathcal{Q}} q_2\}$. Then $\mathcal{T}_{\mathcal{Q}_{(q_1,q_2)}, \leq_{\mathcal{Q}}} = \eta$.*

In other words, the "open interval (q_1, q_2)" is similar to $\mathcal{Q}$.

PROOF: Clearly $\langle \mathcal{Q}_{(q_1,q_2)}, \leq_{\mathcal{Q}} \rangle$ satisfies properties (6.2.1), (6.2.2), and (6.2.3).

†EXAMPLE 6.2.3. We illustrate the correspondence of elements of $\mathcal{Q}$ and of A, via H. Suppose that the elements with indices 0, 1, 2, . . . , 12 in $\mathcal{Q}$ and in A are related as indicated:

$$q_4 <_{\mathcal{Q}} q_7 <_{\mathcal{Q}} q_1 <_{\mathcal{Q}} q_3 <_{\mathcal{Q}} q_{12} <_{\mathcal{Q}} q_0 <_{\mathcal{Q}} q_{11} <_{\mathcal{Q}} q_5 <_{\mathcal{Q}} q_8 <_{\mathcal{Q}} q_9 <_{\mathcal{Q}} q_6 <_{\mathcal{Q}} q_2 <_{\mathcal{Q}} q_{10}$$

and

$$a_{12}Ra_1Ra_6Ra_0Ra_5Ra_{11}Ra_8Ra_{10}Ra_9Ra_2Ra_4Ra_3Ra_7.$$

(1) $H(q_0) = a_0$.
(2) $q_1 <_{\mathcal{Q}} q_0$ and $a_1 R a_0$; hence, q_1 is the first element of $\{q_0, q_1\}$ and a_1 is the element of $A \sim \{a_0\}$ with smallest index M for which a_M is the first element of $\{a_0, a_M\}$ $(= \{H(q_0), a_M\})$. Thus, $H(q_1) = a_1$.
(3) $q_1 <_{\mathcal{Q}} q_0 <_{\mathcal{Q}} q_2$; q_2 is the third element of $\{q_0, q_1, q_2\}$. Moreover, $a_1Ra_0Ra_2$; hence, a_2 is the third element of

$$\{a_0, a_1, a_2\} \; (= \{H(q_0), H(q_1), a_2\}).$$

Thus, $H(q_2) = a_2$.

(4) $q_1 <_{\mathcal{Q}} q_3 <_{\mathcal{Q}} q_0 <_{\mathcal{Q}} q_2$; q_3 is the second element of $\{q_0, q_1, q_2, q_3\}$. However, $a_1Ra_0Ra_2Ra_3$; thus, a_3 is the fourth element of

$$\{a_0, a_1, a_2, a_3\} \; (= \{H(q_0), H(q_1), H(q_2), a_3\})$$

and, as such, is disqualified. Similarly, a_4 is disqualified (because $a_1Ra_0Ra_2Ra_4$), and a_5 is disqualified. However, $a_1Ra_6Ra_0Ra_2$; hence, $H(q_3) = a_6$.
(5) $q_4 <_{\mathcal{Q}} q_1 <_{\mathcal{Q}} q_3 <_{\mathcal{Q}} q_0 <_{\mathcal{Q}} q_2$. Also, 12 is the smallest index M for which $a_MRa_1Ra_6Ra_0Ra_2$—i.e., $a_MRH(q_1)RH(q_3)RH(q_0)RH(q_2)$. Thus, $H(q_4) = a_{12}$.

We now characterize the order type of $\mathcal{R}$, the set of real numbers. Recall how the real numbers were defined (as "Dedekind cuts") in terms of the rationals (Section 3.9). We generalize this process of "cutting" to an arbitrary linearly ordered system.

Let $\langle A, R\rangle$ be a linearly ordered system. Then $\langle A_1, A_2\rangle$ is called an ***R*-cut of** A if the following are true:

(1) $A_1 \neq \varnothing, A_2 \neq \varnothing$.
(2) $A_1 \cup A_2 = A, A_1 \cap A_2 = \varnothing$.
(3) If $\alpha_1 \in A_1$ and $\alpha_2 \in A_2$, then $\alpha_1 R \alpha_2$.

If the relation R is understood, we speak of a **cut in** A.

In the case of the $\leq_{\mathscr{Q}}$-cuts, $\langle L, U\rangle$, of the rationals we also specified that L had no last element. There were cuts in which U had a first element (these corresponded to rational numbers)—and "irrational" cuts in which U had no first element.

We consider three cases for cuts $\langle A_1, A_2\rangle$ relative to an arbitrary linearly ordered system.

Case 1: If A_1 has a last element and A_2 has a first element, then $\langle A_1, A_2\rangle$ is called a **jump**.

Case 2: If either A_1 has a last element or (excl.) A_2 has a first element, then $\langle A_1, A_2\rangle$ is called a **continuous cut**.

Case 3: If A_1 has no last element and A_2 has no first element, then such a cut is called a **gap**.

Thus, in the case of $\langle \mathscr{Q}, \leq_{\mathscr{Q}}\rangle$, every "rational cut" is continuous; every "irrational cut" is a gap.

Observe that a continuous cut $\langle A_1, A_2\rangle$ in A uniquely determines an element of A—the last element of A_1 or (excl.) the first element of A_2. We let "$\widehat{\langle A_1, A_2\rangle}$" denote the element of A so determined.

EXAMPLE 6.2.4

(1) Consider $\langle N, \in\rangle$, where N is a natural number ≥ 2. Every $\in_N$-cut is a jump because every nonempty subset of a finite ordered set has both a first and a last element.

(2) Consider $\langle \omega 2, \in\rangle$. Let $A_1 = \omega$ and let $A_2 = \{\alpha : \omega \leq \alpha < \omega 2\}$. Then $\langle A_1, A_2\rangle$ is continuous. Every other $\in_{\omega 2}$-cut is a jump.

Let $\langle A, R\rangle$ be a linearly ordered system and let $\langle D, R\rangle$ be a subsystem. We say that D is **dense in** A **(with respect to** R) if whenever $a_1, a_2\,(\neq a_1) \in A$ and $a_1 R a_2$, then there is some d in $D \sim \{a_1, a_2\}$ that satisfies $a_1 R d R a_2$. Clearly, if D is dense in A, then D is dense; the converse is false, however.

EXAMPLE 6.2.5. Consider $\langle \mathscr{R}, \leq_{\mathscr{R}}\rangle$.

(1) The set of positive reals is dense, but it is not dense in $\mathscr{R}$; between any two negative reals there is no positive real.

(2) The isomorphic image of the rationals is dense and, according to Exercise 3.10.6, is also dense in $\mathscr{R}$.

(3) Every densely ordered set is dense in itself.

We denote "$\mathscr{T}_{\mathscr{R}, \leq \mathscr{R}}$" by "$\lambda$".

Theorem 6.2.3. *Let $\langle A, R\rangle$ be a linearly ordered system. Then $\mathcal{T}_{A,R} = \lambda$ if and only if the following conditions hold:*

(6.2.1) *A has neither a first nor a last element (with respect to R).*

(6.2.6) *A includes a denumerable subset that is dense in A.*

(6.2.7) *Every R-cut in A is continuous.*

Thus, $\langle A, R\rangle$ satisfies (6.2.1), (6.2.6), and (6.2.7) *if and only if $\langle A, R\rangle \cong \langle \mathcal{R}, \leq_{\mathcal{R}}\rangle$.*

PROOF: We first show that $\langle \mathcal{R}, \leq_{\mathcal{R}}\rangle$ has properties (6.2.1), (6.2.6), and (6.2.7). Clearly, $\mathcal{R}$ has property (6.2.1); moreover, the isomorphic image of $\mathcal{Q}$ in $\mathcal{R}$ is denumerable and is dense in $\mathcal{R}$. It remains to demonstrate property (6.2.7).

Let $\langle \mathcal{R}_1, \mathcal{R}_2\rangle$ be a $\leq_{\mathcal{R}}$-cut in $\mathcal{R}$. Then $\mathcal{R}_1 \neq \varnothing$; let $\Gamma \in \mathcal{R}_1$. Then Γ is a lower bound for $\mathcal{R}_2$; by the Corollary of Theorem 3.10.6, $\mathcal{R}_2$ has a greatest lower bound, Γ_0. Surely, Γ_0 is in (exactly) one of the sets $\mathcal{R}_1$ or $\mathcal{R}_2$. It is easy to show that if $\Gamma_0 \in \mathcal{R}_1$, then Γ_0 is the last element of $\mathcal{R}_1$, whereas if $\Gamma_0 \in \mathcal{R}_2$, it is the first element of $\mathcal{R}_2$. Moreover, in the first case, if $\Gamma_2 \in \mathcal{R}_2$, then there is some Γ_3 in $\mathcal{R}_2$ such that $\Gamma_0 <_{\mathcal{R}} \Gamma_3 <_{\mathcal{R}} \Gamma_2$; otherwise, $\Gamma_2\ (\neq \Gamma_0)$ would be another greatest lower bound for $\mathcal{R}_2$. Thus, in this case $\mathcal{R}_2$ has no first element. In the second case, if $\Gamma_1 \in \mathcal{R}_1$, there is a Γ_4 in $\mathcal{R}_1$ such that $\Gamma_1 <_{\mathcal{R}} \Gamma_4 <_{\mathcal{R}} \Gamma_0$; here, $\mathcal{R}_1$ has no last element.

Clearly, if $\mathcal{T}_{A,R} = \lambda$, then $\langle A, R\rangle \cong \langle \mathcal{R}, \leq_{\mathcal{R}}\rangle$; hence, $\langle A, R\rangle$ has properties (6.2.1), (6.2.6), and (6.2.7).

Now suppose that $\langle A, R\rangle$ has properties (6.2.1), (6.2.6), and (6.2.7). We must show that $\langle A, R\rangle \cong \langle \mathcal{R}, \leq_{\mathcal{R}}\rangle$.

Let D be a denumerable subset of A that is dense in A. Then D has neither a first nor a last element; because D is dense, by Theorem 6.2.2, D is of type η. Hence, $\langle \mathcal{Q}, \leq_{\mathcal{Q}}\rangle \cong \langle D, R\rangle$; let F be a (fixed) similarity mapping. For each irrational r in $\mathcal{R}$, r is determined by a unique gap $\langle L_r, U_r\rangle$ in $\mathcal{Q}$. Let $\langle \mathcal{L}_r, \mathcal{U}_r\rangle$ be the cut in A in which

$$\mathcal{L}_r = \{a \in A : aRd \text{ for some } d \text{ in } F''(L_r)\}$$

and

$$\mathcal{U}_r = A \sim \mathcal{L}_r.$$

Let H be the isomorphic embedding of $\mathcal{Q}$ in $\mathcal{R}$ (of Chapter 3). Define $G : \mathcal{R} \to A$ by

$$G(r) = \begin{cases} F(H^{-1}(r)), & \text{if } r \in H''(\mathcal{Q}); \\ \widehat{\langle \mathcal{L}_r, \mathcal{U}_r\rangle}, & \text{otherwise.} \end{cases}$$

Then G is the required similarity mapping of $\mathscr{R}$ onto A; the verification of this is left as an exercise.

Corollary. *Let* $\Gamma_1, \Gamma_2 \in \mathscr{R}$ *with* $\Gamma_1 <_{\mathscr{R}} \Gamma_2$. *Let*

$$\mathscr{R}_{(\Gamma_1,\Gamma_2)} = \{\Gamma \in \mathscr{R} : \Gamma_1 <_{\mathscr{R}} \Gamma <_{\mathscr{R}} \Gamma_2\}.$$

Then $\mathscr{T}_{(\Gamma_1,\Gamma_2)}, \leq_{\mathscr{R}} = \lambda$.

[Thus, the "open interval" (Γ_1, Γ_2) is similar to $\mathscr{R}$].

Let $\langle A, R\rangle$ be a linearly ordered system, and let $a, b(\neq a) \in A$ be such that aRb. The **open interval** (a, b) is defined to be $\{x \in A \sim \{a, b\} : aRx \text{ and } xRb\}$. Then if $\langle A, R\rangle$ satisfies condition (6.2.6), it also satisfies the following condition:

(6.2.8). Every set of pairwise disjoint open intervals is countable.

(See Exercise 6.2.12.) In particular, $\langle \mathscr{Q}, \leq_{\mathscr{Q}}\rangle$ and $\langle \mathscr{R}, \leq_{\mathscr{R}}\rangle$ satisfy condition (6.2.8). Theorem 6.2.3 asserted that a linearly ordered system $\langle A, R\rangle$ satisfies (6.2.1), (6.2.6), and (6.2.7) iff $\langle A, R\rangle \cong \langle \mathscr{R}, \leq_{\mathscr{R}}\rangle$. By the above, $\langle \mathscr{R}, \leq_{\mathscr{R}}\rangle$ satisfies (6.2.1), (6.2.7), and (6.2.8). **Souslin's Hypothesis** asserts that if $\langle A, R\rangle$ satisfies (6.2.1), (6.2.7), and (6.2.8), then $\langle A, R\rangle \cong \langle \mathscr{R}, \leq_{\mathscr{R}}\rangle$. It has recently been determined that Souslin's Hypothesis is undecidable, it can neither be proved nor disproved in Zermelo–Fraenkel set theory. (See Section 6.12.)

Exercises

6.2.1. Determine $\mathscr{T}_5$, the order type of $\langle 5, \in\rangle$.

6.2.2. Classify the following linearly ordered systems according to order type:

(a) $\langle \omega, \in\rangle$.

(b) $\langle \omega + 1, \in\rangle$.

(c) $\langle \omega 2, \in\rangle$.

(d) $\langle \omega, R_1\rangle$, where MR_1N iff either $M + N$ is even and $M \leq N$—or M is odd and N is even.

(e) $\langle \omega, R_2\rangle$, where R_2 is indicated by $1, 2, \ldots, N, N^+, \ldots; 0$.

(f) $\langle \omega, R_3\rangle$, where MR_3N iff $M \geq N$.

(g) $\langle \mathscr{Z}, \leq_{\mathscr{Z}}\rangle$.

(h) $\langle \mathscr{N}, \leq_{\mathscr{Z}}\rangle$, where $\mathscr{N}$ is the set of negative integers.

(i) $\langle \mathscr{P}, \leq_{\mathscr{Z}}\rangle$, where $\mathscr{P}$ is the set of positive integers.

(j) $\langle \mathscr{Z} \sim \{0\}, \leq_{\mathscr{Z}}\rangle$.

★6.2.3. Consider the proof of Theorem 6.2.2.

(a) Prove that if $\langle A, R\rangle \cong \langle \mathscr{Q}, \leq\rangle$, then $\langle A, R\rangle$ must have properties (6.2.1), (6.2.2), and (6.2.3).

(b) Give a detailed account of how properties (6.2.1), (6.2.2), and (6.2.3) of $\langle A, R\rangle$ assure that for every finite set $\{a_{I_0}, a_{I_1}, \ldots, a_{I_K}\}$ with $I_0 < I_1 < \cdots < I_K$, and for every J with $1 \leq J \leq K^{++}$, there is an M in $\omega \sim \{I_0, I_1, \ldots, I_K\}$ such that a_M is the Jth element of $\{a_{I_0}, a_{I_1}, \ldots, a_{I_K}, a_M\}$.

(c) Prove that for $N \geq 0$ and for all $M \leq N$, q_M is the Ith element of $\{q_0, q_1, \ldots, q_N\}$ iff $H(q_M)$ is the Ith element of $\{H(q_0), H(q_1), \ldots, H(q_N)\}$, $1 \leq I \leq N^+$.

6.2.4. Consider Example 6.2.3.
(a) Determine $H(q_5)$.
(b) Determine $H(q_6)$.
(c) Show that $H(q_7) = a_M$ for $M > 12$.

6.2.5. Let $q_0, q_1, \ldots, q_M \in \mathscr{Q}$ and let $q_0 <_{\mathscr{Q}} q_1 <_{\mathscr{Q}} \cdots <_{\mathscr{Q}} q_M$. Which of the following subsets $\mathscr{Q}_I$ of $\mathscr{Q}$ are such that $\mathscr{T}_{\mathscr{Q}_I, \leq_{\mathscr{Q}}} = \eta$?
(a) $\mathscr{Q} \sim \{q_0, q_1, \ldots, q_M\}$.
(b) $\mathscr{Q} \sim \{\overline{\overline{N}} : N \in \omega\}$.
(c) $\mathscr{Q} \sim \{\overline{\overline{1}}/\overline{\overline{N}} : N \in \omega\}$.
(d) $\{q \in \mathscr{Q} : q < q_0\}$.
(e) $\{q \in \mathscr{Q} : q < q_0 \text{ or } q > q_1\}$.
(f) $\{q \in \mathscr{Q} : q < q_0 \text{ or } q \geq q_1\}$.
(g) $\{q \in \mathscr{Q} : q \leq q_0 \text{ or } q \geq q_1\}$.

6.2.6. Give an example of an R-cut $\langle A_1, A_2\rangle$ of a linearly ordered set A in which A_1 has a last element but A_2 has no first element.

6.2.7. (a) Show that no cut of a well-ordered system is a gap.
(b) Characterize the continuous $\in_\alpha$-cuts of an ordinal system $\langle \alpha, \in\rangle$.
(c) Now characterize the continuous R-cuts of an arbitrary well-ordered system $\langle A, R\rangle$.

6.2.8. With reference to the proof of Theorem 6.2.3, show (a) and (b):
(a) If $\langle A, R\rangle$ satisfies (6.2.1), (6.2.6), and (6.2.7), then $\langle A, R\rangle \cong \langle \mathscr{R}, \leq_{\mathscr{R}}\rangle$.
(b) G is a similarity mapping of $\mathscr{R}$ onto A.
(c) Consider the sentence:

(6.2.9) No R-cut in A is a gap.

Show that $\langle A, R\rangle$ satisfies (6.2.6) and (6.2.7) iff it satisfies (6.2.6) and (6.2.9).

6.2.9. Which of the following subsets A of the reals are such that $\mathscr{T}_{A, \leq_{\mathscr{R}}} = \lambda$?
(a) $\{\Gamma \in \mathscr{R} : \Gamma > 1\}$.
(b) $\{\Gamma \in \mathscr{R} : \Gamma < 1 \text{ or } \Gamma > 2\}$.
(c) $\{\Gamma \in \mathscr{R} : \Gamma \leq 1 \text{ or } \Gamma > 2\}$.
(d) $\{\Gamma \in \mathscr{R} : \Gamma \leq 1 \text{ or } \Gamma \geq 2\}$.
(e) $\mathscr{R} \sim \{\mathscr{R}\text{-integers}\}$.
(f) $\mathscr{I}$ (= {irrationals}).
(g) $\{\Gamma \in \mathscr{R} : 0 < \Gamma < 1 \text{ or } 2 < \Gamma < 3\}$.
(h) $\{\Gamma \in \mathscr{R} : 0 < \Gamma \leq 1 \text{ or } 2 < \Gamma < 3\}$.
(i) $\{\Gamma \in \mathscr{R} : \text{for some } M \text{ (in } \omega),\ 2M < \Gamma \leq 2M + 1\}$.

★**6.2.10.** Let $\langle A, R\rangle$ be a linearly ordered system. Prove that the following conditions are necessary and sufficient for $\mathscr{T}_{A,R}$ to be $\mathscr{T}_\omega$:
(1) A has a first, but no last element.
(2) Every R-cut in A is a jump.

6.2.11. Prove that the collection of order types is not a set.

★**6.2.12.** Prove that if a linearly ordered system satisfies (6.2.6), then it also satisfies (6.2.8).

6.3 Arithmetic of Order Types

Let $\langle A, R\rangle$ be a linearly ordered system and let $\mu = \mathscr{T}_{A,R}$. Then R^{-1} is a linear ordering on A. We denote "$\mathscr{T}_{A,R^{-1}}$" by "$\mathscr{T}_{A,R}{}^{-1}$" as well as by "μ^{-1}"; μ^{-1} is called the **inverse type of** μ. Clearly, $(\mu^{-1})^{-1} = \mu$ *for every order type* μ.

EXAMPLE 6.3.1

(1) For every natural number N, $\mathscr{T}_N{}^{-1} = \mathscr{T}_N$.
(2) $\mathscr{T}_\omega{}^{-1} = \mathscr{T}_{\mathscr{N},\leq_{\mathscr{I}}}$, where $\mathscr{N}$ is the set of negative integers.
(3) $\mathscr{T}_{\omega+1}{}^{-1} = \mathscr{T}_{\omega,R}$, where for M, N in ω, MRN iff either $M = 0$ or $M \geq N > 0$.
(4) $\eta^{-1} = \eta$ because $\langle \mathscr{Q}, \geq_{\mathscr{Q}}\rangle$ satisfies (6.2.1), (6.2.2), and (6.2.3).
(5) $\lambda^{-1} = \lambda$ because $\langle \mathscr{R}, \geq_{\mathscr{R}}\rangle$ satisfies (6.2.1), (6.2.6), and (6.2.7).

Addition and multiplication of ordinal numbers were defined by transfinite induction. We then showed that the sums (or products) of the ordinals α and β could be obtained by considering "representative" well-ordered systems, $\langle A, R_A\rangle$ and $\langle B, R_B\rangle$, and by defining the appropriate well-orderings on $A \cup B$ (or $A \times B$). Definition by Transfinite Induction is not applicable for defining arithmetic operations on order types; however, the latter "representation" procedure yields suitable definitions.

Let μ and ν be arbitrary order types. Let $\langle A, R\rangle$ and $\langle B, S\rangle$ be linearly ordered systems for which $\mathscr{T}_{A,R} = \mu$, $\mathscr{T}_{B,S} = \nu$, and $A \cap B = \varnothing$. Recall the definition of the linear ordering $R_{A,B}$ on $A \cup B$ (Section 5.5). Define "$\mu +_{\mathscr{T}} \nu$" to be "$\mathscr{T}_{A\cup B, R_{A,B}}$". It is a simple exercise to verify that the *definition of* "$\mu +_{\mathscr{T}} \nu$" *is independent of the choice of the representative linearly ordered systems. Order type addition is associative*; the following example indicates that *it is noncommutative.*

†EXAMPLE 6.3.2

(1) $\mathscr{T}_\omega +_{\mathscr{T}} \mathscr{T}_N$, $N < \omega$: Let $\langle A, R\rangle = \langle \omega, \in\rangle$ and let

$$\langle B, S\rangle = \langle (\omega + N) \sim \omega, \mathscr{T}_{\omega+N}\rangle;$$

then $\mathscr{T}_\omega +_{\mathscr{T}} \mathscr{T}_N = \mathscr{T}_{A\cup B, R_{A,B}} = \mathscr{T}_{\omega+N}$.

(2) $\mathcal{T}_N +_{\mathcal{T}} \mathcal{T}_\omega$, $N < \omega$: Here, let $\langle A, R\rangle = \langle N, \in\rangle$ and let $\langle B, S\rangle = \langle \omega \sim N, \in_\omega\rangle$. Then $\mathcal{T}_N +_{\mathcal{T}} \mathcal{T}_\omega = \mathcal{T}_{A\cup B, R_{A,B}} = \mathcal{T}_\omega$.
(3) $\mathcal{T}_\omega^{-1} +_{\mathcal{T}} \mathcal{T}_\omega = \mathcal{T}_{\mathcal{Z}, \leq_{\mathcal{Z}}} \neq \mathcal{T}_\omega +_{\mathcal{T}} \mathcal{T}_\omega^{-1}$.

Parts (1) and (2) of Example 6.3.2 can be generalized to show that addition of order types "preserves ordinal addition".

Theorem 6.3.1. *Let α and β be ordinal numbers. Then*

$$\mathcal{T}_\alpha +_{\mathcal{T}} \mathcal{T}_\beta = \mathcal{T}_{\alpha+\beta}.$$

PROOF: Choose $\langle A, R\rangle = \langle \alpha, \in\rangle$ and $\langle B, S\rangle = \langle (\alpha + \beta) \sim \alpha, \in_{\alpha+\beta}\rangle$. Then $A \cup B = \alpha + \beta$ and $R_{A,B} = \in_{\alpha+\beta}$.

Let μ and ν be order types. Choose linearly ordered systems $\langle A, R\rangle$ and $\langle B, S\rangle$ such that $\mathcal{T}_{A,R} = \mu$ and $\mathcal{T}_{B,S} = \nu$. Recall the definition of the linearly ordering $R_{A\times B}$ (Section 5.6). Define "$\mu \cdot_{\mathcal{T}} \nu$" to be "$\mathcal{T}_{A\times B, R_{A\times B}}$". Again, *the choice of $\langle A, R\rangle$ and $\langle B, S\rangle$ does not affect the definition of "$\mu \cdot_{\mathcal{T}} \nu$". Multiplication of order types is associative, but noncommutative.*

EXAMPLE 6.3.3

(1) Consider $\mathcal{T}_\omega \cdot_{\mathcal{T}} \mathcal{T}_N$, $N < \omega$. Let $\langle A, R\rangle = \langle \omega, \in\rangle$ and $\langle B, S\rangle = \langle N, \in\rangle$. Then $A \times B = \omega \times N$ and $\langle A \times B, R_{A\times B}\rangle \cong \langle \omega N, \in\rangle$. Thus, $\mathcal{T}_\omega \cdot_{\mathcal{T}} \mathcal{T}_N = \mathcal{T}_{A\times B, R_{A\times B}} = \mathcal{T}_{\omega N}$.
(2) Consider $\mathcal{T}_N \cdot_{\mathcal{T}} \mathcal{T}_\omega$, $N < \omega$. Clearly, $\langle N \times \omega, R_{N\times\omega}\rangle \cong \langle \omega, \in\rangle$. Therefore, $\mathcal{T}_N \cdot_{\mathcal{T}} \mathcal{T}_\omega = \mathcal{T}_\omega$.
(3) Suppose that $1 \leq N < \omega$. Then $\eta \cdot_{\mathcal{T}} N = \eta$ because $\langle \mathcal{Q} \times N, R_{\mathcal{Q}\times N}\rangle$ satisfies (6.2.1), (6.2.2), and (6.2.3). (Show this.)
(4) Suppose that $2 \leq N < \omega$. Then $N \cdot_{\mathcal{T}} \eta \neq \eta$ because $\langle N \times \mathcal{Q}, R_{N\times\mathcal{Q}}\rangle$ has jumps—e.g., for each q in $\mathcal{Q}$, $\langle 0, q\rangle$ is the immediate predecessor of $\langle 1, q\rangle$.
(5) $\eta \cdot_{\mathcal{T}} \eta = \eta$ because $\langle \mathcal{Q}^{②}, R_{\mathcal{Q}^{②}}\rangle$ satisfies (6.2.1), (6.2.2), and (6.2.3). (Show this.)

Theorem 6.3.2. *Let α and β be ordinal numbers. Then $\mathcal{T}_\alpha \cdot_{\mathcal{T}} \mathcal{T}_\beta = \mathcal{T}_{\alpha\beta}$.*

PROOF: Let $\langle A, R\rangle = \langle \alpha, \in\rangle$ and $\langle B, S\rangle = \langle \beta, \in\rangle$. Then $\langle \alpha \times \beta, R_{\alpha\times\beta}\rangle \cong \langle \alpha\beta, \in\rangle$.

Theorem 6.3.3. *Let μ and ν be order types.*

(1) $(\mu +_{\mathcal{T}} \nu)^{-1} = \nu^{-1} +_{\mathcal{T}} \mu^{-1}$.
(2) $(\mu \cdot_{\mathcal{T}} \nu)^{-1} = \mu^{-1} \cdot_{\mathcal{T}} \nu^{-1}$.

PROOF: Suppose that $\mathcal{T}_{A,R} = \mu$ and $\mathcal{T}_{B,S} = \nu$. [In part (1) we assume that $A \cap B = \varnothing$.] Then $\mathcal{T}_{A,R^{-1}} = \mu^{-1}$ and $\mathcal{T}_{B,S^{-1}} = \nu^{-1}$. We consider two orderings on each of the sets A and B; thus, the notation "$R_{A,B}$" and "$R_{A\times B}$" is inadequate. Let "$R_{A,R;B,S}$" and "$R_{\langle A,R\rangle \times \langle B,S\rangle}$", respectively, denote "$R_{A,B}$" and "$R_{A\times B}$", when the orderings R on A and S on B are assumed. Let "$R_{B,S^{-1};A,R^{-1}}$" and "$R_{\langle B,S^{-1}\rangle \times \langle A,R^{-1}\rangle}$", respectively, denote "$R_{B,A}$" and "$R_{B\times A}$" when the orderings S^{-1} on B and R^{-1} on A are assumed.

(1) ${R_{A,R;B,S}}^{-1} = R_{B,S^{-1};A,R^{-1}}$. Therefore,

$$\begin{aligned}(\mu +_{\mathcal{T}} \nu)^{-1} &= \mathcal{T}_{A\cup B,{R_{A,R;B,S}}^{-1}} \\ &= \mathcal{T}_{B\cup A,R_{B,S^{-1};A,R^{-1}}} \\ &= \nu^{-1} +_{\mathcal{T}} \mu^{-1}.\end{aligned}$$

(2) ${R_{\langle A,R\rangle \times \langle B,S\rangle}}^{-1} = R_{\langle A,R^{-1}\rangle \times \langle B,S^{-1}\rangle}$ because for $\langle a_1, b_1\rangle, \langle a_2, b_2\rangle$ in $A \times B$, $\langle a_1, b_1\rangle {R_{\langle A,R\rangle \times \langle B,S\rangle}}^{-1} \langle a_2, b_2\rangle$

iff $\langle a_2, b_2\rangle R_{\langle A,R\rangle \times \langle B,S\rangle} \langle a_1, b_1\rangle$
iff either $b_2 \neq b_1$ and $b_2 S b_1$—or else $b_2 = b_1$ and $a_2 R a_1$
iff either $b_1 \neq b_2$ and $b_1 S^{-1} b_2$—or else $b_1 = b_2$ and $a_1 R^{-1} a_2$
iff $\langle a_1, b_1\rangle R_{\langle A,R^{-1}\rangle \times \langle B,S^{-1}\rangle} \langle a_2, b_2\rangle$.

Therefore,

$$\begin{aligned}(\mu \cdot_{\mathcal{T}} \nu)^{-1} &= \mathcal{T}_{A\times B,{R_{\langle A,R\rangle \times \langle B,S\rangle}}^{-1}} \\ &= \mathcal{T}_{A\times B,R_{\langle A,R^{-1}\rangle \times \langle B,S^{-1}\rangle}} \\ &= \mu^{-1} \cdot_{\mathcal{T}} \nu^{-1}.\end{aligned}$$

Finally, we briefly consider a generalization of addition of order types. We would like to define addition so as to have it apply to arbitrary "linear orderings of (not necessarily distinct) order types". The same order type can appear at various places in the ordering, and we must distinguish between these occurrences. For example, we want to define addition in each of the following cases:

(1) $3 + \omega + \omega + 5 + \omega + 3 + \cdots$ (the ordering is that of $\leq$).
(2) $\cdots + 3 + \omega + 5 + \omega + \omega + 3$ (the ordering is that of $\geq$).
(3) $\cdots + 3 + \omega + \omega + 5 + \omega + 3 + \cdots$ (the ordering is that of $\leq_{\mathcal{Z}}$).

Let $\langle X, R_X\rangle$ be a linearly ordered system in which each element x of X is an ordered pair $\langle t_x, u_x\rangle$, the first coordinate of which is an order type. (The u_x serve to index the t_x.) We would like to define the "ordered sum",

$$\sum_{\substack{x\in X\\ R_X}} t_x.$$

We use the Axioms of Substitution and Choice to show (in Exercise 6.3.10) that there exists a linearly ordered system $\langle Y, R_Y\rangle$ with the following properties:

(6.3.1) Y is a set of ordered pairs $\langle y, R_y\rangle$, where R_y is a linear ordering on y.
(6.3.2) If $\langle y_1, R_{y_1}\rangle$ and $\langle y_2, R_{y_2}\rangle$ are distinct elements of Y, then $y_1 \cap y_2 = \varnothing$.
(6.3.3) $\langle X, R_X\rangle \cong \langle Y, R_Y\rangle$. Moreover, there is a similarity map f of X onto Y (with respect to R_X, R_Y) that is such that for each $\langle t_x, u_x\rangle$ in X, if $\langle y, R_y\rangle = f(\langle t_x, u_x\rangle)$, then $\mathscr{T}_{y,R_y} = t_x$.

EXAMPLE 6.3.4. We reconsider the sum (1), given above. Let

$$\langle t_0, u_0\rangle = \langle 3, 0\rangle; \langle t_1, u_1\rangle = \langle \omega, 0\rangle; \langle t_2, u_2\rangle = \langle \omega, 1\rangle;$$
$$\langle t_3, u_3\rangle = \langle 5, 0\rangle; \langle t_4, u_4\rangle = \langle \omega, 3\rangle; \langle t_5, u_5\rangle = \langle 3, 1\rangle.$$

Let $\langle X, R_X\rangle$ be a linearly ordered system in which the first few elements of X are the above ordered pairs; let these be arranged so that

$$\langle 3,0\rangle R_X\langle \omega,0\rangle R_X\langle \omega,1\rangle R_X\langle 5,0\rangle R_X\langle \omega,3\rangle R_X\langle 3,1\rangle \cdots.$$

Let

$$\langle y_0, R_{y_0}\rangle = \langle 3, \in\rangle;$$
$$\langle y_1, R_{y_1}\rangle = \langle \omega 2 \sim \omega, \in_{\omega 2}\rangle;$$
$$\langle y_2, R_{y_2}\rangle = \langle \omega 3 \sim \omega 2, \in_{\omega 3}\rangle;$$
$$\langle y_3, R_{y_3}\rangle = \langle 8 \sim 3, \in_8\rangle;$$
$$\langle y_4, R_{y_4}\rangle = \langle \omega 4 \sim \omega 3, \in_{\omega 4}\rangle;$$
$$\langle y_5, R_{y_5}\rangle = \langle 11 \sim 8, \in_{11}\rangle.$$

Let $\langle Y, R_Y\rangle$ be a linearly ordered system that is similar to $\langle X, R_X\rangle$ and whose first few elements, together with their arrangement, are indicated by

$$\langle 3, \in\rangle R_Y\langle \omega 2 \sim \omega, \in_{\omega 2}\rangle R_Y\langle \omega 3 \sim \omega, \in_{\omega 3}\rangle R_Y\langle 8 \sim 3, \in_8\rangle$$
$$R_Y\langle \omega 4 \sim \omega 3, \in_{\omega 4}\rangle R_Y\langle 11 \sim 8, \in_{11}\rangle.$$

Let Y_1 be the set of first coordinates of elements of Y. We define a linear ordering on $\cup Y_1$ as follows. For z_1, z_2 in $\cup Y_1$, $z_1 \in y_1 \in Y_1$ and $z_2 \in y_2 \in Y_1$ for unique y_1 and y_2.

Let $z_1 R z_2$ iff either $y_1 \neq y_2$ and $y_1 R_Y y_2$—or else $y_1 = y_2$ and $z_1 R_{y_1} z_2$. We define "$\sum_{\substack{x \in X \\ R_X}} t_x$" to be "$\mathscr{T}_{\cup Y_1, R}$".

This definition is independent of the choice of the linearly ordered system $\langle Y, R_Y \rangle$ *corresponding to* $\langle X, R_X \rangle$ [Exercise 6.3.11(2)].

EXAMPLE 6.3.5

(1) Let $X = \{\langle \mathscr{T}_1, q \rangle : q \in \mathscr{Q}\}$ and let R_X be defined by

$$\langle \mathscr{T}_1, q_1 \rangle R_X \langle \mathscr{T}_1, q_2 \rangle \qquad \text{iff } q_1 \leq_{\mathscr{Q}} q_2.$$

Let $Y = \{\langle \{q\}, I_{\{q\}} \rangle : q \in \mathscr{Q}\}$ and let

$$\langle \{q_1\}, I_{\{q_1\}} \rangle R_Y \langle \{q_2\}, I_{\{q_2\}} \rangle \qquad \text{iff } q_1 \leq_{\mathscr{Q}} q_2.$$

Let $f : \langle X, R_X \rangle \to \langle Y, R_Y \rangle$ be defined by

$$f(\langle \mathscr{T}_1, q \rangle) = \langle \{q\}. I_{\{q\}} \rangle.$$

Thus, if $t_x = \mathscr{T}_1$ for all x in X, then

$$\sum_{\substack{x \in X \\ R_X}} t_x = \mathscr{T}_{\cup Y_1, R} = \mathscr{T}_{\mathscr{Q}, \leq_{\mathscr{Q}}} = \eta.$$

(2) Let $X = \{\langle \mathscr{T}_2, q \rangle : q \in \mathscr{Q}\}$ and let $\langle \mathscr{T}_2, q_1 \rangle R_X \langle \mathscr{T}_2, q_2 \rangle$ iff $q_1 \leq_{\mathscr{Q}} q_2$. Then $\sum_{\substack{x \in X \\ R_X}} (\mathscr{T}_2)_x \neq \eta$ because $\langle \cup Y_1, R \rangle$ must have jumps. We can let $Y = \{\langle 2 \times \{q\}, R_q \rangle : q \in \mathscr{Q}\}$, where for each q in $\mathscr{Q}$,

$$\langle M, q \rangle R_q \langle N, q \rangle \qquad \text{iff } 0 \leq M \leq N \leq 1.$$

Let

$$\langle 2 \times \{q_1\}, R_{q_1} \rangle R_Y \langle 2 \times \{q_2\}, R_{q_2} \rangle \qquad \text{iff } q_1 \leq_{\mathscr{Q}} q_2,$$

and let $f : X \to Y$ be the similarity mapping defined by

$$f(\langle \mathscr{T}_2, q \rangle) = 2 \times \{q\}, \qquad q \in \mathscr{Q}.$$

The ordering R on $\cup Y_1$ is such that

$$\langle M, q_1 \rangle R \langle N, q_2 \rangle \qquad \text{iff either } q_1 <_{\mathscr{Q}} q_2\text{—or else } q_1 = q_2 \text{ and } M \leq N.$$

Note that $\langle \cup Y_1, R \rangle \cong \langle 2 \times \mathscr{Q}, R_{2 \times \mathscr{Q}} \rangle$ (with respect to the orderings $\in_2$ and $\leq_{\mathscr{Q}}$). Thus, if $t_x = \mathscr{T}_2$ for all x in X, then

$$\sum_{\substack{x \in X \\ R_X}} t_x = \mathscr{T}_{\cup Y, R} = \mathscr{T}_{2 \times \mathscr{Q}, R_{2 \times \mathscr{Q}}} = \mathscr{T}_2 \cdot_{\mathscr{T}} \eta.$$

We now show that *generalized addition of order types extends our previous definition of* "$\mu +_{\mathscr{T}} \nu$".

Let $X = \{\langle \mu, 0\rangle, \langle \nu, 1\rangle\}$; let $\langle U, R_U\rangle$ and $\langle V, R_V\rangle$ be such that $U \cap V = \varnothing$, $\mathscr{T}_{U,R_U} = \mu$, and $\mathscr{T}_{V,R_V} = \nu$. Let $Y = \{\langle U, R_U\rangle, \langle V, R_V\rangle\}$. Let $R_X = I_X \cup \{\langle\langle \mu, 0\rangle, \langle \nu, 1\rangle\rangle\}$ and let $R_Y = I_Y \cup \{\langle\langle U, R_U\rangle, \langle V, R_V\rangle\rangle\}$. Let f be the similarity mapping of $\langle X, R_X\rangle$ onto $\langle Y, R_Y\rangle$ defined by $f(\langle \mu, 0\rangle) = \langle U, R_U\rangle$ and $f(\langle \nu, 1\rangle) = \langle V, R_V\rangle$. Then

$$\sum_{\substack{x \in X \\ R_X}} t_x = \mathscr{T}_{\cup Y_1, R} = \mathscr{T}_{U \cup V, R_{U,V}} = \mu +_{\mathscr{T}} \nu.$$

Exercises

6.3.1. Let $\langle A_I, R_I\rangle$ be linearly ordered systems, $1 \le I \le 4$, where $A_1 \cap A_2 = A_3 \cap A_4 = \varnothing$.

(a) If $\langle A_3, R_3\rangle \cong \langle A_1, R_1\rangle$ and $\langle A_4, R_4\rangle \cong \langle A_2, R_2\rangle$, show that

$$\langle A_1 \cup A_2, R_{A_1,A_2}\rangle \cong \langle A_3 \cup A_4, R_{A_3,A_4}\rangle.$$

(b) Show that (a) implies that the definition of the sum of order types is independent of the representative linearly ordered systems.

6.3.2. Prove that **(a)** addition and **(b)** multiplication of order types are associative.

6.3.3. For N in ω_+, simplify, if possible.

(a) $\mathscr{T}_\omega^{-1} +_{\mathscr{T}} \mathscr{T}_N$. **(b)** $\mathscr{T}_N +_{\mathscr{T}} \mathscr{T}_\omega^{-1}$. **(c)** $\mathscr{T}_\omega^{-1} +_{\mathscr{T}} \mathscr{T}_N +_{\mathscr{T}} \mathscr{T}_\omega^{-1}$.

6.3.4. Let $N \ge 1$. Categorize the following order types (according to equality):

(a) $\mathscr{T}_\omega +_{\mathscr{T}} \mathscr{T}_\omega^{-1}$.

(b) $\mathscr{T}_\omega +_{\mathscr{T}} \mathscr{T}_N +_{\mathscr{T}} \mathscr{T}_\omega^{-1}$.

(c) $\mathscr{T}_\omega +_{\mathscr{T}} \mathscr{T}_\omega +_{\mathscr{T}} \mathscr{T}_N$.

(d) $\mathscr{T}_N +_{\mathscr{T}} \mathscr{T}_\omega +_{\mathscr{T}} \mathscr{T}_\omega^{-1}$.

(e) $\mathscr{T}_N +_{\mathscr{T}} \mathscr{T}_\omega^{-1} +_{\mathscr{T}} \mathscr{T}_\omega$.

(f) $\mathscr{T}_N +_{\mathscr{T}} \mathscr{T}_\omega^{-1} +_{\mathscr{T}} \mathscr{T}_N +_{\mathscr{T}} \mathscr{T}_\omega$.

(g) $\mathscr{T}_\omega^{-1} +_{\mathscr{T}} \mathscr{T}_\omega +_{\mathscr{T}} \mathscr{T}_N$.

(h) $\mathscr{T}_N +_{\mathscr{T}} \mathscr{T}_\omega +_{\mathscr{T}} \mathscr{T}_N +_{\mathscr{T}} \mathscr{T}_\omega^{-1}$.

6.3.5. **(a)** Prove that the following order types are all distinct: $\eta, \eta +_{\mathscr{T}} \mathscr{T}_1, \mathscr{T}_1 +_{\mathscr{T}} \eta, \mathscr{T}_1 +_{\mathscr{T}} \eta +_{\mathscr{T}} \mathscr{T}_1$.

(b) Let $\langle A, R\rangle$ be a linearly ordered system in which A is denumerable and R densely orders A. Prove that $\mathscr{T}_{A,R}$ is one of the types in (a).

★6.3.6 (By a **denumerable type** we mean the order type of a linearly ordered system $\langle A, R\rangle$, in which A is denumerable.)

(a) Prove that $\eta \cdot_{\mathscr{T}} \mu = \eta$ for every denumerable type η.

(b) Determine denumerable types μ_1 and μ_2 such that $\mu_1 \cdot_{\mathscr{T}} \eta = \eta$, but such that $\mu_2 \cdot_{\mathscr{T}} \eta \ne \eta$.

6.3.7. **(a)** Prove that for any order types μ_1, μ_2, and μ_3,

$$\mu_1 \cdot_{\mathscr{T}} (\mu_2 +_{\mathscr{T}} \mu_3) = \mu_1 \cdot_{\mathscr{T}} \mu_2 +_{\mathscr{T}} \mu_1 \cdot_{\mathscr{T}} \mu_3.$$

(b) Find a counterexample to show that, in general,

$$(\mu_1 +_{\mathscr{T}} \mu_2) \cdot_{\mathscr{T}} \mu_3 \neq \mu_1 \cdot_{\mathscr{T}} \mu_3 +_{\mathscr{T}} \mu_2 \cdot_{\mathscr{T}} \mu_3.$$

(c) Are any of the following equal?

$$\eta \cdot_{\mathscr{T}} (\eta +_{\mathscr{T}} \mathscr{T}_1), \eta \cdot_{\mathscr{T}} (\mathscr{T}_1 +_{\mathscr{T}} \eta), \eta \cdot_{\mathscr{T}} (\mathscr{T}_1 +_{\mathscr{T}} \eta +_{\mathscr{T}} \mathscr{T}_1).$$

6.3.8. Show that **(a)** $\lambda +_{\mathscr{T}} \lambda \neq \lambda$; but **(b)** $\lambda +_{\mathscr{T}} \mathscr{T}_1 +_{\mathscr{T}} \lambda = \lambda$.

6.3.9. Let $\mathscr{I} = \{\text{irrational numbers}\}$ and let $\xi = \mathscr{T}_{\mathscr{I}, \leq_{\mathscr{R}}}$. Prove the following:

(a) $\xi +_{\mathscr{T}} \xi = \xi$.

(b) $\xi +_{\mathscr{T}} \mathscr{T}_1 +_{\mathscr{T}} \xi = \xi$.

(c) $\mathscr{T}_1 +_{\mathscr{T}} \xi \neq \xi$.

★6.3.10. Let $\langle X, R_X \rangle$ be a linearly ordered system in which every x in X is an ordered pair, $\langle t_x, u_x \rangle$, where t_x is an order type. Use the Axioms of Substitution and Choice to prove that there exists an ordered system $\langle Y, R_Y \rangle$ that has the properties (6.3.1), (6.3.2), and (6.3.3).

★6.3.11. (A.C.) In the definition of generalized addition of order types show the following:

(a) $\cup Y_1$ is linearly ordered by R, where for $z_1, z_2 \in \cup Y_1$, $z_1 R z_2$ iff either $z_1 \in y_1 \in Y_1$, $z_2 \in y_2 \in Y_1$, $y_1 \neq y_2$, and $y_1 R_y y_2$—or else $z_1, z_2 \in y_1 \in Y_1$, and $z_1 R_{y_1} z_2$.

(b) The definition of "$\sum_{\substack{x \in X \\ R_X}} t_x$" is independent of the choice of the ordered system $\langle Y, R_Y \rangle$, corresponding to $\langle X, R_X \rangle$ (as in Exercise 6.3.10).

★6.3.12. (A.C.) Let $\langle X, R_X \rangle$ be a well-ordered system in which every x in X is an ordered pair $\langle t_x, u_x \rangle$, where $t_x = \mathscr{T}_{\Gamma_x}$ for some ordinal Γ_x. Prove the following:

(a) $\sum_{\substack{x \in X \\ R_X}} t_x = \mathscr{T}_\delta$ for some ordinal δ.

(b) $\delta = \sum_{\gamma < \rho} \Gamma_\gamma$ (the ordinal sum), where $\langle \rho, \in \rangle \cong \langle X, R_X \rangle$.

6.4 Cardinal Numbers

We associate with every set a certain "cardinal number" in such a way that two sets correspond to the same cardinal number if and only if they are in one-one correspondence with each other. The collection of sets in one-one correspondence with any nonempty set is not a set. The Axiom of Rank enables us to define cardinal numbers as sets without resorting to the Axiom of Choice.

Recall that we write "$X \approx Y$" in case there is a one–one correspondence between the sets X and Y.

For any set A, the **cardinal (number) of** A (the **cardinality of** A), written "$|A|$", is defined to be the collection of all sets x of lowest rank for which $x \approx A$. The Axiom of Rank implies that *every set has a cardinality*. Furthermore, $|A|$ *is always a set*; in fact, *if* $\rho(A) = \alpha$, *then*

$$|A| = \{x \in \mathbf{R}_{\alpha+1} : x \approx A \text{ and for all } y, \text{ if } y \approx A, \text{ then } \rho(x) \leq \rho(y)\}.$$

The elements of $|A|$ can (all) be of rank $< \alpha$ [Examples 6.4.1(5) and 6.4.2(2)]. A set c is a **cardinal number** provided there is some set A for which $c = |A|$.

A cardinal number c is said to be a **finite cardinal** provided that every set A for which $|A| = c$ is finite; otherwise, c is said to be an **infinite cardinal**. *If c is a cardinal, then c is a finite cardinal if and only if c is finite* (Exercise 6.4.2).

The definition of "cardinal number" is reminiscent of that of "order type". In fact, *let X be an N-element set*. Then $|X|$ is the set of N-element sets of lowest rank. Every N-element set can be linearly ordered; moreover, if X_1 and X_2 are both N-element sets and if R_1 and R_2 are linear orderings on X_1 and X_2, respectively, then $\langle X_1, R_1\rangle \cong \langle X_2, R_2\rangle$. It follows that $|X|$ *is the set of first coordinates of the elements of* $\mathcal{T}_N$.

EXAMPLE 6.4.1. We utilize the above remarks together with Example 6.2.1 to compute $|N|$ for $N \leq 4$. This gives us the cardinality of A, whenever A is an N-element set for $N \leq 4$.

(1) $|0| = \{x : \text{for some } y, \langle x, y\rangle \in \mathcal{T}_0\} = \{0\} = 1$.

(2) $|1| = \{1\}$.

(3) $|2| = \{2\}$.

(4) $\mathcal{T}_3 = \{\langle A, R\rangle : \rho(A) = 3,\ A \text{ is a 3-element set, and } R \text{ is a linear ordering on } A\}$. Hence, $|3|$ is the set of 3-element sets of rank 3; i.e., $|3| = \{3, \{0, 1, \{1\}\}, \{0, 2, \{1\}\}, \{1, 2, \{1\}\}\}$.

(5) $\mathbf{R}_3$ is the unique 4-element set of lowest rank; thus, $|4| = \{\mathbf{R}_3\}$.

Note that, in general, if κ is a cardinal number, $|\kappa|$ need not be κ; for example, $\|2\| = |1|$. However, *if* $K \in \kappa$, *then* $|K| = \kappa$.

†EXAMPLE 6.4.2

(1) A denumerable set has infinite rank. ω has rank ω. Thus, $|\omega| = \{x \in \mathbf{R}_{\omega+1} : x \text{ is denumerable}\}$.

(2) $|\omega + 1|$ is the set of all sets x of lowest rank for which $x \approx \omega + 1$. Also, $\omega \approx \omega + 1$, and if $\rho(y) < \omega$, $y \not\approx \omega + 1$. Thus, $|\omega + 1| = \{x \in \mathbf{R}_{\omega+1} : x \text{ is denumerable}\} = |\omega|$.

(3) Similarly, for every denumerable ordinal α, $|\alpha| = |\omega|$; in fact, if X is a denumerable set, $|X| = |\omega|$.

Let c and c' be cardinal numbers. We write "$c \preceq c'$" or "$c' \succeq c$" provided there exist sets A and A' with $|A| = c$, $|A'| = c'$ for which there is a one-one mapping of A into A'. Clearly, $c \preceq c$. It is easy to show that $c \preceq c'$ iff for *all* sets A and A' that are such that $|A| = c$ and $|A'| = c'$, there is a one–one mapping of A into A'. In other words, $c \preceq c'$ *iff for all sets A and A' that are such that $|A| = c$ and $|A'| = c'$, there is a subset B of A such that $B \approx A'$*. We write "$c \prec c'$" or "$c' \succ c$" in case $c \preceq c'$ and $c \neq c'$. Clearly, $c \prec c'$ iff for all sets A of cardinality c and all sets A' of cardinality c', the following are true:

(1) There is a one-one mapping of A *into* A'.
(2) There is no one-one mapping of A *onto* A'.

†EXAMPLE 6.4.3

(1) If M and N are natural numbers, then $M \leq N$ iff $|M| \preceq |N|$, and $M < N$ iff $|M| \prec |N|$. Moreover, $|M| \prec |\omega|$ because $M \subseteq \omega$ and there is no one-one map of M onto ω.
(2) $|\omega| \prec |\mathscr{R}|$: In Chapter 3 we showed that there is an isomorphic image of ω in $\mathscr{R}$; thus, $|\omega| \preceq |\mathscr{R}|$. We also showed that $\mathscr{R}$ is uncountable; thus, $\mathscr{R} \not\approx \omega$. Note that every subset of ω is countable; hence, $\mathscr{R} \not\preceq \omega$.

If c, c', and c'' are cardinal numbers and are such that $c \preceq c' \preceq c''$, then it is immediate that $c \preceq c''$ [Exercise 6.4.6(a)]. Thus, on every *set* of cardinal numbers $\preceq$ is a transitive (and reflexive) relation. Surprisingly, it is far more difficult to show that $c \prec c' \prec c''$ implies $c \prec c''$. In fact, it is also not obvious that $c \prec c'$ and $c' \prec c$ cannot hold simultaneously. These results will follow from the Schröder–Bernstein Theorem, our next goal.

One might also have expected a Law of Trichotomy for $\prec$, i.e., that for any 2 cardinals c and c', (exactly) one of the following should hold:

$$c \prec c', \qquad c = c', \qquad c' \prec c.$$

We will need the Axiom of Choice to establish this; in fact, in Section 6.9 we will show that the Axiom of Choice is equivalent to the Law of Trichotomy for $\prec$. We now consider some laws governing the comparability of cardinals that do not depend upon the Axiom of Choice.

For every finite set X and infinite set Y, clearly there is a one–one map of X into Y but no one-one map of X onto Y; thus, $|X| \prec |Y|$. In particular, for every natural number N, $|N| \prec |Y|$. But we required the Axiom of Choice in order to establish that every infinite set Y has a denumerable subset (Theorem 4.2.1). Thus, we cannot as yet establish that for *every* infinite set Y, $|\omega| \preceq |Y|$. However, even without the Axiom of Choice we know that there exist (infinite) sets X for which $|\omega| \prec |X|$. For example, we did not utilize the Axiom of Choice to establish $|\omega| \prec |\mathscr{R}|$.

The question arises as to whether or not there exists a cardinal number κ satisfying

(6.4.1) $$|\omega| \prec \kappa \prec |\mathscr{R}|.$$

The assertion: "There is no cardinal number satisfying (6.4.1)." is known as the **Continuum Hypothesis**. ($\mathscr{R}$ is often called the **linear continuum**). The determination of the undecidability of the Continuum Hypothesis was one of the outstanding achievements in research in set theory. Gödel proved that the truth of the Continuum Hypothesis is consistent with the axioms of set theory, including the Axiom of Choice, whereas Cohen demonstrated that the falsehood of the Continuum Hypothesis is also consistent with the axioms of set theory (including the Axiom of Choice). Thus, the Continuum Hypothesis amounts to an "undecidable" sentence within the framework of the axioms of set theory—i.e., the Continuum Hypothesis is "independent" of the axioms of set theory.

Cantor's Theorem (4.1.7) indicates that

(6.4.2) $$\text{for every set } Y,\ |Y| \prec |\mathscr{P}(Y)|.$$

Let $Y = \omega$ in (6.4.2); then $|\omega| \prec |\mathscr{P}(\omega)|$. Now let $Y = \mathscr{P}(\omega)$ in (6.4.2); then $|\mathscr{P}(\omega)| \prec |\mathscr{P}\mathscr{P}(\omega)|$. The Axiom of Substitution assures the existence of the set

$$A = \{\mathscr{P}^N(\omega) : N < \omega\},$$

where $\mathscr{P}^0(\omega) = \omega$ and $\mathscr{P}^{K^+}(\omega) = \mathscr{P}(\mathscr{P}^K(\omega))$, $K < \omega$. Substitution also yields the existence of a set of "increasing" cardinals, $\{\kappa_N : N < \omega\}$, where $\kappa_N = |\mathscr{P}^N(\omega)|$; we have

$$\kappa_0 \prec \kappa_1 \prec \kappa_2 \prec \cdots \prec \kappa_N \prec \cdots.$$

Before proving the Schröder–Bernstein Theorem, our main result of this section, we first consider a "fixed-point" argument (due, in various stages, to Banach, Knaster, Tarski, and Whittaker) that will be utilized in that proof.

Let $f : X \to X$. Then x_0 ($\in X$) is called a **fixed-point of** f if $f(x_0) = x_0$.

Theorem 6.4.1. *Let A be any set. Let $h : \mathscr{P}(A) \to \mathscr{P}(A)$ be such that*

(6.4.3) $$x_1 \subseteq x_2 \subseteq A \text{ implies } h(x_1) \subseteq h(x_2).$$

Then there is a subset $A_0 \subseteq A$ such that $h(A_0) = A_0$.

(A_0 is a fixed point of the function h.)

PROOF: Let $A_1 = \{X \in \mathscr{P}(A) : X \subseteq h(X)\}$. Then $\cup A_1 \subseteq A$; we claim that

$$h(\cup A_1) = \cup A_1. \tag{6.4.4}$$

For let $X \in A_1$. Then $X \subseteq \cup A_1$. Therefore, by (6.4.3), $h(X) \subseteq h(\cup A_1)$—hence, $X \subseteq h(X) \subseteq h(\cup A_1)$. It follows that $\cup A_1 \subseteq h(\cup A_1)$.

Again by (6.4.3), $h(\cup A_1) \subseteq h(h(\cup A_1))$. Therefore, $h(\cup A_1) \in A_1$. But then $h(\cup A_1) \subseteq \cup A_1$. We have established (6.4.4).

Corollary. *Let $h : \mathscr{P}(A) \to \mathscr{P}(A)$, and let h satisfy* (6.4.3). *Let*

$$A_1 = \{X \in \mathscr{P}(A) : X \subseteq h(X)\}.$$

Then $h(\cup A_1) = \cup A_1$.

Theorem 6.4.2. Schröder–Bernstein Equivalence Theorem. *Let c and c' be cardinal numbers. Then $c \preceq c'$ and $c' \preceq c$ together imply $c = c'$.*

PROOF: Let A and B be arbitrary sets satisfying $|A| = c$ and $|B| = c'$. Assume that $c \preceq c'$ and $c' \preceq c$. Then for some subsets $A_0 \subseteq A$ and $B_0 \subseteq B$, $A \approx B_0$ and $B \approx A_0$. Let f be a one-one correspondence of A onto B_0 and let g be a one-one correspondence of B onto A_0. We must find a one-one correspondence, F, of A onto B.

Let $h : \mathscr{P}(A) \to \mathscr{P}(A)$ be defined as follows: for $X \subseteq A$, let $h(X) = A \sim g''(B \sim f''(X))$. If $X_1 \subseteq X_2 \subseteq A$, then the following inclusions hold:

$$f''(X_1) \subseteq f''(X_2);$$

$$B \sim f''(X_2) \subseteq B \sim f''(X_1);$$

$$g''(B \sim f''(X_2)) \subseteq g''(B \sim f''(X_1));$$

$$A \sim g''(B \sim f''(X_1)) \subseteq A \sim g''(B \sim f''(X_2)).$$

Thus,

$$X_1 \subseteq X_2 \subseteq A \text{ implies } h(X_1) \subseteq h(X_2).$$

Consequently, h satisfies (6.4.3). Let

$$A_1 = \{X \in \mathscr{P}(A) : X \subseteq h(X)\}.$$

According to the (corollary of the) preceding "fixed-point" theorem,

$$\cup A_1 = h(\cup(A_1)) = A \sim g''(B \sim f''(\cup A_1)).$$

It follows that $g''(B \sim f''(\cup A_1)) = A \sim \cup A_1$. Define $F : A \to B$ by $F(a) = f(a)$ if $a \in \cup A_1$ and $F(a) = g^{-1}(a)$ if $a \in A \sim \cup A_1$. Then F is one-one onto B. (See Figure 6.4.1.)

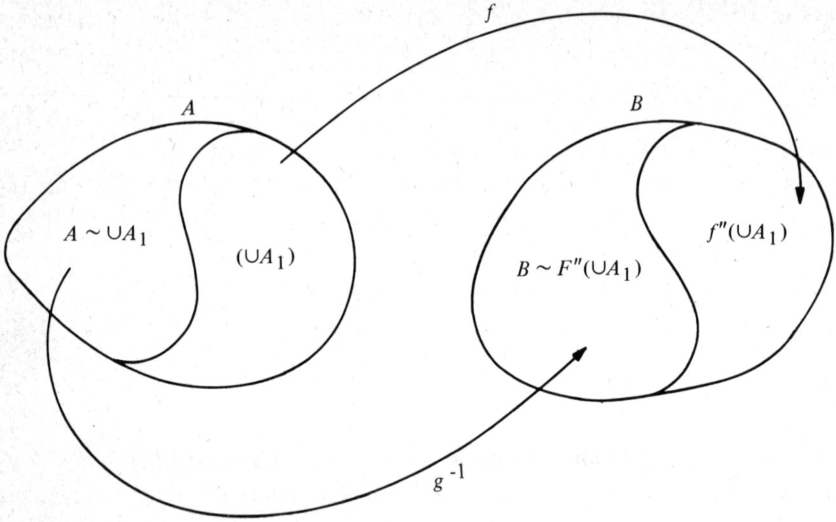

Fig. 6.4.1

EXAMPLE 6.4.4. For I = 1, 2, 3, 4, let $X \subseteq \omega$ and let $h_I : \mathscr{P}(\omega) \to \mathscr{P}(\omega)$ be defined as follows:

(1) $h_1(X) = \begin{cases} X \cup \{\min(\omega \sim X)\}, & \text{if } X \subset \omega; \\ \omega, & \text{if } X = \omega. \end{cases}$

(2) $h_2(X) = \begin{cases} X \sim \{\min(X)\}, & \text{if } X \neq \varnothing; \\ \varnothing, & \text{if } X = \varnothing. \end{cases}$

(3) $h_3(X) = \begin{cases} X \cup \{10\}, & \text{if } X \subseteq 10; \\ \omega, & \text{otherwise.} \end{cases}$

(4) $h_4(X) = \begin{cases} X \cup \{9\}, & \text{if } X \subseteq 10; \\ (X \cup \{9\}) \sim \{\min(X \sim 10)\}, & \text{otherwise.} \end{cases}$

All four functions h_I satisfy condition (6.4.3). We list the fixed point $\cup A_1$ (of the corollary) for each function.

(1) ω, with respect to h_1.
(2) $\varnothing$, with respect to h_2.
(3) ω, with respect to h_3.
(4) 10, with respect to h_4.

In parts (1), (2), and (3), these are the only fixed points of the function in question. Every subset of 10 that contains 9 is a fixed point of h_4.

Corollary 1. *Let c, c', and c'' be cardinal numbers.*
(1) $c \prec c'$ and $c' \prec c$ *cannot simultaneously hold.*
(2) $c \prec c' \prec c''$ *implies* $c \prec c''$.

PROOF: (1) If $c \prec c'$ and $c' \prec c$, then $c \preceq c'$ and $c' \preceq c$. According to the Schröder–Bernstein Theorem, $c = c'$. But this contradicts the assertion that $c \prec c'$.

(2) Suppose that $c \prec c' \prec c''$. Then $c \preceq c' \preceq c''$; hence, $c \preceq c''$. Suppose that $c = c''$. Then $c \prec c'$ and $c' \prec c$, in contradiction of part (1). Thus, $c \prec c''$.

Corollary 2. *Let c and c' be cardinal numbers. Each of the following is an equivalent definition of "$c \prec c'$":*
(1) $c \preceq c'$ *and* $c' \not\preceq c$.
(2) *There exist sets A and A' with $|A| = c$ and $|A'| = c'$, such that*
   (a) *there is a subset B' of A' satisfying $A \approx B'$ and*
   (b) *there is no subset B of A satisfying $A' \approx B$.*
(3) *There exist sets A and A' with $|A| = c$ and $|A'| = c'$, and there is a one-one mapping of A into A' but no one-one mapping of A' into A.*
(4) *For all sets A and A' with $|A| = c$ and $|A'| = c'$,* (2a) and (2b) *hold.*
(5) *For all sets A and A' with $|A| = c$ and $|A'| = c'$, there is a one-one mapping of A into A' but no one-one mapping of A' into A.*

Exercises

6.4.1. Determine $|5|$.

†6.4.2. Prove that for every cardinal c, c is a finite cardinal if c is finite.

6.4.3. **(a)** Suppose that $|A| \prec |A'|$. Use the definition of "$\prec$" to show that there is a proper subset $B' \subset A'$ such that $A \approx B'$.
(b) Suppose that $A \approx B'$ and B' is a proper subset of A'. Give a counterexample to show that we cannot conclude that $|A| \prec |A'|$.

6.4.4. Let s be the sentence, "A is an infinite set and $|\omega| \not\preceq |A|$." Show that the truth of s implies the falsehood of the Axiom of Choice.

6.4.5. Let c be any cardinal number. Prove that there exists a denumerable sequence of increasing cardinals $[\kappa_0, \kappa_1, \kappa_2, \ldots]$ beginning with $\kappa_0 = c$.

†6.4.6. Let c, c', and c'' be cardinal numbers. Show the following:
(a) $c \preceq c'$ and $c' \preceq c''$ together imply $c \preceq c''$.
(b) $c \prec c'$ and $c' \preceq c''$ together imply $c \prec c''$.
(c) $c \preceq c'$ and $c' \prec c''$ together imply $c \prec c''$.

6.4.7. Let $h : \mathscr{P}(\omega 2) \to \mathscr{P}(\omega 2)$ be defined by

$$h(X) = \begin{cases} \omega, & \text{if } X \subseteq \omega; \\ \omega^+, & \text{otherwise.} \end{cases}$$

(a) Determine all fixed points of h.
(b) Does h satisfy condition (6.4.3)? If so, determine $\cup A_1$.

6.4.8. Suppose that $b_0 \in B \subset A$. Let $h : \mathscr{P}(A) \to \mathscr{P}(A)$ be defined by

$$h(X) = \begin{cases} X \cup \{b_0\}, & \text{if } X \cap B = \varnothing; \\ X \sim B, & \text{if } X \cap B \neq \varnothing. \end{cases}$$

(a) Determine all fixed points of h.
(b) Does h satisfy condition (6.4.3)? If so, determine $\cup A_1$.

***6.4.9.** In the proof of the Schröder–Bernstein Equivalence Theorem, suppose that $A = \{1, 3, 5, 7, \ldots\}$, $B = \{0, 2, 4, 6, \ldots\}$, $B_0 = \{0, 4, 8, 12, \ldots\}$, and $A_0 = \{3, 7, 11, \ldots\}$. Furthermore, suppose that $f : A \to B_0$ is such that $f(2N + 1) = 4N$, and $g : B \to A_0$ is such that $g(2N) = 4N + 3$.
(a) Determine $h(\{1, 5, 7, 13\})$.
(b) Show that $A \sim A_0 \subseteq \cup A_1$.
(c) Determine $\cup A_1$.

6.4.10. Prove Corollary 2 of the Schröder–Bernstein Theorem; i.e., prove the equivalence of each of the five sentences to "$c \prec c'$".

†6.4.11. Let X, Y, and Z be sets for which $X \subset Y \subset Z$. Suppose that $X \approx Z$. Prove that $X \approx Y$.

⋆,†6.4.12. Prove that the collection of cardinal numbers is not a set.

6.5 Cardinal Addition and Multiplication

Our method of introducing sums and products of cardinal numbers will be by means of representative sets of the cardinalities in question.

In our development of Sections 6.5–6.8, we will consider only cardinal arithmetical results that can be obtained without utilizing the Axiom of Choice. Many of the important theorems of cardinal arithmetic depend upon the Axiom of Choice; some of these are even equivalent to the Axiom of Choice. These will be considered in Sections 6.9–6.12.

Let c and c' be cardinal numbers. We know that there are sets A and A' that satisfy $|A| = c$, $|A'| = c'$, and $A \cap A' = \varnothing$. Let $c + c' = |A \cup A'|$. *This definition is independent of the representative sets A and A'.* Moreover, by the laws of set operations (Theorem 1.8.1), $A \cup B = B \cup A$ and $(A \cup B) \cup C = A \cup (B \cup C)$ for all sets A, B, and C. It immediately follows that *cardinal addition*, unlike the addition of ordinals and of order types, *is commutative as well as associative.*

†EXAMPLE 6.5.1

(1) For $M, N < \omega$, $|M| + |N| = |M + N|$. For let $A = M$ and let $A' = (M + N) \sim M$. Then $|A| = |M|$, $|A'| = |N|$, $A \cap A' = \varnothing$, and $A \cup A' = M + N$. Thus, $|M| + |N| = |A \cup A'| = |M + N|$.
(2) For M in ω, $|M| + |\omega| = |\omega|$. Here, take $A = M$ and take $A' = \omega \sim M$. Then $|A'| = |\omega|$ and $A \cup A' = \omega$.
(3) $|\omega| + |\omega| = |\omega|$; let $A = \{\text{odd natural numbers}\}$ and let $A' = \{\text{even natural numbers}\}$.
(4) $|\mathscr{R}| + |\mathscr{R}| = |\mathscr{R}|$. Let $A = \{\Gamma \in \mathscr{R} : 0 < \Gamma \leq 1\}$ and let
$$A' = \{\Gamma \in \mathscr{R} : \ 1 < \Gamma \leq 2\}.$$
Then $A \cup A' = \{\Gamma \in \mathscr{R} : 0 < \Gamma \leq 2\}$ and $|A \cup A'| = |\mathscr{R}|$.

Theorem 6.5.1. *Let $\kappa_1, \kappa_2, \kappa_3$, and κ_4 be cardinal numbers that satisfy $\kappa_1 \precsim \kappa_3$ and $\kappa_2 \precsim \kappa_4$. Then $\kappa_1 + \kappa_2 \precsim \kappa_3 + \kappa_4$.*

PROOF: Let A_1, A_2, A_3, A_4 be sets for which $|A_I| = \kappa_I$, $I = 1, 2, 3, 4$, and $A_1 \cap A_2 = A_3 \cap A_4 = \varnothing$. There exists a subset B_3 of A_3 and a one-one mapping, f, of A_1 onto B_3, and there is a subset B_4 of A_4 together with a one-one mapping, g, of A_2 onto B_4. The disjointness of A_1 and A_3 implies that $f \cup g$ is a mapping of $A_1 \cup A_3$ onto $B_2 \cup B_4 \subseteq A_2 \cup A_4$. Finally, because f and g are each one-one and $B_2 \cap B_4 \subseteq A_2 \cap A_4 = \varnothing$, $f \cup g$ is one-one.

Let c and c' be cardinal numbers. Then $c \cdot c'$ is defined to be $|A \times B|$, where A and B are any sets such that $|A| = c$ and $|B| = c'$. Again, *the definition is independent of the choice of the representative sets A and B.*

For all sets A and B, there is obviously a one-one correspondence, F, between $A \times B$ and $B \times A$; if A and B are nonempty, we need only let $F(\langle a, b\rangle) = \langle b, a\rangle$, for a in A and b in B. Thus, $|A \times B| = |B \times A|$. Hence, *for all cardinal numbers c and c', $c \cdot c' = c' \cdot c$—i.e., cardinal multiplication is commutative.* Similarly, for every set C, $(A \times B) \times C$ is in one-one correspondence with $A \times (B \times C)$; here, if A, B, and C are nonempty, let $g(\langle\langle a, b\rangle, c\rangle) = (\langle a, \langle b, c\rangle\rangle)$, $a \in A$, $b \in B$, and $c \in C$. It follows that *cardinal multiplication is associative*; i.e., *for all cardinal c, c', and c″*,

$$(c \cdot c') \cdot c'' = c \cdot (c' \cdot c'').$$

†EXAMPLE 6.5.2

(1) Let $M, N < \omega$.

$$|M| \cdot |N| = |M \times N| = |M \cdot N|.$$

(2) For $N(\neq 0)$ in ω, $|N| \cdot |\omega| = |\omega|$. Here, let $A = N$ and let $B = \omega$. Then $A \times B = \{\langle K, L\rangle : K < N \text{ and } L < \omega\}$. Every natural number has a unique representation of the form $L \cdot N + K$ for some $K < N$ and L in ω. Hence, this defines a one-one correspondence between ω and $A \times B$.

(3) $|\omega| \cdot |\omega| = |\omega|$, according to Theorem 4.3.2.

(4) $|\mathscr{R}| \cdot |\omega| = |\mathscr{R}|$. Here, let $A = \{\Gamma \in \mathscr{R} : \overset{\equiv}{0} <_{\mathscr{R}} \Gamma \leq_{\mathscr{R}} \overset{\equiv}{1}\}$ and let $B = \omega$. Then $A \times B = \{\langle \Gamma, N\rangle : \Gamma \in \mathscr{R} \text{ and } \overset{\equiv}{0} <_{\mathscr{R}} \Gamma \leq_{\mathscr{R}} \overset{\equiv}{1}, \text{ and } N < \omega\}$. Let $C = \{\text{positive reals}\}$ and let $F : A \times B \to C$ be defined by $F(\langle \Gamma, N\rangle) = \overset{\equiv}{N} +_{\mathscr{R}} \Gamma$. Then F is a one-one mapping of $A \times B$ onto C. Therefore, $|\mathscr{R}| \cdot |\omega| = |A \times B| = |C| = |\mathscr{R}|$.

Theorem 6.5.2. *Cardinal multiplication distributes over cardinal addition. Thus, let κ_1, κ_2, and κ_3 be any cardinal numbers.*

(1) $\kappa_1 \cdot (\kappa_2 + \kappa_3) = \kappa_1 \cdot \kappa_2 + \kappa_1 \cdot \kappa_3$.
(2) $(\kappa_2 + \kappa_3) \cdot \kappa_1 = \kappa_2 \cdot \kappa_1 + \kappa_3 \cdot \kappa_1$.

PROOF: We need only demonstrate (1); (2) will follow from the commutativity of multiplication. Let A_1, A_2, and A_3 be sets that satisfy $|A_I| = \kappa_I$, $I = 1, 2, 3$, and $A_2 \cap A_3 = \varnothing$. Then $\kappa_1 \cdot (\kappa_2 + \kappa_3) = |A_1 \times (A_2 \cup A_3)|$, $\kappa_1 \cdot \kappa_2 = |A_1 \times A_2|$, and $\kappa_1 \cdot \kappa_3 = |A_1 \times A_3|$. Also, $(A_1 \times A_2) \cap (A_1 \times A_3) = \varnothing$, for if

$$\langle u, v\rangle \in A_1 \times A_2 \quad \text{and} \quad \langle x, y\rangle \in A_1 \times A_3,$$

then $v \neq y$ because $A_2 \cap A_3 = \varnothing$; hence, $\langle u, v\rangle \neq \langle x, y\rangle$, by Theorem 2.1.1. Thus,

$$\kappa_1 \cdot \kappa_2 + \kappa_1 \cdot \kappa_3 = |(A_1 \times A_2) \cup (A_1 \times A_3)|.$$

But for any sets A_1, A_2, and A_3, $(A_1 \times A_2) \cup (A_1 \times A_3) = A_1 \times (A_2 \cup A_3)$. Part (1) follows.

Theorem 6.5.3. *For every cardinal number c and every positive natural number N,*
$$c \cdot |N| = c + c + \cdots + c \ (N \text{ times}).$$

PROOF: Let $|A| = c$. Then

$$A \times N = \{\langle a, M\rangle : a \in A \text{ and } M < N\} = \cup\{A \times \{M\} : M < N\}.$$

For $M_1, M_2\ (\neq M_1) \in N$, $(A \times \{M_1\}) \cap (A \times \{M_2\}) = \varnothing$. Hence,

$$\{A \times \{M\} : M < N\}$$

is pairwise disjoint, and

$$c \cdot |N| = |A \times N| = |\cup\{A \times \{M\} : M < N\}|$$
$$= c + c + \cdots + c \text{ (N times).}$$

Our last theorem is the multiplicative analogue of Theorem 6.5.1; we leave its proof as an exercise.

Theorem 6.5.4. *Let $\kappa_1, \kappa_2, \kappa_3$, and κ_4 be cardinal numbers that satisfy $\kappa_1 \precsim \kappa_3$ and $\kappa_2 \precsim \kappa_4$. Then $\kappa_1 \cdot \kappa_2 \precsim \kappa_3 \cdot \kappa_4$.*

We briefly consider **cardinal subtraction and division.**

Let c_1, c_2, and κ be cardinal numbers. We write "$\kappa = c_1 - c_2$" provided the following conditions hold:

(1) $\kappa + c_2 = c_1$.
(2) If κ' is any cardinal satisfying $\kappa' + c_2 = c_1$, then $\kappa' = \kappa$.

In other words, $\kappa = c_1 - c_2$ *if it is the unique cardinal solution, c_3, of the equation*

$$c_3 + c_2 = c_1.$$

Let c_1 and c_2 be arbitrary cardinals with $c_2 \precsim c_1$. There is always some cardinal κ such that $\kappa + c_2 = c_1$. For take any C_1, C_2 with $|C_1| = c_1$ and $|C_2| = c_2$. There is a one-one map of C_2 into C_1. Let $\kappa = |C_1 \sim C_2|$. However, κ is not necessarily unique; hence, $c_1 - c_2$ need not necessarily exist.

EXAMPLE 6.5.3

(1) If N_1 and N_2 are natural numbers and $N_2 \leq N_1$, then $|N_1| - |N_2| = |N_1 \dot{-} N_2|$.
(2) For N in ω, $|\omega| - |N|$ exists and equals $|\omega|$. In fact, $|\omega| + |N| = |\omega|$; moreover, if $\kappa + |N| = |\omega|$, κ cannot be finite. If K is a set for which $K \cap N = \varnothing$ and $|K| = \kappa$, then K can be mapped one-one into ω. Hence, K is countable, and therefore denumerable. Thus, $\kappa = |\omega|$.
(3) $|\omega| - |\omega|$ does not exist because for every natural number N, $|N| + |\omega| = |\omega|$. Also, $|\omega| + |\omega| = |\omega|$.

The Axiom of Choice is required to show that $c_2 \prec c_1$ implies the existence of $c_1 - c_2$. (See Exercise 6.9.4.)

We say that "$\kappa = c_1/c_2$" provided the following conditions hold:

(1) $\kappa \cdot c_2 = c_1$.
(2) If κ' is a cardinal number satisfying $\kappa' \cdot c_2 = c_1$, then $\kappa' = \kappa$.

The Axiom of Choice implies that if $c_2 \prec c_1$ and c_1 is an infinite cardinal, then c_1/c_2 exists. (See the remarks following Exercise 6.9.4.)

EXAMPLE 6.5.4

(1) Let $N\ (\neq 0) < \omega$. Then $|\omega|/|N| = |\omega|$ because

(a) $|\omega| \cdot |N| = |\omega|$, and
(b) if $\kappa \cdot |N| = |\omega|$ and $|K| = \kappa$, then $K \times N \approx \omega$.

Thus, $\kappa = |K| \preceq |K \times N| = |\omega|$. Because κ cannot be finite, $\kappa = \omega$.

(2) $|\omega|/|\omega|$ is undefined because for all nonzero natural numbers N, $|N| \cdot |\omega| = |\omega| \cdot |\omega| = |\omega|$.

Exercises

†6.5.1. Prove the assertion made at the beginning of this section concerning the existence of disjoint sets A and A' having the specified cardinalities c and c', respectively.

†6.5.2. Prove that the definitions of cardinal addition and multiplication are independent of the representative sets. Thus let $|A| = |B| = c$ and $|A'| = |B'| = c'$.

(a) Show that if $A \cap A' = B \cap B' = \varnothing$, then $|A \cup A'| = |B \cup B'|$.
(b) Show that $|A \times A'| = |B \times B'|$.

†6.5.3. Let c and c' be arbitrary cardinals. Prove that for any set X for which $|X| = c + c'$, there are disjoint subsets A and A' of X that satisfy $|A| = c$, $|A'| = c'$, and $A \cup A' = X$.

†6.5.4. Prove that $|0|$ is the unique cardinal number c' satisfying $c + c' = c$ for every cardinal c.

6.5.5. Prove that $|\mathscr{R}| + |\omega| = |\mathscr{R}|$.

★6.5.6. Prove that $|\mathscr{R}| + |\mathscr{P}(\mathscr{R})| = |\mathscr{P}(\mathscr{R})|$.

6.5.7. Let c and c' be arbitrary cardinal numbers. Prove that $c \prec c' + |1|$ implies $c \preceq c'$.

†6.5.8. Prove that $|1|$ is the unique cardinal number c' with the property that for all cardinal numbers c, $c \cdot c' = c$.

6.5.9. Prove that $|\mathscr{R}| \cdot |\mathscr{R}| = |\mathscr{R}|$.

6.5.10. Prove Theorem 6.5.4.

★6.5.11. Prove that $|\mathscr{P}(\mathscr{R})| \cdot |\mathscr{P}(\mathscr{R})| = |\mathscr{P}(\mathscr{R})|$.

6.5.12. Prove that for every nonzero natural number N, $|\mathscr{R}| + |N| = |\mathscr{R}| \cdot |N| = |\mathscr{R}|$.

†6.5.13. Prove that if κ_1 and κ_2 are cardinal numbers that satisfy $|1| \prec \kappa_1$ and $|1| \prec \kappa_2$, then $\kappa_1 + \kappa_2 \preceq \kappa_1 \cdot \kappa_2$.

★,†**6.5.14.** Let α and β be ordinals.

(a) Prove that $|\alpha + \beta| = |\alpha| + |\beta|$.

(b) Prove that $|\alpha \cdot \beta| = |\alpha| \cdot |\beta|$.

(c) Show that $\alpha \leq \beta$ does not imply the existence of $|\beta| - |\alpha|$.

(d) Show that if $|\beta| - |\alpha|$ exists, then it equals $|\beta \dot{-} \alpha|$.

6.5.15. Prove that for every infinite cardinal c and for every N in ω, $c - \|N\|$ exists.

6.5.16. Prove that $|\mathscr{R}| / |\mathscr{R}|$ is undefined.

6.6 Cardinal Exponentiation

Let c and κ be any cardinal numbers and let A and B be any sets for which $|A| = \kappa$ and $|B| = c$. Let "$\mathscr{F}(A, B)$" denote the set of all functions f from A into B. More precisely, we let

$$\mathscr{F}(A, B) = \{f \in \mathscr{P}(A \times B) : f \text{ is a function with domain } A\}.$$

Then we define "c^κ" to be "$|\mathscr{F}(A, B)|$"; *this definition is independent of the representative sets A and B.*

The elements of $\mathscr{F}(A, B)$ are functions from A into B. Thus two functions f_1 and f_2 of $\mathscr{F}(A, B)$ will be distinct if they differ on (at least) a single element of A; i.e., if for some a_0 in A, $f_1(a_0) \neq f_2(a_0)$. Intuitively, each a_0 in A determines $|B|$ different functions. Moreover, if $a_1 (\neq a_0) \in A$, then equality of two functions at a_1 is independent of equality at a_0. Hence, the "number" of functions in $\mathscr{F}(A, B)$ should be $|B|$ multiplied by itself $|A|$ times; i.e., $|B|^{|A|} (= c^\kappa)$. In particular, *if $c = |M|$, $\kappa = |N|$, and M and N are positive natural numbers, then $c^\kappa = |\mathscr{F}(N, M)| = |M^N|$*, and our argument indicates that $\mathscr{F}(N, M)$ is, indeed, an M^N-element set.

If $\kappa = |0|$ and c is arbitrary, then $c^\kappa = |1|$ because $A = \varnothing$, and $\varnothing$ is the unique function whose domain is empty. If $c = |0|$ and $\kappa \neq |0|$, then $A \neq \varnothing$ and

$$\begin{aligned}\mathscr{F}(A, 0) &= \{f \in \mathscr{P}(0) : f \text{ is a function with domain } A\} \\ &= \{f \in 1 : f \text{ is a function with domain } A\}.\end{aligned}$$

But 0 is the unique element of 1 and dom $0 = 0 \neq A$. Thus, $\mathscr{F}(A, 0) = \varnothing$, and $|0|^\kappa = |0|$. Clearly, $c \neq |0|$ *implies that for all cardinals κ, $c^\kappa \neq 0$.*

Observe that $|0|^{|0|}$ *is defined* (and is $|1|$), *whereas in ordinal exponentiation, we did not define* "0^0". A more crucial difference between cardinal and ordinal exponentiation will be discussed at the end of the section.

We now show that *for every cardinal number c and for every $N \geq 1$,*

$$c^{|N|} = c \cdot c \cdot \ldots \cdot c \ (N \text{ times}).$$

Let C be a set of cardinality c; then $c^{|N|} = |\mathscr{F}(N, C)|$. To each f in $\mathscr{F}(N, C)$ there corresponds the unique N-tuple, $\langle f(0), f(1), \ldots, f(N \dot{-} 1)\rangle$ in $C^{\circledN}$; to each N-tuple $\langle a_0, a_1, \ldots, a_{N \dot{-} 1}\rangle$ in $C^{\circledN}$ there corresponds the unique function f in $\mathscr{F}(N, C)$ defined by $f(K) = a_K$, $K < N$. Thus, $\mathscr{F}(N, C) \approx C^{\circledN}$; hence,

$$c^{|N|} = |\mathscr{F}(N, C)| = |C^{\circledN}| = c \cdot c \cdot \ldots \cdot c \,(N \text{ times}).$$

(In Section 6.10, when we generalize cardinal multiplication so that it is defined for arbitrarily many factors, we will extend this result.)

We now establish the usual laws of exponentiation. We first note that *for all cardinal c,*

$$c^{|1|} = c$$

because if A is a unit set, $\{a\}$, and if $|B| = c$, then

$$\mathscr{F}(A, B) = \{f \in \mathscr{P}(A \times B) : f(a) = b \text{ for some } b \text{ in } B\}.$$

Clearly,

$$|\mathscr{F}(A, B)| = |B| = c.$$

Theorem 6.6.1. *Let κ_1, κ_2, and κ_3 be any cardinal numbers.*

(1) $\kappa_1^{\kappa_2+\kappa_3} = \kappa_1^{\kappa_2} \cdot \kappa_1^{\kappa_3}$.

(2) $(\kappa_1 \cdot \kappa_2)^{\kappa_3} = \kappa_1^{\kappa_3} \cdot \kappa_2^{\kappa_3}$.

(3) $(\kappa_1^{\kappa_2})^{\kappa_3} = \kappa_1^{\kappa_2 \cdot \kappa_3}$.

PROOF: We let A, B, and C be sets for which $|C| = \kappa_1$, $|B| = \kappa_2$, and $|A| = \kappa_3$. In part (1) we will also assume that $A \cap B = \varnothing$.

(1) Let $f : B \to C$ and let $g : A \to C$. Define $H_1(\langle f, g\rangle) = f \cup g$; then $f \cup g : A \cup B \to C$. We leave as an exercise [6.6.2(a)] the verification that H_1 is a one-one map of $\mathscr{F}(B, C) \times \mathscr{F}(A, C)$ onto $\mathscr{F}(A \cup B, C)$. Thus,

$$\kappa_1^{\kappa_2+\kappa_3} = |\mathscr{F}(A \cup B, C)| = |\mathscr{F}(B, C) \times \mathscr{F}(A, C)| = \kappa_1^{\kappa_2} \cdot \kappa_1^{\kappa_3}.$$

(2) Let $f : A \to C \times B$. Define $f_1 : A \to C$ and $f_2 : A \to B$ by letting $f_1(a)$ be the first, and $f_2(a)$ the second, coordinate of $f(a)$ for each a in A. Define

$$H_2 : \mathscr{F}(A, C \times B) \to \mathscr{F}(A, C) \times \mathscr{F}(A, B)$$

by $H_2(f) = \langle f_1, f_2\rangle$. Then H_2 is one-one onto $\mathscr{F}(A, C) \times \mathscr{F}(A, B)$ [Exercise 6.6.2(b)]. Thus,

$$(\kappa_1 \cdot \kappa_2)^{\kappa_3} = |\mathscr{F}(A, C \times B)| = |\mathscr{F}(A, C) \times \mathscr{F}(A, B)| = \kappa_1^{\kappa_3} \cdot \kappa_2^{\kappa_3}.$$

(3) Let $f \in \mathscr{F}(A, \mathscr{F}(B, C))$. Then for every a in A, $f(a) : B \to C$. Define

$$H_3 : \mathscr{F}(A, \mathscr{F}(B, C)) \to \mathscr{F}(B \times A, C)$$

by $H_3(f) = h$, where $h: B \times A \to C$ and for all a in A, b in B, $h(\langle b, a\rangle) = (f(a))(b)$. Then H_3 is the required one-one function onto $\mathscr{F}(B \times A, C)$ [Exercise 6.6.2(*c*)], and

$$(\kappa_1{}^{\kappa_2})^{\kappa_3} = |\mathscr{F}(A, \mathscr{F}(B, C))| = |\mathscr{F}(B \times A, C)|$$
$$= \kappa_1{}^{\kappa_2 \cdot \kappa_3}.$$

Theorem 6.6.2. *Let κ_1, κ_2, κ_3, and κ_4 be cardinal numbers that satisfy*

(6.6.1) $$\kappa_1 \preceq \kappa_3, \qquad \kappa_2 \preceq \kappa_4,$$

and

(6.6.2) $$\textit{if } \kappa_2 + \kappa_3 = |0|, \textit{ then } \kappa_4 = |0|.$$

It follows that

$$\kappa_1{}^{\kappa_2} \preceq \kappa_3{}^{\kappa_4}.$$

PROOF: First suppose that $\kappa_3 = |0|$. If $\kappa_4 = |0|$, then, by (6.6.1), $\kappa_1 = \kappa_2 = |0|$; hence

$$\kappa_1{}^{\kappa_2} = \kappa_3{}^{\kappa_4} = |1|.$$

If $\kappa_4 \succ |0|$, then, by (6.6.1) and (6.6.2), $\kappa_1 = |0|$ and $\kappa_2 \succ |0|$; hence,

$$\kappa_1{}^{\kappa_2} = \kappa_3{}^{\kappa_4} = |0|.$$

Now suppose that $\kappa_3 \succ |0|$. Let D have cardinality κ_3 and let C have cardinality κ_4. Then there is a subset B of D of cardinality κ_1 as well as a subset A of C of cardinality κ_2. By our assumption about κ_3, $D \neq \varnothing$; let $d_0 \in D$. For each f in $\mathscr{F}(A, B)$, let $\hat{f} = f \cup \{\langle c, d_0\rangle : c \in C \sim A\}$. Then the function $G: \mathscr{F}(A, B) \to \mathscr{F}(C, D)$—defined by $G(f) = \hat{f}$—is one-one; hence

$$\kappa_1{}^{\kappa_2} = |\mathscr{F}(A, B)| \preceq |\mathscr{F}(C, D)| = \kappa_3{}^{\kappa_4}.$$

Examples[1] are known of cardinal numbers κ_1, κ_2, κ_3, and κ_4 that satisfy

$$\kappa_1 \prec \kappa_3, \kappa_2 \prec \kappa_4, \text{ and } \kappa_1{}^{\kappa_2} = \kappa_3{}^{\kappa_4}.$$

†EXAMPLE 6.6.1

(1) $|\omega|^{|1|} = |\omega|^{|2|}$; in fact, $|\omega|^{|N|} = |\omega|$ for all nonzero natural numbers N. Thus, $\kappa_2 \prec \kappa_4$ and $\kappa_1{}^{\kappa_2} = \kappa_1{}^{\kappa_4}$ can hold simultaneously.
(2) In Example 6.6.2, we will see that $|2|^{|\omega|} = |\mathscr{R}|^{|\omega|}$. Thus, $\kappa_1 \prec \kappa_3$ and $\kappa_1{}^{\kappa_2} = \kappa_3{}^{\kappa_2}$ can simultaneously hold.

[1] See Sierpiński (1958), pp. 153–154.

(3) On the other hand, Example 6.6.2 will indicate that $|\omega|^{|2|} \prec |\omega|^{|\omega|}$.
(4) Note that $|0|^{|0|} \succ |0|^{|\mathscr{R}|}$; for this reason we added condition (6.6.2) in Theorem 6.6.2.

Theorem 6.6.3. *For* $2 \leq N < \omega$,

$$|N|^{|\omega|} = |2|^{|\omega|}.$$

PROOF: By Theorem 6.6.2, $|2|^{|\omega|} \preceq |N|^{|\omega|}$.

Let $f : \omega \to N$. Define $g_f : \omega \to 2$ as follows: For each L in ω, L has a unique representation of the form $L = KN + R$ for $K \geq 0$ and $0 \leq R < N$.

Let

$$g_f(L) = \begin{cases} 1, & \text{if } f(K) \geq R; \\ 0, & \text{if } f(K) < R. \end{cases}$$

Then if $G : \mathscr{F}(\omega, N) \to \mathscr{F}(\omega, 2)$ is defined by $G(f) = g_f$, surely, G is one-one. (Verify.) Thus,

$$|N|^{|\omega|} = |\mathscr{F}(\omega, N)| \preceq |\mathscr{F}(\omega, 2)| = |2|^{|\omega|}.$$

The asserted equality follows.

Corollary. *For* $2 \leq M < \omega$ *and* $2 \leq N < \omega$,

$$|M|^{|\omega|} = |N|^{|\omega|}.$$

Theorem 6.6.4. $|\mathscr{R}| = |2|^{|\omega|}$.

PROOF: By Theorem 6.6.3, it suffices to show that $|\mathscr{R}| = |10|^{|\omega|}$. Thus, we must show that $\mathscr{R} \approx \mathscr{F}(\omega, 10)$. Recall that in Section 4.6, we let $\mathscr{T}$ be the set of $\mathscr{R}$-natural numbers < 10. (See p. 157 for our notational conventions.) We showed that $\mathscr{R} \approx \mathscr{D}$, where $\mathscr{D}$ is the set of denumerable sequences $\langle d_0, d_1, d_2, \ldots, d_k, \ldots\rangle$ that satisfy the following conditions:

(4.6.1) $d_0 = 0$

and for each $N = 1, 2, 3, \ldots,$

(4.6.2) $d_N \in \mathscr{T}$

and

(4.6.3) there is a natural number M, $M \geq N$, such that $d_M > 0$.

For each d in $\mathscr{D}$, let $d' : \omega \to 10$ be such that $\overline{\overline{d'_N}} = d_{N^+}$, $N = 0, 1, 2, \ldots$; let $\mathscr{D}' = \{d' : d \in \mathscr{D}\}$. Clearly, $\mathscr{D}' \approx \mathscr{D} \approx \mathscr{R}$. Let $\mathscr{E} = \mathscr{F}(\omega, 10) \sim \mathscr{D}'$. It suffices to show that $|\mathscr{E}| = |\omega|$ because then

$$|10|^{|\omega|} = |\mathscr{F}(\omega, 10)| = |\mathscr{D}'| + |\mathscr{E}| = |\mathscr{R}| + |\omega| = |\mathscr{R}|.$$

(The last equality is by Exercise 6.5.5.)

$\mathscr{E} = \{f \in \mathscr{F}(\omega, 10) :$ there is an M in ω such that $f(N) = 0$ for all $N \geq M\}$.

For each N in ω, let

$$f_N(M) = \begin{cases} 1, & \text{if } M = N; \\ 0, & \text{if } M \neq N. \end{cases}$$

Then $\{f_N : N < \omega\}$ is a denumerable subset of $\mathscr{E}$; hence,

$$|\omega| \precsim |\mathscr{E}|.$$

For the reverse inequality we note that $\mathscr{E}$ is the disjoint union,

$\mathscr{E} = \cup\{\mathscr{E}_K : K < \omega\}$, where $\mathscr{E}_K$ is the set of those functions f in $\mathscr{E}$ for which K is the smallest index M with the property that $f(N) = 0$ for all $N \geq M$. Clearly, $|\mathscr{E}_K| \precsim |10|^{|K|}$. Let E_K be the natural number for which $|\mathscr{E}_K| = |E_K|$; we can enumerate the elements of $\mathscr{E}_K$ as

$$\{e_{0,K}, e_{1,K}, \ldots, e_{E_K \dot{-} 1,K}\}.$$

We define $g : \mathscr{E} \to \omega$ by

$$g(e_{J,K}) = \begin{cases} 0, & \text{if } K = 0; \\ \displaystyle\sum_{L=0}^{K \dot{-} 1} 10^L + J, & \text{if } K > 0. \end{cases}$$

Then g is a one-one map of $\mathscr{E}$ into ω; thus,

$$|E| \precsim |\omega|.$$

By the Schröder–Bernstein Theorem,

$$|E| = |\omega|,$$

and our theorem follows.

Corollary. *For* $2 \leq N < \omega$, $|N|^{|\omega|} = |\mathscr{R}|$.

Finally, we extend Theorem 4.1.4, which asserted that if X is an N-element set, $N < \omega$, then $\mathscr{P}(X)$ is a 2^N-element set.

Theorem 6.6.5. *For every set* X, $|\mathscr{P}(X)| = |2|^{|X|}$.

PROOF: First, $|2|^{|X|} = |\mathscr{F}(X, 2)|$. Define $H : \mathscr{F}(X, 2) \to \mathscr{P}(X)$ as follows: If $f: X \to 2$, let $H(f) = f^{-1}(1)$. Then for each such f, $H(f)$ is the subset of X consisting of those x for which $f(x) = 1$. If f and g are distinct functions in $\mathscr{F}(X, 2)$, then for some x_0 in X, $f(x_0) \neq g(x_0)$. Suppose that $f(x_0) = 1$ and $g(x_0) = 0$. Then $x_0 \in f^{-1}(1)$, but $x_0 \notin g^{-1}(1)$. It follows that $x_0 \in H(f)$, but $x_0 \notin H(g)$. Similarly, if $f(x_0) = 0$ but $g(x_0) = 1$, $x_0 \in H(g) \sim H(f)$. Consequently, H is one-one. $\mathscr{F}(X, 2)$ is the set of characteristic functions of X; if A is any subset of X, $\chi_A : X \to 2$ and $\chi_A{}^{-1}(1) = A$. Thus, H is onto $\mathscr{P}(X)$.

Corollary. $|\mathscr{R}| = |\mathscr{P}(\omega)|$.

†EXAMPLE 6.6.2

(1) Let $N > 0$. Then $|\mathscr{R}|^{|N|} = (|2|^{|\omega|})^{|N|} = 2^{|\omega|\cdot|N|} = |2|^{|\omega|} = |\mathscr{R}|$.
(2) $|\mathscr{R}|^{|\omega|} = (|2|^{|\omega|})^{|\omega|} = |2|^{|\omega|\cdot|\omega|} = |2|^{|\omega|} = |\mathscr{R}|$.
(3) $|\omega|^{|\omega|} \preceq |\mathscr{R}|^{|\omega|} = |\mathscr{R}|$; $|\mathscr{R}| = |2|^{|\omega|} \preceq |\omega|^{|\omega|}$. Thus, $|\omega|^{|\omega|} = |\mathscr{R}|$.

Part (3) indicates that, in general, *for ordinal numbers* α *and* β,

$$|\alpha^{\beta}| \neq |\alpha|^{|\beta|}.$$

In fact, ω^{ω} is denumerable.

In the next section we will consider the relationship between $c^{|N|}$ and $|2|^{c}$ for arbitrary infinite cardinals c and finite cardinals $|N|$.

Exercises

†6.6.1. Suppose that $|A| = |A'|$ and $|B| = |B'|$. Show that $|\mathscr{F}(A, B)| = |\mathscr{F}(A', B')|$.

***6.6.2.** In the proof of Theorem 6.6.1, verify the following:

(a) H_1 is a one-one map of $\mathscr{F}(B, C) \times \mathscr{F}(A, C)$ onto $\mathscr{F}(A \cup B, C)$.

(b) H_2 is a one-one map of $\mathscr{F}(A, C \times B)$ onto $\mathscr{F}(A, C) \times \mathscr{F}(A, B)$.

(c) H_3 is a one-one map of $\mathscr{F}(A, \mathscr{F}(B, C))$ onto $\mathscr{F}(B \times A, C)$.

***6.6.3.** **(a)** In the proof of Theorem 6.6.3, verify that the function G from $\mathscr{F}(\omega, N)$ into $\mathscr{F}(\omega, 2)$, defined by $G(f) = g_f$, is one-one.

(b) Suppose that $N = 5$. Let

$$f(L) = \begin{cases} 0, & \text{if } N \neq 93; \\ 1, & \text{if } N = 93. \end{cases}$$

Let

$$F(L) = \begin{cases} 0, & \text{if } N \neq 93; \\ 2, & \text{if } N = 93. \end{cases}$$

On which numbers do g_f and g_F differ?

6.6.4. Prove, by induction on N, that for all $N \geq 1$, $|\omega|^{|N|} = |\omega|$.

6.6.5. Prove that for every cardinal c, other than $|0|$ and $|1|$, $|\omega| \prec c|^{\omega}|$.

6.6.6. In Example 6.6.2(1), prove, by induction on N, that $|\mathscr{R}| = |\mathscr{R}|^{|N|}$ for $N \geq 1$.

6.6.7. **(a)** Prove that for all $N \geq 2$, $|N|^{|\mathscr{R}|} = |\omega|^{|\mathscr{R}|} = |\mathscr{R}|^{|\mathscr{R}|}$.

(b) What does this tell us about $\mathscr{F}(\mathscr{R}, 2)$ and $\mathscr{F}(\mathscr{R}, \mathscr{R})$?

6.6.8. Let $\mathscr{F}_0(\omega, \omega)$ be the set of all functions $f : \omega \to \omega$ that are such that $f(N) = 0$ except for a finite number of N. Determine $|\mathscr{F}_0(\omega, \omega)|$.

6.7 Alephs

In this section we consider the cardinal numbers of well-orderable sets. We do not assume the Axiom of Choice; thus not every set is necessarily well-orderable. Many arithmetical results hold for cardinals of well-orderable sets—but not for cardinals in general (without the Axiom of Choice). For example, if c is an arbitrary infinite cardinal, we cannot conclude that $c + |N| = c$ for all N in ω, because we require the Axiom of Choice to obtain a denumerable subset of a set of cardinality c. But *if κ is the cardinal number of an infinite well-orderable set, then every set of cardinality κ is well-orderable and*, hence, *has a denumerable subset*. Thus, *for all N in ω, $\kappa + |N| = \kappa$.*

If c is an arbitrary cardinal number, we let "$\mathbf{Z}(c)$" denote the collection of ordinal numbers α for which $|\alpha| = c$. We call $\mathbf{Z}(c)$ the **number class of** c. Thus, $\mathbf{Z}(|N|) = \{N\}$ *for N in ω, and $\mathbf{Z}(|\omega|)$ is the second number class*, which is a set, according to Theorem 5.12.1.

Theorem 6.7.1. *For every cardinal number c, $\mathbf{Z}(c)$ is a set.*

PROOF: If $\mathbf{Z}(c) = \varnothing$, $\mathbf{Z}(c)$ is a set; otherwise, let $C \in c$, and let

$$A_C = \{x \in \{C\} \times C^{②} : x = \langle C, R\rangle \text{ and } R \text{ is a well-ordering on } C\}.$$

(We proceed as in the proof of Theorem 5.12.1.)

By the Counting Theorem, each such $\langle C, R\rangle$ in A_C is similar to a unique ordinal system, $\langle \alpha, \in\rangle$, in which (clearly) $|\alpha| = c$. Moreover, every ordinal system, $\langle \beta, \in\rangle$, in which $|\beta| = c$ corresponds to at least one such $\langle C, R\rangle$. For let $C \approx \beta$; with respect to any one-one mapping f of C onto β, we can define R_β on C by $x_1 R_\beta x_2$ iff $f(x_1) < f(x_2)$, $x_1, x_2 \in C$. Thus, R_β is the required well-ordering on C. Let s be the bimonadic formula, "$x \in A_C$ and y is the first coordinate of an ordinal system that is similar to x." By the Axiom of Substitution, there exists a set

$$S = \{c_2 : \text{for some } c_1 \text{ in } A_C, \text{ "}s\begin{pmatrix} x & y \\ c_1 & c_2 \end{pmatrix}\text{" is true}\}. \text{ Clearly, } S = \mathbf{Z}(c).$$

Observe that $\mathbf{Z}(c) \neq \varnothing$ *iff there is a well-ordered set of cardinality c*. In fact, $\mathbf{Z}(c) \neq \varnothing$ *iff every set of cardinality c is well-orderable*. For if $\alpha \in \mathbf{Z}(c)$ and $|C| = c$, then any one-one correspondence between α and C induces a well-ordering on C. Conversely, if $|C| = c$ implies C is well-ordered by R, then $\langle C, R\rangle$ is similar to an ordinal system $\langle \beta, \in\rangle$; hence, $\beta \in \mathbf{Z}(c)$.

Because $\mathbf{Z}(c)$ is a set of ordinals, it is well-ordered by $\in_{\cup\mathbf{Z}(c)}$. For each infinite cardinal number c for which $\mathbf{Z}(c) \neq \varnothing$, the smallest ordinal of $\mathbf{Z}(c)$ is called the **initial ordinal of** $\mathbf{Z}(c)$ or the **initial ordinal corresponding to** c. An ordinal number α is called an **initial ordinal** if for some infinite cardinal c, α is the initial ordinal of $\mathbf{Z}(c)$. *An infinite ordinal is an initial ordinal iff for every ordinal* β, $\beta < \alpha$ *iff* $|\beta| \prec |\alpha|$. Clearly, *an initial ordinal is a limit ordinal* (Exercise 6.7.1).

For every cardinal number c, let $\mathbf{Y}(c)$ be the collection of ordinal numbers α for which $|\alpha| \prec c$. *If c is a finite cardinal,* $|N|$, *then* $\mathbf{Y}(c) = N$. *If c is an infinite cardinal for which* $\mathbf{Z}(c) \neq \varnothing$, *then* $\mathbf{Y}(c)$ *is the initial ordinal of* $\mathbf{Z}(c)$. (Verify!) Thus, in these 2 cases $\mathbf{Y}(c)$ is a set. Finally, *if c is an infinite cardinal for which* $\mathbf{Z}(c) = \varnothing$, $\mathbf{Y}(c)$ *is also a set*—but the verification of this is more difficult and involves a modification of the proof of Theorem 6.7.1 [Exercise 6.7.2(b)].

For every cardinal c we let "$\mathbf{X}(c)$" denote "$\mathbf{Y}(c) \cup \mathbf{Z}(c)$"; thus, $\mathbf{X}(c) = \{\alpha : |\alpha| \preceq c\}$. Finally, we let "$\aleph(c)$" denote "$|X(c)|$".

†Example 6.7.1

(1) $\mathbf{Z}(\omega)$ is the second number class; ω is the initial number of $\mathbf{Z}(\omega)$.
(2) $\mathbf{Y}(\omega)$ is the set of finite ordinals (the "first number class"). Thus, $\mathbf{Y}(\omega) = \omega$.
(3) $\mathbf{X}(\omega) = \mathbf{Y}(\omega) \cup \mathbf{Z}(\omega) = \omega_1$, the first uncountable ordinal.
(4) $\aleph(\omega) = |\mathbf{X}(\omega)| = |\omega_1|$.

†Example 6.7.2

(1) $\mathbf{Z}(\omega_1)$ is the set of all ordinals that are in one-one correspondence with ω_1; ω_1 is the initial number of $\mathbf{Z}(\omega_1)$.
(2) $\mathbf{Y}(\omega_1)$ is the set of countable ordinals.
(3) $\mathbf{X}(\omega_1) = \mathbf{Y}(\omega_1) \cup \mathbf{Z}(\omega_1)$; it is the first uncountable ordinal that is not in one-one correspondence with ω_1.

Theorem 6.7.2. Hartog's Theorem. *For every cardinal number c,* $\mathbf{X}(c)$ *is an ordinal number that satisfies*

$$\aleph(c) = |\mathbf{X}(c)| \npreceq c.$$

PROOF: Let $\beta_1 \in \beta_2 \in \mathbf{X}(c)$. Then β_1 and β_2 are ordinals, $\beta_1 < \beta_2$, and $|\beta_1| \preceq |\beta_2| \preceq c$. It follows that $|\beta_1| \preceq c$ and that $\beta_1 \in \mathbf{X}(c)$. Thus, $\mathbf{X}(c)$, as a transitive set of ordinals, is itself an ordinal.

Suppose that $\aleph(c) \preceq c$. Then $\mathbf{X}(c) \in \mathbf{X}(c)$. But $\mathbf{X}(c)$ is ordinary. This contradiction yields $\aleph(c) \npreceq c$.

Corollary 1. *For every set A there is an ordinal number α for which $|\alpha| \npreceq |A|$.*

Corollary 2. *For every infinite cardinal number, c, $\mathbf{X}(c)$ is an initial number.*

PROOF: $\aleph(c)$ is an infinite cardinal because for every finite cardinal κ, $\kappa \prec c$. For any ordinal β, if $\beta < \mathbf{X}(c)$, then $\beta \in \mathbf{X}(c)$; consequently, $|\beta| \preceq c$ and $|\beta| \neq \aleph(c)$. Thus, if β is an ordinal of cardinality $\aleph(c)$, then $\mathbf{X}(c) \leq \beta$.

Corollary 3. *For every infinite cardinal number c, $\mathbf{Y}(c)$ is an initial number.*

(The proof is left as an exercise.)

Hartog's Theorem has some important consequences; in order to discuss these, we introduce the notion of an aleph.

An **aleph** is the cardinal number of an infinite well-orderable set. Alephs are generally denoted by "$\aleph$".[2] Thus, $|\omega|$ and $|\omega_1|$ are each alephs. *For every infinite cardinal c, $\aleph(c)$ is an aleph.*

If $\aleph$ is an aleph and κ is a cardinal satisfying $\kappa \prec \aleph$, then κ, too, must be an aleph. For if $|A| = \aleph$ and $|K| = \kappa$, then there exists a one-one mapping, F, of K into A. Range F, as a subset of A, is well-orderable; hence, K, too, is well-orderable.

Hartog's Theorem asserts that for every cardinal c, $\aleph(c) \npreceq c$. This can be strengthened.

Theorem 6.7.3. *For every cardinal number c and aleph $\aleph$, if $\aleph \npreceq c$, then $\aleph(c) \preceq \aleph$.*

[*$\aleph(c)$ is the smallest aleph $\aleph$ satisfying $\aleph \npreceq c$.*]

PROOF: Suppose that $\aleph \npreceq c$. Let $|A| = \aleph$ and let R be a well-ordering on A. For every section $\langle Y, \in_{\mathbf{X}(c)} \rangle$ of $\langle \mathbf{X}(c), \in \rangle$, $|Y| \preceq c$. Thus, $\langle A, R \rangle$ is not similar to any such section. It follows that either $\langle \mathbf{X}(c), \in \rangle \cong \langle A, R \rangle$ or $\langle \mathbf{X}(c), \in \rangle$ is similar to a section of $\langle A, R \rangle$. Thus,

$$\aleph(c) = |\mathbf{X}(c)| \preceq |A| = \aleph.$$

[2] $\aleph$ (aleph) is the first letter of the Hebrew alphabet.

[There may be cardinals κ that are not alephs and that satisfy $\kappa \not\preceq c$ as well as $\aleph(c) \not\preceq \kappa$.]

We now consider the relationship between alephs and initial ordinals.

Let $\aleph$ be an aleph. Then $\mathbf{Z}(\aleph)$ *is the set of ordinals of cardinality* $\aleph$; $\mathbf{Y}(\aleph)$ *is the initial ordinal of* $\mathbf{Z}(\aleph)$. Thus, $|\mathbf{Y}(\aleph)| = \aleph$.

Let α be an initial ordinal. Then $|\alpha|$ is an aleph, $\aleph$. By the above, $\alpha = \mathbf{Y}(\aleph)$.

The following theorem is easily verified.

Theorem 6.7.4. *Let $\aleph$ and $\aleph^*$ be alephs. Then*

$$\aleph \prec \aleph^* \textit{ iff } \mathbf{Y}(\aleph) < \mathbf{Y}(\aleph^*).$$

Corollary 1. *Every set A of alephs is well-ordered by the relation R—defined by $\aleph R \aleph^*$ iff $\aleph \preceq \aleph^*$, $\aleph, \aleph^* \in A$.*

PROOF: Consider the corresponding set of initial ordinals and apply the theorem.

Corollary 2. Trichotomy for Alephs. *For any two alephs, $\aleph$ and $\aleph^*$, exactly one of the following is true:*

(1) $\aleph \prec \aleph^*$.

(2) $\aleph = \aleph^*$.

(3) $\aleph^* \prec \aleph$.

Corollary 2, together with Theorems 6.4.2 (Schröder–Bernstein) and 6.7.3, yield the following corollary.

Corollary 3. *If κ is an aleph, then $\kappa \prec \aleph(\kappa)$; moreover, if $\aleph^*$ is any aleph such that $\kappa \prec \aleph^*$, then $\aleph(\kappa) \preceq \aleph^*$.*

[$\aleph(\kappa)$ *is the smallest aleph "greater than"* κ; because κ is itself an aleph, $\aleph(\kappa)$ *can be considered its "immediate successor"*.]

Note that if κ is an aleph, conceivably there may be a cardinal c that is not an aleph and that satisfies $\kappa \prec c$ and $\aleph(\kappa) \not\preceq c$.

Corollary 4. *If β is any ordinal, then* $\mathbf{Y}(\aleph(|\beta|))$ *is the smallest initial ordinal containing β.*

Let s be the strictly bimonadic formula, "x is an ordinal number and y is the smallest initial ordinal containing x—or x is not an ordinal number and $y = 0$." In Definition by Transfinite Induction (Theorem 5.4.1), let $c_0 = \omega$.

Then for every ordinal number $\alpha > 0$ there is a unique function i_α with domain α that satisfies the following:

(1) $i_\alpha(0) = \omega$.
(2) $i_\alpha(\gamma^+) = \beta$, where β is the smallest initial ordinal containing $i_\alpha(\gamma)$, if $\gamma^+ < \alpha$.
(3) $i_\alpha(\gamma) = \cup\{i_\alpha(\delta) : \delta < \gamma\}$, if γ is a limit ordinal in α.

In (2), β is the unique set for which

$$S\begin{pmatrix} x & y \\ i_\alpha(\gamma) & \beta \end{pmatrix}$$

is true. For any ordinal numbers γ, α_1, and α_2 such that $\gamma \in \alpha_1 \cap \alpha_2$,

$$i_{\alpha_1}(\gamma) = i_{\alpha_2}(\gamma).$$

For every ordinal number α, we define "ω_α" to be "$i_{\alpha^+}(\alpha)$" and we define "$\aleph_\alpha$" to be "$|\omega_\alpha|$".

Thus, *we have indexed the alephs* (*as well as the corresponding initial ordinals*) *by means of the ordinals.*

By transfinite induction, one shows that

$$\alpha < \beta$$

iff both

$$\omega_\alpha < \omega_\beta$$

and

$$\aleph_\alpha \prec \aleph_\beta.$$

For every ordinal α, $\aleph_{\alpha+1}$ *is the aleph successor of* $\aleph_\alpha$. Moreover, $\aleph_{\alpha+1}$ *is the cardinal successor of* $\aleph_\alpha$; *i.e., there is no cardinal number* κ *satisfying* $\aleph_\alpha \prec \kappa \prec \aleph_{\alpha+1}$. For we previously showed that any cardinal κ—such that $\kappa \prec \aleph_{\alpha+1}$—must be an aleph.

We note that *for all ordinals* α,

$$\omega_\alpha = \mathbf{Y}(\aleph_\alpha)$$

and that

$$\omega_{\alpha+1} = \mathbf{Y}(\aleph_{\alpha+1}) = \mathbf{X}(\aleph_\alpha).$$

Thus,

$$|\mathbf{Y}(\aleph_\alpha)| = \aleph_\alpha$$

and

$$|\mathbf{X}(\aleph_\alpha)| = \aleph(\aleph_\alpha) = \aleph_{\alpha+1}.$$

Exercises

†6.7.1. Prove that an initial ordinal is a limit ordinal.

*6.7.2. †(a) Prove that if c is an infinite cardinal and $\mathbf{Z}(c) \neq \varnothing$, then $\mathbf{Y}(c)$ is the initial ordinal of $\mathbf{Z}(c)$.

(b) Prove that $\mathbf{Y}(c)$ is a set for the case in which c is an infinite cardinal and $\mathbf{Z}(c) = \varnothing$.

(c) Prove Corollary 3 of Theorem 6.7.2; i.e., prove that for every infinite cardinal c, $\mathbf{Y}(c)$ is an initial number.

6.7.3. Prove that $\omega_1 \leq \mathbf{Y}(|\mathscr{R}|)$. (Useful for Section 6.12.)

6.7.4. Suppose that C is an infinite set that has no denumerable subset. Determine $\mathbf{Y}(|C|)$, $\mathbf{Z}(|C|)$, and $\mathbf{X}(|C|)$.

6.7.5. (a) Prove that for every natural number $N, \aleph(|N|) = |N^+|$.

(b) Prove that for every infinite cardinal c, either $\omega \in \mathbf{Y}(c)$ or (excl.) $\omega = \mathbf{Y}(c)$.

6.7.6. Let c be a cardinal number. Prove that c is an aleph iff $\mathbf{Z}(c)$ is infinite.

6.7.7. (a) Prove Theorem 6.7.4.

(b) Prove Corollary 4 of Theorem 6.7.4.

6.7.8. Prove that if A is any set of ordinals, there is a smallest ordinal β with the property that $|\alpha| \prec |\beta|$ for every α in A.

†6.7.9. Let γ, α_1, and α_2 be ordinal numbers that obey $\gamma \in \alpha_1 \cap \alpha_2$. Prove that $i_{\alpha_1}(\gamma) = i_{\alpha_2}(\gamma)$.

†6.7.10. Let α and β be ordinal numbers. Prove that $\alpha < \beta$ iff both $\omega_\alpha < \omega_\beta$ and $\aleph_\alpha \prec \aleph_\beta$.

†6.7.11. Let β be an ordinal number. Suppose that c is an infinite cardinal number and $c \neq \aleph_\alpha$ for $\alpha < \beta$. What conclusion can be drawn?

6.7.12. (a) Prove that the collection of alephs is not a set.

(b) Prove that the collection of cardinal numbers is not a set.

(c) Prove that the collection of initial ordinals is not a set.

†6.7.13. Use Definition by Induction and the Axiom of Substitution to define an ordinal γ such that $\omega_\gamma = \gamma$. (Used in Exercise 6.11.7.)

6.7.14. †(a) Prove that for every ordinal α, $|\alpha| \preceq \aleph_\alpha$.

†(b) If $\alpha = 0$ or if α is a successor ordinal, prove that $|\alpha| \prec \aleph_\alpha$.

(c) Determine limit ordinals α_1 and α_2 that satisfy

$$(1)\ |\alpha_1| \prec \aleph_{\alpha_1}; \qquad (2)\ |\alpha_2| = \aleph_{\alpha_2}.$$

*6.7.15. Prove that every initial ordinal ω_α is a power of ω.

*6.7.16. Determine the Cantor normal form of $\omega^{\omega^\omega} \cdot \omega_1$.

6.8 Aleph Arithmetic

We proceed to show that addition and multiplication of alephs are rather trivial operations; in each case the maximal aleph is preserved. Moreover, raising alephs to the power $\aleph_\beta$ is the same for all alephs $\preceq \aleph_\beta$; in fact, this amounts to determining the cardinality of $\mathscr{P}(\aleph_\beta)$.

Recall that for every denumerable ordinal α, α^+, $\alpha + \alpha$, $\alpha \cdot \alpha$, and α^α are each denumerable. We consider a classical result connecting ordinal and cardinal numbers; the present proof is due to Kopperman.

Theorem 6.8.1. *For every infinite ordinal number* α, $|\alpha|^{|2|} = |\alpha|$.

PROOF: Clearly, for every ordinal α,

$$|\alpha| = |\alpha \times 1| \preceq |\alpha \times \alpha| = |\alpha| \cdot |\alpha| = |\alpha|^{|2|}.$$

Assume there are infinite ordinals α that satisfy $|\alpha| \prec |\alpha|^{|2|}$; let α_0 be the first such infinite ordinal. Surely $\alpha_0 > \omega$.

α_0 must be an initial ordinal, for if $\beta < \alpha_0$ and $|\beta| = |\alpha_0|$, then

$$|\alpha_0|^{|2|} = |\beta|^{|2|} = |\beta| = |\alpha_0|.$$

It follows that α_0 is a limit ordinal. Consequently,

$$\begin{aligned} |\alpha_0| \prec |\alpha_0|^{|2|} &= |\alpha_0| \cdot |\alpha_0| = |\alpha_0 \times \alpha_0| \\ &= |\cup\{\beta \times \beta : \beta < \alpha_0\}| \\ &= |\cup\{(\{\beta \times \beta\} \times \{\beta\}) : \beta < \alpha_0\}| \\ &= |\sum_{\beta < \alpha_0} \beta^2| \\ &= |\cup\{\sum_{\beta < \gamma} \beta^2 : \gamma < \alpha_0\}| \end{aligned}$$

Thus, $\alpha_0 < \cup\{\sum_{\beta<\gamma} \beta^2 : \gamma < \alpha_0\}$; hence, for some infinite $\gamma < \alpha_0$, $\alpha_0 < \sum_{\beta<\gamma} \beta^2$. But $\sum_{\beta<\gamma} \beta^2 \leq \sum_{\beta<\gamma} \gamma_\beta$, where each $\gamma_\beta = \gamma^2$. Therefore,

$$\alpha_0 < \sum_{\beta<\gamma} \beta^2 \leq \sum_{\beta<\gamma} \gamma_\beta = \gamma^2 \cdot \gamma,$$

and because α_0 is an initial number,

$$|\alpha_0| \preceq |\gamma^2 \cdot \gamma| = |\gamma^2| \cdot |\gamma| = |\gamma| \prec |\alpha_0|.$$

This is a contradiction. Thus, for every infinite α, $|\alpha|^{|2|} = |\alpha|$.

Corollary. *For every infinite ordinal number* α,

$$|\alpha| = |\alpha^+| = |\alpha| \mathbin{\pmb{+}} |\alpha| = |\alpha| \cdot |\alpha|.$$

PROOF:

$$|\alpha| = |\alpha + 0| \preceq |\alpha + 1| = |\alpha| + |1| \preceq |\alpha| + |\alpha| = |\alpha| \cdot |2|$$

$$\preceq |\alpha| \cdot |\alpha| = |\alpha|^{|2|} = |\alpha|.$$

Thus,

$$|\alpha| = |\alpha + 1| = |\alpha^+| = |\alpha| + |\alpha| = |\alpha| \cdot |\alpha|.$$

Theorem 6.8.2. *Let α and β be ordinal numbers and let $\alpha \leq \beta$. Then the following are true:*

(1) $\aleph_\alpha + \aleph_\beta = \aleph_\alpha \cdot \aleph_\beta = \aleph_\beta$.

(2) *If κ is any cardinal number that is such that $\aleph_\alpha \preceq \kappa$, then $\aleph_\alpha + \kappa = \kappa$.*

(3) $\aleph_\alpha^{\aleph_\beta} = |2|^{\aleph_\beta}$.

(4) *For N in ω, if $\alpha + N \leq \beta$, then*

$$\aleph_{\alpha+N}^{\aleph_\beta} = \aleph_\alpha^{\aleph_\beta} \cdot \aleph_{\alpha+N}.$$

PROOF: We utilize the preceding corollary together with the Schröder–Bernstein Theorem.

(1) $\aleph_\beta = |0| + \aleph_\beta \preceq \aleph_\alpha + \aleph_\beta = |\omega_\alpha| + |\omega_\beta| \preceq |\omega_\beta| + |\omega_\beta| = |\omega_\beta| = \aleph_\beta$.

Also,

$$\aleph_\beta = |1| \cdot \aleph_\beta \preceq \aleph_\alpha \cdot \aleph_\beta = |\omega_\alpha| \cdot |\omega_\beta| \preceq |\omega_\beta| \cdot |\omega_\beta| = |\omega_\beta| = \aleph_\beta.$$

Thus,

$$\aleph_\alpha + \aleph_\beta = \aleph_\alpha \cdot \aleph_\beta = \aleph_\beta.$$

(2) Because $\aleph_\alpha \preceq \kappa$, there exists a cardinal c such that $\kappa = \aleph_\alpha + c$. (See the remark preceding Example 6.5.3.) Hence,

$$\aleph_\alpha + \kappa = \aleph_\alpha + (\aleph_\alpha + c) = (\aleph_\alpha + \aleph_\alpha) + c = \aleph_\alpha + c = \kappa.$$

(3) By Theorems 6.6.1, 6.6.2, and part (1) above,

$$|2|^{\aleph_\beta} \preceq \aleph_\alpha^{\aleph_\beta} \preceq (|2|^{\aleph_\alpha})^{\aleph_\beta} = |2|^{\aleph_\alpha \cdot \aleph_\beta} = |2|^{\aleph_\beta}.$$

Thus, $\aleph_\alpha^{\aleph_\beta} = |2|^{\aleph_\beta}$.

(4) Let $N \in \omega$ and let α and β be ordinals that obey $\alpha + N \leq \beta$. Then

$$\begin{aligned}
\aleph_{\alpha+N}{}^{\aleph_\beta} &= |2|^{\aleph_\beta} \\
&\preceq \aleph_\alpha{}^{\aleph_\beta} \cdot \aleph_{\alpha+N} \\
&= |2|^{\aleph_\beta} \cdot \aleph_{\alpha+N} \\
&\preceq |2|^{\aleph_\beta} \cdot \aleph_\beta \\
&\preceq |2|^{\aleph_\beta} \cdot |2|^{\aleph_\beta} \\
&= (|2|^{\aleph_\beta})^{|2|} \\
&= |2|^{\aleph_\beta \cdot |2|} \\
&= |2|^{\aleph_\beta}.
\end{aligned}$$

The inequalities must be equalities and (4) follows by induction.

EXAMPLE 6.8.1

(1) For every ordinal α, $\aleph_\alpha + |2|^{\aleph_\alpha} = |2|^{\aleph_\alpha} = \aleph_\alpha{}^{\aleph_\alpha}$.
(2) $\aleph_5{}^{\aleph_\omega} = \aleph_0{}^{\aleph_\omega} \cdot \aleph_5$.
(3) $\aleph_1{}^{\aleph_0} \preceq (|2|^{\aleph_1})^{\aleph_0} = |2|^{\aleph_1 \cdot \aleph_0} = |2|^{\aleph_1}$, or, better, $\aleph_1{}^{\aleph_0} \preceq \aleph_1{}^{\aleph_1} = |2|^{\aleph_1}$.

Whereas the sum and product of alephs are alephs, exponentiation of alephs need not necessarily yield an aleph. For example,

$$\aleph_0{}^{\aleph_0} = |2|^{\aleph_0} = |\mathscr{R}|,$$

and it is consistent with the axioms of set theory to assume that the set of real numbers is not well-orderable. Without the Axiom of Choice we do not know that $\aleph_\alpha{}^{\aleph_\beta}$ is an aleph; hence, we cannot further simplify the formula in Theorem 6.8.2(4).

Part (4) of Theorem 6.8.2 is a special case of a recursion formula for exponentiation of alephs. The case of $\beta < \alpha + N$ utilizes the Axiom of Choice and will be treated in Section 6.11.

Our next theorem is not about alephs, but its proof utilizes several theorems about alephs. Theorem 6.8.3 is due to Specker; the somewhat strengthened version presented here is essentially that of Kuratowski and Mostowski. From Theorem 6.8.1 we deduce that for every aleph $\aleph$ and natural number N, $\aleph^{|N|} = \aleph$ and thus

$$\aleph^{|N|} \prec |2|^{\aleph}. \tag{6.8.1}$$

For arbitrary infinite cardinals the inequality (6.8.1) is replaced by a weaker formula.

Theorem 6.8.3. *For every infinite cardinal number c and natural number N,*

$$|2|^c \not\preceq c^{|N|}.$$

PROOF: For every infinite cardinal c,

$$c^{|0|} = |1| \prec |2|^c;$$

thus,

$$|2|^c \not\preceq c^{|0|}.$$

Let $N > 0$, let $|C| = c$, and suppose that $|2|^c \preceq c^{|N|}$. Then there exists a one-one mapping

$$F : \mathscr{P}(C) \to C^{\textcircled{N}}.$$

We will derive a contradiction of Hartog's Theorem.

Recall that $\mathbf{X}(c)$ is the initial number of $\aleph(c)$. Thus, if $\alpha < \mathbf{X}(c)$, then $|\alpha| \preceq c$. In Exercise 3.2.15, we showed that

$$\text{for each } N \geq 0, \text{ if } K \geq 2^{N+2}, \text{ then } K^N < 2^K.$$

Let

$$A_1 = \{\text{one-one } \alpha\text{-sequences in } C \text{ for } \alpha < 2^{N+2}\},$$

$$A_2 = \{\text{one-one } \alpha\text{-sequences in } C \text{ for } 2^{N+2} \leq \alpha < \omega\},$$

$$A_3 = \{\text{one-one } \alpha\text{-sequences in } C \text{ for } \omega \leq \alpha < \mathbf{X}(c)\}.$$

Let $A = A_1 \cup A_2 \cup A_3$. We now construct a function G from A into C that is such that for each f in A, $G(f) \in C \sim \text{range} f$.

Let $f \in A_1$. Then $|\text{range} f| = |\alpha| \prec |2|^{|N+2|} \prec c$. With respect to any (fixed) 2^{N+2}-element subset of C, $\{c_1, c_2, \ldots, c_{2^{N+2}}\}$, for each f in A_1, let

$$G(f) = c_{I_f}, \qquad \text{where } I_f \text{ is the smallest subscript } I \text{ for which } C_I \notin \text{range} f.$$

Let $f \in A_2$. Then for some $K \geq 2^{N+2}$,

$$|(\text{range} f)^{\textcircled{N}}| = |K|^{|N|} \prec |2|^{|K|} = |\mathscr{P}(\text{range} f)|.$$

Because F is a one-one map of $\mathscr{P}(C)$ into $C^{\textcircled{N}}$ and range $f \subset C$, there must be a subset $Z \subseteq \text{range} f$ for which

(6.8.2) $$F(Z) \in C^{\textcircled{N}} \sim (\text{range} f)^{\textcircled{N}}.$$

Furthermore, $\mathscr{P}(\text{range } f)$ can be well-ordered as follows:

for $B_1, B_2 \subseteq \text{range } f$, let $B_1 R_f B_2$
if either $B_1 = B_2$ or if the element
b of $B_1 \triangle B_2$ for which $f^{-1}(b)$ is
smallest is in B_1.

Let Z_0 be the R_f-first subset Z of range f that satisfies (6.8.2), and let

$G(f)$ be the first coordinate of $F(Z_0)$
that is in $C \sim \text{range } f$.

For f in A_3, dom f is an infinite ordinal, α. By Theorem 6.8.1, $|\alpha| = |\alpha|^{|N|} = |\alpha^{\circledN}|$; hence, there is a one-one mapping, h, of α onto $\alpha^{\circledN}$. For $\beta < \alpha$, $h(\beta)$ is thus an N-tuple in α, which we denote by

(6.8.3) $$\text{“}\langle \beta_{(1)}, \beta_{(2)}, \ldots, \beta_{(N)} \rangle\text{”}.$$

We define a mapping H from α into $\mathscr{P}(C)$ as follows: For $\beta < \alpha$, let

$$H(\beta) = \begin{cases} F^{-1}(\langle f(\beta_{(1)}), f(\beta_{(2)}), \ldots, f(\beta_{(N)}) \rangle), \\ \qquad \text{if } \langle f(\beta_{(1)}), f(\beta_{(2)}), \ldots, f(\beta_{(N)}) \rangle \in \text{range } F; \\ \varnothing, \qquad \text{otherwise.} \end{cases}$$

Let

$$C_f = \{x \in C : \text{for some } \beta < \alpha, x = f(\beta) \notin H(\beta)\},$$

and consider $F(C_f)$ (in $C^{\circledN}$).

Suppose that $F(C_f) \in (\text{range } f)^{\circledN}$; then for $\langle \beta_1, \beta_2, \ldots, \beta_N \rangle \in \alpha^{\circledN}$,

$$F(C_f) \text{ is of the form } \langle f(\beta_1), f(\beta_2), \ldots, f(\beta_N) \rangle.$$

$\langle \beta_1, \beta_2, \ldots, \beta_N \rangle = h(\hat{\beta})$ for some unique $\hat{\beta} < \alpha$; thus, by (6.8.3),

$$\langle \beta_1, \beta_2, \ldots, \beta_N \rangle = \langle \hat{\beta}_{(1)}, \hat{\beta}_{(2)}, \ldots, \hat{\beta}_{(N)} \rangle,$$

and

$$F(C_f) = \langle f(\hat{\beta}_{(1)}), f(\hat{\beta}_{(2)}), \ldots, f(\hat{\beta}_{(N)}) \rangle.$$

Thus, $H(\hat{\beta}) = C_f$. But then

$$f(\hat{\beta}) \in H(\hat{\beta}) \qquad \text{iff } f(\hat{\beta}) \notin H(\hat{\beta}).$$

This contradiction guarantees that $F(C_f) \in C^{\circledN} \sim (\text{range } f)^{\circledN}$. Let

$G(f)$ be the first coordinate of $F(C_f)$ in $C \sim \text{range } f$.

We have defined

$$G: \{\text{one-one } \alpha\text{-sequences in } C, \text{ for all } \alpha < \mathbf{X}(c)\} \to C.$$

By Theorem 5.4.2, there exists an $\mathbf{X}(c)$-sequence, Φ in C, defined by

$$\Phi(\beta) = G(\Phi \upharpoonright \beta) \qquad \text{for } \beta < \mathbf{X}(c).$$

For $\beta_1 < \beta_2 < \mathbf{X}(c)$, $\Phi(\beta_2) = G(\Phi \upharpoonright \beta_2) \in C \sim \text{range}\,(\Phi \upharpoonright \beta_1)$; thus, $\Phi(\beta_2) \neq \Phi(\beta_1)$, and Φ is a one-one function from $\mathbf{X}(c)$ into C. Hence,

$$\aleph(c) = |\mathbf{X}(c)| \preceq |C| = c.$$

This contradicts Hartog's Theorem.

From the corollary of Theorem 6.8.1, we know that for every aleph, $\aleph$,

$$\aleph + |1| = \aleph \prec |2|^{\aleph}.$$

Corollary 1. *For every cardinal c other than $|0|$ and $|1|$,*

$$c + |1| \prec |2|^c.$$

PROOF: For finite cardinals, $|N|$, $2 \le N < \omega$, the result is immediate; for example, it can be proved by induction on N.

Let c be an infinite cardinal and let $|C| = c$. Then because $c \prec |2|^c$, there exists a one-one mapping $f : C \to \mathscr{P}(C)$ and a subset C_0 (of C) in $\mathscr{P}(C) \sim \text{range}\, f$. Hence, $f' = f \cup \{\langle C, C_0\rangle\}$ is a one-one mapping of $C \cup \{C\}$ into $\mathscr{P}(C)$. From $C \notin C$, it follows that

$$c + |1| = |C \cup \{C\}| \preceq |\mathscr{P}(c)| = |2|^c.$$

If $|2|^c = c + |1|$, then

$$|2|^c = c + |1| \preceq c + c = |2| \cdot c \preceq c \cdot c = c^{|2|},$$

in contradiction of the theorem.

Corollary 2. *For every infinite cardinal c and finite cardinal $|N|$,*

$$c + |N| \prec |2|^c.$$

Let c be an arbitrary cardinal. In Theorem 6.7.3, we characterized $\aleph(c)$ as the smallest aleph $\aleph$ satisfying $\aleph \not\preceq c$. We consider another result of this kind.

Theorem 6.8.4. *Let c be a cardinal number.*

(1) $c \prec c + \aleph(c)$.

(2) *If c is infinite, then there is no cardinal number κ satisfying $c \prec \kappa \prec c + \aleph(c)$.*

PROOF: (1) $c \preceq c + \aleph(c)$. If $c = c + \aleph(c)$, then $\aleph(c) \preceq c$, in contradiction of Hartog's theorem. Thus,

$$c \prec c + \aleph(c)).$$

(2) Let c be infinite. Suppose that κ is a cardinal satisfying $c \prec \kappa \prec c + \aleph(c)$. Then if $|K| = \kappa$, $|C| = c$, and $C \cap \mathbf{X}(c) = \varnothing$, K is in one-one correspondence with a subset of $C \cup \mathbf{X}(c)$. Thus, $\kappa = c_1 + \aleph$, where $c_1 \preceq c$ and $\aleph \preceq \aleph(c)$. Furthermore,

$$c + \aleph \preceq \kappa + \aleph = \kappa$$

because $\aleph \preceq \kappa$. Thus, $\aleph \neq \aleph(c)$ because $\kappa \prec c + \aleph(c)$, and therefore $\aleph \prec \aleph(c)$. By Theorem 6.7.3, $\aleph \preceq c$. Consequently,

$$\kappa = c_1 + \aleph \preceq c + \aleph = c,$$

contradicting $c \prec \kappa$.

We note that part (2) does *not* say that for every cardinal κ, $c \prec \kappa$ implies $c + \aleph(c) \preceq \kappa$; for it is conceivable that κ is incomparable with $c + \aleph(c)$.

Theorem 6.8.5. *Let c be an infinite cardinal number. Then a sufficient condition for c to be an aleph is that*

$$c + \aleph(c) = c \cdot \aleph(c).$$

PROOF: Let $|C| = c$. Because $c + \aleph(c) = c \cdot \aleph(c)$, there is a partition $\{A, B\}$ of $C \times \mathbf{X}(c)$ for which $|A| = c$ and $|B| = \aleph(c)$. Thus, B is well-orderable; let R be a fixed well-ordering on B.

For each a in A, $|\{a\} \times B| = |B| = \aleph(c)$. By Hartog's theorem, $\aleph(c) \npreceq c$; thus, $\{a\} \times B \nsubseteq A$. In other words, for each a in A, $B \cap (\{a\} \times B) \neq \varnothing$. Let $F : A \to B$ be such that $F(a)$ is the R-first element of $B \cap (\{a\} \times B)$. Clearly, F is one-one because if a_1 and a_2 are distinct elements of A, then for b_1, b_2 in B,

$$F(a_1) = \langle a_1, b_1 \rangle \neq \langle a_2, b_2 \rangle = F(a_2).$$

Thus, A is in one-one correspondence with a subset of B; hence, A is well-orderable, and c $(= |A|)$ is an aleph.

Note, that, by Theorem 6.8.2, this condition on c is also necessary for it to be an aleph.

Exercises

6.8.1. In the proof of Theorem 6.8.1, show that for a limit ordinal α_0, the following are true:

(a)
$$\alpha_0 \times \alpha_0 = \cup \{\beta \times \beta : \beta < \alpha_0\}.$$

(b)
$$|\cup\{(\{\beta \times \beta\} \times \{\beta\}) : \beta < \alpha_0\}| = |\sum_{\beta < \alpha_0} \beta^2|.$$

(c) Does (a) or (b) hold for successor ordinals α_0?

6.8.2. What is wrong with the following "proof" of Theorem 6.8.2(4)?
Let $\alpha + N \leq \beta$. Then $\aleph_{\alpha+N} \preceq \aleph_\beta \prec |2|^{\aleph_\beta}$.
Thus,

$$\aleph_\alpha^{\aleph_\beta} \cdot \aleph_{\alpha+N} = |2|^{\aleph_\beta} \cdot \aleph_{\alpha+N} = |2|^{\aleph_\beta}$$
$$= \aleph_{\alpha+N}{}^{\aleph_\beta}.$$

†6.8.3. Prove that for all $N \geq 1$ and all ordinal numbers α,

$$\aleph_\alpha^{|N|} = \aleph_\alpha.$$

6.8.4. Determine a cardinal number c satisfying each.
(a) $c \prec c^{\aleph_0}$. **(b)** $c = c^{\aleph_0}$.

6.8.5. Prove that for every ordinal number α and every cardinal $\kappa \prec \aleph_\alpha, \aleph_\alpha - \kappa = \aleph_\alpha$.

†6.8.6. Prove that if $\aleph$ and $\aleph'$ are alephs and $\aleph^{|2|} = \aleph'^{|2|}$, then $\aleph = \aleph'$.

★6.8.7. Prove that for every aleph, $\aleph$, and for all cardinal numbers c and κ, $\aleph \preceq c \cdot \kappa$ implies either $\aleph \preceq c$ or $\aleph \preceq \kappa$. (Used in Theorem 6.9.3.)

6.8.8. In the proof of Theorem 6.8.3, verify that R_f is a well-ordering of $\mathscr{P}$ (range f), f in A_2.

6.8.9. Simplify the proof of Theorem 6.8.3 for cardinals c satisfying $\aleph_0 \prec c$.

6.8.10. Find all cardinals c satisfying each.
(a) $|2|^c \not\preceq c^{|2|}$. **(b)** $|2|^c \not\preceq c^{|3|}$.

★6.8.11. **†(a)** Show that for every cardinal c, $c \preceq c + |1|$.
(b) Show that if $c = \kappa^{\aleph_0}$ for some cardinal $\kappa \succeq |2|$, then $c + |1| = c$.
(c) State some other conditions on c that are sufficient for the equality $c = c + |1|$.

6.8.12. Let $\aleph$ be an aleph and let κ be an arbitrary cardinal. What can be said about $\aleph + \kappa$?

†6.8.13. Prove that for every cardinal c,

$$c \prec c \cdot \aleph(c).$$

6.9 Cardinal Numbers and the Axiom of Choice

It is clear that *were we to assume the Axiom of Choice among our axioms, then the arithmetic rules established for alephs would be true for arbitrary infinite cardinal numbers*. For the Axiom of Choice is equivalent to the Well-ordering Theorem, which asserts that every set is well-orderable. Thus, *every infinite cardinal would be an aleph*. The following theorem is easily established.

Theorem 6.9.1. (*A.C.*)

(1) *Every set A of cardinal numbers is well-ordered by the relation R—defined by* $c_1 R c_2$ *iff* $c_1 \preceq c_2, c_1, c_2 \in A$.

(2) *For any two cardinal numbers,* c_1 *and* c_2, *exactly one of the following is true:*

$$\text{(a) } c_1 \prec c_2; \quad \text{(b) } c_1 = c_2; \quad \text{(c) } c_2 \prec c_1.$$

(3) *For every set A,* $|A| \prec \aleph(|A|)$.

(4) *Every cardinal number c has an "immediate successor", i.e., there is a cardinal number c′ such that* $c \prec c'$ *and such that if* κ *is a cardinal number satisfying* $c \prec \kappa$, *then* $c' \prec \kappa$.

(5) *For every infinite cardinal c,*

$$c = c + |1| = |2| \cdot c = c^{|2|}.$$

(6) *If c and* κ *are infinite cardinals and* $c \preceq \kappa$, *then*

$$c + \kappa = c \cdot \kappa = \kappa$$

and

$$c^{\kappa} = |2|^{\kappa}.$$

(7) *For cardinals numbers c and* κ, $c^{|2|} = \kappa^{|2|}$ *implies* $c = \kappa$.

(8) *If* c_1, c_2, c_3, *and* c_4 *are any cardinal numbers that satisfy*

$$c_1 \prec c_3 \quad \textit{and} \quad c_2 \prec c_4,$$

then

$$c_1 + c_2 \prec c_3 + c_4$$

and

$$c_1 \cdot c_2 \prec c_3 \cdot c_4.$$

(9) *If* c_1, c_2, *and* c_3 *are any cardinal numbers such that*

$$c_1 + c_3 \prec c_2 + c_3 \quad \textit{or} \quad c_1 \cdot c_3 \prec c_2 \cdot c_3,$$

then

$$c_1 \prec c_2.$$

(10) *For every aleph,* $\aleph$, $|2|^{\aleph}$ *is also an aleph.*

PROOF: Because the Axiom of Choice implies that every infinite cardinal is an aleph, part (10) is immediate and parts (1), (2), (4), (5), and (6) follow from theorems on alephs together with well-known facts about finite cardinals. Part (3) follows from (2) together with Hartog's Theorem; (7) follows from (2) together with the observation that

$$c \prec \kappa \text{ or } \kappa \prec c$$

would imply

$$c^{|2|} \prec \kappa^{|2|} \text{ or } \kappa^{|2|} \prec c^{|2|}$$

respectively.

In (8), if all four cardinals are infinite, then

$$c_1 + c_2 = c_1 \cdot c_2 = \max(c_1, c_2) \prec \max(c_3, c_4) = c_3 + c_4 = c_3 \cdot c_4.$$

The other cases are easily established using (5), (6), and the comparable rules for natural number arithmetic.

Part (9) is left as an exercise.

The rather amazing fact is that each of the sentences (1)–(10) implies the Axiom of Choice. Thus, each is an equivalent of the Axiom of Choice. We proceed to demonstrate several of these equivalencies; the others will be left as exercises. In some cases, even weaker statements imply the Axiom of Choice.

Most of the equivalents discussed in this section were discovered by Tarski. Theorem 6.9.5 is due to H. Rubin; the equivalent of Exercise 6.9.4 is due to Sierpiński. The most comprehensive collection of these equivalents is found in Rubin and Rubin.

Theorem 6.9.2. *Each of the sentences* (1), (2), *and* (3) *of Theorem* 6.9.1. *is an equivalent of the Axiom of Choice.*

PROOF: Figure 6.9.1 indicates what must be proved.

(*W.O.*) → (*Choice*): See Section 4.8.

(*Choice*) → (*1*): This is Theorem 6.9.1(1).

(*1*) → (*2*): Let c_1 and c_2 be any cardinals. Then by (1), $\{c_1, c_2\}$ is linearly ordered by $\preceq$; thus, $c_1 \preceq c_2$ or $c_2 \preceq c_1$, and at least one of the relationships—$c_1 \prec c_2$, $c_1 = c_2$, $c_2 \prec c_1$—holds. Also, $c_1 \preceq c_2$ and $c_2 \preceq c_1$ together imply

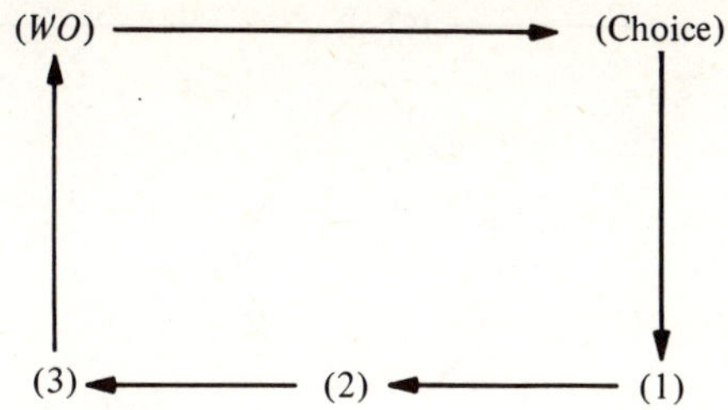

Fig. 6.9.1

$c_1 = c_2$; this together with the definition of "$\prec$" indicate that at most one of the above relationships holds.

(2) → *(3)*: Suppose Trichotomy holds for cardinals. Then for any set A, $|A| \prec \aleph(|A|)$ or (excl.) $\aleph(|A|) \preceq |A|$. But by Hartog's Theorem, $\aleph(|A|) \npreceq |A|$; thus, $A \prec \aleph(|A|)$.

(3) → *(W.O.)*: Suppose that for every set A, $|A| \prec \aleph(|A|)$. Then A is in one-one correspondence with a subset of the well-ordered set $\mathbf{X}(|A|)$. Therefore, A is itself well-orderable.

Theorem 6.9.3. *Let* (5′) *and* (6′) *be the sentences:*

(5′) *For every infinite cardinal number c,*

$$c = c^{|2|}.$$

(6′) *For all infinite cardinal numbers c and κ,*

$$c \cdot \kappa = \max(c, \kappa).$$

Then each of the sentences (4), (5), (5′), (6), (6′), *and* (7) *is an equivalent of the Axiom of Choice.*

PROOF: Figure 6.9.2 indicates what must be proved.

(W.O.) → *(Choice)*.

(Choice) → *(4)*, *(5)*, *and (6)*: This follows by Theorem 6.9.1(4), (5), and (6).

(5) → *(5′) and (6)* → *(6′)*: These are trivial.

(6′) → *(5′)*: For any infinite cardinal, c, $c = \max(c, c)$; thus,

$$c^{|2|} = c \cdot c = c.$$

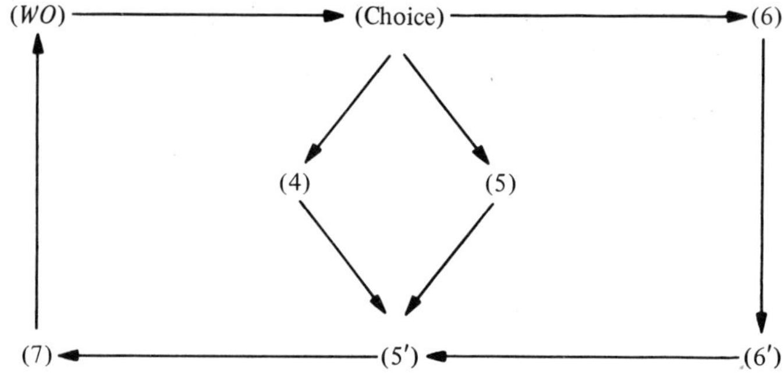

Fig. 6.9.2

(4) → *(5′)*: Let c be an infinite cardinal. Then by (4), there is a cardinal c' such that (a) $c \prec c'$ and (b) if κ is any cardinal such that $c \prec \kappa$, then $c' \preceq \kappa$. By Theorem 6.8.4, $c' = c + \aleph(c)$. Now if $c \neq c^{|2|}$, then $c \prec c^{|2|}$, because

$$c = c \cdot |1| \preceq c \cdot c = c^{|2|}.$$

Therefore

$$\aleph(c) \preceq c + \aleph(c) \preceq c^{|2|} = c \cdot c.$$

But by Exercise 6.8.7, this implies $\aleph(c) \preceq c$, contradicting Hartog's Theorem.

(5′) → *(7)*: Suppose that c and κ are any cardinal numbers and $c^{|2|} = \kappa^{|2|}$. Clearly, c and κ are both finite or both infinite. If c is finite, then the corresponding theorem of arithmetic indicates that $c = \kappa$. If c is infinite, then by (5′),

$$c = c^{|2|} = \kappa^{|2|} = \kappa.$$

(7) → *(W.O.)*: Let c be an arbitrary infinite cardinal and consider the following cardinal equalities:

$$\begin{aligned}
(c^{\aleph_0} + \aleph(c^{\aleph_0}))^{|2|} &= (c^{\aleph_0})^{|2|} + |2| \cdot c^{\aleph_0} \cdot \aleph(c^{\aleph_0}) + (\aleph(c^{\aleph_0}))^{|2|} \\
&= c^{\aleph_0 \, |2|} + c^{\aleph_0} \cdot (|2| \cdot \aleph(c^{\aleph_0})) + \aleph(c^{\aleph_0}) \\
&= c^{\aleph_0} + (c^{\aleph_0} \cdot \aleph(c^{\aleph_0}) + |1| \cdot \aleph(c^{\aleph_0})) \\
&= c^{\aleph_0} + (c^{\aleph_0} + |1|) \cdot \aleph(c^{\aleph_0}) \\
&= c^{\aleph_0} \cdot |1| + c^{\aleph_0} \cdot \aleph(c^{\aleph_0}) \\
&= c^{\aleph_0} \cdot (|1| + \aleph(c^{\aleph_0})) \\
&= c^{\aleph_0 \cdot |2|} \cdot \aleph(c^{\aleph_0}) \\
&= (c^{\aleph_0})^{|2|} \cdot (\aleph(c^{\aleph_0}))^{|2|} \\
&= (c^{\aleph_0} \cdot \aleph(c^{\aleph_0}))^{|2|}.
\end{aligned}$$

[The transition from line 4 to line 5 depends upon Exercise 6.8.11(b).]

By (7), $c^{\aleph_0} + \aleph(c^{\aleph_0}) = c^{\aleph_0} \cdot \aleph(c^{\aleph_0})$, and, by Theorem 6.8.5, $c^{\aleph_0}$ is an aleph. Clearly, $c \preceq c^{\aleph_0}$; thus, c, which is an arbitrary infinite cardinal, is an aleph. It follows that every infinite set—hence every set— is well-orderable.

Theorem 6.9.4. *Let* (8′), (8″), (9′), and (9″) *be the sentences:*

(8′) *If* c_1, c_2, c_3, *and* c_4 *are any cardinal numbers that satisfy* $c_1 \prec c_3$ *and* $c_2 \prec c_4$, *then* $c_1 + c_2 \prec c_3 + c_4$.

(8″) *If* c_1, c_2, c_3, *and* c_4 *are any cardinal numbers that satisfy* $c_1 \prec c_3$ *and* $c_2 \prec c_4$, *then* $c_1 \cdot c_2 \prec c_3 \cdot c_4$.

(9′) *If* c_1, c_2, *and* c_3 *are any cardinal numbers such that* $c_1 + c_3 \prec c_2 + c_3$, *then* $c_1 \prec c_2$.

(9″) *If* c_1, c_2, *and* c_3 *are any cardinal numbers such that* $c_1 \cdot c_3 \prec c_2 \cdot c_3$, *then* $c_1 \prec c_2$.

Each of the sentences (8), (8′), (8″), (9), (9′), and (9″) *are equivalents of the Axiom of Choice.*

PROOF: The diagrams of Figure 6.9.3 indicate what must be proved; we verify only the first of these, and leave the others as exercises.

(W.O.) → *(Choice).*

(Choice) → *(8)*: This follows by Theorem 6.9.1(8).

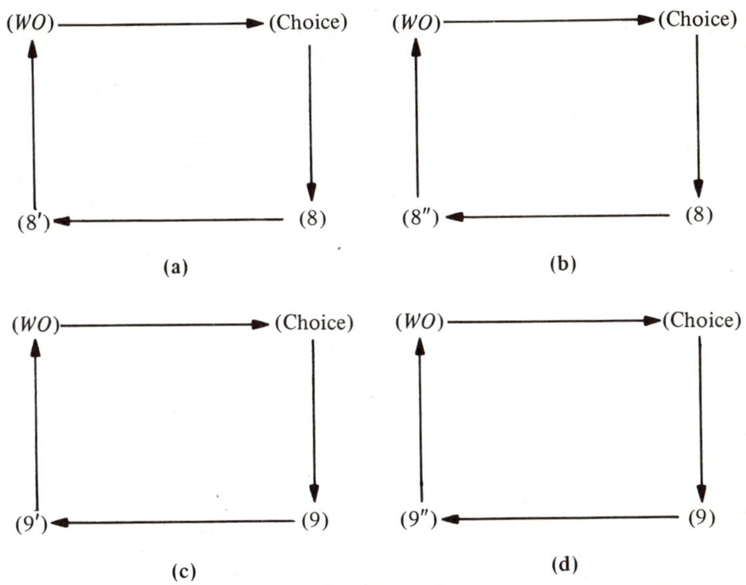

Fig. 6.9.3

(8) → (8′): This is trivial.

(8′) → (*W.O.*): Let c be any infinite cardinal. Then

$$c \cdot \aleph_0 \prec c \cdot \aleph_0 + \aleph(c \cdot \aleph_0),$$

by Theorem 6.8.4(1), and

$$\aleph(c \cdot \aleph_0) \preceq c \cdot \aleph_0 + \aleph(c \cdot \aleph_0).$$

Suppose that $\aleph(c \cdot \aleph_0) \prec c \cdot \aleph_0 + \aleph(c \cdot \aleph_0)$. Then, by (8′),

$$\begin{aligned} c \cdot \aleph_0 + \aleph(c \cdot \aleph_0) &\prec |2| \cdot (c \cdot \aleph_0 + \aleph(c \cdot \aleph_0)) \\ &= c \cdot (|2| \cdot \aleph_0) + |2| \cdot \aleph(c \cdot \aleph_0) \\ &= c \cdot \aleph_0 + \aleph(c \cdot \aleph_0). \end{aligned}$$

This is a contradiction. Thus,

$$\aleph(c \cdot \aleph_0) = c \cdot \aleph_0 + \aleph(c \cdot \aleph_0) \qquad \text{and} \qquad c \preceq c \cdot \aleph_0 \preceq \aleph(c \cdot \aleph_0).$$

Consequently, every set of cardinality c is in one-one correspondence with a subset of a well-orderable set. Every infinite cardinal is an aleph, and every set is well-orderable.

Theorem 6.9.5. *Let* (10′) *be the sentence*:

(10′) *For every set* X, *if* X *is well-orderable, then* $\mathscr{P}(X)$ *is well-orderable. Then* (10′) *is equivalent to the Axiom of Choice.*

PROOF: It suffices to show that (10′) is equivalent to the Well-ordering Theorem. Clearly, the Well-ordering Theorem implies that for every set X, $\mathscr{P}(X)$ is well-orderable.

Suppose that (10′) is true and let A be an arbitrary set. By the Axiom of Rank, A has rank α for some (unique) ordinal α. Thus, $A \in \mathbf{R}_{\alpha+1}$—or equivalently, $A \subseteq \mathbf{R}_\alpha$. The ordinal $\mathbf{X}(|\mathbf{R}_{\alpha+1}|)$ is well-orderable; hence, so is $\mathscr{P}(\mathbf{X}(|\mathbf{R}_{\alpha+1}|))$. Let W^* be a fixed well-ordering on $\mathscr{P}(\mathbf{X}(|\mathbf{R}_{\alpha+1}|))$. Note that A is a set of sets all of which are of rank $<\alpha$. We proceed to define a specific well-ordering, $\hat{W}_\beta$, on each R_β, $\beta \leq \alpha$.

Suppose that for some $\beta < \alpha$, W_β is a well-ordering on $\mathbf{R}_\beta$. Then $\langle \mathbf{R}_\beta, W_\beta \rangle$ is similar to a unique ordinal system, $\langle \gamma, \in \rangle$; let f_β be the similarity map. Clearly,

$$\mathbf{R}_\beta| \preceq |\mathbf{R}_\alpha| \prec |\mathscr{P}(\mathbf{R}_\alpha)| = |\mathbf{R}_{\alpha+1}|;$$

thus, $\gamma < \mathbf{X}(|\mathbf{R}_{\alpha+1}|)$. Define $g_\beta : \mathbf{R}_{\beta+1} \to \mathscr{P}(\mathbf{X}(|\mathbf{R}_{\alpha+1}|))$ by

$$g_\beta(B) = f_\beta''(B), \qquad B \subseteq \mathbf{R}_\beta;$$

because f_β is one-one, so is g_β. (Verify.)

We define the relation W_{β^+} on $\mathbf{R}_{\beta^+}$ (in terms of W_β) as follows: For x_1, x_2 in $\mathbf{R}_{\beta^+}$, let $x_1 W_{\beta^+} x_2$ if

(1) $x_1, x_2 \in \mathbf{R}_\beta$ and $x_1 W_\beta x_2$,
(2) $x_1 \in \mathbf{R}_\beta$, $x_2 \in \mathbf{R}_{\beta^+} \sim \mathbf{R}_\beta$, or
(3) $x_1, x_2 \in \mathbf{R}_{\beta^+} \sim \mathbf{R}_\beta$ and $g_\beta(x_1) W^* g_\beta(x_2)$.

Then W_{β^+} well-orders $\mathbf{R}_{\beta^+}$. (Verify.) Let s be the formula,

"$x = \langle \mathbf{R}_\beta, W_\beta \rangle$ for some $\beta \le \alpha$ and some well-ordering W_β on $\mathbf{R}_\beta$ and $y = \langle \mathbf{R}_{\beta^+}, W_{\beta^+} \rangle$—or else x is not of the above form and $y = 0$."

Then s is strictly bimonadic. Let $c_0 = \langle 0, 0 \rangle = \langle \mathbf{R}_0, \varnothing \rangle$. According to Definition by Transfinite Induction (Theorem 5.4.1), there is a unique α^+-sequence F that satisfies the following conditions:

(1) $F(0) = \langle \mathbf{R}_0, \varnothing \rangle$.
(2) For every successor ordinal $\beta^+ \le \alpha$, if $F(\beta) = \langle \mathbf{R}_\beta, \hat{W}_\beta \rangle$, then $F(\beta^+) = \langle \mathbf{R}_{\beta^+}, \hat{W}_{\beta^+} \rangle$, where $\hat{W}_{\beta^+}$ is the well-ordering corresponding to $\hat{W}_\beta$, as previously defined.
(3) For every limit ordinal $\delta \le \alpha$, $F(\delta) = \langle \mathbf{R}_\delta, \hat{W}_\delta \rangle$, where

$$\hat{W}_\delta = \cup\{\hat{W}_\beta : \beta < \delta\}.$$

It is easily verified that in case (3), $\hat{W}_\delta$ well-orders $\mathbf{R}_\delta$.

Clearly, $\hat{W}_\alpha$ well-orders $\mathbf{R}_\alpha$; hence, A, as a subset of a well-orderable set, is well-orderable.

Corollary. (10) *is equivalent to the Axiom of Choice.*

We note that the Axiom of Choice implies that for every infinite cardinal c, $c = |2| \cdot c$. However, the sentence "For every infinite cardinal number $c, c = |2| \cdot c$" does not imply the Axiom of Choice.[3] It is not known whether the consequent of the Axiom of Choice, "For every infinite cardinal c, $c^{|2|} = c^{|3|}$" is an equivalent.

We consider a theorem on cardinal exponentiation that utilizes the Axiom of Choice.

[3] See (Sageev).

For any set X and any cardinal $\kappa \preceq |X|$, let "$\mathscr{P}_\kappa(X)$" denote the set of subsets, of cardinality κ, of X. If X and Y are any two sets of cardinality c and if $\kappa \preceq c$, then

(6.9.1) $$|\mathscr{P}_\kappa(X)| = |\mathscr{P}_\kappa(Y)|;$$

to see this, let g be any one-one mapping of X onto Y. Then g induces a mapping h of $\mathscr{P}_\kappa(X)$ into $\mathscr{P}_\kappa(Y)$, defined by

$$h(A) = g''(A), \qquad A \in \mathscr{P}_\kappa(X).$$

Clearly, h is one-one onto $\mathscr{P}_\kappa(Y)$; hence, (6.9.1) follows.

If X is an N-element set and $\kappa = |K|$ for $K \leq N < \omega$, then $|\mathscr{P}_\kappa(X)| = \left|\binom{N}{K}\right|$. In general, $\left|\binom{N}{K}\right| \neq |N|^{|K|}$.

Theorem 6.9.6. (*A.C.*) *If c is any aleph—hence any infinite cardinal number, then for every cardinal $\kappa \preceq c$ and every set X of cardinality c,*

$$c^\kappa = |\mathscr{P}_\kappa(X)|.$$

PROOF: Let K be a fixed set of cardinality κ; we must establish a one-one correspondence between $\mathscr{F}(K, X)$ and $\mathscr{P}_\kappa(X)$.

Let $f \in \mathscr{F}(K, X)$; then f is a subset, of cardinality κ, of $K \times X$. Hence, $\mathscr{F}(K, X) \subseteq \mathscr{P}_\kappa(K \times X)$. Also,

$$|K \times X| = \kappa \cdot c = c;$$

thus, by (6.9.1),

$$c^\kappa = |\mathscr{F}(K, X)| \preceq |\mathscr{P}_\kappa(K \times X)| = |\mathscr{P}_\kappa(X)|.$$

Suppose that $A \in \mathscr{P}_K(X)$. Then $F_A = \{f \in \mathscr{F}(K, X) : f \text{ is one-one onto } A\} \neq \varnothing$. Use the Axiom of Choice to select a function f_A from each element of

$$\{F_A : A \in \mathscr{P}_\kappa(X)\}.$$

In fact, $\{F_A : A \in \mathscr{P}_\kappa(X)\}$ is pairwise disjoint because if $A, B\,(\neq A) \in \mathscr{P}_\kappa(X)$, then a function f cannot be a one-one mapping of K onto both A and B. Thus,

$$|\mathscr{P}_\kappa(X)| \preceq |\mathscr{F}(K, X)| = c^\kappa.$$

The asserted equality follows.

†EXAMPLE 6.9.1. (A.C.) The set A of all denumerable subsets of $\mathscr{R}$ is of cardinality

$$|\mathscr{R}|^{\aleph_0} = (|2|^{\aleph_0})^{\aleph_0} = |2|^{\aleph_0 \cdot \aleph_0} = |2|^{\aleph_0} = |\mathscr{R}|.$$

Thus, $A \approx R$. (The Axiom of Choice is not necessary for this result.)

Exercises

6.9.1. Prove part (9) of Theorem 6.9.1.

6.9.2. Prove that (8″) implies the Well-ordering Theorem.

6.9.3. **(a)** Prove that (9′) implies the Well-ordering Theorem.

(b) Prove that (9″) implies the Well-ordering Theorem.

★6.9.4. Prove that the sentence, "For all cardinal numbers c and κ with $c \prec \kappa$, $\kappa - c$ exists." is an equivalent of the Axiom of Choice. [The sentence, "For all cardinal numbers c and κ where κ is an infinite cardinal and $|0|c \prec \kappa$, κ/c exists." is also an equivalent of the Axiom of Choice. The proof of this is somewhat more difficult, and can be found in (Rubin and Rubin, pp. 60–61).]

★6.9.5. Prove that each of the following is equivalent to the Axiom of Choice:

(a) "For every infinite cardinal number κ, whenever c_1 and c_2 are cardinals such that $c_1 + c_2 = \kappa$, either $c_1 \not\prec \kappa$ or $c_2 \not\prec \kappa$."

(b) "For every pair of infinite cardinals $\{c, \kappa\}$, either $c + \kappa = c$ or $c + \kappa = \kappa$."

(c) "For every pair of infinite cardinals $\{c, \kappa\}$, either $c \cdot \kappa = c$ or $c \cdot \kappa = \kappa$."

6.9.6. Prove that the Axiom of Choice implies that for every infinite cardinal c, $c^{|2|} = c^{|3|}$.

6.9.7. In the proof of Theorem 6.9.5, verify the following for each $\beta < \alpha$, and for each limit ordinal $\delta \leq \alpha$:

(a) g_β is one-one.

(b) $\hat{W}_{\beta+}$ well-orders $\mathbf{R}_{\beta+}$.

(c) $\hat{W}_\delta$ well-orders $\mathbf{R}_\delta$.

6.9.8. What is wrong with the following "proof" of (the second half of) Theorem 6.9.5? First, $\mathbf{R}_0$ $(= \varnothing)$ is well-orderable. Suppose that for all $\beta < \alpha$, $\mathbf{R}_\beta$ is well-orderable. If $\alpha = \gamma^+$, then by (10′), $\mathbf{R}_\alpha$ is well-orderable. If α is a limit ordinal, then for each $\beta < \alpha$, let W_β well-order $\mathbf{R}_\beta$. Define W_α on $\mathbf{R}_\alpha$ as follows: For $x, y \in \mathbf{R}_\alpha$, let $xW_\alpha y$ if (1) $\rho(x) < \rho(y)$ or (2) $\rho(x) = \rho(y)$ and $xW_{\beta+}y$. Thus, every $\mathbf{R}_\alpha$ is well-orderable; by the Axiom of Rank, every set is well-orderable.

6.9.9. Prove the result of Example 6.9.1 without utilizing the Axiom of Choice.

6.10 Transfinite Addition and Multiplication of Cardinal Numbers

It will be convenient to call an ordinal α an **initial number**[4] if α is either an initial ordinal or a natural number. Thus, *if we assume the Axiom of Choice, then every cardinal number c corresponds to a unique initial number. Moreover, this initial number is of cardinality c.* If the cardinal c corresponds to the initial number α, we will say that α is the **initial number of** c. In other words, α *is the*

[4] In set theory with the Axiom of Choice, cardinal numbers are usually *defined* to be (what we call) initial numbers.

initial number of c if and only if α is the smallest ordinal of cardinality c. Thus, N is the initial number of the cardinal $|N|$. and ω_α is the initial number of $\aleph_\alpha$.

We defined the sum of two cardinals, κ_1 and κ_2, in terms of representative disjoint sets A and B whose cardinalities are κ_1 and κ_2, respectively. Thus, $\kappa_1 + \kappa_2 = |A \cup B|$. In the case of infinitely many cardinals, we require a choice function in order to pick such representative sets, in general. We will therefore assume the Axiom of Choice among our axioms (throughout this section). The infinite cardinals we discuss are therefore alephs.

A second difficulty we encounter in defining infinite sums of cardinals is the fact that some of the cardinals to be added may be repeated—perhaps even infinitely many times. The Axiom of Choice implies that every set is in one-one correspondence with a unique initial number. We can utilize ordinal sums in defining transfinite cardinal sums.

Let α be any ordinal number. By an α-**cardinal-sequence** we mean a function, κ, whose domain is α and whose range is a set of cardinals.

Let $[\kappa_\gamma]_{\gamma<\alpha}$ be an α-cardinal-sequence. Let Γ_γ be the initial number of κ_γ, $\gamma < \alpha$. We associate the α-ordinal-sequence $[\Gamma_\gamma]_{\gamma<\alpha}$ with $[\kappa_\gamma]_{\gamma<\alpha}$, and define

$$\text{“}\sum_{\gamma<\alpha} \kappa_\gamma\text{” to be “}\Big|\sum_{\gamma<\alpha} \Gamma_\gamma\Big|\text{”}.$$

According to Theorem 5.9.5,

$$\sum_{\gamma<\alpha} \kappa_\gamma = |\cup\{\Gamma_\gamma \times \{\gamma\} : \gamma < \alpha\}|.$$

We use this fact to prove the following **commutativity law**.

Theorem 6.10.1. (*A.C.*) *Let α and β be ordinal numbers of the same cardinality; let f be any one-one mapping of α onto β. Let $[c_\gamma]_{\gamma<\alpha}$ and $[\kappa_\delta]_{\delta<\beta}$ be cardinal-sequences for which $c_\gamma = \kappa_{f(\gamma)}, \gamma < \alpha$. Then*

$$\sum_{\gamma<\alpha} c_\gamma = \sum_{\delta<\beta} \kappa_\delta.$$

PROOF: For $\gamma < \alpha$ and for $\delta < \beta$, let Γ_γ and Δ_δ be the initial numbers of c_γ and κ_δ, respectively. By the above remark,

$$\sum_{\gamma<\alpha} c_\gamma = |\cup\{\Gamma_\gamma \times \{\gamma\} : \gamma < \alpha\}|.$$

The one-one mapping f of α onto β induces a one-one mapping g of

$$\cup\{\Gamma_\gamma \times \{\gamma\} : \gamma < \alpha\} \text{ onto } \cup\{\Delta_\delta \times \{\delta\} : \delta < \beta\},$$

defined by $g(\langle\Gamma_\gamma, \gamma\rangle) = \langle\Delta_{f(\gamma)}, f(\gamma)\rangle, \gamma < \alpha$. Thus,

$$\begin{aligned}\sum_{\gamma<\alpha} c_\gamma &= |\cup\{\Gamma_\gamma \times \{\gamma\} : \gamma < \alpha\}| \\ &= |\cup\{\Delta_\delta \times \{\delta\} : \delta < \beta\}| \\ &= \sum_{\delta<\beta} \kappa_\delta.\end{aligned}$$

Theorem 6.10.1 suggests that, for the most part, we need only consider α-sequences with reference to initial numbers α.

Our new definition of cardinal sums preserves finite sums in the sense that for any natural number N and for any cardinal numbers $\kappa_0, \kappa_1, \ldots, \kappa_N$,

$$\sum_{\gamma<N^+} \kappa_\gamma = \kappa_0 + \kappa_1 + \kappa_1 + \cdots + \kappa_N.$$

Example 6.10.1 (A.C.)

(1) Let $\kappa_\gamma = |1|$ and let $\Gamma_\gamma = 1$ for $\gamma < \omega$. Then

$$\sum_{\gamma<\omega} \kappa_\gamma = |\sum_{\gamma<\omega} \Gamma_\gamma| = |\omega_0| = \aleph_0.$$

(2) $$\sum_{\gamma<\omega 2} \aleph_\gamma = |\sum_{\gamma<\omega 2} \omega_\gamma| = |\omega_{\omega 2}| = \aleph_{\omega 2} = \sum_{\gamma<\omega} c_\gamma,$$

where for $\gamma < \omega$,

$$c_\gamma = \begin{cases} \aleph_N, & \text{if } \gamma = 2N; \\ \aleph_{\omega+N}, & \text{if } \gamma = 2N+1. \end{cases}$$

An associative law for transfinite sums of cardinals follows from Theorem 5.9.3. To this end, we first extend our original definition of transfinite sums (of cardinals), and we modify the corresponding discussion of Section 5.9.

Let ω_α be an initial ordinal and let $[\kappa_\gamma]_{\gamma<\omega_\alpha}$ be an ω_α-cardinal-sequence; let $[\Gamma_\gamma]_{\gamma<\omega_\alpha}$ be the associated ordinal-sequence in which each Γ_γ is the initial number of κ_γ. If δ_1 and δ_2 are initial numbers and if $\delta_1 \le \delta_2 \le \omega_\alpha$, define

$$\text{“}\sum_{\delta_1\le\gamma<\delta_2} \kappa_\gamma\text{” to be “}|\sum_{\delta_1\le\gamma<\delta_2} \Gamma_\gamma|\text{”.}$$

Let $\delta \le \omega_\alpha$ and let $[\Lambda_\mu]_{\mu<\delta^+}$ be an increasing sequence of ordinals in $\omega_\alpha{}^+$ that is such that each Λ_μ is a nonzero initial number and $\Lambda_\delta = \omega_\alpha$. Recall that "$\Lambda_{0^-}$" represents "0" and for $\mu < \delta^+$—"$\Lambda_{(\mu^+)^-}$" is "Λ_μ", and "Λ_{μ^-}" is "$\cup\{\Lambda_\gamma : \gamma < \mu\}$" if μ is a limit ordinal. (Note that in the last case, Λ_{μ^-} is an initial ordinal.) For $\mu < \delta^+$, define

$$K_\mu = \sum_{\Lambda_{\mu^-}\le\gamma<\Lambda_\mu} \kappa_\gamma.$$

We call k_μ the **(associated) μth interval sum**. Observe that $K_\mu = |0|$ if $\mu = \mu^-$.

Theorem 6.10.2. *(A.C.) Let ω_α be an initial ordinal and let $[\kappa_\gamma]_{\gamma<\omega_\alpha}$ be any ω_α-cardinal-sequence. Let $\delta \leq \omega_\alpha$ and let $[\Lambda_\mu]_{\mu<\delta^+}$ be an increasing ordinal-sequence in ω_α^+ that is such that each Λ_μ is a nonzero initial number and $\Lambda_\delta = \omega_\alpha$. Let $[K_\mu]_{\mu<\delta^+}$ be the associated sequence of μth interval sums. Then*

$$\sum_{\gamma<\omega_\alpha} \kappa_\gamma = \sum_{\mu<\delta^+} K_\mu.$$

(The proof is left as an exercise.)

Theorem 6.10.3. *(A.C.) Let $[\kappa_\gamma]_{\gamma<\omega_\alpha}$ be a cardinal-sequence in which, for a fixed cardinal κ, $\kappa_\gamma = \kappa$ for every $\gamma < \omega_\alpha$. Then*

$$\sum_{\gamma<\omega_\alpha} \kappa_\gamma = \kappa \cdot \aleph_\alpha.$$

PROOF: Let β be the initial number of κ. Then if $\Gamma_\gamma = \beta$ for $\gamma < \omega_\alpha$, according to Exercise 6.5.14(b),

$$\sum_{\gamma<\omega_\alpha} \kappa_\gamma = \Big|\sum_{\gamma<\omega_\alpha} \Gamma_\gamma\Big| = |\beta \cdot \omega_\alpha| = |\beta| \cdot |\omega_\alpha| = \kappa \cdot \aleph_\alpha.$$

Theorem 6.10.4. *(A.C.) Let $[c_\gamma]_{\gamma<\omega_\alpha}$ and $[\kappa_\gamma]_{\gamma<\omega_\alpha}$ be ω_α-cardinal-sequences in which $c_\gamma \precsim \kappa_\gamma$ for each $\gamma < \omega_\alpha$. Then*

$$\sum_{\gamma<\omega_\alpha} c_\gamma \precsim \sum_{\gamma<\omega_\alpha} \kappa_\gamma.$$

(The proof is left as an exercise.)

†EXAMPLE 6.10.2. (A.C.) We use Theorems 6.10.3 and 6.10.4 to determine $\sum_{N<\omega} |N|$. Let $c_N = |1|$ and $\kappa_N = \omega$, $N < \omega$. Then

$$\begin{aligned} \aleph_0 &= |\omega_0| = |1| \cdot |\omega_0| \\ &= \sum_{N<\omega} c_N \precsim \sum_{N<\omega} |N| \precsim \sum_{N<\omega} \kappa_N \\ &= |\omega_0| \cdot |\omega_0| = \aleph_0 \cdot \aleph_0 = \aleph_0. \end{aligned}$$

Thus, $\sum_{N<\omega} |N| = \aleph_0$.

We next consider a relationship between transfinite addition of cardinals and the exponentiation of alephs.

Theorem 6.10.5. *(A.C.) If α and β are ordinal numbers that satisfy $0 \leq \alpha \leq \beta$, then*

$$|2|^{\aleph_\beta} = \aleph_\alpha^{\aleph_\beta} = \sum_{\gamma<\alpha} \aleph_\gamma^{\aleph_\beta}.$$

PROOF: Theorem 6.8.2(3) asserts that if $\alpha \leq \beta$, then $|2|^{\aleph_\beta} = \aleph_\alpha^{\aleph_\beta}$. For each $\gamma < \alpha$, define κ_γ to be $\aleph_\alpha$. Then

$$\begin{aligned}\aleph_\alpha^{\aleph_\beta} &\preceq (|2|^{\aleph_\alpha})^{\aleph_\beta} = |2|^{\aleph_\alpha \cdot \aleph_\beta} = |2|^{\aleph_\beta} \\ &\preceq \aleph_0^{\aleph_\beta} \preceq \sum_{\gamma<\alpha} \aleph_\gamma^{\aleph_\beta} \preceq \sum_{\gamma<\alpha} \kappa_\gamma^{\aleph_\beta} \\ &= \aleph_\alpha^{\aleph_\beta} \cdot |\alpha| \preceq \aleph_\alpha^{\aleph_\beta} \cdot \aleph_\alpha \\ &= \aleph_\alpha^{\aleph_\beta + |1|} = \aleph_\alpha^{\aleph_\beta}.\end{aligned}$$

[The last inequality is by virtue of Exercise 6.7.14(a).] Thus, each inequality can be replaced by equality.

Another case of the equation "$\aleph_\alpha^{\aleph_\beta} = \sum_{\gamma<\alpha} \aleph_\gamma^{\aleph_\beta}$" will be considered in the next section.

EXAMPLE 6.10.3. (A.C.)

$$\aleph_\omega^{\aleph_\omega} = \sum_{N<\omega} \aleph_N^{\aleph_\omega}.$$

We cannot readily modify the above discussion to define transfinite products of cardinals in terms of transfinite products of ordinals. For we want the product of a transfinite sequence of cardinals to have the cardinality of the Cartesian product of a sequence of representative sets with these cardinalities. (See Exercise 5.10.8.) If $\kappa_\gamma = \kappa$ for each $\gamma < \omega_\alpha$, then "$\prod_{\gamma<\omega_\alpha} \kappa_\gamma$" should be "$\kappa^{\aleph_\alpha}$". But if, for example, $\kappa_\gamma = \aleph_0$ and if $\Gamma_\gamma = \omega_0$ for $\gamma < \omega_0$, then "$\prod_{\gamma<\omega_0} \kappa_\gamma$" should be "$\aleph_0^{\aleph_0}$", whereas

$$|\prod_{\gamma<\omega_0} \Gamma_\gamma| = |\omega_0^{\omega_0}| = \aleph_0 \prec |2|^{\aleph_0} = \aleph_0^{\aleph_0}.$$

Let α be an ordinal number and let $[\kappa_\gamma]_{\gamma<\alpha}$ be an α-cardinal-sequence. Let Γ_γ be the initial number of κ_γ, $\gamma < \alpha$. Consider the α-ordinal-sequence $f = [\Gamma_\gamma]_{\gamma<\alpha}$. We define

$$\text{"}\prod_{\gamma<\alpha} \kappa_\gamma\text{" to be "}|\bigtimes_{\alpha,f} \{\Gamma_\gamma : \gamma < \alpha\}|\text{".}$$

Observe that if $\alpha = 0$, then $\prod_{\gamma<\alpha} \kappa_\gamma = 0$. (See Section 4.7.)

If, in this definition, we let α be a nonzero natural number, we see that *we have extended the definition of cardinal multiplication so as to allow infinitely* many "*Cartesian factors*". Recall that $\bigtimes_{\varnothing,\varnothing} \varnothing = 1$; hence, *for any* 0-*cardinal-sequence* $[\kappa_\gamma]_{\gamma<0}$,

$$f = [\Gamma_\gamma]_{\gamma<0} = \{\Gamma_\gamma : \gamma < 0\} = 0,$$

and

$$\prod_{\gamma<0} \kappa_\gamma = |1|.$$

If $\{B_\gamma : \gamma < \alpha\}$ *is such that* $|B_\gamma| = \kappa_\gamma$ *for each* $\gamma < \alpha$, *and if* $h = [B_\gamma]_{\gamma<\alpha}$, *then*

$$\prod_{\gamma<\alpha} \kappa_\gamma = |\bigtimes_{\alpha,h} \{B_\gamma : \gamma < \alpha\}|.$$

Clearly, *transfinite multiplication of cardinals is commutative in the sense that if* α *and* β *are any ordinals of the same cardinality, then with respect to any one-one mapping* F *of* α *onto* β, *if* $c_\gamma = \kappa_{F(\gamma)}$, $\gamma < \alpha$, then

$$\prod_{\gamma<\alpha} c_\gamma = \prod_{\delta<\beta} \kappa_\delta.$$

Again, we will generally consider α-sequences with reference to initial numbers α.

†EXAMPLE 6.10.4. (A.C.)

(1) Let $\kappa_\gamma = |1|$ and let $\Gamma_\gamma = 1$, $\gamma < \omega_\alpha$. Let $f : \omega_\alpha \to \{1\}$. Then $\langle \omega_\alpha, f\rangle$ is an indexing pair for $\{\Gamma_\gamma : \gamma < \omega_\alpha\}$. Furthermore, $1 = \cup\{1\} = \cup\{\Gamma_\gamma : \gamma < \omega_\alpha\}$. The unique function f from ω_α into 1 such that $f(\gamma) \in \Gamma_\gamma$ is the constant function, given by $f(\gamma) = 0$, $\gamma < \omega_\alpha$. Thus,

$$\prod_{\gamma<\omega_\alpha} \kappa_\gamma = |\bigtimes_{\omega_\alpha, f} \{\Gamma_\gamma : \gamma < \omega_\alpha\}| = |1|.$$

(2) Let $\kappa_\gamma = |2|$ and let $\Delta_\gamma = 2$ for $\gamma < \omega$. If $g : \omega \to 2$, then

$$\prod_{\gamma<\omega} \kappa_\gamma = |\bigtimes_{\omega,g} \{\Delta_\gamma : \gamma < \omega\}|$$
$$= |\mathscr{F}(\omega, 2)| = |2|^{|\omega|} = |\mathscr{R}|.$$

We next consider the multiplicative analogues of Theorems 6.10.2–6.10.4. If δ_1 and δ_2 are initial numbers and if $\delta_1 \le \delta_2 \le \omega_\alpha$, we let

$$I = \{\gamma : \delta_1 \le \delta < \delta_2\} \qquad \text{and} \qquad g = f \upharpoonright I.$$

We define

$$\text{“}\prod_{\delta_1 \le \gamma < \delta_2} \kappa_\gamma\text{” to be “}|\bigtimes_{I,g} \{\Gamma_\gamma : \delta_1 \le \gamma < \delta_2\}|\text{”.}$$

Let ω_α be an initial ordinal and let $[\kappa_\gamma]_{\gamma<\omega_\alpha}$ be an ω_α-cardinal-sequence. Let $\delta \le \omega_\alpha$ and let $[\Lambda_\mu]_{\mu<\delta^+}$ be an increasing sequence of ordinals in $\omega_\alpha{}^+$ that is such that each Λ_μ is a nonzero initial number and $\Lambda_\delta = \omega_\alpha$. For $\mu \le \delta$, let

$$P_\mu = \prod_{\Lambda_{\mu^-} \le \gamma < \Lambda_\mu} \kappa_\gamma .$$

Here, P_μ is called the **μth interval product**. Note that $P_\mu = |1|$ if $\mu = \mu^-$.

Theorem 6.10.6. (*A.C.*) *Let ω_α be an initial ordinal and let $[\kappa_\gamma]_{\gamma<\omega_\alpha}$ be any ω_α-cardinal sequence. Let $\delta \le \omega_\alpha$ and let $[\Lambda_\mu]_{\mu<\delta^+}$ be an increasing ordinal-sequence in $\omega_\alpha{}^+$ that is such that each Λ_μ is a nonzero initial number and $\Lambda_\delta = \omega_\alpha$. Let $[P_\mu]_{\mu<\delta^+}$ be the associated sequence of μth interval products. Then*

$$\prod_{\gamma<\omega_\alpha} \kappa_\gamma = \prod_{\mu<\delta^+} P_\mu . \tag{6.10.1}$$

PROOF: Let Γ_γ be the initial number of κ_γ, $\gamma < \omega_\alpha$; let Δ_μ be the initial number of P_μ, $\mu < \delta^+$. Let $f = [\Gamma_\gamma]_{\gamma<\omega_\alpha}$ and let $g = [\Delta_\mu]_{\mu<\delta^+}$. In order to show (6.10.1), it suffices to show

$$\mathop{\times}_{\omega_\alpha, f} \{\Gamma_\gamma : \gamma < \omega_\alpha\} \approx \mathop{\times}_{\delta^+, g} \{\Delta_\mu : \mu < \delta^+\}. \tag{6.10.2}$$

For $\mu < \delta^+$, let $I_\mu = \{\gamma : \Delta_{\mu^-} \le \gamma < \Delta_\mu\}$, and let $f_\mu = f \restriction I_\mu$. Thus, $I_\mu = f_\mu = \varnothing$ if $\mu^- = \mu$. For each $\mu < \delta^+$, $\mathop{\times}_{I_\mu, f_\mu} \{\Gamma_\gamma : \gamma \in I_\mu\} \approx \Delta_\mu$; let h_μ be any such one-one correspondence.

If $F \in \mathop{\times}_{\omega_\alpha, f} \{\Gamma_\gamma : \gamma < \omega_\alpha\}$, then $F(\gamma) < \Gamma_\gamma$ for all $\gamma < \omega_\alpha$. Consequently, F determines a sequence of functions $[F_\mu]_{\mu<\delta^+}$, where each $F_\mu \in \mathop{\times}_{I_\mu, f_\mu} \{\Gamma_\gamma : \gamma \in I_\mu\}$. In fact, $F_\mu(\gamma) = F(\gamma)$, $\gamma \in I_\mu$, if $\mu^- < \mu$; and $F_\mu = \varnothing$ for $\mu^- = \mu$. Define $G : \mathop{\times}_{\omega_\alpha, f} \{\Gamma_\gamma : \gamma < \omega_\alpha\} \to \mathop{\times}_{\delta^+, g} \{\Delta_\mu : \mu < \delta^+\}$ by

$$(G(F))(\mu) = h_\mu(F_\mu), \qquad \mu < \delta^+.$$

The reader can check that G is a one-one correspondence by means of which (6.10.2) is established.

Note that in (6.10.3) the arguments of h_μ are functions; thus, we do not mean the composite function.

Theorem 6.10.7. (*A.C.*) *Let $[\kappa_\gamma]_{\gamma<\omega_\alpha}$ be an ω_α-cardinal-sequence in which, for a fixed cardinal κ, $\kappa_\gamma = \kappa$ for every $\gamma < \omega_\alpha$. Then*

$$\prod_{\gamma<\omega_\alpha} \kappa_\gamma = \kappa^{\aleph_\alpha}.$$

Theorem 6.10.8. *(A.C.) Let $[c_\gamma]_{\gamma<\omega_\alpha}$ and $[\kappa_\gamma]_{\gamma<\omega_\alpha}$ be ω_α-cardinal-sequences in which $c_\gamma \preceq \kappa_\gamma$ for each $\gamma < \omega_\alpha$. Then*

$$\prod_{\gamma<\omega_\alpha} c_\alpha \preceq \prod_{\gamma<\omega_\alpha} \kappa_\gamma.$$

(The proofs of the preceding theorems are left as exercises.)

There is an important relationship between cardinal addition and cardinal multiplication; before presenting it in its generality, we consider a particular case.

EXAMPLE 6.10.5. (A.C.) By Examples 6.10.2 and 6.10.4(2),

$$\sum_{N<\omega} |N| = \aleph_0 \prec |\mathscr{R}| = \prod_{N<\omega} c_N,$$

where each $c_N = |2|$. Let $\kappa_N = \aleph_0$, $N < \omega$. Then

$$|\mathscr{R}| = \prod_{N<\omega} c_N \preceq \prod_{1<N<\omega} |N^+| \preceq \prod_{N<\omega} \kappa_N = |\mathscr{R}|,$$

by Theorems 6.10.7 and 6.10.8 and Example 6.6.2(3). By the Schröder–Bernstein Theorem,

$$\prod_{N<\omega} |N^+| = \prod_{1\le N<\omega} |N^+| = |\mathscr{R}|.$$

Thus,

$$\sum_{N<\omega} |N| \prec \prod_{N<\omega} |N^+|.$$

Theorem 6.10.9. König–Zermelo Inequality. *(A.C.) Let α be a nonzero initial number. Let $[c_\gamma]_{\gamma<\alpha}$ and $[\kappa_\gamma]_{\gamma<\alpha}$ be α-cardinal-sequences in which $c_\gamma \prec \kappa_\gamma$ for each $\gamma < \alpha$. Then*

$$\sum_{\gamma<\alpha} c_\gamma \prec \prod_{\gamma<\alpha} \kappa_\gamma.$$

PROOF: Let β_γ be the initial number of κ_γ, and let $B_\gamma = \beta_\gamma \times \{\gamma\}$, $\gamma < \alpha$. Let $G_1 = [\beta_\gamma]_{\gamma<\alpha}$ and let $G_2 = [B_\gamma]_{\gamma<\alpha}$. Then

$$\prod_{\gamma<\alpha} \kappa_\gamma = |\bigtimes_{\alpha,G_1} \{\beta_\gamma : \gamma < \alpha\}| = |\bigtimes_{\alpha,G_2} \{B_\gamma : \gamma < \alpha\}|.$$

Furthermore, $c_\gamma \prec \kappa_\gamma = |B_\gamma|$, $\gamma < \alpha$; we can use the Axioms of Choice and Substitution to obtain a set $\{C_\gamma : \gamma < \alpha\}$, where $|C_\gamma| = c_\gamma$ and $C_\gamma \subset B_\gamma$, $\gamma < \alpha$.

Moreover, $\{C_\gamma : \gamma < \alpha\}$ is pairwise disjoint because $\{B_\gamma : \gamma < \alpha\}$ is. Thus,

$$\sum_{\gamma<\alpha} c_\gamma = |\cup\{C_\gamma : \gamma < \alpha\}|.$$

We next utilize the Axioms of Choice and Substitution to obtain a set $\{b_\gamma : \gamma < \alpha\}$, where each b_γ is in $B_\gamma \sim C_\gamma$. With each element x in $\cup\{C_\gamma : \gamma < \alpha\}$ we can associate the function f in $\bigtimes_{\alpha,G_2} \{B_\gamma : \gamma < \alpha\}$, defined by

$$f(\gamma) = \begin{cases} x, & \text{if } x \in C_{\gamma_0} \text{ and } \gamma = \gamma_0; \\ b_\gamma, & \text{otherwise.} \end{cases}$$

This mapping is clearly one-one; hence

$$\sum_{\gamma<\alpha} c_\gamma = |\cup\{C_\gamma : \gamma < \alpha\}| \preceq |\bigtimes_{\alpha,G_2} \{B_\gamma : \gamma < \alpha\}|$$
$$= \prod_{\gamma<\alpha} \kappa_\gamma.$$

Suppose that $\sum_{\gamma<\alpha} c_\gamma = \prod_{\gamma<\alpha} \kappa_\gamma$. Then

$$\cup\{C_\gamma : \gamma < \alpha\} \approx \bigtimes_{\alpha,G_2} \{B_\gamma : \gamma < \alpha\};$$

let F be any such one-one correspondence. Then

(6.10.5) $\{F''(C_\gamma) : \gamma < \alpha\}$ is pairwise disjoint and its union is $\bigtimes_{\alpha,G_2} \{B_\gamma : \gamma < \alpha\}$.

For each $\gamma < \alpha$, let $A_\gamma = \{x \in B_\gamma : x = h(\gamma)$ for some h in $F''(C_\gamma)\}$. Clearly $A_\gamma \subseteq B_\gamma$. Furthermore,

$$|A_\gamma| \preceq |F''(C_\gamma)| = |C_\gamma| = c_\gamma \prec \kappa_\gamma = |B_\gamma|.$$

Thus, $B_\gamma \sim A_\gamma \neq \varnothing$ for each $\gamma < \alpha$. We obtain a set $\{d_\gamma : \gamma < \alpha\}$ consisting of a single element, d_γ, from each such $B_\gamma \sim A_\gamma$. Define $g : \alpha \to \cup\{B_\gamma : \gamma < \alpha\}$ by $g(\gamma) = d_\gamma$ for each $\gamma < \alpha$. Then $g \in \bigtimes_{\alpha,G_2} \{B_\gamma : \gamma < \alpha\}$. But for each $\gamma < \alpha$, $g(\gamma) = d_\gamma \notin A_\gamma$; hence, $g \notin F''(C_\gamma)$ for any such γ. This contradicts (6.10.5); (6.10.4) follows.

Exercises

†6.10.1. Let α be an ordinal and let $[\kappa_\gamma]_{\gamma<\alpha}$ be an α-cardinal-sequence. Suppose that $\{C_\gamma : \gamma < \alpha\}$ is such that $|C_\gamma| = \kappa_\gamma$ for each $\gamma < \alpha$.

(a) Prove that $|\bigcup_{\gamma<\alpha} C_\gamma| \preceq \sum_{\gamma<\alpha} \kappa_\gamma$.

(b) Prove that if $\{C_\gamma : \gamma < \alpha\}$ is pairwise disjoint, then $|\bigcup_{\gamma<\alpha} C_\gamma| = \sum_{\gamma<\alpha} \kappa_\gamma$.

6.10.2. Let $[\Lambda_\mu]_{\mu<\delta_0}$ be an increasing ordinal-sequence in which each Λ_μ is a nonzero initial number. Prove that if μ is a limit ordinal $< \delta_0$, then Λ_{μ^-} is an initial ordinal.

6.10.3. Prove Theorem 6.10.2.

6.10.4. Prove Theorem 6.10.4.

6.10.5. In Theorem 6.10.6, prove that G is a one-one mapping of $\underset{\omega_\alpha, f}{\times} \{\Gamma_\gamma : \gamma < \omega_\alpha\}$ onto $\underset{\delta^+, g}{\times} \{\Delta_\mu : \mu < \delta^+\}$.

6.10.6. Prove Theorem 6.10.7.

6.10.7. Prove Theorem 6.10.8.

6.10.8. (A.C.) Show, by means of counterexamples, that in each of Theorems 6.10.4 and 6.10.8, the inequalities cannot both be strengthened:

(a) Determine cardinal-sequences $[c_\gamma]_{\gamma<\omega_\alpha}$ and $[\kappa_\gamma]_{\gamma<\omega_\alpha}$ in which $c_\gamma \prec \kappa_\gamma$ for each $\gamma < \omega_\alpha$ and $\sum_{\gamma<\omega_\alpha} c_\gamma = \sum_{\gamma<\omega_\alpha} \kappa_\gamma$.

(b) Determine cardinal-sequences $[k_\gamma]_{\gamma<\omega_\beta}$ and $[l_\gamma]_{\gamma<\omega_\beta}$ in which $k_\gamma \prec l_\gamma$ for each $\gamma < \omega_\beta$ and $\prod_{\gamma<\omega_\beta} k_\gamma = \prod_{\gamma<\omega_\beta} l_\gamma$.

6.10.9. (A.C.) For each K in ω, let N_K be a natural number ≥ 2.

(a) Prove that $\sum_{K<\omega_0} |N_K| = \aleph_0$.

(b) Prove that $\prod_{K<\omega_0} |N_K| = |\mathscr{R}|$.

6.10.10. In the proof of the König–Zermelo Inequality, show that there is a set $\{C_\gamma : \gamma < \alpha\}$, where $|C_\gamma| = c_\gamma$ and $C_\gamma \subset B_\gamma$, $\gamma < \alpha$.

⋆,†6.10.11. (A.C.) Prove that $|2|^{\aleph_0} \neq \aleph_\omega$.

6.11. Ordinal Cofinality

We consider a relationship among ordinal numbers that will have important applications in the subsequent development of cardinal arithmetic.

Let $\alpha \leq \beta$. Then β is said to be **cofinal with** α if there exists an increasing α-sequence, $f = [\Gamma_\gamma]_{\gamma<\alpha}$, with the properties:

(1) Each Γ_γ, $\gamma < \alpha$, is either 0 or a successor ordinal.
(2) $\cup$ range $f = \beta$.

We refer to such a sequence f as an **increasing α-sequence for** β. When β is cofinal with α, we write "$\beta \hat{C} \alpha$". If α and β are ordinals and β is not cofinal with α, we write "$\beta \hat{\not{C}} \alpha$".

†Example 6.11.1

(1) Every successor ordinal, β^+, is cofinal with 1 because the increasing 1-sequence $f = \{\langle 0, \beta^+\rangle\}$ is such that $\cup$ range $f = \cup\{\beta^+\} = \beta^+$.

(2) For every ordinal α, $\alpha\hat{C}\alpha$ because the increasing α-sequence, $\{\langle\gamma, \gamma^+\rangle : \gamma < \alpha\}$, satisfies (1) and $\cup\{\gamma^+ : \gamma < \alpha\} = \alpha$.
(3) ω is not cofinal with any natural number N. For if $[\Gamma_\gamma]_{\gamma<N}$ is a sequence in which each $\Gamma_\gamma < \omega$, then $\cup$ range $f = \max\{\Gamma_\gamma : \gamma < N\} < \omega$. By (1), no Γ_γ can equal ω. If $\Gamma_{\gamma_0} > \omega$ for some $\gamma_0 < N$, then $\cup$ range $f \geq \cup\{\Gamma_{\gamma_0}\} = \Gamma_{\gamma_0} > \omega$.
(4) $\omega 2\hat{C}\omega$. Consider the ω-sequence given by $\Gamma_N = \omega + N^+, N < \omega$.
(5) $\omega^2\hat{C}\omega$. Consider the ω-sequence given by $\Gamma_N = (\omega N)^+$, $N < \omega$.

Note that the specification that $\alpha \leq \beta$ in the above definition was not necessary.

Also, part (3) of Example 6.11.1 suggests that

(6.11.1) *if f $(= [\Gamma_\gamma]_{\gamma<\alpha})$ is an increasing α-sequence for β, then $\Gamma_\gamma \leq \beta$ for each $\gamma < \alpha$.*

In fact,

$$\Gamma_\gamma \subseteq \cup \text{ range } f = \beta;$$

hence,

$$\Gamma_\gamma \leq \beta.$$

Our first theorem shows that a necessary condition for β to be cofinal with α is for both ordinals to be of "the same kind".

Theorem 6.11.1. *Let α and β be ordinal numbers and let $\beta\hat{C}\alpha$. Then (exactly) one of the following is true:*
(1) *Both α and β are* 0.
(2) *Both α and β are successor ordinals.*
(3) *Both α and β are limit ordinals.*

PROOF: Assume $\beta\hat{C}\alpha$ and let $f = [\Gamma_\gamma]_{\gamma<\alpha}$ be an increasing α-sequence for β. If $\beta = 0$, then because $\alpha \leq \beta, \alpha = 0$; if $\beta \neq 0$, then α cannot be 0, because otherwise,

$$\cup \text{ range } f = \cup\varnothing = 0 < \beta.$$

Suppose that β is a successor ordinal and α is a limit ordinal. Then by the Division Algorithm for Ordinal Numbers (Theorem 5.7.4), $\beta = \omega\delta + N$ for some ordinal δ and nonzero natural number N. If each $\Gamma_\gamma < \omega\delta$, then $\cup$ range $f \leq \omega\delta$. (Contradiction.) If for some $\gamma_0 < \alpha$, $\Gamma_{\gamma_0} \geq \omega\delta$, then $\gamma_0 + N^+ < \alpha$ and, because f is increasing, a simple inductive argument shows that $\Gamma_{\gamma_0+N^+} > \omega\delta + N = \beta$, in contradiction of (6.11.1).

Suppose that β is a limit ordinal but α is a successor ordinal. Then $\alpha = \delta^+$ for some ordinal δ. Moreover, $\Gamma_\delta = \max\{\Gamma_\gamma : \gamma < \alpha\}$; hence, $\cup$ range $f = \cup\{\Gamma_\delta\} = \Gamma_\delta$. But Γ_δ is zero or a successor ordinal—hence cannot equal β.

Corollary

$$\{\beta : \beta \textit{ is an ordinal number and } \beta\hat{C}0\}$$

$$= \{\alpha : \alpha \textit{ is an ordinal number and } 0\hat{C}\alpha\} = \{0\}.$$

Theorem 6.11.2. *Let α and β be successor ordinals and let $\alpha \leq \beta$. Then $\beta\hat{C}\alpha$.*

PROOF: Let $\alpha = \alpha_0{}^+$. Define $f\,(= [\Gamma_\gamma]_{\gamma<\alpha})$ on α by

$$\Gamma_\gamma = \begin{cases} \gamma^+, & \gamma < \alpha_0; \\ \beta, & \gamma = \alpha_0. \end{cases}$$

Thus, f is increasing, each Γ_γ is a successor ordinal, and range $f = \cup\{\beta\} = \beta$.

Corollary

(1) *Let β be a successor ordinal. Then*

$\{\alpha : \alpha$ *is an ordinal number and* $\beta\hat{C}\alpha\} = \{\alpha : \alpha$ *is a successor ordinal and* $\alpha \leq \beta\}$.

(2) *Let α be a successor ordinal. Then $\beta\hat{C}\alpha$ iff—β is a successor ordinal and $\alpha \leq \beta$.*

Thus, the only interesting case of the relationship $\beta\hat{C}\alpha$ is the one in which both α and β are limit ordinals.

Theorem 6.11.3. *Let α and β be ordinal numbers such that $\beta\hat{C}\alpha$, and let $f = [\Gamma_\gamma]_{\gamma<\alpha}$ be any increasing α-sequence for β. Then for every $\delta < \beta$ there is some $\gamma < \alpha$ for which $\Gamma_\gamma > \delta$.*

PROOF: Let $\delta < \beta = \cup$ range $f = \cup\{\Gamma_\gamma : \gamma < \alpha\}$. Then for some $\gamma < \alpha$, $\delta < \Gamma_\gamma$.

EXAMPLE 6.11.2. Let $\beta = \omega 2$, let $\alpha = \omega$, and, as in Example 6.11.1(4), let f be the increasing ω-sequence for $\omega 2$ given by $\Gamma_N = \omega + N^+$, $N < \omega$. Let $\delta < \omega 2$. If $\delta \leq \omega$, take $\gamma = 0$; if $\delta = \omega + N$ for $N < \omega$, take $\gamma = N$.

Theorem 6.11.4. *Let α and β be ordinal numbers and let $\beta\hat{C}\alpha$. Then for every $\delta \leq \alpha$, $\alpha\hat{C}\delta$ iff $\beta\hat{C}\delta$.*

PROOF: If β is 0 or a successor ordinal, this follows by the corollaries of Theorems 6.11.1 and 6.11.2.

Assume that β is a limit ordinal. Let $\beta\hat{C}\alpha$ and let $f(=[\Gamma_\gamma]_{\gamma<\alpha})$ be an increasing α-sequence for β.

If $\alpha\hat{C}\delta$ and if g is an increasing δ-sequence for α, then one easily verifies [Exercise 6.11.5(a)] that the function h, defined by

$$h(\gamma) = f \circ g(\gamma), \qquad \gamma < \delta,$$

is an increasing δ-sequence for β. Therefore, $\beta\hat{C}\delta$.

Now suppose that $\beta\hat{C}\delta$, where $\delta \leq \alpha$, and let $[\Lambda_\mu]_{\mu<\delta}$ be an increasing δ-sequence for β. Then, by Theorem 6.11.3, for every $\mu < \delta$ there is a $\gamma < \alpha$ such that

$$\Gamma_\gamma > \Lambda_\mu. \tag{6.11.2}$$

We can define H_1 by letting $H_1(\mu) = \gamma_0$, $\mu < \delta$, where γ_0 is the smallest subscript γ for which (6.11.2) is satisfied. We can show that $\cup$ range $H_1 = \alpha$ [Exercise 6.11.5(b)]. Next, we apply the second form of Definition by Transfinite Induction to remedy the defects of H_1. Let $B = \{\mu\text{-sequences in } \alpha : \mu < \delta\}$. For f in B with domain μ, let

$$G(f) = \max\{H_1(\mu), \cup \text{ range } f\} + 1.$$

Then $G : B \to \alpha$, and, according to Theorem 5.4.2, there is a unique δ-sequence, H_2, in α that is such that $H_2(\mu) = G(H_2 \upharpoonright \mu)$, $\mu < \delta$. Then H_2 is an increasing δ-sequence for α [Exercise 6.11.5(c)]; consequently, $\alpha\hat{C}\delta$.

Corollary. *Let α, β, and δ be ordinal numbers. Then $\beta\hat{C}\alpha$ and $\alpha\hat{C}\delta$ together imply $\beta\hat{C}\delta$.*

Thus, if A is any set of ordinals and if the relation $\hat{C}_A$ is defined by

$$\beta\hat{C}_A\alpha \text{ iff—}\beta, \alpha \in A \text{ and } \beta\hat{C}\alpha,$$

then C_A is a partial ordering on A.

Suppose that $\alpha < \beta$ and $|\alpha| = |\beta|$. Clearly, β need not be cofinal with α; but if it is not, β is cofinal with some ordinal $\rho < \alpha$.

Theorem 6.11.5. *Let α and β be ordinal numbers for which $\alpha < \beta$ and $|\alpha| = |\beta|$. Then there exists an ordinal number ρ that is such that $\rho \leq \alpha$ and $\beta\hat{C}\rho$.*

PROOF: Clearly, α and β must be infinite. If β is a successor ordinal, then let $\rho = 1$.

Let β be a limit ordinal and let f be a one-one map of α onto β. For $\delta < \beta$, $\delta = f(\gamma)$ for some unique $\gamma < \alpha$. Say that δ is "special" provided (1) δ is a successor ordinal and (2) for every $\mu < \gamma$ for which $f(\mu)$ is a successor ordinal,

$f(\mu) < \delta$. Let A be the set of special ordinals in β. Then A is well-ordered by $\in_\beta$; hence, $\langle A, \in_\beta\rangle$ is similar to a unique ordinal system, $\langle \rho, \in\rangle$. Let F be the (unique) similarity map of $\langle \rho, \in\rangle$ onto $\langle A, \in_\beta\rangle$. We claim that F is an increasing ρ-sequence for β.

Clearly, F is a ρ-ordinal-sequence; it is increasing because it is a similarity mapping. For $\mu < \rho$, $F(\mu) \in A$; hence $F(\mu)$ is a successor ordinal. Also, $\cup$ range $F = \cup A \subseteq \beta$. In order to show that $\beta \subseteq \cup$ range $F = \cup A$, it suffices to show that for each δ in β, there is some ξ such that $\delta \in \xi \in A$.

Let $\delta < \beta$. Then $\delta^+ < \beta$. If δ^+ is special, then $\delta \in \delta^+ \in A$. Otherwise, $\delta^+ = f(\gamma)$ for some γ, and there exists $\mu < \gamma$ such that

(6.11.3) $$f(\mu) \text{ is a successor ordinal and } f(\mu) > \delta^+.$$

Let μ_0 be the smallest ordinal μ satisfying (6.11.3). Then $f(\mu_0)$ is special because $f(\mu_0)$ is a successor ordinal and for $\mu < \mu_0, f(\mu) < \delta^+ < f(\mu_0)$. Thus $\delta \in f(\mu_0) \in A$.

Consequently, F is a ρ-sequence for β, and $\rho \hat{C} \beta$.

Corollary. *Let α, β, and γ be ordinal numbers and suppose that $\gamma \hat{C} \beta$, $\alpha \leq \beta$, and $|\alpha| = |\beta|$. Then there exists an ordinal number ρ such that $\rho \leq \alpha$ and $\gamma \hat{C} \rho$.*

PROOF: This follows from Theorems 6.11.4 and 6.11.5.

For every ordinal number β, the **cofinality character of** β, written "cf(β)", is defined to be the smallest ordinal number α such that $\beta \hat{C} \alpha$. Clearly, $\text{cf}(\beta) \leq \beta$. Furthermore, β is said to be **regular** if $\text{cf}(\beta) = \beta$—and **singular** if $\text{cf}(\beta) < \beta$.

†EXAMPLE 6.11.3

(1) $\text{cf}(0) = 0$ because $\alpha \hat{C} \alpha$ for every ordinal α and because 0 is the smallest ordinal. Thus, 0 is regular.
(2) $\text{cf}(5) = 1$ because $5 \not\hat{C} 0$ (by Theorem 6.11.1) and because $5 \hat{C} 1$, by Example 6.11.1(1). Consequently, 5 is singular.
(3) $\text{cf}(\omega) = \omega$, by Example 6.11.1(2) and (3); thus, ω is regular.
(4) $\text{cf}(\omega 2) = \omega$ because of Example 6.11.1(4) and a modification of Example 6.11.1(3); $\omega 2$ is singular.

Theorem 6.11.6. *Let α and β be ordinal numbers. Then $\beta \hat{C} \alpha$ implies* $\text{cf}(\beta) = \text{cf}(\alpha)$.

PROOF: Let $\beta \hat{C} \alpha$. By the Corollary of Theorem 6.11.4, $\beta \hat{C} \alpha$ and $\alpha \hat{C}\, \text{cf}(\alpha)$ together imply $\beta \hat{C}\, \text{cf}(\alpha)$. By the definitions of $\text{cf}(\beta)$ and of $\hat{C}$,

$$\text{cf}(\beta) \leq \text{cf}(\alpha) \leq \alpha.$$

But Theorem 6.11.4 guarantees that

$$\beta \hat{C} \alpha, \ \mathrm{cf}(\beta) \leq \alpha, \text{ and } \beta \hat{C}\ \mathrm{cf}(\beta)$$

together imply $\alpha \hat{C}\ \mathrm{cf}(\beta)$. Consequently,

$$\mathrm{cf}(\alpha) \leq \mathrm{cf}(\beta)$$

and

$$\mathrm{cf}(\beta) = \mathrm{cf}(\alpha).$$

EXAMPLE 6.11.4. Note that $\beta > \alpha$ and $\mathrm{cf}(\beta) = \mathrm{cf}(\alpha)$ do not imply $\beta \hat{C} \alpha$. For we have

$$\mathrm{cf}(\omega^2 + \omega) = \mathrm{cf}(\omega^2) = \omega;$$

we will show that $\omega^2 + \omega \not\hat{C} \omega^2$. Let $F = [\Gamma_\gamma]_{\gamma < \omega^2}$ be any increasing ω^2-sequence. If for each $\gamma < \omega^2$, $\Gamma_\gamma < \omega^2$, then surely $\cup$ range $F \neq \omega^2 + \omega$. Otherwise, let $\gamma < \omega^2$ be such that $\Gamma_\gamma \geq \omega^2$. Then for some N, K in ω, $\gamma = \omega N + K$. For $L_1 > L_2 > K$, $\Gamma_{\omega N + L_1} > \Gamma_{\omega N + L_2} > \Gamma_{\omega N + K}$, and it can easily be shown that

$$\Gamma_{\omega N^+} \geq \omega^2 + \omega$$

and

$$\Gamma_{(\omega N^+)^+} \geq \omega^2 + \omega + 1.$$

Consequently,

$$\cup \text{ range } \Gamma > \omega^2 + \omega.$$

Theorem 6.11.7. *Let β be any ordinal number. Then*

$$\mathrm{cf}(\beta) = \begin{cases} 0, & \text{if } \beta = 0; \\ 1, & \text{if } \beta \text{ is a successor ordinal}; \\ \omega_\delta, & \text{for some ordinal number } \delta, \text{ if } \beta \text{ is a limit ordinal.} \end{cases}$$

PROOF: First, $\mathrm{cf}(0) = 0$, according to Example 6.11.3(1).

Let β be a successor ordinal. Then $\beta \not\hat{C} 0$, by Theorem 6.11.1; $\beta \hat{C} 1$, by Example 6.11.1(1). Thus, $\mathrm{cf}(\beta) = 1$.

Let β be a limit ordinal. Then if $\beta \hat{C} \alpha$, by Theorem 6.11.1, α must be a limit ordinal. Hence, $\mathrm{cf}(\beta)$ is a limit ordinal and is consequently infinite. Let ω_δ be the initial number of $|\mathrm{cf}(\beta)|$. Then

$$\omega_\delta \leq \mathrm{cf}(\beta).$$

Furthermore, $|\omega_\delta| = |\text{cf}(\beta)|$; by Theorem 6.11.5, there is some $\rho \leq \omega_\delta$ for which $\text{cf}(\beta)\hat{C}\rho$. Because $\beta\hat{C}\,\text{cf}(\beta)$, it follows that $\beta\hat{C}\rho$, and hence that $\text{cf}(\beta) \leq \rho$. Thus,

$$\rho \leq \omega_\delta \leq \text{cf}(\beta) \leq \rho,$$

and $\omega_\delta = \text{cf}(\beta)$.

Up until this point, we have not needed the Axiom of Choice in our discussion of cofinality. However, part (2) of Theorem 6.11.8 utilizes transfinite addition of cardinals.

Theorem 6.11.8

(1) *ω_0 is regular.*

(2) *(A.C.) If β is a successor ordinal, then ω_β is regular.*

PROOF: (1) By Example 6.11.3(3), ω_0 $(= \omega)$ is regular.

(2) Let $\beta = \alpha^+$ and suppose that ω_β is singular. Then by Theorem 6.11.7, $\text{cf}(\omega_\beta) = \omega_\delta$ for some $\delta \leq \alpha$. Let $f\,(= [\Gamma_\gamma]_{\gamma<\omega_\delta})$ be an increasing ω_δ-sequence for ω_β. For each $\gamma < \omega_\delta$, Γ_γ is either 0 or else a successor ordinal $< \omega_\beta$. Thus,

$$\begin{aligned}\aleph_{\alpha+1} &= \aleph_\beta = |\omega_\beta| = |\cup \text{range} f| = |\cup\{\Gamma_\gamma : \gamma < \omega_\delta\}| \\ &\preceq \sum_{\gamma<\omega_\delta} |\Gamma_\gamma| \preceq |\omega_\alpha| \cdot |\omega_\delta| = \aleph_\alpha \cdot \aleph_\delta \\ &= \aleph_\alpha.\end{aligned}$$

This contradiction establishes the regularity of ω_β.

It is instructive to consider where the argument of part (2) of the proof fails if β is a limit ordinal. In this case, if ω_β is singular, we can only conclude that $\text{cf}(\omega_\beta) = \omega_\delta$ for some $\delta < \beta$. As before, if $f\,(= [\Gamma_\gamma]_{\gamma<\omega_\delta})$ is an increasing ω_δ-sequence for ω_β, then for each $\gamma < \omega_\delta$, $\Gamma_\gamma < \omega_\beta$. But we cannot find a uniform $\alpha < \beta$ for which $|\Gamma_\gamma| \leq \aleph_\alpha$. Hence, at best we can say that

$$\begin{aligned}\aleph_\beta &= |\cup \text{range} f| \preceq \sum_{\gamma<\omega_\delta} |\Gamma_\gamma| \preceq |\omega_\beta| \cdot |\omega_\delta| \\ &= \aleph_\beta \cdot \aleph_\delta = \aleph_\beta.\end{aligned}$$

We have to leave open the possibility that $|\cup \text{range} f| = \aleph_\beta$; no contradiction is produced.

We consider some limit ordinals β for which ω_β is singular.

EXAMPLE 6.11.5

(1) We claim that $\text{cf}(\omega_\omega) = \omega$, and hence that ω_ω is singular. For the increasing ω-sequence for ω_ω defined by $\Gamma_N = \omega_N{}^+$, $N < \omega$, estab-

lishes that $\mathrm{cf}(\omega_\omega) \leq \omega$. Theorem 6.11.7 guarantees that $\mathrm{cf}(\omega_\omega) \geq \omega_0 = \omega$.

(2) $\mathrm{cf}(\omega_{\omega 2}) = \omega$ and $\omega_{\omega 2}$ is singular. Here, we let $\Gamma_N = \omega_{\omega+N}{}^+$, $N < \omega$, and we conclude that $\mathrm{cf}(\omega_{\omega 2}) \leq \omega$; again, we use Theorem 6.11.7 to obtain cf $(\omega_{\omega 2}) \geq \omega$.

(3) $\mathrm{cf}(\omega_{\omega 2}) = \omega$ and $\omega_{\omega 2}$ is singular. Here, let $\Gamma_N = \omega_{\omega N}{}^+$, $N < \omega$.

The existence of limit ordinals β for which ω_β is regular cannot be proved on the basis of our axioms. On the other hand, no contradiction has been found by assuming that such regular ordinals exist. In the concluding section we will formulate this notion in another way.

We have already shown that limit ordinals are those ordinals with cofinality > 1. We present another characterization of limit ordinals in terms of cofinality.

Theorem 6.11.9. *Let β be a limit ordinal. Then the following sentences are equivalent:*
(1) *β is a limit ordinal.*
(2) *ω_β is cofinal with β.*
(3) $\mathrm{cf}(\omega_\beta) = \mathrm{cf}(\beta)$.

PROOF:

(1) → (2): Let β be a limit ordinal and let $[\Gamma_\gamma]_{\gamma<\beta}$ be defined by $\Gamma_\gamma = \omega_\gamma{}^+$. Then $[\Gamma_\gamma]_{\gamma<\beta}$ is an increasing β-sequence for ω_β.

(2) → (3): This follows by Theorem 6.11.6.

(3) → (1): Suppose that $\mathrm{cf}(\omega_\beta) = \mathrm{cf}(\beta)$. If β were 0 or a successor ordinal, then $\mathrm{cf}(\omega_\beta) = \mathrm{cf}(\beta) \leq 1$. But ω_β is a limit ordinal; hence, $\mathrm{cf}(\omega_\beta) \geq \omega > 1$. Thus, β must be a limit ordinal.

EXAMPLE 6.11.6

$$\mathrm{cf}(\omega_{\omega_1}) = \mathrm{cf}(\omega_1) = \omega_1,$$

according to Theorems 6.11.9 and 6.11.8.

We now consider some important applications of ordinal cofinality to the exponentiation of alephs. We were able to prove the formula

$$\aleph_{\alpha+N}{}^{\aleph_\beta} = \aleph_\alpha{}^{\aleph_\beta} \cdot \aleph_{\alpha+N}$$

when $\alpha + N \leq \beta$ without utilizing the Axiom of Choice [Theorem 6.8.2(4)]. We now use the Axiom of Choice to generalize this result.

Theorem 6.11.10. The Haudorff Recursion Formula (*A.C.*) *For all ordinal numbers α and β,*

$$\aleph_{\alpha+1}{}^{\aleph_\beta} = \aleph_\alpha{}^{\aleph_\beta} \cdot \aleph_{\alpha+1}.$$

PROOF: For $\alpha + 1 \leq \beta$, this follows from Theorem 6.8.2(4). Assume that $\beta < \alpha + 1$. We first show that

$$\mathscr{F}(\omega_\beta, \omega_{\alpha+1}) = \cup\{\mathscr{F}(\omega_\beta, \gamma) : \gamma < \omega_{\alpha+1}\}. \tag{6.11.4}$$

Clearly, if $f : \omega_\beta \to \gamma$ for some $\gamma < \omega_{\alpha+1}$, then $f : \omega_\beta \to \omega_{\alpha+1}$; hence,

$$\cup\{\mathscr{F}(\omega_\beta, \gamma) : \gamma < \omega_{\alpha+1}\} \subseteq \mathscr{F}(\omega_\beta, \omega_{\alpha+1}).$$

Let $f \in \mathscr{F}(\omega_\beta, \omega_{\alpha+1})$. Define g on ω_β by

$$g(\gamma) = \max\{f(\delta) : \delta \leq \gamma\} + \gamma + 1, \qquad \text{for } \gamma < \omega_\beta.$$

Then $g \in \mathscr{F}(\omega_\beta, \omega_{\alpha+1})$ and

$$\cup \text{ range } f \subseteq \cup \text{ range } g \subseteq \omega_{\alpha+1}.$$

Also, $\cup$ range f and $\cup$ range g, as transitive sets of ordinals, are ordinals. Moreover, g is an increasing ω_β-sequence for $\cup$ range g. Because $\omega_{\alpha+1}$ is regular, it follows that

$$\cup \text{ range } f \leq \cup \text{ range } g < \omega_{\alpha+1}.$$

Thus, $f \in \mathscr{F}(\omega_\beta, \cup \text{ range } f)$ and, consequently,

$$\mathscr{F}(\omega_\beta, \omega_{\alpha+1}) \subseteq \cup\{\mathscr{F}(\omega_\beta, \gamma) : \gamma < \omega_{\alpha+1}\}.$$

(6.11.4) follows.

Therefore,

$$\begin{aligned}
\aleph_{\alpha+1}{}^{\aleph_\beta} &= |\mathscr{F}(\omega_\beta, \omega_{\alpha+1})| \\
&= |\cup\{\mathscr{F}(\omega_\beta, \gamma) : \gamma < \omega_{\alpha+1}\}| \\
&\preceq \sum_{\gamma < \omega_{\alpha+1}} |\mathscr{F}(\omega_\beta, \gamma)| \\
&= \sum_{\gamma < \omega_{\alpha+1}} |\gamma|^{\aleph_\beta} \\
&\preceq \aleph_{\alpha+1}{}^{\aleph_\beta} \cdot \aleph_{\alpha+1} \\
&= \aleph_{\alpha+1}{}^{\aleph_\beta + |1|} \\
&= \aleph_{\alpha+1}{}^{\aleph_\beta}.
\end{aligned}$$

Hence,

$$\aleph_{\alpha+1}{}^{\aleph_\beta} = \aleph_\alpha{}^{\aleph_\beta} \cdot \aleph_{\alpha+1},$$

as was to be shown.

Corollary. The Generalized Hausdorff Recursion Formula. (*A.C.*) *For all ordinal numbers* α *and* β *and for all natural numbers* N,

$$\aleph_{\alpha+N}{}^{\aleph_\beta} = \aleph_\alpha{}^{\aleph_\beta} \cdot \aleph_{\alpha+N}.$$

(The simple inductive proof is left as an exercise, as is the proof of the following.)

Theorem 6.11.11. The Bernstein Formula. (*A.C.*) *For all ordinal numbers* β *and for all natural numbers* N,

$$\aleph_N{}^{\aleph_\beta} = |2|^{\aleph_\beta} \cdot \aleph_N.$$

EXAMPLE 6.11.7

(1)
$$\aleph_{\omega 2+5}{}^{\aleph_\omega} = \aleph_{\omega 2}{}^{\aleph_\omega} \cdot \aleph_{\omega 2+5},$$

by the Generalized Hausdorff Recursion Formula.

(2)
$$\aleph_2{}^{\aleph_0} = |2|^{\aleph_0} \cdot \aleph_2 = |\mathscr{R}| \cdot \aleph_2,$$

by the Bernstein Formula.

If $\alpha \leq \beta$, then $\aleph_\alpha{}^{\aleph_\beta} = \sum_{\gamma<\alpha} \aleph_\gamma{}^{\aleph_\beta}$, by Theorem 6.10.5. We now extend this result.

Theorem 6.11.12. (*A.C.*) *If* β *is an ordinal number and if* α *is a limit ordinal for which* $\mathrm{cf}(\alpha) > \omega_\beta$, *then*

$$\aleph_\alpha{}^{\aleph_\beta} = \sum_{\gamma<\alpha} \aleph_\gamma{}^{\aleph_\beta}.$$

PROOF: According to Exercise 6.7.14(a), $|\alpha| \leq \aleph_\alpha$. Thus, as in the proof of Theorem 6.10.5,

$$\begin{aligned}\sum_{\gamma<\alpha} \aleph_\gamma{}^{\aleph_\beta} &\preceq \aleph_\alpha{}^{\aleph_\beta} \cdot |\alpha| \preceq \aleph_\alpha{}^{\aleph_\beta} \cdot \aleph_\alpha \\ &= \aleph_\alpha{}^{\aleph_\beta + |1|} = \aleph_\alpha{}^{\aleph_\beta}.\end{aligned}$$

By Theorem 6.9.6, in order to show that $\aleph_\alpha{}^{\aleph_\beta} \preceq \sum_{\gamma<\alpha} \aleph_\gamma{}^{\aleph_\beta}$, it suffices to show that

$$|\mathscr{P}_{\aleph_\beta}(\omega_\alpha)| \preceq \sum_{\gamma<\alpha} \aleph_\gamma{}^{\aleph_\beta}.$$

Let $A \in \mathscr{P}_{\aleph_\beta}(\omega_\alpha)$ and let $A^* = \{\gamma^+ : \gamma \in A\}$. Then $\langle A^*, \in_{\omega_\alpha}\rangle$ is similar to a unique ordinal system $\langle \rho, \in\rangle$, where $|\rho| = \aleph_\beta$. The similarity mapping F of $\langle \rho, \in\rangle$ onto $\langle A^*, \in_{\omega_\alpha}\rangle$ is increasing. Thus $\cup$ range $F < \omega_\alpha$, because otherwise, F would be an increasing ρ-sequence for ω_α. However, $\text{cf}(\omega_\alpha) = \text{cf}(\alpha) \geq \omega_{\beta+1} > \rho$. Because α is a limit ordinal and $\cup$ range $F < \omega_\alpha$, $A \subseteq \cup$ range $F \subseteq \omega_\gamma$ for some $\gamma < \alpha$. Thus,

$$\mathscr{P}_{\aleph_\beta}(\omega_\alpha) \subseteq \cup\{\mathscr{P}_{\aleph_\beta}(\omega_\gamma) : \beta \leq \gamma < \alpha\}$$

and

$$\begin{aligned} |\mathscr{P}_{\aleph_\beta}(\omega_\alpha)| &\preceq |\cup\{\mathscr{P}_{\aleph_\beta}(\omega_\gamma) : \beta \leq \gamma < \alpha\}| \\ &\preceq \sum_{\beta \leq \gamma < \alpha} |\mathscr{P}_{\aleph_\beta}(\omega_\gamma)| \\ &\preceq \sum_{\gamma < \alpha} \aleph_\gamma^{\aleph_\beta}. \end{aligned}$$

EXAMPLE 6.11.8. Let $\alpha = \omega_1$ and let $\beta = 0$. Then $\text{cf}(\alpha) = \omega_1 > \omega_0 = \omega_\beta$. Thus,

$$\aleph_{\omega_1}^{\aleph_0} = \sum_{\gamma < \omega_1} \aleph_\gamma^{\aleph_0}.$$

Exercises

6.11.1. Let $N > 0$ be a natural number. Determine all ordinals α such that $\omega N \hat{C} \alpha$.

6.11.2. **(a)** Prove that for all N in ω_+, $\omega^\omega \hat{C} \omega N$.

(b) Prove that $\omega^\omega \hat{C} \omega^2$.

6.11.3. Prove that in the definition "β is cofinal with α" the specification "$\alpha \leq \beta$" is unnecessary.

6.11.4. Consider the increasing ω-sequence for ω^2 given by $\Gamma_N = (\omega N)^+$, $N < \omega$ [as in Example 6.11.1(5)]. For every $\delta < \omega^2$, determine an $N < \omega$ for which $\delta < \Gamma_N$.

6.11.5. Consider the proof of Theorem 6.11.4. Verify the following:

(a) h is an increasing δ-sequence for β.

(b) $\cup$ range $H_1 = \alpha$.

(c) H_2 is an increasing δ-sequence for α.

6.11.6. Prove that ω is the only regular denumerable ordinal.

6.11.7. **(a)** Consider the ordinal γ of Exercise 6.7.13. Determine cf(γ).

(b) Determine cf(ε_0) where ε_0 is the first ε-number. (See the remark following Corollary 1 of Theorem 5.12.6: also see Exercise 5.12.6.)

6.11.8. Give examples of singular ordinals β of cardinality $\aleph_1$ for each case.

(a) $\text{cf}(\beta) = \omega_1$. **(b)** $\text{cf}(\beta) = \omega_0$.

6.11.9. Prove that for every ordinal β, cf(β) is regular.

6.11.10. **(a)** Prove the Generalized Hausdorff Recursion Formula from the Hausdorff Recursion Formula.

(b) Verify the Bernstein Formula.

★,†**6.11.11.** **(a)** For which cardinals does the Bernstein Formula depend upon the Axiom of Choice.

(b) Use the Bernstein Formula to evaluate $\aleph_1^{\aleph_0}$.

(c) Now evaluate $\aleph_1^{\aleph_0}$ using the Axiom of Choice, but without using the Bernstein Formula.

★**6.11.12.** Prove the following **Tarski Recursion Formula**: (A.C.) Let α be a limit ordinal, let $f = [\Gamma_\gamma]_{\gamma<\alpha}$ be an increasing α-ordinal-sequence, and let $\delta = \cup$ range f. Then, for any ordinal number β for which $\mathrm{cf}(\alpha) > \omega_\beta$,

$$\aleph_\delta^{\aleph_\beta} = \sum_{\gamma<\alpha} \aleph_{\Gamma_\gamma}^{\aleph_\beta}.$$

6.12 Generalized Continuum Hypothesis

The Continuum Hypothesis is the statement: There is no cardinal number κ satisfying

$$\aleph_0 \prec \kappa \prec |2|^{\aleph_0}.$$

We consider the relationship between the Continuum Hypothesis (C.H.), the Axiom of Choice (Choice), and other sentences of set theory. Unless otherwise specified, we do not assume the Axiom of Choice.

Theorem 6.12.1

(1) $|2|^{\aleph_0} = \aleph_1$ *implies the Continuum Hypothesis.*

(2) *The Continuum Hypothesis implies either* $|2|^{\aleph_0} = \aleph_1$ or $|2|^{\aleph_0}$ *and* $\aleph_1$ *are incomparable.*

(3) *The Continuum Hypothesis together with the Axiom of Choice imply* $|2|^{\aleph_0} = \aleph_1$.

PROOF: (1) Suppose that $|2|^{\aleph_0} = \aleph_1$ and that κ is a cardinal satisfying $\aleph_0 \prec \kappa \prec |2|^{\aleph_0}$. Then κ is an aleph because $\kappa \prec |2|^{\aleph_0} = \aleph_1$. Our indexing procedure of Section 6.7 assures that

(6.12.1) there is no aleph, $\aleph$, satisfying $\aleph_0 \prec \aleph \prec \aleph_1$.

(2) Assume (C.H.) and $|2|^{\aleph_0} \neq \aleph_1$. Obviously, $\aleph_0 \prec \aleph_1$; hence by (C.H.), $\aleph_1 \not\prec |2|^{\aleph_0}$. Also,

$$\aleph_0 \prec |2|^{\aleph_0}.$$

Because $|2|^{\aleph_0} \prec \aleph_1$ would imply that $|2|^{\aleph_0}$ is an aleph, the assumed inequality would contradict (6.12.1).

(3) (Choice) implies that $|2|^{\aleph_0}$ is an aleph. Thus, (3) follows from Trichotomy for alephs, (C.H.), and (2).

It will sometimes be convenient to use

"$\rightarrow$" for "implies",

"$\nrightarrow$" for "does not imply",

and

"$\leftrightarrow$" for "iff".

There are many facts about real numbers that can be expressed in terms of formlas relating $\aleph_1$ to $|2|^{\aleph_0}$. We present one such example.

Theorem 6.12.2. *Let A be the set of all denumerable sequences in $\mathscr{R}$. Then $\aleph_1 \precsim |2|^{\aleph_0}$ if and only if*

(6.12.2) *there exists a function $F : A \rightarrow \mathscr{R}$ that has the property that for each f in A, $F(f) \notin$ range f.*

PROOF: $\aleph_1 \precsim |2|^{\aleph_0} \rightarrow$ (6.12.2):

Suppose that $\aleph_1 \precsim |2|^{\aleph_0}$. Then there is a subset B, of cardinality $\aleph_1$, of $\mathscr{R}$. For any f in A, $|\text{range } f| \precsim \aleph_0 \prec \aleph_1$. Clearly, B is well-orderable; with respect to any well-ordering W on B, let b_f be the first element of $B \sim$ range f. Let $F : A \rightarrow \mathscr{R}$ be such that $F(f) = b_f$, $f \in A$.

(6.12.2) $\rightarrow \aleph_1 \precsim |2|^{\aleph_0}$:

Suppose that $F_A : A \rightarrow \mathscr{R}$ and $F_A(f) \notin$ range f for each f in A. We define a subset C, of cardinality $\aleph_1$, of $\mathscr{R}$. Let B be the set of finite sequences in $\mathscr{R}$. Define $F_B : B \rightarrow \mathscr{R}$ by $F_B(f) = \overline{\overline{\overline{N}}}$, where N is the smallest natural number such that $\overline{\overline{\overline{N}}} \notin$ range f. Then $F_A \cup F_B : \{f : \text{for some } \beta < \omega_1, f : \beta \rightarrow \mathscr{R}\} \rightarrow \mathscr{R}$. By Theorem 5.4.2, there is a unique ω_1-sequence in $\mathscr{R}$, $\hat{F}$, defined by $\hat{F}(\beta) = (F_A \cup F_B)(\hat{F} \upharpoonright \beta)$, $\beta < \omega_1$. We claim that $\hat{F}$ is one-one and that, consequently, $|\text{range } F| = |\omega_1| = \aleph_1$.

First we show that for $N < \omega, \hat{F}(N) = \overline{\overline{\overline{N}}}$. In fact, $\hat{F}(0) = (F_A \cup F_B)(\hat{F} \upharpoonright 0) = F_B(\varnothing) = \overline{\overline{\overline{0}}}$. Suppose that $\hat{F}(N) = \overline{\overline{\overline{N}}}$ for all $N < K$, $K \geq 1$. Then $\hat{F}(K) = (F_A \cup F_B)(\hat{F} \upharpoonright K) = F_B(\{\langle N, \overline{\overline{\overline{N}}} \rangle : N < K\}) = \overline{\overline{\overline{K}}}$.

For $\omega \leq \beta_1 < \beta_2 < \omega_1$, $\hat{F}(\beta_1) = (F_A \cup F_B)(\hat{F} \upharpoonright \beta_1) = F_A(\hat{F} \upharpoonright \beta_1)$ and $\hat{F}(\beta_2) = F_A(\hat{F} \upharpoonright \beta_2)$. Hence, $\hat{F}(\beta_1) \notin$ range $\hat{F} \upharpoonright \beta_1$ and $\hat{F}(\beta_2) \notin$ range $\hat{F} \upharpoonright \beta_2$. Therefore, $\hat{F}(\beta_1) \in \mathscr{R} \sim \{\overline{\overline{\overline{N}}} : N < \omega\}$ and $\hat{F}(\beta_2) \in \mathscr{R} \sim (\{\overline{\overline{\overline{N}}} : N < \omega\} \cup \{\hat{F}(\beta_1)\})$.

Corollary. *The Continuum Hypothesis together with the Axiom of Choice imply* (6.12.2).

It is known to be consistent with the axioms of set theory, including the Axiom of Choice, to assume that $|2|^{\aleph_0} = \aleph_N$ for any $N \geq 1$ (in ω), or even that

$|2|^{\aleph_0} \succ \aleph_\omega$; however, the Axiom of Choice implies $|2|^{\aleph_0} \neq \aleph_\omega$ (Exercise 6.10.11). In Exercise 6.12.3 we consider other restrictions on $|2|^{\aleph_0}$.

We next consider some more general versions of the foregoing sentences. **The Generalized Continuum Hypothesis** (G.C.H.) states:

For every infinite cardinal number c there is no cardinal κ satisfying

$$c \prec \kappa \prec |2|^c.$$

The Aleph Hypothesis (A.H.) states:

For every ordinal number α,

$$|2|^{\aleph_\alpha} = \aleph_{\alpha+1}.$$

We were only able to prove the implication "$|2|^{\aleph_0} = \aleph_1 \rightarrow$ (C.H.)", but had to utilize the Axiom of Choice for the opposite direction—i.e., "((C.H.) and (Choice)) $\rightarrow |2|^{\aleph_0} = \aleph_1$". In contrast with this, we have the following:

Theorem 6.12.3

(1) *The Generalized Continuum Hypothesis is equivalent to the Aleph Hypothesis.*

$$\text{(G.C.H.)} \leftrightarrow \text{(A.H.)}$$

(2) *The Generalized Continuum Hypothesis implies the Axiom of Choice; the Aleph Hypothesis implies the Axiom of Choice.*

$$\text{((G.C.H.) or (A.H.))} \rightarrow \text{(Choice)}$$

PROOF: We prove the following implications:

(a) (G.C.H.) $\rightarrow$ (Choice). (b) (G.C.H.) $\rightarrow$ (A.H.).
(c) (A.H.) $\rightarrow$ (Choice). (d) (A.H.) $\rightarrow$ (G.C.H.).

Statements (1) and (2) follow from these. The proofs we give of (a) and (c) are due to Sierpiński and H. Rubin, respectively. Note that we utilize (a) in proving (b)—and (c) in proving (d).

(G.C.H.) $\rightarrow$ *(Choice)*: Let c be an arbitrary infinite cardinal. Then $c + |1| \prec |2|^c$, by Corollary 1 of Theorem 6.8.3; hence,

$$|c| \preceq |c| + |1| \prec |2|^c.$$

By (G.C.H.), $c = c + |1|$. Therefore,

$$c \cdot |2| \preceq |2|^c \cdot |2| = |2|^{c+|1|} = |2|^c.$$

Also, $c \cdot |2| = |2|^c$ would imply

$$|2|^c = c \cdot |2| \preceq c \cdot c = c^{|2|},$$

in contradiction of Theorem 6.8.3. Thus,

$$c \preceq c \cdot |2| \prec |2|^c.$$

By (G.C.H.), $c \cdot |2| = c$; therefore,

$$c^{|2|} \preceq (|2|^c)^{|2|} = |2|^{c \cdot |2|} = |2|^c.$$

Again by Theorem 6.8.3, $c^{|2|} \neq |2|^c$. Consequently,

$$c \preceq c^{|2|} \prec |2|^c,$$

and again (G.C.H.) implies

$$c = c^{|2|}.$$

Thus, for every infinite cardinal c, $c^{|2|} = c$; Theorem 6.9.3 asserts that this is equivalent to the Axiom of Choice.

(b) *(G.C.H.)* $\rightarrow$ *(A.H.)*: Let α be an ordinal number. By (G.C.H.), there is no cardinal κ satisfying $\aleph_\alpha \prec \kappa \prec |2|^{\aleph_\alpha}$. Furthermore, $\aleph_\alpha \prec \aleph_{\alpha+1}$; thus $\aleph_{\alpha+1} \nprec |2|^{\aleph_\alpha}$. But by (a), (G.C.H.) $\rightarrow$ (Choice); hence, the Law of Trichotomy for cardinals holds. Thus,

$$|2|^{\aleph_\alpha} \preceq \aleph_{\alpha+1}.$$

Also, $|2|^{\aleph_\alpha}$ must be an aleph. Because $\aleph_{\alpha+1}$ is the successor of $\aleph_\alpha$,

$$\aleph_{\alpha+1} \preceq |2|^{\aleph_\alpha};$$

consequently,

$$|2|^{\aleph_\alpha} = \aleph_{\alpha+1}.$$

(c) *(A.H.)* $\rightarrow$ *(Choice)*: Let $\aleph_\alpha$ be an arbitrary aleph. By (A.H.), $|2|^{\aleph_\alpha} = \aleph_{\alpha+1}$; thus, $|2|^{\aleph_\alpha}$ is also an aleph. By the Corollary of Theorem 6.9.5, the Axiom of Choice is true.

(d) *(A.H.)* $\rightarrow$ *(G.C.H.)*: Let c be any infinite cardinal. By (A.H.) and (c), $c = \aleph_\alpha$ for some ordinal α. By (A.H.), $|2|^c = |2|^{\aleph_\alpha} = \aleph_{\alpha+1}$. Furthermore, every infinite cardinal is an aleph, and there is no aleph, $\aleph$, satisfying

$$c = \aleph_\alpha \prec \aleph \prec \aleph_{\alpha+1} = |2|^c.$$

Notice that *the Axiom of Rank was indirectly used in proving* (c); even had we taken a different definition of "cardinal number", we would have needed to know that every set has rank in order to invoke Theorem 6.9.5.

Intuitively speaking, a **model of set theory** is an "ordered pair" $\langle M, R \rangle$, in which M is a collection and R is a "relation on M". (Neither M nor R need be a set.) The collection M represents the "universe of all sets"; R is a stand-in for $\in$. Sets a are identified with "elements" m_a of M. An **atomic sentence of set theory**, $a \in b$, **is true in the model** iff $m_a R m_b$. The **truth (in the model) of "molecular sentences"** is defined in terms of component atomic sentences. For example, "$s_1 \vee s_2$" is true in the model iff one of the sentences, "s_1", "s_2" is true in the model. If x is the unique free variable in the formula $s(x)$, then "for some x, $s(x)$" is true in the model iff there is some set a for which "$s(m_a)$" is true in the model. A precise definition of a "model" would require a formal development of logic. [See (Margaris, Chaps. 2 and 3) or (Mendelson, pp. 45–70)]. There, one defines "theories"—such as number theory, group theory, set theory—and "models" for these theories. Gödel's celebrated Completeness Theorem states that a "set" of sentences of a theory is consistent if and only if it has a model. Another formulation of the Completeness Theorem asserts that a sentence s is provable from a "set" of sentences S if and only if s is true in every model in which all the sentences of S are true. Sentence s is true in a model iff $\sim s$ is false in that model. Thus, $\sim s$ *cannot* be proved from S iff s is true in *some* model in which all the sentences of S are true. Frequently, the Completeness Theorem is utilized to obtain "relative consistency" results. We suppose that S is consistent. Then S has a model. We utilize this model to construct another model in which S together with other sentences $s_1, s_2, \ldots, s_N$ are all true. Then the consistency of S implies the consistency of $S \cup \{s_1, s_2, \ldots, s_N\}$. The construction of models is the key tool in determining consistency questions. Gödel constructed a model in which the axioms of Zermelo–Fraenkel set theory as well as the Generalized Continuum Hypothesis (and hence the Continuum Hypothesis and the Axiom of Choice) are true. Thus, (G.C.H.) is relatively consistent with the axioms of Zermelo–Fraenkel set theory. Cohen, constructed a model in which the Axioms of Zermelo–Fraenkel set theory are true but the Continuum Hypothesis (hence, the Generalized Continuum Hypothesis) and the Axiom of Choice are false. Thus, the negation of the Continuum Hypothesis is also relatively consistent with the axioms of Zermelo–Fraenkel set theory. The Continuum Hypothesis, the Generalized Continuum Hypothesis, and the Axiom of Choice are all "undecidable" sentences. In each case, we can prove neither the sentence nor its negation on the basis of the axioms of set theory. We will describe (a reformation of) Gödel's model.

Let A be a set and let $a \in A$. Let $F(x)$ be a formula (of set theory) with exactly one free variable, x, in which all constants that occur in $F(x)$ are elements of A. We say that a **satisfies** $F(x)$ if $F(a)$ is true. A subset B of A **is** said to be **defined over** A **by** $F(x)$ if for every a in A,

$$a \in B \text{ iff } F(a) \text{ is true.}$$

A subset B of A is called **definable over** A if it is defined over A by some formula (of the above type). Let "$\mathscr{D}(A)$" denote the set of definable subsets of A.

EXAMPLE 6.12.1. Let $A = \omega$ and let $F(x)$ be "$x \in 10$". Then $0, 1, 2, \ldots, 9$ are all the elements of ω that satisfy $F(x)$. Consequently, 10 is the subset of ω that is defined over ω by $F(x)$.

†EXAMPLE 6.12.2

(1) If A is finite, then every subset of A is definable. Thus $\mathscr{D}(A) = \mathscr{P}(A)$.

(2) If A is denumerable, then there are only denumerable many formulas $F(x)$ with constants in A. There are denumerably many definable subsets of A. Thus, $\mathscr{D}(A) \subset \mathscr{P}(A)$.

(3) If $|A| = \aleph_\alpha$, then there are $\aleph_\alpha$ formulas $F(x)$ with constants in A. Again, $\mathscr{D}(A) \subset \mathscr{P}(A)$.

By transfinite induction, we define the following hierarchy of sets:

$$L_0 = \varnothing;$$

$$L_{\alpha+1} = \mathscr{D}(L_\alpha).$$

If δ is a limit ordinal,

$$L_\delta = \cup\{L_\gamma : \gamma < \delta\}.$$

Observe that for $\alpha \le \omega$, $L_\alpha = \boldsymbol{R}_\alpha$; for $\alpha > \omega$, $L_\alpha \subset \mathscr{R}_\alpha$.

A set x is said to be **constructible** if $x \in L_\alpha$ for some ordinal number α. All ordinal and cardinal numbers are constructible. Let L be the collection of constructible sets. Then it can be shown that all the axioms of Zermelo–Fraenkel set theory (including the Axiom of Rank) are true if we substitute "constructible set" for "set" in each axiom. For example, the relativized Power Set Axiom would say:

If A is a constructible set, then the collection

of all constructible subsets of A is constructible.

Furthermore, (G.C.H.) [hence (C.H.) and (Choice)] are true in this model. The **Axiom of Constructibility** (Construct) asserts:

Every set is constructible.

Let (Z.F.) denote Zermelo–Fraenkel set theory, i.e., formalized versions of the axioms of Chapter 1, of Rank, and of Substitution. Then

$$((\text{Z.F.}) \text{ and } (\text{Construct})) \to ((\text{Z.F.}) \text{ and } (\text{G.C.H.})) \to ((\text{Z.F.}) \text{ and } (\text{Choice})).$$

On the other hand, Cohen showed that

$$(\text{Z.F.}) \nrightarrow (\text{Choice}),$$

$$((\text{Z.F.}) \text{ and } (\text{Choice})) \nrightarrow (\text{C.H.}),$$

and

$$((\text{Z.F.}) \text{ and } (\text{G.C.H})) \nrightarrow (\text{Construct}).$$

It follows that

$$(\text{Z.F.}) \nrightarrow (\text{G.C.H.})$$

and that

$$((\text{Z.F.}) \text{ and } (\text{Choice})) \nrightarrow (\text{G.C.H.}).$$

(Cohen's models are presented in (Cohen, Chap. IV)).

Let β be an ordinal number and let g be any increasing β-ordinal-sequence that is such that for all $\alpha < \beta$,

$$\text{cf}(\omega_{g(\alpha)}) > \text{cf}(\omega_\alpha).$$

Easton proved that it is consistent with the axioms of set theory including the Axiom of Choice to assume that for any such ordinal β and any such β-sequence g,

$$|2|^{\aleph_\alpha} = \aleph_{g(\alpha)}, \qquad \alpha < \beta \text{ and } \alpha \text{ regular.}$$[5]

In particular, it is consistent with the axioms of set theory to assume that $|2|^{\aleph_0}$ is any aleph $\aleph_\gamma$ for which

$$\text{cf}(\omega_\gamma) > \text{cf}(\omega_0).$$

The other values are exempted by means of the König–Zermelo Inequality, as was $\aleph_\omega$ (Example 6.10.10). Solovay showed that

$$((\text{Z.F.}) \text{ and } (\text{C.H.})) \nrightarrow (\text{Choice})$$

and that

$$((\text{Z.F.}) \text{ and } (\text{Choice}) \text{ and } (\text{C.H.})) \nrightarrow (\text{G.C.H.}).$$

Souslin's Hypothesis (Section 6.2) is intrinsically related to the above questions. [See the survey article (Jech)]. Tennenbaum, Jech, and Solovay have shown that Souslin's Hypothesis (Souslin) is undecidable. Let S be a "set" of sentences and let "Consists S" stand for "S is consistent." Then

$$\text{Consist (Z.F.)} \rightarrow \text{Consist } ((\text{Z.F.}) \cup \{(\text{Choice}), (\text{Souslin})\});$$

$$\text{Consist (Z.F.)} \rightarrow \text{Consist } ((\text{Z.F.}) \cup \{(\text{G.C.H}) \sim (\text{Souslin})\});$$

$$\text{Consist (Z.F.)} \rightarrow \text{Consist } ((\text{Z.F.}) \cup \{(\text{Choice}), \sim(\text{Souslin}), |2|^{\aleph_0} = \aleph_2\}).$$

The Generalized Continuum Hypothesis enables us to completely determine the exponentiation of alephs (and hence of cardinals).

---

[5] In class-set theory, in which one can consider "sequences" that are not sets, Easton's result has the stronger formulation: Let g be any increasing ordinal "sequence" that is such that for all α,

$$\text{cf}(\omega_{g(\alpha)}) > \text{cf}(\omega_\alpha).$$

Then it is consistent with the axioms of (Gödel–Bernays class-) set theory including (the class form of) the Axiom of Choice to assume that for regular ordinals α, $|2|^{\aleph_\alpha} = \aleph_{g(\alpha)}$. (In the class form the "choice function" is not necessarily a set.)

Theorem 6.12.4. *Let α and β be ordinal numbers. Then the Generalized Continuum Hypothesis implies that*

$$\aleph_\alpha^{\aleph_\beta} = \begin{cases} \aleph_{\beta+1}, & \text{if } \alpha \le \beta; \\ \aleph_{\alpha+1}, & \text{if } \alpha > \beta \text{ and } \operatorname{cf}(\omega_\alpha) \le \omega_\beta; \\ \aleph_\alpha, & \text{if } (\alpha > \beta \text{ and}) \operatorname{cf}(\omega_\alpha) > \omega_\beta. \end{cases}$$

PROOF: Assume (G.C.H.)

(1) Let $\alpha \le \beta$. Then by Theorem 6.8.2(c), and (G.C.H.),

$$\aleph_\alpha^{\aleph_\beta} = |2|^{\aleph_\beta} = \aleph_{\beta+1}.$$

(2) Suppose that $\alpha > \beta$ and $\operatorname{cf}(\omega_\alpha) \le \omega_\beta$. By case (1),

$$\aleph_\alpha^{\aleph_\beta} \preceq \aleph_\alpha^{\aleph_\alpha} = \aleph_{\alpha+1}. \tag{6.12.3}$$

Furthermore, $\operatorname{cf}(\omega_\alpha) = \omega_\gamma$ for some $\gamma \le \beta$. Let $f = [\Gamma_\mu]_{\mu<\omega_\gamma}$ be an increasing ω_γ-sequence for ω_α. Define $g = [\Lambda_\mu]_{\mu<\omega_\gamma}$ by

$$\Lambda_\mu = \Gamma_\mu \sim \cup\{\Gamma_\nu : \nu < \mu\}.$$

Then $\omega_\alpha = \cup \operatorname{range} f = \cup \operatorname{range} g$, and $\{\Lambda_\mu : \mu < \omega_\gamma\}$ is pairwise disjoint. We utilize the König-Zermelo Inequality to obtain

$$\aleph_\alpha = \sum_{\mu<\omega_\gamma} |\Lambda_\mu| \preceq \sum_{\mu<\omega_\gamma} |\Gamma_\mu| \prec \prod_{\mu<\omega_\gamma} \kappa_\mu, \qquad \text{where each } \kappa_\mu = \aleph_\alpha.$$

This, together with (6.12.3), yields

$$\aleph_\alpha \prec \aleph_\alpha^{\aleph_\gamma} \preceq \aleph_\alpha^{\aleph_\beta} \preceq \aleph_{\alpha+1}.$$

Because $\aleph_\alpha^{\aleph_\beta}$ is an aleph, $\aleph_\alpha^{\aleph_\beta} = \aleph_{\alpha+1}$.

(3) Suppose that $\operatorname{cf}(\omega_\alpha) > \omega_\beta$. If α is a successor ordinal, γ^+, then, because $\alpha > \beta$, it follows that $\gamma \ge \beta$ and that

$$\begin{aligned} \aleph_\alpha^{\aleph_\beta} &= \aleph_{\gamma+1}^{\aleph_\beta} = (|2|^{\aleph_\gamma})^{\aleph_\beta} \\ &= |2|^{\aleph_\gamma \cdot \aleph_\beta} = |2|^{\aleph_\gamma} \\ &= \aleph_{\gamma+1} = \aleph_\alpha. \end{aligned}$$

If α is a limit ordinal, then for all $\delta < \alpha$,

$$|2|^{\aleph_\delta} = \aleph_{\delta+1} \prec \aleph_\alpha.$$

Moreover, $\mathrm{cf}(\alpha) = \mathrm{cf}(\omega_\alpha) > \omega_\beta$. Therefore, we use Theorem 6.11.12 to establish that

$$\begin{aligned}
\aleph_\alpha^{\aleph_\beta} &= \sum_{\gamma<\alpha} \aleph_\gamma^{\aleph_\beta} \\
&\preceq \sum_{\gamma<\alpha} (|2|^{\aleph_\gamma})^{\aleph_\beta} \\
&= \sum_{\gamma<\alpha} |2|^{\aleph_\gamma \cdot \aleph_\beta} \\
&= |2|^{\aleph_\beta} \cdot |\beta| + \sum_{\beta \le \gamma < \alpha} |2|^{\aleph_\gamma} \\
&\preceq \aleph_\alpha \cdot |\beta| + \aleph_\alpha \cdot |\alpha \dot{-} \beta| \\
&= \aleph_\alpha \cdot (|\beta| + |\alpha \dot{-} \beta|) \\
&= \aleph_\alpha \cdot |\alpha| \\
&= \aleph_\alpha .
\end{aligned}$$

EXAMPLE 6.12.3. Assume (G.C.H.)

(1) $\aleph_\omega^{\aleph_{\omega^2}} = \aleph_{\omega^2+1}$.
(2) $\aleph_{\omega^2}^{\aleph_0} = \aleph_{\omega^2+1}$ because $\mathrm{cf}(\omega_{\omega^2}) = \omega_0$.
(3) $\aleph_{\omega_3}^{\aleph_2} = \aleph_{\omega_3}$ because $\mathrm{cf}(\omega_3) = \omega_3 > \omega_2$.

The Generalized Continuum Hypothesis has an interesting application in the study of "large" cardinals. The definition of the cardinals that we now consider utilizes transfinite addition and multiplication and thus depends upon the Axiom of Choice. Hereafter we assume (Choice) among our axioms; thus, all infinite cardinals will be alephs. The existence of these "inaccessible cardinals" cannot be proved on the basis of our axioms [including (Choice)]. It is not known whether the assumption of their existence leads to a contradiction. However, the addition of an axiom asserting their existence has significant consequences in more advanced topics in set theory as well as in various other fields of mathematics.

We note the following properties of the cardinal number $\aleph_0$:

(6.12.4) For every cardinal number $\kappa_1 \prec \aleph_0$ there is a cardinal κ_2 such that $\kappa_1 \prec \kappa_2 \prec \aleph_0$.

(6.12.5) Whenever $[\kappa_\gamma]_{\gamma<\delta}$ is a δ-cardinal-sequence, for $|\delta| \prec \aleph_0$, in which each $\kappa_\gamma \prec \aleph_0$, then $\sum_{\gamma<\delta} \kappa_\gamma \prec \aleph_0$.

(6.12.6) Whenever $[\kappa_\gamma]_{\gamma<\delta}$ is a δ-cardinal-sequence, for $|\delta| \prec \aleph_0$, in which each $\kappa_\gamma \prec \aleph_0$, then $\prod_{\gamma<\delta} \kappa_\gamma \prec \aleph_0$.

(6.12.7) For all cardinals $\kappa_1, \kappa_2 \prec \aleph_0$, $\kappa_1^{\kappa_2} \prec \aleph_0$.

Every $\aleph_\alpha$ such that α is a successor ordinal satisfies (6.12.5) with "$\aleph_\alpha$" replacing each occurrence of "$\aleph_0$" (Exercise 6.12.1).

Every $\aleph_\alpha$ for which α is a limit ordinal satisfies (6.12.4) with "$\aleph_\alpha$" replacing "$\aleph_0$". Moreover, if we were to assume (G.C.H.), then every $\aleph_\alpha$ for which α is a limit ordinal would also satisfy (6.12.7) with "$\aleph_\alpha$" replacing "$\aleph_0$". But we cannot produce examples of limit ordinals α for which (6.12.5) or (6.12.6) are satisfied with "$\aleph_\alpha$" in lieu of "$\aleph_0$".

An aleph, $\aleph_\beta$, is said to be **weakly inaccessible**[6] if it satisfies the following conditions:

(6.12.8) β is a limit ordinal.

(6.12.9) Whenever $\alpha < \beta$ and $[\kappa_\gamma]_{\gamma<\omega_\alpha}$ is an ω_α-cardinal-sequence in which $\kappa_\gamma \prec \aleph_\beta$ for each $\gamma < \omega_\alpha$, then $\sum_{\gamma<\omega_\alpha} \kappa_\gamma \prec \aleph_\beta$.

Thus, if $\aleph_\beta$ *is weakly inaccessible, then it satisfies* (6.12.4) *and* (6.12.5) *with* "$\aleph_\beta$" *in lieu of* "$\aleph_0$".

Recall that we could not give an example of a regular initial ordinal ω_α in which α is a limit ordinal. In fact, the existence of such an ordinal cannot be proved from our axioms, according to the following.

Theorem 6.12.5. (*A.C.*) *Let β be a limit ordinal. Then $\aleph_\beta$ is weakly inaccessible if and only if ω_β is regular.*

PROOF: Let β be a limit ordinal.

Suppose that $\aleph_\beta$ is weakly inaccessible. Let $f\ (= [\Gamma_\gamma]_{\gamma<\mathrm{cf}(\omega_\beta)})$ be an increasing $\mathrm{cf}(\omega_\beta)$-sequence for ω_β. Define $g\ (= [\Delta_\gamma]_{\gamma<\mathrm{cf}(\omega_\beta)})$ by $\Delta_\gamma = \Gamma_\gamma \sim \cup\{\Gamma_\delta : \delta < \gamma\}$, $\gamma < \mathrm{cf}(\omega_\beta)$. Then for each such γ, $|\Delta_\gamma| \preceq |\Gamma_\gamma| \prec \aleph_\beta$ and $\{\Delta_\gamma : \gamma < \beta\}$ is pairwise disjoint. Thus,

$$\sum_{\gamma<\mathrm{cf}(\omega_\beta)} |\Delta_\gamma| = |\cup\{\Delta_\gamma : \gamma < \beta\}| = |\cup \text{ range } g|$$

$$= |\cup \text{ range } f| = \aleph_\beta.$$

Because $\aleph_\beta$ is weakly inaccessible, $\mathrm{cf}(\omega_\beta) \nless \omega_\beta$. Therefore, $\mathrm{cf}(\omega_\beta) = \omega_\beta$, and ω_β is regular.

Suppose that ω_β is regular. For any $\alpha < \beta$, let $[\kappa_\gamma]_{\gamma<\omega_\alpha}$ be an ω_α-cardinal-sequence in which $\kappa_\gamma \prec \aleph_\beta$. Then

$$\sum_{\gamma<\omega_\alpha} \kappa_\gamma \preceq \aleph_\beta \cdot \aleph_\alpha = \aleph_\beta.$$

[6] Many authors include $\aleph_0$ among the weakly (and strongly) inaccessible alephs.

Suppose that $\sum_{\gamma<\omega_\alpha} \kappa_\gamma = \aleph_\beta$. Let $\delta\,(\leq \omega_\alpha)$ be the smallest ordinal such that $\sum_{\gamma<\delta} \kappa_\gamma = \aleph_\beta$. For $\gamma < \delta$, let ρ_γ be the initial number of κ_γ, and let $\Delta_\gamma = (\sum_{\mu<\gamma} \rho_\mu)^+$. Then $h = [\Delta_\gamma]_{\gamma<\delta}$ is an increasing sequence of successor ordinals. It is easily verified that $\cup$ range $h = \omega_\beta$. Consequently, h is an increasing δ-sequence for ω_β. But $\delta \leq \omega_\alpha < \omega_\beta$, and this contradicts the regularity of ω_β. It follows that $\sum_{\gamma<\omega_\alpha} \kappa_\gamma \prec \aleph_\beta$ and that $\aleph_\beta$ is weakly inaccessible.

Corollary 1. (*A.C.*) *For every limit ordinal β, if $\aleph_\beta$ is weakly inaccessible, then $\omega_\beta = \beta$.*

Corollary 2. (*A.C.*) *$\aleph_\beta$ is weakly inaccessible if (and only if) β is a regular limit ordinal.*

EXAMPLE 6.12.4. There are initial ordinals β such that $\omega_\beta = \beta$—yet $\aleph_\beta$ is not weakly inaccessible. Let $\beta_0 = \omega$, and for each N (in ω) let $\beta_{N^+} = \omega_{\beta_N}$. Let $\beta = \cup\{\beta_N : N < \omega\}$. Then β is an initial ordinal, $\omega_\beta = \beta$, and $\mathrm{cf}(\beta) = \omega$.

An aleph, $\aleph_\beta$, is said to be **strongly inaccessible** if the following are true:

(6.12.8) β is a limit ordinal.

(6.12.10) Whenever $\alpha < \beta$ and $[\kappa_\gamma]_{\gamma<\omega_\alpha}$ is an ω_α-cardinal-sequence in which $\kappa_\gamma \prec \aleph_\beta$ for each $\gamma < \omega_\alpha$, then $\prod_{\gamma<\omega_\alpha} \kappa_\gamma \prec \aleph_\beta$.

Thus, *strongly inaccessible alephs satisfy conditions* (6.12.4) *and* (6.12.6), *with "$\aleph_\beta$" in place of "$\aleph_0$"*. The next theorem will show that *strongly inaccessible cardinals also satisfy* (6.12.5) *and* (6.12.7) (*with the same replacement*).

Theorem 6.12.6. (*A.C.*) *An aleph, $\aleph_\beta$, is strongly inaccessible if and only if the following are true*:

(6.12.11) *$\aleph_\beta$ is weakly inaccessible.*

(6.12.12) *Whenever κ_1 and κ_2 are cardinals satisfying $\kappa_1 \prec \aleph_\beta$ and $\kappa_2 \prec \aleph_\beta$, then ${\kappa_1}^{\kappa_2} \prec \aleph_\beta$.*

PROOF: Suppose that $\aleph_\beta$ is strongly inaccessible. Then, by definition, β is a limit ordinal. Let $\alpha < \beta$ and let $[\kappa_\gamma]_{\gamma<\omega_\alpha}$ be an ω_α-cardinal-sequence in which $\kappa_\gamma \prec \aleph_\beta$ for each $\gamma < \omega_\alpha$. Define $[c_\gamma]_{\gamma<\omega_\alpha}$ as follows:

$$c_\gamma = \begin{cases} \aleph_0, & \text{if } \kappa_\gamma \prec \aleph_0; \\ \aleph_{\mu+1}, & \text{if } \kappa_\gamma = \aleph_\mu. \end{cases}$$

Then $\kappa_\gamma \prec c_\gamma \prec \aleph_\beta$ for each $\gamma < \omega_\alpha$. Therefore, by the König-Zermelo Inequality (Theorem 6.10.9) together with (6.12.10),

$$\sum_{\gamma<\omega_\alpha} \kappa_\gamma \prec \prod_{\gamma<\omega_\alpha} c_\gamma \prec \aleph_\beta.$$

We have established that $\aleph_\beta$ is weakly inaccessible.

Let κ_1 and κ_2 be cardinals $\prec \aleph_\beta$. If both κ_1 and κ_2 are finite, then ${\kappa_1}^{\kappa_2} \prec \aleph_0 \prec \aleph_\beta$. If κ_2 is finite, but κ_1 is infinite, then ${\kappa_1}^{\kappa_2} \preceq \kappa_1 \prec \aleph_\beta$. Otherwise, let ω_δ be the initial number of κ_2 and let $\kappa_\gamma = \kappa_1$ for $\gamma < \omega_\delta$. Then

$${\kappa_1}^{\kappa_2} = \prod_{\gamma<\omega_\delta} \kappa_\gamma \prec \aleph_\beta.$$

Suppose that $\aleph_\beta$ satisfies (6.12.11) and (6.12.12). Then β is a limit ordinal, by the definition of "weakly inaccessible". Let $[\kappa_\gamma]_{\gamma<\omega_\alpha}$ be an ω_α-cardinal-sequence, for $\alpha < \beta$, in which each $\kappa_\gamma \prec \aleph_\beta$. Because $\aleph_\beta$ is weakly inaccessible, for each $\gamma < \omega_\alpha$,

$$\kappa_\gamma \preceq \sum_{\mu<\omega_\alpha} \kappa_\mu \prec \aleph_\beta.$$

According to (6.12.12),

$$\prod_{\gamma<\omega_\alpha} \kappa_\gamma \preceq \Big(\sum_{\mu<\omega_\alpha} \kappa_\mu\Big)^{\aleph_\alpha} \prec \aleph_\beta.$$

Theorem 6.12.7. *Assume (G.C.H.). Then an aleph is strongly inaccessible if and only if it is weakly inaccessible.*

PROOF: Assume (G.C.H.). Theorem 6.12.6 indicates that we need only show that for every weakly inaccessible aleph, $\aleph_\beta$, if $\kappa_1, \kappa_2 \prec \aleph_\beta$, then ${\kappa_1}^{\kappa_2} \prec \aleph_\beta$. If κ_1 and κ_2 are both finite, then ${\kappa_1}^{\kappa_2} \prec \aleph_0 \prec \aleph_\beta$. If κ_2 is finite and κ_1 is infinite, then ${\kappa_1}^{\kappa_2} \preceq \kappa_1 \prec \aleph_\beta$. Otherwise, κ_2 is infinite. If $\kappa_1 \preceq \kappa_2$, then $\kappa_2 = \aleph_\gamma$ for some $\gamma < \beta$ and ${\kappa_1}^{\kappa_2} \preceq \aleph_{\gamma+1} \prec \aleph_\beta$. If $\kappa_2 \prec \kappa_1$, then $\kappa_1 = \aleph_\delta$ for some $\delta < \beta$, and (G.C.H.) implies ${\kappa_1}^{\kappa_2} \preceq \aleph_{\delta+1} \prec \aleph_\beta$.

Exercises

6.12.1. Prove that $|2|^{\aleph_0} = \aleph_1 \leftrightarrow {\aleph_1}^{\aleph_0} = \aleph_1$.

6.12.2. **(a)** Prove that $|2|^{\aleph_0} = \aleph_1 \rightarrow {\aleph_1}^{\aleph_0} \prec {\aleph_2}^{\aleph_0}$.

(b) Prove that ((Choice) and ${\aleph_1}^{\aleph_0} \prec {\aleph_2}^{\aleph_0}$) $\rightarrow |2|^{\aleph_0} = \aleph_1$.

(c) (A.C.) Prove that (C.H.)$\leftrightarrow {\aleph_1}^{\aleph_0} \prec {\aleph_2}^{\aleph_0}$.

★6.12.3. (A.C.)

(a) Show that for each N in ω, $|2|^{\aleph_0} \neq \aleph_{\omega N}$.

(b) Show that for each M, N in ω, $|2|^{\aleph_M} \neq \aleph_{\omega N}$.

(c) Show that for each M in ω, $|2|^{\aleph_M} \neq \aleph_{\omega^2}$.

6.12.4. Show that (G.C.H.) implies that the following holds true for all cardinals c and κ:
(a) $c = \kappa$ iff $|2|^c = |2|^\kappa$.
(b) $c \prec \kappa$ iff $|2|^c \prec |2|^\kappa$.

6.12.5. Assume (G.C.H.). Prove that if α is a successor ordinal, then for every ordinal β, $\aleph_\alpha^{\aleph_\beta} = \aleph_{\max\{\alpha,\beta+1\}}$.

6.12.6. Assume (G.C.H.). Calculate each of the following:
(a) $|5|^{\aleph_0}$. **(b)** $\aleph_0^{\aleph_0}$. **(c)** $\aleph_0^{\aleph_\omega}$.
(d) $\aleph_1^{\aleph_2}$. **(e)** $\aleph_\omega^{\aleph_\omega}$.

6.12.7. Assume (G.C.H.). Calculate each of the following:
(a) $\aleph_9^{\aleph_7}$. **(b)** $\aleph_\omega^{\aleph_{\omega 2}}$. **(c)** $\aleph_{\omega 2}^{\aleph_\omega}$.
(d) $\aleph_{\omega_1}^{\aleph_{\omega 2}}$. **(e)** $\aleph_{\omega 2+\omega}^{\aleph_0}$. **(f)** $\aleph_{\omega_\omega}^{\aleph_{\omega 2}}$.

★6.12.8. We call $[\Gamma_\alpha]_{\alpha<\omega_1}$ a **Jensen sequence** if it satisfies the following:
(a) For each $\alpha < \omega_1$, $\Gamma_\alpha : \alpha \to \alpha$.
(b) Whenever $f : \omega_1 \to \omega_1$, there is some $\alpha < \omega_1$ for which $\Gamma_\alpha = f|\alpha$.
Prove that the existence of a Jensen sequence implies (C.H.).

★6.12.9. Let (S) be the sentence: "For every infinite set X, there is an ordinal β and a β-sequence in $\mathscr{P}(X)$, $\{F_\gamma\}_{\gamma<\beta}$, that is such that the following are true:
(1) For each $\gamma < \beta$, $|F_\gamma| = X$.
(2) For $\gamma_1 < \gamma_2 < \beta$, $F_{\gamma_1} \subset F_{\gamma_2}$.
(3) $\mathscr{P}(X) = \cup\{F_\gamma : \gamma < \beta\}$".
(a) Prove that (A.H.) → (S).
(b) Prove that [(S) and (Choice)] → (A.H.).

6.12.10. Determine the set of all finite cardinals $|N|$ that satisfy each of the following:
(a) For every cardinal $\kappa_1 \prec |N|$ there is a cardinal $\kappa_2 \prec |N|$ such that $\kappa_1 \prec \kappa_2 \prec |N|$.
(b) Whenever $[\kappa_\gamma]_{\gamma<M}$ is an M-cardinal-sequence, for $M < N$, in which each $\kappa_\gamma \prec |N|$, then $\sum_{\gamma<M} \kappa_\gamma \prec |N|$.
(c) Whenever $[\kappa_\gamma]_{\gamma<M}$ is an M-cardinal-sequence, for $M < N$, in which each $\kappa_\gamma \prec |N|$, then $\prod_{\gamma<M} \kappa_\gamma \prec |N|$.
(d) If $\kappa_1, \kappa_2 \prec |N|$, then $\kappa_1^{\kappa_2} \prec |N|$.

6.12.11. (A.C.) Let α be a successor ordinal. Prove that whenever $[\kappa_\gamma]_{\gamma<\delta}$ is a δ-cardinal-sequence, for $|\delta| \prec \aleph_\alpha$, in which each $\kappa_\gamma \prec \aleph_\alpha$, then $\sum_{\gamma<\delta} \kappa_\gamma \prec \aleph_\alpha$.

6.12.12. (A.C.)
(a) In Theorem 6.12.5, show that $\cup$ range $h = \omega_\beta$.
(b) Prove the Corollaries of Theorem 6.12.5.

6.12.13. (A.C.) Prove that an aleph, $\aleph_\beta$, is strongly inaccessible iff (1) and (2) are true:
(1) $\aleph_\beta$ is weakly inaccessible.
(2) Whenever $c \prec \aleph_\beta$, then $|2|^c \prec \aleph_\beta$.

6.12.14. Prove the cardinality assertions of Example 6.12.2.

Bibliography

We generally list readily available texts rather than the original journal articles. This list is by no means exhaustive. For further sources the reader is referred to bibliographies given in Bachmann, *Transfinite Zahlen*; Rubin, *Set Theory for the Mathematician*; and Sierpiński, *Cardinal and Ordinal Numbers.*

Abian, A. *The Theory of Sets and Transfinite Arithmetic.* Philadelphia: W. B. Saunders Company, 1965.

———. "On the Cofinality of Ordinal Numbers", *Math. Ann.*, 179 (1969), 142–152.

Ayres, F., Jr. *Schaum's Outline of Theory and Problems of Modern Algebra.* New York: McGraw-Hill Book Company, 1965.

Bachmann, H. *Transfinite Zahlen.* Berlin: Springer-Verlag, 1967.

Bernays, P., and A. A. Fraenkel. *Axiomatic Set Theory.* Amsterdam: North-Holland Publishing Company, 1958.

Beth, E. *The Foundations of Mathematics.* Amsterdam: North-Holland Publishing Company, 1959.

Birkhoff, G., and S. MacLane. *A Survey of Modern Algebra*, 3rd ed. New York: Macmillan Publishing Co., Inc., 1965.

Cantor, G. *Contributions to the Theory of Transfinite Numbers.* New York: Dover Publications, Inc., 1952.

Church, A. *Introduction to Mathematical Logic*, Vol. 1. Princeton, N.J.: Princeton University Press, 1956.

Churchill, R. V. *Complex Variables and Applications*, 2nd ed. New York: McGraw-Hill Book Company, 1960.

Cohen, P. J. *Set Theory and the Continuum Hypothesis.* Menlo Park, Calif.: W. A. Benjamin, Inc., 1966.

Copi, I. M. *Symbolic Logic*, 2nd ed. New York: Macmillan Publishing Co., Inc., 1965.

Davis, M. *Mathematical Logic.* New York: New York University Courant Institute Lecture Notes, 1959.

Devidé, V. "A Proof of Zermelo's Theorem", *J. Symbolic Logic*, 32 (1967), 366.

Dinkines, F. *Introduction to Mathematical Logic.* New York: Appleton-Century-Crofts, 1964.

Easton, W. B. "Powers of Regular Cardinals", *Ann. Math. Logic*, 1 (1970), 139–178.

Eisenberg, M. *Axiomatic Theory of Sets and Classes.* New York: Holt, Rinehart and Winston, Inc., 1971.

Feferman, S. *The Number Systems: Foundations of Algebra and Analysis.* Reading, Mass.: Addison-Wesley Publishing Co., Inc., 1964.

Fraenkel, A. A. *Abstract Set Theory*, 2nd ed. Amsterdam: North-Holland Publishing Company, 1961.

——— and Y. Bar-Hillel. *Foundations of Set Theory.* Amsterdam: North-Holland Publishing Company, 1958.

Freilich, G. "A Denumerability Formula for the Rationals", *Amer. Math. Monthly*, 72 (1965), 1013–1014.

Gödel, K. *The Consistency of the Axiom of Choice and of the Generalized Continuum-Hypothesis with the Axioms of Set Theory* (Annals of Math. Studies 3). Princeton, N.J.: Princeton University Press, 1940.

———. "What Is Cantor's Continuum Problem?" *Amer. Math. Monthly*, 54 (1947), 515–525.

Goffman, C. *Real Functions.* Boston: Prindle, Weber & Schmidt, 1953.

Halmos, P. R. *Measure Theory.* New York: Van Nostrand Reinhold Company, 1950.

———. *Naïve Set Theory.* New York: Van Nostrand Reinhold Company, 1960.

———. *Lectures on Boolean Algebras.* New York: Van Nostrand Reinhold Company, 1963.

Hatcher, W. S. *Foundations of Mathematics.* Philadelphia: W. B. Saunders Company, 1968.

Hausdorff, F. *Set Theory.* New York: Chelsea Publishing Company, 1962.

Hayden, S., and J. F. Kennison. *Zermelo–Fraenkel Set Theory.* Columbus, Ohio: Charles E. Merrill Publishers, 1968.

Herstein, I. N. *Topics in Algebra.* Lexington, Mass.: Xerox Corporation, 1964.

Hilbert, D., and W. Ackermann. *Principles of Mathematical Logic.* New York: Chelsea Publishing Company, 1950.

Jech, T. J. "Trees", *J. Symbolic Logic*, 36 (1971), 1–14.

Kamke, E. *Theory of Sets.* New York: Dover Publications, Inc., 1950.

Kelley, J. L. *General Topology.* New York: Van Nostrand Reinhold Company, 1955.

Kleene, S. C. *Introduction to Metamathematics.* New York: Van Nostrand Reinhold Company, 1952.

———. *Mathematical Logic.* New York: John Wiley & Sons, Inc., 1967.

Kopperman, R. D. "Idempotence in Cardinal Arithmetic", *Amer. Math. Monthly*, 77 (1970), 873–874.

Kuratowski, K., and A. Mostowski. *Set Theory.* Amsterdam: North-Holland Publishing Company, 1968.

Landau, E. *Foundations of Analysis.* New York: Chelsea Publishing Company, 1951.

Levy, A., "Parameters in Comprehension Axiom Schemas of Set Theory", *Proc. Symposia Pure Math.*, 25 (1974), 309–324.

Lightstone, A. H. *Symbolic Logic and the Real Number System.* New York: Harper & Row, Publishers, 1965.

Lipschutz, S. *Schaum's Outline of Theory and Problems of Set Theory and Related Topics.* New York: McGraw-Hill Book Company, 1964.

Mac Lane, S., and G. Birkhoff. *Algebra.* Macmillan Publishing Co., Inc., 1967.

Margaris, A. *First Order Mathematical Logic.* Lexington, Mass.: Xerox Corporation, 1967.

Mendelson, E. *Introduction to Mathematical Logic.* New York: Van Nostrand Reinhold Company, 1964.

Monk, J. D. *Introduction to Set Theory.* New York: McGraw-Hill Book Company, 1969.

Morse, A. P. *A Theory of Sets.* New York: Academic Press, Inc., 1965.

Niven, I., and H. Zuckerman. *An Introduction to the Theory of Numbers.* New York: John Wiley & Sons, Inc., 1960.

Pinter, C. C. *Set Theory.* Reading, Mass.: Addison-Wesley Publishing Co., Inc., 1971.

Quigley, F. D. *Manual of Axiomatic Set Theory.* New York: Appleton-Century-Crofts, 1970.

Quine, W. V. *Mathematical Logic*, rev. ed. New York: Harper & Row, Publishers, 1951.

———. *Set Theory and Its Logics.* Cambridge, Mass.: Harvard University Press, 1969.

Ramsey, F. P. "On a Problem in Formal Logic", *Proc. London Math. Soc.* (2), 30 (1930), 264–286.

Rosenbloom, P. C., *Elements of Mathematical Logic.* New York: Dover Publications, Inc., 1950.

Rosser, J. B. *Logic for Mathematicians.* New York: McGraw-Hill Book Company, 1953.

———. *Simplified Independence Proofs.* New York: Academic Press, Inc., 1969.

Royden, H. L. *Real Analysis*, 2nd ed. New York: Macmillan Publishing Co., Inc., 1968.

Rubin, H., and J. Rubin. *Equivalents of the Axiom of Choice.* Amsterdam: North-Holland, 1963.

Rubin, J. E. *Set Theory for the Mathematician.* San Francisco: Holden-Day, Inc., Publisher, 1967.

Sageev, G., "An Independence Result Concerning the Axiom of Choice". Ph.D. Thesis, Hebrew University, Jerusalem, 1973.

Schoenfield, J. R. *Mathematical Logic.* Reading, Mass.: Addison-Wesley Publishing Co., Inc., 1967.

Scott, D. "Definitions by Abstraction in Axiomatic Set Theory", *Bull. Amer. Math. Soc.*, 61 (1955), 442.

Sierpiński, W. "Sur un Problème conduisant à un ensemble non mesurable", *Fund. Math.*, 10 (1927), 177–179.

———. "Sur les Séries infinies de nombres ordinaux", *Fund. Math.*, 36 (1949), 248–253.

———. *Algèbre des ensembles.* Warsaw: Nakadem Polskiego Towarzystwa Matematycznego z Subwencji Ministerstwa Szkol Wyzszych i Nauki, 1951.

———. *Hypothèse du continue*, 2nd ed. New York: Chelsea Publishing Company, 1956.

———. *Cardinal and Ordinal Numbers*, 2nd ed. Warsaw: Państowe Wydawnictwo Naukowe; New York: Hafner Publishing Company, Inc., 1958.

Sigler, L. E. *Exercises in Set Theory*. New York: Van Nostrand Reinhold Company, 1966.
Sikorski, R. *Boolean Algebras*. Berlin: Springer-Verlag, 1960.
Smullyan, R. M. *Lecture Notes on Set Theory* (unpublished). Yeshiva University, 1965.
———. *First-Order Logic*. New York: Springer-Verlag, 1968.
Solovay, R. "Independence Results in the Theory of Cardinals" (abstract). *Notices Amer. Math. Soc.*, 10 (1963), 595.
———. "$2^{\aleph_0}$ Can Be Anything It Ought to Be". In *The Theory of Models, Studies in Logic and the Foundations of Mathematics*. Amsterdam: North-Holland Publishing Company, 1965, 435.
Stoll, R. R. *Set Theory and Logic*. San Francisco: W. H. Freeman & Co., Publishers, 1961.
Suppes, P. *Introduction to Logic*. New York: Van Nostrand Reinhold Company, 1957.
———. *Axiomatic Set Theory*. New York: Van Nostrand Reinhold Company, 1960.
Takeuti, G., and W. M. Zaring. *Introduction to Axiomatic Set Theory*. New York: Springer-Verlag, 1971.
Wilder, R. L. *Introduction to the Foundations of Mathematics*. New York: John Wiley & Sons, Inc., 1952.

Selected Answers and Hints

Chapter 1: The Basic Axioms of Set Theory

Section 1.1

1.1.1. **(c)**, **(d)**, **(e)**, **(g)**. In **(h)**, there is insufficient information; for example, conceivably $A = d$.

1.1.3. **(a)** Suppose that $A \subset B$ and $B \subseteq C$. Then $A \subseteq B$ and $B \subseteq C$; hence, $A \subseteq C$. Because A is properly included in B, there is some b in B for which $b \notin A$. Also, $b \in C$ because $B \subseteq C$. Thus, there is an element of C that is not in A. It follows that $A \neq C$ and, therefore, that $A \subset C$.

(b) The proof here is similar.

(c) No.

(d) (1), (3).

1.1.5. **(a)** Transitivity. **(b)** Reflexivity, transitivity.

(c) (None). **(d)** Reflexivity.

(e) (None).

1.1.6. **(a)**, **(c)**, **(e)**; insufficient information for **(d)**.

Section 1.2

1.2.1. **(a)** For every x, if $x \in A$, then $x \in B$.
(b) A is a subset of B and not $A = B$.

1.2.2. **(a)** $x^2 \leq 0$.
(b) $x + 3 < 5$.
(c) $1 < x$ and $x < 2$.
(d) Not $1 < x$, or not $x < 2$; i.e., $1 \geq x$ or $x \geq 2$.

1.2.4. **(a)** $s\begin{pmatrix} x \\ c \end{pmatrix}$: $c \in A$ and $y \in B$.

$s\begin{pmatrix} y \\ c \end{pmatrix}$: $x \in A$ and $c \in B$.

(b) $s'\begin{pmatrix} x \\ c \end{pmatrix} \equiv s'$ (x has no free occurrence in s'.)

$s'\begin{pmatrix} y \\ c \end{pmatrix}$: For some x, $x = c$; and for all x, $x \in A$.

(c) $s''\begin{pmatrix} x \\ c \end{pmatrix} \equiv s''\begin{pmatrix} y \\ c \end{pmatrix} \equiv s''$.

$s''\begin{pmatrix} z \\ c \end{pmatrix}$: For every x there is a y such that if $x \in A$ and $y \in B$, then either $c \in C$ or $c = x$.

1.2.5. **(a)** 0, 2, 3, 4, 5, 6, 7.
(c) 5.

Section 1.3

1.3.1. **(b)**, **(c)**, **(d)**, **(f)**. There is insufficient information for determining the truth of **(e)**.
1.3.2. **(a)**, **(b)**, **(d)**. In fact, **(d)** is true because "$A = B$" is false.
1.3.3. All but **(a)**.
1.3.5. **(a)**, **(c)**, **(f)**, **(h)**.
1.3.6. **(a)**, **(c)**, **(f)**.
1.3.7. **(c)**, **(d)**, **(e)**, **(f)**.
1.3.8. **(A)**: **(a)**, **(b)**; **(B)**: **(a)**, **(c)**, **(f)**; **(C)**: **(c)**, **(d)**, **(e)**, **(f)**.
1.3.9. **(A)**: **(a)**, **(c)**, **(f)**; **(B)**: **(a)**, **(c)**, **(f)**; **(C)**: **(a)**, **(c)**, **(f)**.

1.3.10. **(d)**

s_1	s_2	s_3	s_4	s_1 or s_2	Not s_3	Both not s_3 and s_4	s_1 or s_2 —implies— both not s_3 and s_4.
t	t	t	t	t	f	f	f
t	t	t	f	t	f	f	f
t	t	f	t	t	t	t	t
t	t	f	f	t	t	f	f
t	f	t	t	t	f	f	f
t	f	t	f	t	f	f	f
t	f	f	t	t	t	t	t
t	f	f	f	t	t	f	f
f	t	t	t	t	f	f	f
f	t	t	f	t	f	f	f
f	t	f	t	t	t	t	t
f	t	f	f	t	t	f	f
f	f	t	t	f	f	f	t
f	f	t	f	f	f	f	t
f	f	f	t	f	t	t	t
f	f	f	f	f	t	f	t

Section 1.4

1.4.1. **(a)** $\{a\} = \{x \in A : x = a\}$.

(b) $\{a, b\} = \{x \in A : x = a \text{ or } x = b\}$.

(c) $\{a, b, c\} = \{x \in A : x \neq d\}$.

1.4.2. **(a)** By the Axiom of Subsets, S is a set; furthermore, $x \in S$ implies $x \in A$, by the definition of S.

(b) Suppose that $A \in S$. Then $A \in A$ and $s\begin{pmatrix} x \\ A \end{pmatrix}$; i.e., $A \in A$ and $A \notin A$. This contradiction establishes **(b)**.

(c) is true by parts **(a)** and **(b)** and rule (5), p. 18.

(d) Either $S \in A$ or (excl.) $S \notin A$. We assume that $S \in A$ and derive a contradiction.

Either $S \in S$ or (excl.) $S \notin S$. Suppose $S \in S$. Then $S \in A$; by the definition of S, $S \notin S$. (Contradiction.) Now suppose that $S \notin S$. By our assumption that $S \in A$ and by the definition of S, we have $S \in S$. (Contradiction.)

Thus, the assumption "$S \in A$" contradicts the true sentence "$S \in S$ or (excl.) $S \notin S$".

1.4.3. Suppose there is a set, X, that contains every set. Then by the Axiom of Subsets, there is a set $Y = \{A \in X : A \notin A\}$. But because X contains every set, "$A \in X$" is automatically true. Hence, $Y = \{A : A \notin A\}$; i.e., Y is the set of all ordinary sets. This contradicts the Russell Paradox.

Section 1.5

1.5.1. Let A be an arbitrary set. Two sets are equal iff they contain the same elements. Thus, if $x \in A$ never holds for any set x, $A = \varnothing$; otherwise, for some set a, $a \in A$. In this case $A \neq \varnothing$.

1.5.3. Suppose that the empty set exists. Then surely there exists a set. Conversely, suppose there exists a set A. By the Axiom of Subsets, $S = \{B \in A : B \notin A\}$ exists (as a set). Thus, S is a set that contains no elements.

1.5.4. All.

1.5.5. Insufficient information. Conceivably, we could have $B \in A$, $A \in A$, and for any set X different from A and B, $X \notin A$; here, A would equal $\{A, B\}$.

1.5.9. **(a)** $\{a, b, c, d\}$. **(b)** C.

(c) $\{a, b, c, d, e\}$. **(d)** $\{a, b, c, d, \{a\}\}$.

(e) Insufficient information. (We do not know the elements of a or of b. We do know that $a \in \cup D$ because $a \in \{a\} \in D$.)

(f) D.

Section 1.6

1.6.1. By Pairing, $\{A\}$ is a set. By Pairing and Union, A^+ $(= A \cup \{A\} = \cup\{A, \{A\}\})$ is also a set.

1.6.2. **(c)**, **(d)**, **(g)**, **(j)**.

1.6.4. **(b)** Assume the Axiom of Infinity; i.e., assume there is a set, ω, consisting of all the natural numbers. By Theorem 1.6.2(1), every natural number is transitive; because $0 \in 1$ and $0 \notin \{1\}$, $\{1\}$ cannot be a natural number.

Suppose $\{1\}$ is not a natural number. Then there must exist some inductive set A for which $\{1\} \notin A$. Let $W = \{x \in A : x \text{ is a natural number}\}$; then W is a set, by the Axiom of Subsets. Because every natural number is a member of A, it follows that $x \in W$ iff x is a natural number. Thus, W equals the set that we have denoted by ω.

1.6.7. **(a)** Let $A = \{X \in \omega : X \text{ is ordinary}\}$. Surely, $0 \in A$, because $y \notin 0$ for every set y. Suppose that $N \in A$, but $N^+ \notin A$. Then $N^+ \in N^+ = N \cup \{N\}$, i.e., either $N^+ \in N$ or $N^+ \in \{N\}$.

Suppose that $N^+ \in N$. Then by the transitivity of N, $N^+ \subseteq N$; hence, $N \in N$, because $N \in N^+$. This contradicts the ordinariness of N.

Now suppose that $N^+ \in \{N\}$; thus, $N^+ = N$. But then $N \in N^+ = N$. (Contradiction.)

Thus, $N \in A$ implies $N^+ \in A$; A, as an inductive subset of ω, must equal ω. Consequently, every natural number is ordinary.

(b) Suppose that ω is a natural number. Then $\omega \in \omega$. Hence, by **(a)**, ω is ordinary. Thus, $\omega \notin \omega$. (Contradiction.)

(c) By **(b)**, $\omega \notin \omega$; thus, ω is ordinary.

1.6.8. **(a)** $4 = \{0, \{0\}, \{0, \{0\}\}, \{0, \{0\}, \{0, \{0\}\}\}\}$.

(b) $\cup 4 = \cup\{0, 1, 2, 3\} = \{x : x \in 0 \text{ or } x \in 1 \text{ or } x \in 2 \text{ or } x \in 3\} = \{0, 1, 2\} = 3$.

(c) 0 is in every inductive set; hence, $1(=0^+)$ is in every inductive set. Similarly, $2(=1^+)$, $3(=2^+)$, and $4(=3^+)$ are in every inductive set.

(d) $4=\{0,1,2,3\}$.
$0=\varnothing\subseteq 4$. (In fact, $\varnothing$ is a subset of every set.)
$1=\{0\}\subseteq 4$. $2=\{0,1\}\subseteq 4$. $3=\{0,1,2\}\subseteq 4$.

(e) $4=\{0,1,2,3\}$. As indicated in the proof of **(d)**, $3\in 4$, but 3 is not an element of any element of 4. Thus, $4\neq 0, 4\neq 1, 4\neq 2, 4\neq 3$; i.e., $4\notin 4$.

1.6.9. **(b)** The proof is by induction. Let A be the set of all natural numbers N that have this property. Let K be an arbitrary set. Then $0\in A$ because "$K\in 0$" is false. Suppose that $N\in A$; i.e., suppose that if $K\in N$, then either $K^+\in N$ or (excl.) $K^+=N$. We must show that if $X\in N^+$, then either $X^+\in N^+$ or (excl.) $X^+=N^+$. Suppose that $X\in N^+$. Then because $N^+=N\cup\{N\}$, it follows that $X\in N$ or (excl.) $X=N$. By the inductive hypothesis, if $X\in N$, then $X^+\in N$ or (excl.) $X^+=N$. But if $X^+\in N$, then $X^+\in N^+$ because $N\subset N^+$. If $X=N$, then $X^+=N^+$. Because N^+ is ordinary, either $X^+\in N^+$ or (excl.) $X^+=N^+$. It follows that $N^+\in A$ and that $A=\omega$.

1.6.13. Say that a set A is "★-inductive" if (1) $0\in A$ and (2) whenever $x\in A$; then $x^\star\in A$. Proceed as in the development leading up to the Axiom of Infinity.

Section 1.7

1.7.2. The elements of $\mathcal{P}(A)$ are the following: $\varnothing$; $\{1\}$, $\{\{1,2\}\}$, $\{2\}$, $\{\{\{1,2\},\{1\}\}\}$; $\{1,\{1,2\}\}$, $\{1,2\}$, $\{1,\{\{1,2\},\{1\}\}\}$, $\{\{1,2\},2\}$, $\{\{1,2\},\{\{1,2\},\{1\}\}\}$, $\{2,\{\{1,2\},\{1\}\}\}$; $\{1,\{1,2\},2\}$, $\{1,\{1,2\},\{\{1,2\},\{1\}\}\}$, $\{1,2,\{\{1,2\},\{1\}\}\}$, $\{\{1,2\},2,\{\{1,2\},\{1\}\}\}$; A.

1.7.3. $\mathcal{P}(A)=\{\varnothing,\{a\},\{b\},\{a,b\}\}$. The following are the elements of $\mathcal{P}(\mathcal{P}(A))$: $\varnothing$; $\{\varnothing\}$, $\{\{a\}\}$, $\{\{b\}\}$, $\{\{a,b\}\}$; $\{\varnothing,\{a\}\}$, $\{\varnothing,\{b\}\}$, $\{\varnothing,\{a,b\}\}$, $\{\{a\},\{b\}\}$, $\{\{a\},\{a,b\}\}$, $\{\{b\},\{a,b\}\}$; $\{\varnothing,\{a\},\{b\}\}$, $\{\varnothing,\{a\},\{a,b\}\}$, $\{\varnothing,\{b\},\{a,b\}\}$, $\{\{a\},\{b\},\{a,b\}\}$; $\mathcal{P}(A)$.

1.7.4. First, $0\ (=\varnothing)\subseteq X$; thus, $0\in\mathcal{P}(X)$, $1=\{0\}\subseteq\mathcal{P}(X)$; thus, $1\in\mathcal{P}(\mathcal{P}(X))$. Next, $0\subseteq\mathcal{P}(X)$; hence, $0\in\mathcal{P}(\mathcal{P}(X))$. Therefore, $2=\{0,1\}\subseteq\mathcal{P}(\mathcal{P}(X))$; consequently, $2\in\mathcal{P}(\mathcal{P}(\mathcal{P}(X)))$.

1.7.6. **(c)** alone. For $X\subseteq X$. Thus, $X\in\mathcal{P}(X)$ and $\{X\}$ is a (unit) subset of $\mathcal{P}(X)$. The fact that $X\in\mathcal{P}(X)$ means that **(d)** is false for every set X. "$X=\{1\}$" is a counterexample to **(a)** and **(b)**; here $\mathcal{P}(X)=\{\varnothing,\{1\}\}$.

1.7.7. **(a)** Suppose that $A\subseteq B$. Let $x\in\mathcal{P}(A)$. Then $x\subseteq A\subseteq B$, and $x\in\mathcal{P}(B)$. Therefore, $\mathcal{P}(A)\subseteq\mathcal{P}(B)$. Now suppose that $\mathcal{P}(A)\subseteq\mathcal{P}(B)$. Then $A\in\mathcal{P}(A)$; hence, $A\in\mathcal{P}(B)$. Therefore, $A\subseteq B$.

1.7.9. **(b)** $\mathcal{P}(A)\cup\mathcal{P}(B)=\mathcal{P}(A\cup B)$ iff either $A\subseteq B$ or $B\subseteq A$.

(c) Suppose that $A\subseteq B$ or $B\subseteq A$. Then $A\cup B=A$ or $A\cup B=B$. Thus, $\mathcal{P}(A\cup B)=\mathcal{P}(A)$ or $\mathcal{P}(A\cup B)=\mathcal{P}(B)$. In either case,

$$\mathcal{P}(A\cup B)\subseteq\mathcal{P}(A)\cup\mathcal{P}(B)$$

and, by **(a)**, equality follows.

If $A \nsubseteq B$ and $B \nsubseteq A$, there are sets a and b such that $a \in A$, $a \notin B$, $b \notin A$, $b \in B$. Then $\{a, b\} \in \mathcal{P}(A \cup B)$ and $\{a, b\} \notin \mathcal{P}(A) \cup \mathcal{P}(B)$.

1.7.13. **(a)** Say that a set A has property E if $0 \in A$ and if whenever $x \in A$, then $(x^+)^+ \in A$. (Clearly, ω has property E.) Say that a set is even if it is in every set that has property E. Let $\mathbf{E} = \{N \in \omega : N \text{ is even}\}$.

Section 1.8

1.8.1. Suppose that $\cap\varnothing$ is in

$$\{b : \text{for every } x \text{ in } \varnothing, b \in x\}.$$

$$\text{For every } x \text{ in } \varnothing, b \in x$$

means

$$\text{For every } x, \text{ if } x \in \varnothing, \text{ then } b \in x.$$

Let $s(x)$ be "If $x \in \varnothing$, then $b \in x$". For every set a, $s(a)$ is true because "$a \in \varnothing$" is false. Thus, "For every x, $s(x)$" is true. It follows that for every set b, $b \in \cap\varnothing$. Hence $\cap\varnothing$ is the set of all sets; this contradicts Cantor's Paradox (Exercise 1.4.3).

1.8.2. **(a)** $\{8, 10\}$. **(b)** $\{8\}$.
(c) $\varnothing$. **(d)** $\varnothing$.
(e) $\{1, 2, 4, 6, 7\}$. **(f)** $\{1, 6, 7\}$.
(g) $\{1, 2, 6, 7\}$. **(h)** $\{1, 6\}$.

1.8.4. **(a)** $\{2, 3\}$. **(b)** $\varnothing$. **(c)** $\{1\}$.

1.8.7. **(a)** Suppose that $a \in A$. Then surely $a \in A$ or $a \in B$; thus, $A \subseteq A \cup B$.
(b) Suppose that $a \in A \cap B$. Then $a \in A$ and $a \in B$. Thus, $a \in A$; hence, $A \cap B \subseteq A$.
(c) Suppose that $X \in \mathcal{P}(A) \cap \mathcal{P}(B)$. Then $X \subseteq A$ and $X \subseteq B$. For every x in X, $x \in A$ and $x \in B$. Thus, $x \in A \cap B$, from which it follows that $X \subseteq A \cap B$, $X \in \mathcal{P}(A \cap B)$, and $\mathcal{P}(A) \cap \mathcal{P}(B) \subseteq \mathcal{P}(A \cap B)$.

Suppose that $Y \in \mathcal{P}(A \cap B)$. Then $Y \subseteq A \cap B$. But $A \cap B \subseteq A$ and $A \cap B \subseteq B$. Therefore, $Y \subseteq A$ and $Y \subseteq B$. Thus, $Y \in \mathcal{P}(A) \cap \mathcal{P}(B)$ and $\mathcal{P}(A \cap B) \subseteq \mathcal{P}(A) \cap \mathcal{P}(B)$. By the first part, these two sets are the same.

1.8.8. **(b)** $A \cap B = \varnothing$ or $A \subseteq B$.
(d) If $A \subseteq B$, then $A \sim B = \varnothing$. Thus, $\mathcal{P}(A \sim B) = \{\varnothing\}$. Furthermore, $\mathcal{P}(A) \subseteq \mathcal{P}(B)$; hence, $\mathcal{P}(A) \sim \mathcal{P}(B) = \varnothing \in \{\varnothing\} = \mathcal{P}(A \sim B)$.

1.8.10. **(1a)** 1, 2, 3. **(4a)** 0.

1.8.11. **(2b)** 7. **(3a)** 1, 4, 5, 6, 7. **(5b)** 1, 5, 7.

1.8.13. **(3c)** $x \in P \cup (\cap Q)$ iff $x \in P$ or $x \in \cap Q$, iff $x \in P$ or for each q in Q, $x \in q$, iff $x \in P \cup q$ for each q in Q, iff $x \in \cap\{P \cup q : q \in Q\}$.
(5c)
$$\begin{aligned}(P \sim Q) \cup R &= (P \cap \tilde{Q}) \cup R = (P \cup R) \cap (\tilde{Q} \cup R)\\ &= (P \cup R) \cap \widetilde{(Q \cap \tilde{R})} = (P \cup R) \cap \widetilde{(Q \sim R)}\\ &= (P \cup R) \sim (Q \sim R).\end{aligned}$$

Chapter 2: Relations

Section 2.1

2.1.3. Sets (1), (5), and (7) are equal, as are sets (2) and (9), sets (3) and (10), and sets (6) and (8). (No other equalities exist.)

2.1.4. $\langle\{a\}, \{a\}\rangle$.

2.1.6. **(a)** By Exercise 1.6.9(a), every nonzero natural number contains 0. For sets x and y, $\{x\} \in \langle x, y\rangle$; hence, $\langle x, y\rangle \neq 0$. Thus, if $\langle x, y\rangle = \{\{x\}, \{x, y\}\} \in \omega$, then either $\{x\} = 0$ or $\{x, y\} = 0$. But $x \in \{x\}$ and $x \in \{x, y\}$, whereas $x \notin 0$. Therefore, $\langle x, y\rangle$ cannot be a natural number.

(b) Suppose that $\langle x, y\rangle \subset \omega$. Then both $\{x\}$ and $\{x, y\}$ must be natural numbers. Thus, x must be 0. Also, $\{x, y\} = \{0, y\}$; consequently, y must be 0 or 1. Thus, $\langle 0, 0\rangle$ and $\langle 0, 1\rangle$ are the only ordered pairs that are subsets of ω.

2.1.7. Use the Axioms of Pairing, Power Set, and Subsets—in this order.

Section 2.2

2.2.1. For a in A and b in B, $\langle a, b\rangle$ $(= \{\{a\}, \{a, b\}\}) \in \mathscr{P}(\mathscr{P}(A \cup B))$. Thus, $A \times B = \{x \in \mathscr{P}(\mathscr{P}(A \cup B)) :$ for some a in A and for some b in B, $x = \langle a, b\rangle\}$.

2.2.2. $A = \{0, 1, 2\}$, $B = \{0, 1\}$.

(a) $A \times B = \{\langle 0,0\rangle, \langle 0,1\rangle, \langle 1,0\rangle, \langle 1,1\rangle, \langle 2,0\rangle, \langle 2,1\rangle\}$.

(b) $B \times A = \{\langle 0,0\rangle, \langle 0,1\rangle, \langle 0,2\rangle, \langle 1,0\rangle, \langle 1,1\rangle, \langle 1,2\rangle\}$.

2.2.4. **(a)** $A \times B = \{\langle a, b\rangle : a \in A \text{ and } b \in B\} = \{\{\{a\}, \{a, b\}\} : a \in A \text{ and } b \in B\}$.

$\cup(A \times B) = \{x : x = \{a\}$ for some a in A, or $x = \{a, b\}$ for some a in A and some b in $B\}$.

$\cup(\cup(A \times B)) = \{x : x \in \{a\}$ for some a in A, or $x \in \{a, b\}$ for some a in A and some b in $B\} = A \cup B$.

(b) Let $B = A$, in **(a)**.

Section 2.3

2.3.2. **(a)** Dom $R_1 = \{x \in \cup\cup R_1 :$ for some y, $xR_1y\} = \{0, 2, 4\}$.

(b) Range $R_1 = \{y \in \cup\cup R_1 :$ for some x, $xR_1y\} = 3$.

(c) Field $R_2 =$ dom $R_2 \cup$ range $R_2 = 2 \cup 4 = 4$.

(d)
$$\begin{aligned} R_2 \circ R_1 &= \{\langle x, z\rangle \in \text{dom } R_1 \times \text{range } R_2 : \text{for some } y, \text{ both } xR_1y \text{ and } yR_2z\} \\ &= \{\langle 0,0\rangle, \langle 0,1\rangle, \langle 0,2\rangle, \langle 0,3\rangle, \langle 2,0\rangle, \langle 2,1\rangle, \langle 2,2\rangle, \langle 2,3\rangle, \\ &\qquad \langle 4,0\rangle, \langle 4,1\rangle, \langle 4,2\rangle, \langle 4,3\rangle\} \\ &= \{0, 2, 4\} \times 4. \end{aligned}$$

(e) $R_1 \circ R_2 = \{\langle 0,0\rangle, \langle 1,0\rangle, \langle 0,1\rangle, \langle 1,1\rangle, \langle 0,2\rangle, \langle 1,2\rangle\} = 2 \times 3$.

(f) $R_1^{-1} = \{\langle 0,0\rangle, \langle 1,2\rangle, \langle 2,2\rangle, \langle 0,4\rangle\}$.

(g) $R_2^{-1} \circ R_1 = \{\langle 0,0\rangle, \langle 0,1\rangle, \langle 2,0\rangle, \langle 2,1\rangle, \langle 4,0\rangle, \langle 4,1\rangle\}$.

(h) $R_1''(4) = \{x :$ for some N in 4, $NR_1x\} = \{0, 1, 2\} = 3$ because $0R_10$, $2R_11$, and $2R_12$.

2.3.3. **(a)** $\langle x, y\rangle \in (R_1^{-1})^{-1}$ iff $\langle y, x\rangle \in R_1^{-1}$ iff $\langle x, y\rangle \in R_1$.

(b) $\langle x, z\rangle \in (R_2 \circ R_1)^{-1}$ iff $zR_2 \circ R_1x$, iff for some y in dom R_2—both zR_1y and yR_2x, iff for some y in range R_2^{-1}—both $xR_2^{-1}y$ and $yR_1^{-1}z$, iff

$$\langle x, z\rangle \in R_1^{-1} \circ R_2^{-1}.$$

2.3.5. **(a)** $\{\langle 0,6\rangle, \langle 1,0\rangle, \langle 1,1\rangle, \langle 1,2\rangle, \langle 1,3\rangle, \langle 1,4\rangle, \langle 1,5\rangle, \langle 1,6\rangle, \langle 2,6\rangle\}$.

(c) $\{\langle 0,0\rangle, \langle 0,1\rangle, \langle 0,6\rangle, \langle 1,1\rangle, \langle 1,4\rangle, \langle 1,6\rangle, \langle 2,1\rangle, \langle 2,6\rangle\}$.

2.3.6. **(c)**

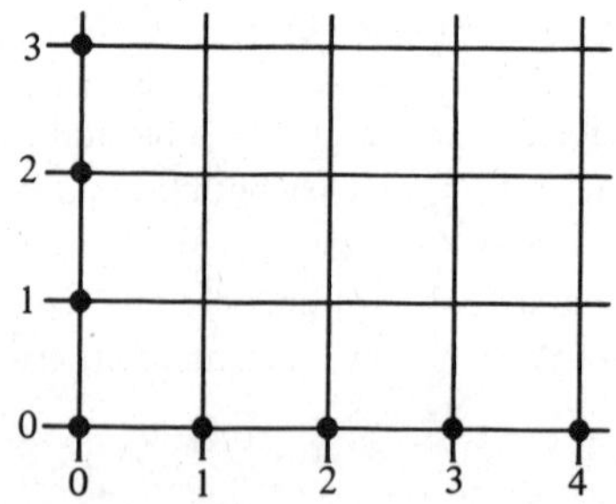

2.3.7. **(a)** $y \in R''(A \cup B)$ iff for some x in $A \cup B$, xRy, iff for some x in A, xRy or for some x in B, xRb, iff $y \in R''(A)$ or $y \in R''(B)$, iff $y \in R''(A) \cup R''(B)$.

(b) Suppose that $y \in R''(A \cap B)$. Then for some x in $A \cap B$, xRy. But for such an x, $x \in A$ and $x \in B$; hence, there is an x in A such that xRy and there is an x in B such that xRy. Thus, $y \in R''(A) \cap R''(B)$.

(c) Let $A_1 = 2$, let $B_1 = \{2, 3\}$, and let $R = \{\langle 0,2\rangle, \langle 2,2\rangle\}$. Then $A_1 \cap B_1 = \varnothing$; hence, $R''(A_1 \cap B_1) = \varnothing$. Also, $0R2$ and $2R2$; thus, $2 \in R''(A_1) \cap R''(B_1)$.

Section 2.4

2.4.1. If $\mathscr{F}$ is the collection of points of the plane that represent the elements of a function F in a graphic diagram, then any vertical line in the plane intersects $\mathscr{F}$ in at most one point. F is one–one iff any horizontal line in the plane intersects $\mathscr{F}$ in at most one point.

(a) Not a function.

(b) Function (neither one-one nor onto 5).

(c) One-one function onto 5.

(d) Function with domain $\{1, 2, 3\}$ (one-one but not onto 5).

2.4.2. R_3; R_5 (assuming a monogamous social structure).

2.4.3. **(a)** $F_2 \circ F_1 = \{\langle 0,2\rangle, \langle 1,0\rangle, \langle 2,2\rangle, \langle 3,3\rangle\}$.

(b) $(F_2 \circ F_1)^{-1} = \{\langle 2,0\rangle, \langle 0,1\rangle, \langle 2,2\rangle, \langle 3,3\rangle\}$. Note that this is not a function.

(c) $F_3 \circ F_4 = \{\langle 0,1\rangle, \langle 2,1\rangle, \langle 3,2\rangle\}$.

(d) $F_1 \upharpoonright 3 = \{\langle 0,2\rangle, \langle 1,5\rangle, \langle 2,2\rangle\}$.

2.4.4. **(a)** $G \circ F = \{\langle a, c\rangle : \text{for some } b \text{ in } B, \text{ both } \langle a, b\rangle \in F \text{ and } \langle b, c\rangle \in G\}$.

For each a in A there is some unique b in B for which $\langle a, b\rangle \in F$ (because F is a function with domain A); similarly, for each b in B there is some unique c in C for which $\langle b, c\rangle \in G$. Thus, to each a in A there is a unique c in C corresponding to a; c is characterized by the fact that $\langle F(a), c\rangle \in G$. Thus, $G \circ F : A \to C$.

(b) Suppose that F and G are each one-one and suppose that $G \circ F$ is not one-one. Then for some c in C, $\langle a, c\rangle \in G \circ F$, $\langle a', c\rangle \in G \circ F$, and $a \neq a'$. It follows from the definition of $G \circ F$ that there exist b, b' in B such that $\langle a, b\rangle \in F$, $\langle a', b'\rangle \in F$, $\langle b, c\rangle \in G$, $\langle b', c\rangle \in G$. Furthermore, $b = b'$ because G is one-one. Therefore, $\langle a, b\rangle \in F$ and $\langle a', b'\rangle \in F$; thus, $a = a'$ because F is one-one. (Contradiction.)

(c) Let $F : 2 \to 2$ and $G : 2 \to 2$ be such that $F(0) = 0$, $F(1) = G(0) = G(1) = 1$. Then $G \circ F : 2 \to 2$, $G \circ F(0) = G(0) = 1$, $G \circ F(1) = G(1) = 1$; thus, $G \circ F$ is not one-one.

2.4.6. (a) Suppose that $x \in F^{-1''}(A \sim B)$. Then for some y in $A \sim B$, $x = F^{-1}(y)$, i.e., for some y, $F(x) = y \in A \sim B \subseteq A$. It follows that $x \in F^{-1''}(A)$. Suppose that $x \in F^{-1''}(B)$, as well. Then for some z in B, $x = F^{-1}(z)$; i.e., for some $z\,(\neq y)$, $F(x) = z$. This would contradict the assumption that F is a function. Thus, $x \in F^{-1}(A) \sim F^{-1}(B)$.

Suppose that $x \in F^{-1''}(A) \sim F^{-1''}(B)$. Then for some y in A, $x = F^{-1}(y)$. Also, $y \notin B$, as otherwise $x \in F^{-1''}(B)$. Thus, $x \in F^{-1''}(A \sim B)$.

2.4.9. (a) It suffices to show that $\chi_{A\cap B}(x) = 1$ iff $\chi_A(x) \cap \chi_B(x) = 1$. $\chi_{A\cap B}(x) = 1$ iff $x \in A \cap B$, iff $x \in A$ and $x \in B$, iff $\chi_A(x) = 1$ and $\chi_B(x) = 1$, iff $\chi_A(x) \cap \chi_B(x) = 1 \cap 1 = 1$.

(c) Suppose that $A \subseteq B$. Let $x \in \mathscr{U}$. Then if $x \notin A$, $\chi_A(x) = 0 \subseteq \chi_B(x)$. If $x \in A$, then $x \in B$ and $\chi_A(x) = \chi_B(x) = 1$; hence, $\chi_A(x) \subseteq \chi_B(x)$.

If $A \not\subseteq B$, there is some x in $\mathscr{U}$ for which $x \in A$ and $x \notin B$. Hence, $\chi_A(x) = 1$, $\chi_B(x) = 0$, and $\chi_A(x) \subseteq \chi_B(x)$.

(d) No. If $\mathscr{U} = \omega$, $A = 1$, $B = 2$, then $A \subset B$; yet $\chi_1(0) = \chi_2(0) = 1$.

2.4.10. $2(N^2)$.

2.4.11. Because a one-one function of N onto A enumerates the elements of A, we can consider the number of different enumerations. For $N \geq 1$, there are N choices for the element a_0, $N - 1$ choices for a_1, and, in general, $N - J$ choices for the element a_J, $J = 0, 1, 2, \ldots N$. There are $N \cdot (N - 1) \cdot (N - 2) \ldots 3 \cdot 2 \cdot 1$ (or $N!$) such one-one correspondences between N and A. For $N = 0$, there is only one such function, the empty set (of ordered pairs); as is customary, we let $0! = 1$.

In other words, for each N in ω the collection B of one-one functions of N onto A is a subset of $(N \times A)$ (by the Axiom of Subsets); there exists a one-one function of $N!$ onto B.

2.4.12. N^M (unless both N and M are 0—in which case there is 1). Note that $Y^{\varnothing} = \{\varnothing\}$; if $X \neq \varnothing$, then $\varnothing^X = \varnothing$.

Section 2.5

2.5.1. (a) None of these.

(b) Symmetric.

(c) Reflexive, symmetric, transitive (equivalence relation).
(d) Symmetric.
(e) (None of these.)
(f) Transitive.

2.5.2. **(a), (b), (d).**

2.5.3. **(a)** Let $R_1 = \{\langle 0,0\rangle, \langle 1,0\rangle, \langle 0,1\rangle, \langle 1,1\rangle\}$.
Note that R_1 is an equivalence relation on 2.
(b) Let $R_2 = \{\langle 0,0\rangle, \langle 1,1\rangle, \langle 2,2\rangle, \langle 0,1\rangle, \langle 1,0\rangle, \langle 1,2\rangle, \langle 2,1\rangle\}$.
(c) Let $R_3 = \{\langle 0,0\rangle, \langle 1,1\rangle, \langle 2,2\rangle, \langle 0,1\rangle, \langle 1,2\rangle, \langle 0,2\rangle\}$.

2.5.4. **(b)** and **(e)** represent equivalence relations. The corresponding partitions are, respectively, $\{\{1,4\}, \{2,3\}\}$, and $\{\{1,2,4\}, \{3\}\}$. **(a)** lacks symmetry ($2R1$ but not $1R2$) as well as transitivity ($2R1$ and $1R3$, but not $2R3$). **(c)** lacks reflexivity (not $2R2$) as well as transitivity ($2R3$ and $3R2$, but not $2R2$). **(d)** lacks transitivity ($2R1$ and $1R3$, but not $2R3$).

2.5.5. **(a)** $R_1 = \{\langle 1,1\rangle, \langle 2,2\rangle, \langle 1,2\rangle, \langle 2,1\rangle, \langle 3,3\rangle, \langle 4,4\rangle, \langle 5,5\rangle, \langle 6,6\rangle, \langle 3,4\rangle, \langle 4,3\rangle, \langle 3,5\rangle, \langle 5,3\rangle, \langle 3,6\rangle, \langle 6,3\rangle, \langle 4,5\rangle, \langle 5,4\rangle, \langle 4,6\rangle, \langle 6,4\rangle, \langle 5,6\rangle, \langle 6,5\rangle, \langle 7,7\rangle\}$.

(b)

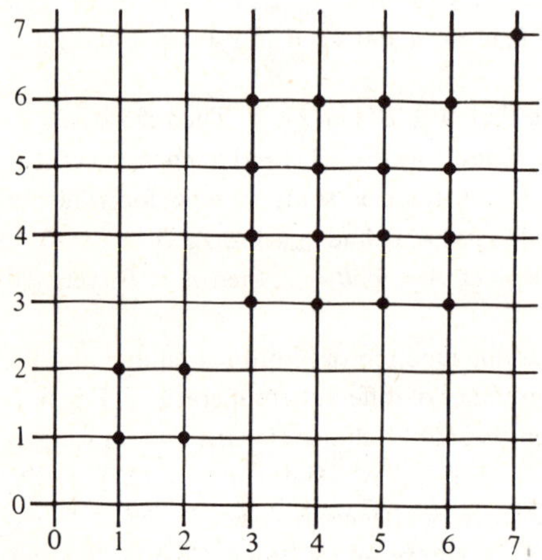

2.5.6. **(a)** We need only show that R is reflexive. If R is empty, then $\operatorname{dom} R = \varnothing$. Hence, for any a in $\varnothing$, aRa. Suppose that $R \neq \varnothing$; then $\operatorname{dom} R \neq \varnothing$. If $a \in \operatorname{dom} R$, then for some b, aRb. By symmetry, bRa, and by transitivity, aRa. Thus, R is reflexive. If $\operatorname{dom} R \subset A$, then R is not reflexive on A.

2.5.12. Let $Y = \{x \cap B : x \in X\} \sim \{\varnothing\}$. By definition, the elements of Y are nonempty. If $x_1 \cap B$ and $x_2 \cap B$ are distinct elements of Y, then $x_1 \neq x_2$. Because x_1 and x_2 are distinct elements of the partition X, they are disjoint. Hence, $(x_1 \cap B) \cap (x_2 \cap B) \subseteq x_1 \cap x_2 = \varnothing$. Finally, $\cup\, Y = \cup(\{x \cap B : x \in X\} \sim \{\varnothing\})$ $= \cup(x \cap B : x \in X\} = B \cap \cup\{x : x \in X\} = B \cap X = B$.

2.5.15. If $K_1 + K_2 + \cdots + K_R = N$, $1 \le K_J \le N$, $J = 1, 2, \ldots, R$, then by a $(K_1, K_2, \ldots, K_R)$-type partition of N we mean a partition that contains a K_1-element cell, a K_2-element cell, ..., a K_R-element cell. Clearly, X is a partition of an N-element set iff X is a $(K_1, K_2, \ldots, K_R)$-type partition, where $K_1 + K_2 + \cdots + K_R = N$. For $N = 3, 4, 5$ we list the various types of partitions as well as the number of distinct partitions of each type.

	Type	Number of each type
(a)	(1, 1, 1)	1
	(1, 2)	3
	(3)	1
		5
(b)	(1, 1, 1, 1)	1
	(1, 1, 2)	6
	(1, 3)	4
	(2, 2)	3
	(4)	1
		15
(c)	(1, 1, 1, 1, 1)	1
	(1, 1, 1, 2)	10
	(1, 1, 3)	10
	(1, 2, 2)	15
	(1, 4)	5
	(2, 3)	10
	(5)	1
		52

For example, in determining the number of (1, 2, 2)-type partitions we note that there are five ways of selecting the 1-element cell; of the remaining four elements a, b, c, and d, there are 3 ways of associating them—$\{a, b\}$, $\{c, d\}$; $\{a, c\}$, $\{b, d\}$; $\{a, d\}$, $\{b, c\}$.

Section 2.6

2.6.2. **(a)**

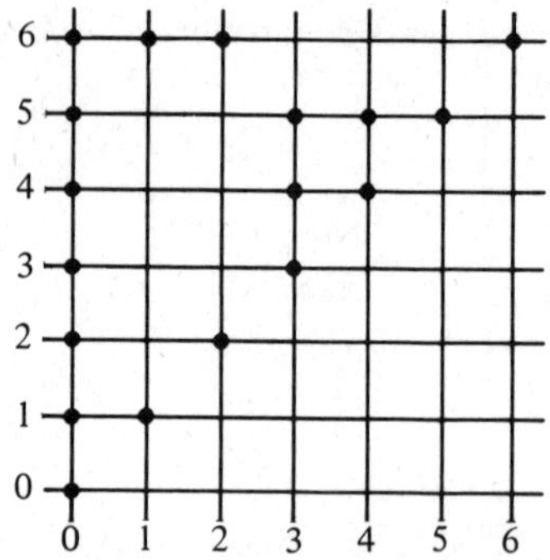

Minimal elements: 0
Maximal elements: 5, 6
First element: 0
Last element: (none)

(d)

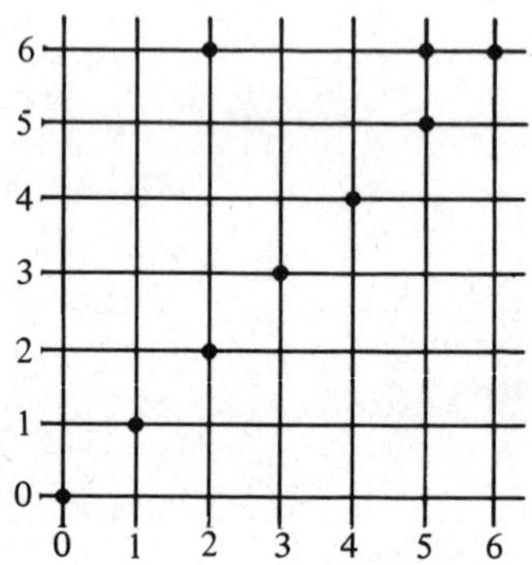

Minimal elements: 0, 1
2, 3, 4, 5
Maximal elements: 0, 1
3, 4, 6
First element: (none)
Last element: (none)

2.6.4. **(a)** Partial ordering.

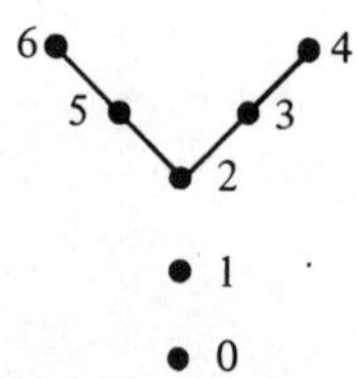

(b) Not partial ordering (not antisymmetric, e.g., $0R6$ and $6R0$).
(c) Not partial ordering (not transitive, e.g., $0R1$, $1R2$, but $0R2$).
(d) Linear ordering (hence partial ordering).

2.6.5. $R_0 = \{\langle N, N\rangle : N \in 5\} \cup \{\langle 0,2\rangle, \langle 2,3\rangle, \langle 2,1\rangle, \langle 3,4\rangle, \langle 0,3\rangle, \langle 0,4\rangle, \langle 0,1\rangle, \langle 2,4\rangle\}$.

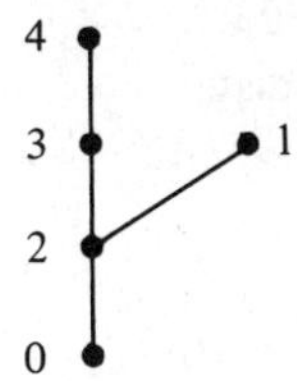

2.6.6. **(b)** Suppose $A = \{1,2\}$, $B = \{2,3\}$, $R = \{\langle 1,1\rangle, \langle 1,2\rangle, \langle 2,2\rangle\}$, and $S = \{\langle 2,2\rangle, \langle 2,3\rangle, \langle 3,3\rangle\}$. Then R is a linear ordering on A, S is a linear ordering on B. However, $R \cup S$ is not a partial ordering on $A \cup B$ because $\langle 1,2\rangle \in R \cup S$, $\langle 2,3\rangle \in R \cup S$, but $\langle 1,3\rangle \notin R \cup S$. (Here, $A \cap B \neq \varnothing$.)

2.6.7. Recall that for $M, N \in \omega$, $M \leq N$ iff either $M \in N$ or $M = N$.

(a) $0 = 0$; hence, $0 \leq 0$. For $N \neq 0$, $0 \in N$; thus, $0 \leq N$. It follows that 0 is a lower bound for ω. But if N is any lower bound for ω, then because $0 \leq N$ and $N \leq 0$, N must equal 0. Thus, 0 is the only lower bound for ω; hence, it is the greatest lower bound for ω.

(b) For each N in $6 (= \{0,1,2,3,4,5\})$, $N \leq 5$; thus, 5 is an upper bound for 6. If M in ω is such that for each N in 6, $N \leq M$, then $6 \subseteq M$ or (excl.) $M = 5$. Also, $5 \in 6 \subseteq M$ implies $5 \leq M$. Thus, 5 is the least upper bound for 6.

(c) For each N in ω, $N^+ \in \omega$ and $N^+ \nleq N$. Hence, no N in ω can be an upper bound for ω.

(d) 0, 1, and 2 are each lower bounds for $\omega \sim \{0, 1, 5, 6\}$; 2 is the greatest lower bound for this set.

Section 2.7

2.7.2. Let R be a well-ordering on A. Let B be any subset of A, and consider the relation $R_B = R \cap B^{②}$. We claim that this is a well-ordering on B.

In fact, the properties of R carry over to R_B. For example, if $B \neq \varnothing$ and if a, b, and c in B are such that aR_Bb, bR_Bc, then aRb and bRc. Because R is transitive, aRc; hence, aR_Bc, and R_B is transitive. Similarly, reflexivity, antisymmetry, and the comparability of any two elements of B carry over from R to R_B. Consequently, R_B linearly orders B. Let B_0 be a nonempty subset of B; then B_0 is a nonempty subset of A. As such, it contains a first element, b_0, with respect to R. Thus, b_0Rb for every b in B; hence, b_0R_Bb for every b in B. It follows that b_0 is a minimal element of B. Because R_B linearly orders B, b_0 must be the first element of B.

Finally, if $B = \varnothing$, each of the properties, reflexivity, antisymmetry, transitivity, and comparability hold vacuously for $R_B (= \varnothing)$. Similarly, B has no nonempty subsets; hence, every nonempty subset of B has a first element with respect to R_B. Thus, R_B is a well-ordering.

2.7.3. Let $\langle X, R\rangle$ be a partially ordered system in which every nonempty subset has a first element. In order to show that X is well-ordered by R, we need only show that any two distinct elements x, y are R-comparable. But $\{x, y\}$ has a first element, say x. Thus, xRy.

The other half is immediate.

Section 2.8

2.8.1. **(a)**
$$A \,\Delta\, B = (A \sim B) \cup (B \sim A)$$
$$= (A \cap \tilde{B}) \cup (B \cap \tilde{A}).$$

(b)
$$(A \cup B) \sim (A \cap B) = (A \cup B) \cap \widetilde{(A \cap B)}$$
$$= (A \cup B) \cap (\tilde{A} \cup \tilde{B})$$
$$= (A \cap (\tilde{A} \cup \tilde{B})) \cup (B \cap (\tilde{A} \cup \tilde{B}))$$
$$= ((A \cap \tilde{A}) \cup (A \cap \tilde{B})) \cup ((B \cap \tilde{A}) \cup (B \cap \tilde{B}))$$
$$= (A \cap \tilde{B}) \cup (B \cap \tilde{A})$$
$$= A \,\Delta\, B.$$

2.8.3. **(a)** A. **(b)** $\tilde{A}$.
(c) $\varnothing$. **(d)** $\mathscr{U}$.

2.8.4. **(a)** $A = B$. **(b)** $A = \tilde{B}$.
(c) $B \subseteq A$. **(d)** $A \subseteq B$.

2.8.8. **(a), (c), (d).**

2.8.10. Define H to be the function $\{\langle x, G(\{x\})\rangle : x \in X\}$. Then $H : X \to X$. By Definition by Induction, there is a unique function F from ω into X that is such that $F(0) = x_0$ and for all N in ω, $F(N^+) = H(F(N)) = G(\{F(N)\})$. Then F is certainly a function with the required properties. Moreover, were there a second such function, F', then F' would also satisfy the requirements for the function in the theorem on Definition by Induction; hence, it would contradict the uniqueness of that function.

2.8.11. **(a)** $F(0) = a$, $F(1) = c$, $F(2) = e$, $F(3) = b$, $F(4) = d$; in general,

$$F(m) = \begin{cases} a, & \text{if } M \equiv 0 \pmod 5; \\ c, & \text{if } M \equiv 1 \pmod 5; \\ e, & \text{if } M \equiv 2 \pmod 5; \\ b, & \text{if } M \equiv 3 \pmod 5; \\ d, & \text{if } M \equiv 4 \pmod 5). \end{cases}$$

Chapter 3: Real and Complex Numbers as Sets

Section 3.2

3.2.1. **(a)** $M + 1 = F_M(1) = F_M(0^+) = F_M(0)^+ = M^+$.

Let $A = \{M \in \omega : 1 + M = M^+\}$. Clearly, $0 \in A$ because $1 + 0 = F_1(0) = 1 = 0^+$. Suppose that $K \in A$. Then $1 + K = K^+$. Hence, $1 + K^+ = F_1(K^+) = F_1(K)^+ = (1 + K)^+ = K^{++}$. By induction, $A = \omega$; thus, $M + 1 = 1 + M = M^+$ for all M in ω.

(b) $M \cdot 1 = G_M(1) = G_M(0^+) = F_M(G_M(0)) = F_M(0) = M$.

Let $B = \{M \in \omega : 1 \cdot M = M\}$. First, $0 \in B$ because $1 \cdot 0 = G_1(0) = 0$. Suppose that $K \in B$; i.e., suppose that $1 \cdot K = K$. Then

$$1 \cdot K^+ = G_1(K^+) = F_1(G_1(K)) = 1 + G_1(K) = 1 + 1 \cdot K = 1 + K = K^+.$$

By induction, $B = \omega$; thus, for all M in ω, $M \cdot 1 = 1 \cdot M = M$.

3.2.2. **(a)** Let $A = \{R \in \omega : \text{for all } M, N \text{ in } \omega, M + (N + R) = (M + N) + R\}$. Then $0 \in A$ because

$$\begin{aligned} M + (N + 0) &= F_M(F_N(0)) = F_M(N) = M + N \\ &= F_{M+N}(0) = (M + N) + 0. \end{aligned}$$

Suppose $R \in A$. Then for all M, N, R in ω, $M + (N + R) = (M + N) + R$; i.e., $F_M(F_N(R)) = F_{M+N}(R)$. Thus,

$$\begin{aligned} M + (N + R^+) &= M + F_N(R^+) = M + F_N(R)^+ = F_M(F_N(R)^+) \\ &= F_M(F_N(R))^+ = F_{M+N}(R)^+ = F_{M+N}(R^+) \\ &= (M + N) + R^+. \end{aligned}$$

By induction, $A = \omega$.

(b) Let $B = \{\langle M, N\rangle \in \omega^{②} : M + N = N + M\}$. We use (a variant of) the Principle of Double Induction (Exercise 2.2.10) to show that $B = \omega^{②}$.

Clearly, $\langle 0, 0\rangle \in B$. Suppose that $\langle M, N\rangle \in B$; we must show that $\langle M, N^+\rangle \in B$ and $\langle M^+, N\rangle \in B$. (We will show only the first of these.)

$$\begin{aligned} M + N^+ &= M + (N + 1) = (M + N) + 1 \\ &= (N + M) + 1 = N + (M + 1) \\ &= N + (1 + M) = (N + 1) + M \\ &= N^+ + M. \end{aligned}$$

Thus, $\langle M, N^+ \rangle \in B$.

3.2.3. Let $A = \{R \in \omega : \text{for all } M, N \text{ in } \omega, \text{ whenever } M + R = N + R, \text{ then } M = R\}$. Then $0 \in A$ because $M + 0 = N + 0$ implies $M = N$. Suppose that $K \in A$ and suppose that for arbitrary M, N in ω, $M + (K + 1) = N + (K + 1)$. By the associativity and commutativity of addition, $(M + 1) + K = (N + 1) + K$. By the inductive hypothesis, $M + 1 = N + 1$. But then $M^+ = M + 1 = N + 1 = N^+$; by Theorem 1.6.4 [Peano Postulate (4)]; $M = N$, Thus, $A = \omega$.

3.2.7. **(a)** True.

(b) False. Let $N = 4$, $M = 5$, $R = 3$. Then $(4 \dot{-} 5) \dot{-} 3 = 0; 4 \dot{-} (5 \dot{-} 3) = 2$.

(c) True.

(d) True.

3.2.10. For each M in ω, let G_M be the unique operation on ω satisfying $G_M(0) = 1$ and $G_M(N^+) = M \cdot G_M(N)$. Let

"$\exp(M, N)$" be defined to be "$G_M(N)$", $\langle M, N \rangle \in \omega^{②} \sim \{\langle 0, 0 \rangle\}$.

3.2.12. **(a)** $1.(H(1, N) = 1, \text{ all } N \in \omega.)$ **(b)** $2^2 (= 4)$.
(c) $2^{(2^{(2^2)})} (= 2^{16})$. **(d)** $3^{(3^3)} (= 3^{27})$.

3.2.15. $N = 0$:

(a) $F(0) = 4$. Clearly, $K^0 = 1 < 2^K$ for $K \geq 4$.

(b) By induction on K (≥ 4), we establish that

$$(K + 1) \cdot 2^0 < 2^{K \dot{-} 1}.$$

Suppose that $N \geq 0$ is such that for all $K \geq F(N)$, **(a)** and **(b)** hold. Let $L \geq F(N + 1) = 2F(N)$.

Case 1: L is even:

Then $L = 2K$, where $K \geq F(N) \geq 4$.

(a)
$$\begin{aligned} L^{N+1} &= (2K)^{N+1} = K^N \cdot 2^{N+1} \cdot K < 2^K \cdot 2^{N+1} \cdot K \\ &= 2^{K+1}(2^N \cdot K) < 2^{K+1}((K + 1)2^N) \\ &< 2^{K+1} \cdot 2^{K \dot{-} 1} = 2^{2K} = 2^L. \end{aligned}$$

(b)
$$\begin{aligned} (L + 1)2^{N+1} &= (2K + 1)2^{N+1} < (K + 1)2^{N+2} \\ &= (K + 1)2^N \cdot 2^2 < 2^{K+1} < 2^L. \end{aligned}$$

Case 2: L is odd:

Then $L = 2K + 1$, where $K \geq F(N) \geq 4$.

(a)
$$\begin{aligned} L^{N+1} &= (2K+1)^{N+1} < (2(K+1))^{N+1} \\ &= 2^{N+1}(K+1)^N(K+1) < 2^{N+1}\cdot 2^{K+1}(K+1) \\ &= ((K+1)2^N)2^{K+2} < 2^{K\dot{-}1}\cdot 2^{K+2} \\ &= 2^{2K+1} = 2^L. \end{aligned}$$

(b)
$$\begin{aligned} (L+1)2^{N+1} &= (2K+2)2^{N+1} = 2^2((K+1)2^N) \\ &< 2^2\cdot 2^{K\dot{-}1} = 2^{K+1} < 2^{L\dot{-}1}. \end{aligned}$$

Section 3.3

3.3.1. The primes are 29, 43, 53, 61.

3.3.2. **(a)** $96 = 2^5\cdot 3^1$.
(b) $124 = 2^2\cdot 31^1$.

3.3.3. **(a)** $1176 = 2^3\cdot 3^1\cdot 7^2$; **(d)** $20{,}449 = 11^2\cdot 13^2$.

3.3.5. $\{3,8\}, \{3,20\}, \{8,15\}$.

3.3.6. Suppose $N = K\cdot M + R = K'\cdot M$. Then $R = (K' \dot{-} K)\cdot M$. If $K' > K$, then $(K' \dot{-} K)\cdot M \geq 1\cdot M > R$. If $K' \leq K$, then $(K' \dot{-} K)\cdot M = 0 < R$.

3.3.9. **(a)** For $N > 1$, consider the prime factorization of N. Extend Exercise 3.3.7(a).

Section 3.4

3.4.1. **(b)**, **(c)**, **(e)**.

3.4.6. **(a)** $-\overline{3}$. **(b)** $\overline{4}$. **(c)** $\overline{2}$. **(d)** $-\overline{2}$.

3.4.11. **(b)** See Section 4.5.

Section 3.5

3.5.1. **(a)** 11. **(b)** 0. **(c)** $M = 7, N = 11$.

3.5.2. **(a)** and **(b)** $\{\langle M, M\rangle : M \in \omega\}$. **(c)** $\omega^{②}$.
(d) $\{\langle 0,0\rangle, \langle 2,2\rangle\}$.

3.5.6. **(a)** 10. **(b)** 4. **(c)** 3.
(d) 4. **(e)** 0. **(f)** 3.

Section 3.6

3.6.1. **(a)** $-\overline{6}$. **(b)** $-\overline{9}$. **(c)** $\overline{25}$.
(d) $\overline{0}$ or $\overline{5}$. **(e)** Every nonzero integer.

3.6.2. $\{\langle\overline{4},\overline{2}\rangle, \langle\overline{6},\overline{3}\rangle, \langle-\overline{2},-\overline{1}\rangle\}$; $\{\langle\overline{3},\overline{4}\rangle, \langle\overline{6},\overline{8}\rangle\}$; $\{\langle\overline{1},-\overline{1}\rangle, \langle-\overline{1},\overline{1}\rangle\}$; $\{\langle\overline{1},\overline{2}\rangle\}$; $\{\langle-\overline{1},\overline{2}\rangle\}$.

3.6.5. No. $\langle\overline{0},\overline{1}\rangle S_2\langle\overline{0},\overline{0}\rangle S_2\langle\overline{1},\overline{0}\rangle$; yet $\langle\overline{0},\overline{1}\rangle \not S_2\langle\overline{1},\overline{0}\rangle$.

Section 3.7

3.7.1. **(a)** $[5/-3]$. **(b)** $[13/3]$.
(c) $[4/-1]$. **(d)** $[-4/9]$.

3.7.6. **(b)** Suppose that $\alpha/\beta = \rho$ and $\gamma/\delta = \sigma$. Then $\alpha = \beta\cdot\rho$ and $\gamma = \delta\cdot\sigma$. Thus, $\alpha\cdot\gamma = (\beta\cdot\rho)\cdot(\delta\cdot\sigma) = (\beta\cdot\delta)\cdot(\rho\cdot\sigma)$. Hence, $(\alpha\cdot\gamma)/(\beta\cdot\delta) = \rho\cdot\sigma = (\alpha/\beta)\cdot(\gamma/\delta)$.

Section 3.8

3.8.1. Zero: **(b)**, **(f)**.
Positive: **(a)**, **(d)**, **(g)**, **(i)**.
Negative: **(c)**, **(e)**, **(h)**, **(j)**.

3.8.2. $[7/-4] < [-2/4] < [0/-3] < [-1/-2] < [7/10]$.

3.8.8. For any $\langle K, L\rangle$ in α with $L > 0$, let $N = L^+$.

3.8.11. R is the identity function on X, by the preceding exercise.

3.8.12. **(a)**, **(b)**, **(d)**.

Section 3.9

3.9.1. $\mathcal{R} = \left\{a \in \mathcal{P}(\mathcal{P}(\mathcal{Q})) : s\begin{pmatrix} x \\ a \end{pmatrix}\right\}$, where $s(x)$ is the formula "x is a partition $\{L, U\}$ of $\mathcal{Q}$ such that with respect to $\leq_{\mathcal{Q}}$, L has no last element and L is an initial segment of $\mathcal{Q}$."

3.9.3. $\{N \in \omega : N$ is a "perfect square"$\} = \{0, 1, 4, 9, 16, \ldots\}$.

3.9.6. **(a)**, **(c)**, **(e)**, **(f)**.

Section 3.10

3.10.1. **(b)** $\{L^*, U^*\}$, where $L^* = \{\alpha \in \mathcal{Q} : \alpha$ is negative or $(\alpha - 1)^2 < 3\}$, and $U^* = \mathcal{Q} \sim L^*$.

(d) $\{L^{**}, U^{**}\}$, where $L^{**} = \{\alpha \in \mathcal{Q} : \alpha$ is negative or $\alpha^2 < 75\}$, and $U^* = \mathcal{Q} \sim L^{**}$.

3.10.6. **(a)** $L_1 \subset L_2$; hence, there is some β in $L_2 \sim L_1 = L_2 \cap U_1$. Surely, β cannot be the last element of L_2; consequently, there is an α in L_2 such that $\beta <_{\mathcal{Q}} \alpha$. Clearly, $\alpha \in U_1$ and α is not the first element of U_1. Thus,

$$L_1 \subset \{\gamma \in \mathcal{Q} : \gamma < \alpha\} \subset L_2; \text{i.e., } \{L_1, U_1\} <_{\mathcal{R}} \overline{\alpha} <_{\mathcal{R}} \{L_2, U_2\}.$$

(b) Let α and β be as in part (a). Then

$$(\overline{\beta} +_{\mathcal{R}} \sqrt{\overline{\overline{\overline{2}}}} \cdot_{\mathcal{R}} \overline{\alpha}) /_{\mathcal{R}} (\overline{\overline{\overline{1}}} +_{\mathcal{R}} \sqrt{\overline{\overline{\overline{2}}}})(= \Gamma)$$

is irrational. (Show this.) Clearly, $\{L_1, U_1\} \leq_{\mathcal{R}} \overline{\beta} <_{\mathcal{R}} \Gamma <_{\mathcal{R}} \overline{\alpha} <_{\mathcal{R}} \{L_2, U_2\}$.

3.10.11. **(a)** l.u.b. $= \overline{\overline{\overline{2}}}$; g.l.b. $= \overline{\overline{\overline{0}}}$.

(b) l.u.b. does not exist; g.l.b. $= \overline{\overline{\overline{0}}}$.

(c) l.u.b. $= \overset{\equiv}{1}$; g.l.b. $= \overset{\equiv}{0}$.

(d) l.u.b. $= \overset{\equiv}{1}$; g.l.b. $= \overset{\equiv}{-1}$.

3.10.15. **(a)** See the proof of Theorem 6.2.3.

Section 3.11

3.11.1. **(a)** $-3 \oplus 6i$. **(b)** $72 \ominus 12i$.
(c) $3 \oplus (1/4)i$. **(d)** $(1/5) \odot (2 \oplus i)$.

3.11.2. **(a)** $4 \oplus 5i$. **(b)** $12 \oplus 19i$.
(c) i. **(d)** $-3/2 \oplus (7/2)i$.

3.11.3. **(a)** $\sqrt{3}i, -\sqrt{3}i$.
(b) $(\sqrt{2}/2) \odot (1 + \sqrt{3}i), (\sqrt{2}/2) \odot (1 - \sqrt{3}i)$.

3.11.4. **(c)** The square roots of $\Gamma + \Delta i$ are

$$\sqrt{\mathrm{P}} \odot (\cos(N\Phi/2) \oplus \sin(N\Phi/2)i) \quad \text{and} \quad \sqrt{\mathrm{P}} \odot (\cos((N\Phi/2) + \pi) \oplus \sin((N\Phi/2) + \pi)i)$$

$$= -\sqrt{\mathrm{P}} \odot (\cos(N\Phi/2) \oplus \sin(N\Phi/2)i).$$

The cube roots of $\Gamma \oplus \Delta i$ are

(1) $\sqrt[3]{\mathrm{P}} \odot (\cos(N\Phi/3) \oplus \sin(N\Phi/3)i)$.

(2) $\sqrt[3]{\mathrm{P}} \odot (\cos((N\Phi + 2\pi)/3) \oplus \sin((N\Phi + 2\pi)/3)i)$.

(3) $\sqrt[3]{\mathrm{P}} \odot (\cos(N\Phi + 4\pi)/3) \oplus \sin((N\Phi + 4\pi)/3)i)$.

3.11.6. The quotient is $\Gamma_3 + \Delta_3 i$, where $\langle \Gamma_3, \Delta_3 \rangle$ is the solution of the equations (3.11.2). These two linear equations (in the unknowns Γ_3 and Δ_3) have a nonzero determinant (on the left side) whose value, $\Gamma_2{}^2 + \Delta_2{}^2$, is nonzero. Cramer's Rule guarantees the existence and uniqueness of a solution. [The existence and uniqueness of the solution of (3.11.2) can also be demonstrated by elementary algebraic techniques.]

Section 3.12

3.12.1. **(a)** $\langle -4, 7 \rangle$. **(b)** $\langle 15/2, 2 \rangle$.
(c) $\langle 9 + \sqrt{2}, \sqrt{2} + \sqrt{3} \rangle$. **(d)** $\langle -16/5, -7/5 \rangle$.

3.12.2. $\langle 2, \sqrt{3} \rangle, \langle 2, -\sqrt{3} \rangle$.

3.12.4. **(b)** The vertices of a regular N-sided polygon inscribed in the "unit circle" (the set of all complex $\mathscr{A}$ whose modulus is 1).

Chapter 4: Denumerability and Choice

Section 4.1

4.1.3. Suppose that X is finite. Let Y ($\neq \varnothing$) be any set of subsets of X. Every element of Y (as a subset of X) is finite. Let $C = \{N \in \omega : \text{there is an } N\text{-element set in } Y\}$.

C, as a nonempty subset of ω, has a least element, K. Every K-element set in Y is clearly $\mathscr{I}_X$-minimal.

4.1.6. Suppose that $\cup X$ is finite. Then $\mathscr{P}(\cup X)$ is finite. Also, $A \in X$ implies $A \subseteq \cup X$. Therefore, X is finite.

4.1.8. A subset A of ω is denumerable iff A has no upper bound with respect to $\leq$.

4.1.12. Let $M \in \omega_+$ and let Q be the smallest prime that does not divide M. Then $\{Q^K : K \in \omega\} \subseteq P_M \subseteq \omega$.

Section 4.2

4.2.1. Let X be any nonempty set of nonempty subsets of ω. Let F be the collection of ordered pairs $\langle Y, y\rangle$ such that $Y \in X$ and y is the first element of Y with respect to $\leq$. Then F is included in $X \times \cup X$; surely, F can be described by a formula of set theory, $s(x)$. (Show this.) By the Axiom of Subsets, F is a set (of ordered pairs). Moreover, if $\langle Y_1, y_1\rangle$ and $\langle Y_2, y_2\rangle$ are distinct elements of F, then $Y_1 \neq Y_2$. Hence, F is a function.

4.2.2. We proceed by induction on the number of elements of X. Let

$$A = \{0\} \cup \{N : N \geq 1 \text{ and for every } N\text{-element set } X \text{ of nonempty sets, a choice function on } X \text{ exists}\}.$$

Then $0 \in A$.

Before proceeding to the inductive step, we consider the case of $N = 1$; this will be utilized in the inductive step. Let $X = \{Y\}$, where Y is nonempty. Then $X \times \cup X$ is a nonempty set, each element, $\langle Y, y\rangle$, of which is such that $\{\langle Y, y\rangle\}$ is a choice function on X.

Suppose that for $N \geq 1$, $N \in A$; let X be any N^+-element set of nonempty sets. Then with respect to some one-one mapping of N^+ onto X, $X = \{Y_1, Y_2, \ldots, Y_N, Y_{N^+}\}$. Hence, $X = \{Y_1, Y_2, \ldots, Y_N\} \cup \{Y_{N^+}\}$; let F be a choice function on $\{Y_1, Y_2, \ldots, Y_N\}$ and let G be a choice function on $\{Y_{N^+}\}$. Then $F(Y_K) \in Y_K$, $K = 1, 2, \ldots, N$, and $G(Y_{N^+}) \in Y_{N^+}$. Thus, $F \cup G$ is a choice function on X. It follows that $N^+ \in A$ and that $A = \omega$. Consequently, every nonempty finite set of nonempty sets has a choice function.

4.2.6. **(a)** Prove this by induction on N. The case of $N = 1$ is trivial. Assume that this is true for $N = L\,(\geq 1)$. For $N = L^+$, consider two cases according to whether or not there exist denumerable subsets $\{x_1, \ldots, x_K, \ldots\}$ of X and $\{X_1, X_2, \ldots, X_K, \ldots\}$ of $\mathscr{P}(X)$ that have the following properties: For all $K = 1, 2, \ldots$,

(1) $x_{K^+} \in X_K$;

(2) $X_{K^+} \subseteq X_K \sim \{x_{K^+}\}$;

(3) $\{\{x_K\} \cup A : A \in \mathscr{P}^{(L)}(X_K)\} \subseteq C_1$.

4.2.7. **(c)** Let X be any nonempty set of 4-element sets; let X^* be the set of 2-element subsets of elements of X. By [2], a choice function, F^*, exists on X^*; we define a choice function F on X by considering the possible choices from each of the six 2-element subsets of Y, for each Y in X. For example, if one element, y_0, of Y is chosen by F^* more than any other element of Y, let $F(Y) = y_0$; if each of three elements are chosen twice by F^*, let $F(Y)$ be the fourth element, etc.

Section 4.3

4.3.1. Every countable set is in one-one correspondence with a subset of ω—hence is well-orderable.

4.3.2. **(a)** Let $A_J = \{N \in \omega : N \geq J\}$, $J \in \omega$. Then $\cap X = \varnothing$.

(b) Let $K \in \omega_+$. Let $A_J = \{N \in \omega : N \geq J \text{ or } N < K\}$, $J \in \omega$. Then $\cap X = K$.

(c) Let $A_J = \{N \in \omega : \text{for some } M \text{ in } \omega, N = 2M, \text{ or } J < N\}$, $J \in \omega$. Then $\cap X = \{N \in \omega : \text{for some } M \text{ in } \omega, N = 2M\}$. Thus $\cap X$ equals the set of even natural numbers.

4.3.5. Let $Q_1 = \{0, 1\} \cup \{P^K : P \text{ prime}, K = 1, 2, \ldots\}$; for $N \geq 2$, let

$$Q_N = \{M \in \omega : M \text{ has exactly } N \text{ distinct prime factors}\}.$$

4.3.8. **(b)** Let $\mathscr{F} = \{\text{one-one correspondences of } \omega \text{ onto } A\}$ and let

$$\mathscr{G} = \{\text{one-one correspondences of } \omega \text{ onto } B\}.$$

Then $\{\mathscr{F}, \mathscr{G}\}$ is a 2-element set of nonempty sets. We do not need the Axiom of Choice to select an f from $\mathscr{F}$ and a g from $\mathscr{G}$. With respect to f we can enumerate A; with respect to g we can enumerate B. Interweave these two enumerations, as in the Cauchy diagonal method.

Section 4.4

4.4.1. **(a), (b), (d), (e)** are the functions.

(a) Domain $= \{\langle 1, 7\rangle, \langle 1, 4\rangle, \langle 2, 4\rangle\}$; range $= \{1, 2, 7\}$.

(b) Domain $= \{\langle 0, 0\rangle\}$; range $= \{0\}$.

(d) Domain $= \{\langle M, 0\rangle : M \in \omega\} = \omega \times 1$; range $= \omega$.

(e) Domain $= 1 \times \omega$, range $= \omega$.

4.4.2. $2^{(K^N)}$.

Section 4.5

4.5.5. (Sketch) Let $\mathscr{A}_N$ be the set of algebraic numbers Γ that satisfy an equation (of degree N)

$$\overline{\overline{\overline{A_N}}}\Gamma^N +_{\mathscr{R}} \overline{\overline{\overline{A_{N \dot{-} 1}}}}\Gamma^{N \dot{-} 1} +_{\mathscr{R}} \cdots +_{\mathscr{R}} \overline{\overline{\overline{A_1}}}\Gamma +_{\mathscr{R}} \overline{\overline{\overline{A_0}}} = \overline{\overline{\overline{0}}},$$

$$A_0, A_1, \ldots, A_{N \dot{-} 1}, A_N \in \omega,$$

but that fail to satisfy any such equation of degree $< N$. Then each $\mathscr{A}_N$ is denumerable, $N \in \omega_+$. Use Theorem 4.3.1(1). (Note that the Axiom of Choice is not needed to assure the denumerability of $\mathscr{A}$; see the comment on p. 150.)

Section 4.6

4.6.1. First show that the collection of all functions F from ω into $\mathscr{T}$ is a set, where $\overset{\equiv}{\omega}$ is the set of $\mathscr{R}$-natural numbers.

4.6.4. **(c)** Begin by letting $\mathscr{D}^*$ be the set of denumerable sequences d (of $\mathscr{D}$) that are such that only a finite number of terms of d are nonzero.

4.6.5. $\mathrm{Tan}(\pi \cdot (\Gamma - 1/2))$.

4.6.7. For each prime P, $\sqrt{P}$ is irrational; a modification of the proof of this shows that $\sqrt[3]{P}$ is irrational. Let $A = \{\sqrt{P} : P \text{ is prime}\}$, let $B = \{\sqrt[3]{P} : P \text{ is prime}\}$, and let $C = \mathscr{I} \sim (A \cup B)$. Then $\{A, B, C\}$ is a partition of $\mathscr{I}$. Because A and B are each denumerable, so is $A \cup B$. Let $\overline{\mathscr{Q}}$ be the set of $\mathscr{R}$-rationals; $\overline{\mathscr{Q}}$ is denumerable. Let f be a one-one correspondence between A and $\overline{\mathscr{Q}}$, and let g be a one-one correspondence between B and $A \cup B$. Then $f \cup g \cup I_c$ is a one-one correspondence between $\mathscr{I}$ and $\mathscr{R}$.

Section 4.7

4.7.1. **(d)** 3^5.

Section 4.8

4.8.5. Well-order $\cup X$.

Chapter 5: Ordinal Numbers

Section 5.1

5.1.3. In the linear ordering counterparts of Theorem 5.1.1 and Corollaries 2 and 3, let $\langle X_1, R_1\rangle = \langle X_2, R_2\rangle = \langle \mathscr{R}, \leq_{\mathscr{R}}\rangle$. Consider $F : \mathscr{R} \to \mathscr{R}$, defined by $F(\Gamma) = \Gamma -_{\mathscr{R}} \overset{\equiv}{\overline{1}}$. In Corollaries 1 and 4, $\langle \mathscr{Z}_{-}, \leq_{\mathscr{Z}}\rangle \cong \langle \mathscr{Z}_{-\overline{N}}, \leq_{\mathscr{Z}}\rangle$, where $\mathscr{Z}_{-\overline{N}}$ consists of the integers $\leq_{\mathscr{Z}} -\overline{N}$.

5.1.4. $\langle X, R\rangle = \langle \omega, \leq\rangle$; $Y = \{\text{even natural numbers}\}$; $F(N) = 2N$.

5.1.5. **(a)** Use induction on N; show that if R is any well-ordering on an N-element set X, then $\langle X, R\rangle \cong \langle N, \leq \cap N^{②}\rangle$.

(b) No. Consider Example 5.1.3.

Section 5.2

5.2.3. **(a)** $\{1\}$. **(b)** $\{1\}$. **(c)** $\{1, 2\}$.
(d) $\{1, 2, \{2\}\}$. **(e)** $\{2\}$. **(f)** $\varnothing$.
(g) $\{2, 4\}$. **(h)** $\{\{2\}\}$.

5.2.4. **(a)** (1), (2). **(b)** None of these. **(c)** (3).
(d) (1), (2). **(e)** (2).

5.2.6. If α is a finite ordinal, then α is in one-one correspondence with some natural number, N. If $\alpha \neq N$, then either $\alpha \in N$ or (excl.) $N \in \alpha$. Thus, $\alpha \subset N$ or (excl.)

$N \subset \alpha$. Neither alternative is possible because a finite set cannot be in one-one correspondence with a proper subset.

5.2.7. $\{0, 2\} \in \mathscr{P}(\omega)$; $\{0, 2\}$ is not an ordinal.

5.2.9. $\omega_+ = \{1, 2, \ldots\}$.

5.2.10. Proof of Corollary 6: Let A be the collection of all ordinal numbers. If A were a set, then, by Corollary 3, $A \subseteq \alpha$ for some ordinal α. By the definition of A and by the transitivity of α, $\alpha \in A \subseteq \alpha$; hence, $\alpha \in \alpha$. But α is ordinary.

5.2.14. $\{0, 1, \{1\}\}$.

(a) $0 \notin \{1\}$; $\{1\} \notin 0$.

(b) $0 \in 1$; $1 \in \{1\}$; thus, $\{1\}$ is not transitive.

5.2.16. **(a)** Suppose that X is an ordinal number and that A is a transitive proper subset of X. By Theorems 5.2.6 and 5.2.7, A is an ordinal number. Either $A \in X$ or (excl.) $X \in A$. Clearly, the first relationship holds.

Suppose that X is a set all of whose transitive proper subsets are elements of X. Then $\{\gamma \in X : \gamma \text{ is an ordinal}\}$ is a set of ordinals; hence, it is properly included in an ordinal, β. Let α be the first ordinal in $\beta \sim X$. Then every element in α (is an ordinal and) is in X; thus, $\alpha \subseteq X$. Furthermore, α is transitive. If $\alpha \subset X$, then $\alpha \in X$—contradicting $\alpha \in \beta \sim X$. Thus $\alpha = X$.

Section 5.3

5.3.1. **(b)** Bimonadic: (2), (3), (4), and (6); of these, all but (4) are strictly bimonadic.

5.3.3. **(a)** For sets x, y, z, say that y is an x-power of z if there is a set A satisfying the following:

(1) $z \in A$.

(2) If $a\,(\neq z) \in A$, then $a = \mathscr{P}(b)$ for some b in A.

(3) $y \in A$ and $\mathscr{P}(y) \notin A$.

(4) $x \in \omega$ and A is an x^+-element set.

Let s be the formula, "y is an x-power of ω." Show that s is bimonadic.

Consider $B = \{c_2 : \text{for some } N \text{ in } \omega, \text{“}s\begin{pmatrix} x & y \\ N & c_2 \end{pmatrix}\text{” is true}\}$.

(b) Let B be as in part **(a)**. Then, by the Axiom of Subsets, $C = \{y \in \cup B : y \text{ is } x\text{-linked to } 0\}$.

5.3.6. **(b)** $\omega2, (\omega2)^+, (\omega2)^{++}$, etc.

5.3.7. Let s_1 be a formula that has x as its one free variable and all of whose constants refer to sets whose existence has been previously guaranteed. Let s_2 be the bimonadic formula, "s_1 and $x = y$". Then "$s_2\begin{pmatrix} x & y \\ c_1 & c_2 \end{pmatrix}$" is true for at most one set c_2. Therefore, for every set A, the collection B—consisting of sets c_2 such that for some c_1 in A, "$s_2\begin{pmatrix} x & y \\ c_1 & c_2 \end{pmatrix}$" is true—is a set. But

$$B = \{c \in A : \text{“}s_1\begin{pmatrix} x \\ c \end{pmatrix}\text{” is true}\}.$$

5.3.13. **(a)** Suppose there were a set, A, consisting of all limit ordinals. Then by Corollary 1 of Theorem 5.3.2, there is a limit ordinal α such that $A \subset \alpha$; hence, $\alpha \in A \subset \alpha$—contradicting the ordinariness of α.

5.3.14. For N in ω_+ and for any ordinal α, let $\mathscr{L}^N(\alpha)$ be the Nth smallest limit ordinal containing α; let $\mathscr{L}^0(\alpha) = \alpha$. For sets x, y, and z, say that y is an x-**denumerable limit of** z if there exists a set A of ordinals satisfying the following:

(1) $z \in A$.

(2) If $\alpha\,(\neq z) \in A$, then for some ordinal β in A—α is the smallest limit ordinal containing $\mathscr{L}^N(\alpha)$ for each N in ω.

(3) $\cup A = y$.

(4) $x \in \omega$ and A is an x^+-element set.

Let s be the formula "y is an x-denumerable limit of 0." The desired set will be

$\{c_2 : \text{for some } N \text{ in } \omega, \text{"}s\begin{pmatrix} x & y \\ N & c_2 \end{pmatrix}\text{" is true}\}$.

Section 5.4

(We utilize the notation of Section 5.3, slightly modified.)

5.4.1. $f = \{\langle 0, 6\rangle\} \cup \{\langle N, \omega \cdot N\rangle : N \in \omega_+\} \cup \{\langle S_N(\omega), \omega^2 + \omega \cdot N\rangle : N \in \omega\}$.

5.4.2. $f = \{\langle N, \omega \cdot N^+\rangle : N \in \omega\} \cup \{\langle S_N(\omega), \omega^2 + \omega \cdot N\rangle : N \in \omega\}$.

5.4.3. **(a)** $f = \{\langle N, \omega^2 \cdot N\rangle : N \in \omega\}$.

(b) $g = f \cup \{\langle S_N(\omega), \omega^3 + \omega^2 \cdot N : N \in \omega\}$.

Section 5.5

5.5.1. Let α and γ be arbitrary ordinals and let β_1 and β_2 be any ordinals that contain γ. Suppose that $f_{\alpha,\beta_1}(\gamma) \neq f_{\alpha,\beta_2}(\gamma)$. Then $A = \{\delta < \gamma^+ : f_{\alpha,\beta_1}(\delta) \neq f_{\alpha,\beta_2}(\delta)\} \neq \varnothing$. Let δ_0 be the first element of A (with respect to $\in_{\gamma^+}$). Thus, if $\delta < \delta_0$, then $f_{\alpha,\beta_1}(\delta) = f_{\alpha,\beta_2}(\delta)$. Clearly, $\delta_0 \neq 0$ because $f_{\alpha,\beta_1}(0) = f_{\alpha,\beta_2}(0) = \alpha$. Furthermore, δ_0 cannot be a successor ordinal because if $\delta_0 = \zeta^+$, then $f_{\alpha,\beta_1}(\delta_0) = (f_{\alpha,\beta_1}(\zeta))^+ = (f_{\alpha,\beta_2}(\zeta))^+ = f_{\alpha,\beta_2}(\delta_0)$. Finally, δ_0 cannot be a limit ordinal because otherwise, $f_{\alpha,\beta_1}(\delta_0) = \cup\{f_{\alpha,\beta_1}(\delta) : \delta < \delta_0\} = \cup\{f_{\alpha,\beta_2}(\delta) : \delta < \delta_0\} = f_{\alpha,\beta_2}(\delta_0)$. This contradicts our original assumption.

(The above proof is, essentially, by transfinite induction because we have shown that for each x in γ^+, if $\text{seg}_{\in_{\gamma^+}}(x) \subseteq \gamma^+ \sim A$, then $x \in \gamma^+ \sim A$).

5.5.2. **(a)** $S_4(\omega)$. **(b)** $S_0(\omega)(= \omega)$. **(c)** $S_5(\omega)$.
(d) $S_3(\omega)$. **(e)** $S_2(\omega)$. **(f)** $S_0(\omega)$.

5.5.3. **(a)** ω^3. **(b)** $\omega^3 + \omega^2$. **(c)** $\omega^2 2$.
(d) $\omega^6 2$. **(e)** ω^ω.

5.5.8. **(a)** $\omega = 1 + 2 + \omega = 2 + 1 + \omega$; $\omega + 1 = 2 + \omega + 1$;
$\omega + 2 = 1 + \omega + 2$; $\omega + 3 = \omega + 1 + 2 = \omega + 2 + 1$.

5.5.9. **(a)** Let $\{\alpha_1, \alpha_2, \alpha_3\} \subset \omega$.
(b) Let $\alpha_1 = 1$, $\alpha_2 = \omega$, $\alpha_3 = \omega + 1$.

Then

$$\alpha_1 + \alpha_3 + \alpha_2 = \alpha_3 + \alpha_1 + \alpha_2 = \omega 2;$$

$$\alpha_1 + \alpha_2 + \alpha_3 = \alpha_3 + \alpha_2 + \alpha_1 = \alpha_2 + \alpha_1 + \alpha_3 = \omega 2 + 1;$$

$$\alpha_2 + \alpha_3 + \alpha_1 = \omega 2 + 2.$$

(c) See Example 5.5.4.

5.5.10. Let $A = \{\gamma < \alpha^+ : 0 + \gamma = \gamma\}$. Clearly, $0 \in A$. Suppose that $\gamma (\neq 0) \in \alpha^+$ and $\gamma \subseteq A$. If $\gamma = \delta^+$, then $\delta \in A$ and

$$0 + \gamma = f_{0,\gamma^+}(\gamma) = (f_{0,\gamma^+}(\delta))^+ = (f_{0,\delta^+}(\delta))^+ = \delta^+ = \gamma.$$

Now let γ be a limit ordinal in α^+. Then each $\delta < \gamma$ is in A, so that for each such δ, $0 + \delta = \delta$. Therefore,

$$\begin{aligned} 0 + \gamma = f_{0,\gamma^+}(\gamma) &= \cup\{f_{0,\gamma^+}(\delta) : \delta < \gamma\} \\ &= \cup\{f_{0,\delta^+}(\delta) : \delta < \gamma\} = \cup\{0 + \delta : \delta < \gamma\} \\ &= \cup\{\delta : \delta < \gamma\} = \cup\gamma = \gamma. \end{aligned}$$

Consequently, $A = \alpha^+$; because $\alpha \in \alpha^+$, it follows that $0 + \alpha = \alpha$.

5.5.14. **(a)** If $\beta < \gamma$, then $\beta \neq \gamma$. By Theorem 5.5.7, $\alpha + \beta \neq \alpha + \gamma$. Thus, $\alpha + \beta < \alpha + \gamma$ or (excl.) $\alpha + \beta > \alpha + \gamma$. Proceed as in the beginning of the proof of Theorem 5.5.7 to show that the former inequality holds.

(b) $\alpha = \omega, \beta = 0, \gamma = 1$.

Section 5.6

5.6.2. **(a)** $\omega 1 (= \omega)$. **(b)** $\omega 5$. **(c)** $\omega 2$.
(d) $\omega 1$. **(e)** $\omega 10$.

5.6.3. **(a)** $(\omega + 1) \cdot 3 = 0 + (\omega + 1) + (\omega + 1) + (\omega + 1) = \omega + (1 + \omega) + (1 + \omega) + 1 = \omega + \omega + \omega + 1 = \omega \cdot 3 + 1$.
(b) $\omega + 3$.
(c) $\omega + 1$.
(d) $(\omega + 2) \cdot \omega = \cup\{(\omega + 2)N : N < \omega\} = \cup\{\omega N + 2 : N < \omega\} = \omega^2$.
(e) $\omega^2 + \omega 2$.
(f) $\omega^2 + \omega + 2$.
(g) $(\omega + 1)(\omega + 2) = (\omega + 1)\omega + (\omega + 1)2 = \omega^2 + \omega 2 + 1$.

5.6.6. **(d)** $\alpha = 0$.

5.6.7. **(d)** Corollary 1 is true because a Cartesian product is empty iff (at least) one of the "factors" is empty.

5.6.9.

$$\omega^2 = 3 \cdot (\omega 2) \cdot \omega = (\omega 2) \cdot 3 \cdot \omega,$$

$$\omega^2 2 = 3 \cdot \omega \cdot (\omega 2) = \omega \cdot 3 \cdot (\omega 2),$$

$$\omega^2 3 = (\omega 2) \cdot \omega \cdot 3,$$

$$\omega^2 6 = \omega \cdot (\omega 2) \cdot 3.$$

5.6.10. For $1 \leq K \leq N$, let $\alpha_K = \omega + K$.
5.6.15. **(c)** Yes.

Section 5.7

5.7.2. **(a)** 7. **(b)** 3. **(c)** $\omega + 8$.
(d) $\omega 2 + 2$. **(e)** $\omega 2 + 2$. **(f)** ω^2.

5.7.4. Let α and β be ordinal numbers with $\alpha \geq \beta$. Then $\alpha \dot{-} \beta = \gamma$ iff the following is true: For all well-ordered systems $\langle A, R_A \rangle$ and $\langle B, R_B \rangle$ similar to $\langle \alpha, \in \rangle$ and $\langle \beta, \in \rangle$, respectively, every well-ordered system $\langle C, R_C \rangle$, similar to $\langle \gamma, \in \rangle$ and for which $B \cap C = \varnothing$, is such that $\langle C, R_C \rangle \cong \langle A \sim A_1, R_A \rangle$, where $\langle A_1, R_A \rangle$ is the initial segment of $\langle A, R_A \rangle$ for which $\langle A_1, R_A \rangle \cong \langle B, R_B \rangle$.

5.7.5. **(a)** $\alpha = \beta = \omega$. Then every natural number is a solution.
(b) $\alpha = \omega + 1, \beta = \omega$.
(c) $\alpha = \omega + 1, \beta = 1$.

5.7.7. **(a)** False. Let $\alpha = \omega + 1, \beta = \gamma = \omega$.
(b) True. (Use the associativity of addition.)

5.7.8. **(a)** $\gamma = \omega, \rho = 3$. **(b)** $\gamma = 2, \rho = 2$.
(c) $\gamma = \omega^\omega, \rho = 0$. **(d)** $\gamma = 1, \rho = \omega 4 + 5$.
(e) $\gamma = \omega^\omega, \rho = \omega^2 + \omega 2 + 1$. **(f)** $\gamma = 0, \rho = \omega^4$.

5.7.9. **(a)** $1, \omega, \omega^2$.
(b) Every nonzero ordinal $\alpha \leq \omega^2$ is a left divisor of ω^2.

5.7.10. $\omega, \omega + 1, \omega^2 + 1$. In fact, for $N \geq 2$, $\omega + N = N(\omega + 1)$, and for $2 \leq K < \omega$, $N < \omega$, $\omega K + N = (\omega + N)K$. Also, $\omega^2 = \omega\omega$. Thus, the third prime is at least $\omega^2 + 1$. If $\alpha < \omega$ and $\beta \leq \omega^2$, $\alpha\beta \leq \omega^2$. If $\omega \leq \alpha < \omega^2$ and $\beta \leq \omega$, then $\alpha\beta \leq \omega^2$. If $\alpha = \omega^2$ and $\beta \leq 1$, then $\alpha\beta \leq \omega^2$. If both $0 < \alpha \leq \omega^2$ and $0 < \beta \leq \omega^2$, then $\alpha\beta \geq \omega^2 + \omega$.

5.7.13. No.

Section 5.8

5.8.2. **(a)** $\omega 5$.

(b)
$$\begin{aligned}(\omega + 1)^3 &= h_{\omega+1,4}(3) = \cdots = h_{\omega+1,4}(0) \cdot (\omega + 1) \cdot (\omega + 1) \cdot (\omega + 1) \\ &= 1 \cdot ((\omega + 1) \cdot (\omega + 1)) \cdot (\omega + 1) \\ &= ((\omega + 1) \cdot \omega + (\omega + 1) \cdot 1) \cdot (\omega + 1) \\ &= (\omega^2 + \omega + 1) \cdot (\omega + 1) = (\omega^2 + \omega + 1) \cdot \omega + (\omega^2 + \omega + 1) \cdot 1 \\ &= \omega^3 + \omega^2 + \omega + 1.\end{aligned}$$

(c) ω^ω.
(d) $\omega^{\omega+1} + \omega^\omega 2$.
(e) ω^ω.
(f) $\omega^3 \cdot 243$.

5.8.4. Use Theorem 5.6.6.

5.8.8. Use Exercise 5.8.7 to show that $\{\delta : \gamma < \alpha^\delta\} \neq \varnothing$. Let δ_0 be the smallest such δ; show that δ_0 is a successor ordinal, β^+. This will establish the existence of such a β.

Section 5.9

5.9.1. **(a)** ω. **(b)** ω. **(c)** ω^2.
(d) ω. **(e)** ω^2. **(f)** ω^2.
(g) ω^ω.

5.9.2. **(a)** $\omega^2 + 3$. **(b)** $\omega^2 + \omega$. **(c)** ω^2.

5.9.3. **(a)** $\omega 2$. **(b)** ω^2. **(c)** ω^2.

5.9.4. **(a)** ω^2. **(b)** $\omega^2 + \omega 6 + 2$. **(c)** $\omega^2 2$.
(d) ω^3.

5.9.6. **(a)** $\alpha = \Gamma_0 = \omega; \alpha_N = 0$ for $1 \leq N < \omega$.

5.9.11. **(b)** $\alpha = \omega; \Gamma_\gamma = 1, \Delta_\gamma = 2$ for all $\gamma < \omega$.

Section 5.10

5.10.1. **(a)** 0. **(b)** ω. **(c)** ω.
(d) ω^ω. **(e)** ω^ω. **(f)** ω^ω.

5.10.2. **(a)** $\omega^\omega 8$. **(b)** $\omega^{\omega+1}$. **(c)** ω^ω.

5.10.3. **(a)** ω^2. **(b)** ω^ω. **(c)** ω^ω.

5.10.4. **(a)** ω^ω.
(b) $\omega^{\omega+3}2 + \omega^{\omega+2}4 + \omega^{\omega+1}2$.
(c) $\omega^{\omega 2}$.
(d) $\omega^{(\omega^2)}$.

5.10.8. Let $X = \{2\}$, let $I = \omega$, and let $f(N) = 2$ for all $N < \omega$. Then $\cup X = 2$ and $\underset{I,f}{\times} X = \{g : \omega \to 2\}$.
Let

$$g_1(N) = \begin{cases} 0, & \text{if } N \text{ is even;} \\ 1, & \text{if } N \text{ is odd.} \end{cases}$$

Let

$$g_2(N) = \begin{cases} 1, & \text{if } N \text{ is even;} \\ 0, & \text{if } N \text{ is odd.} \end{cases}$$

Which function would come first in an antilexicographical ordering?

Section 5.11

5.11.1. **(a)** $\omega 3 + 2$. **(b)** $\omega^2 + \omega 4 + 8$. **(c)** $\omega^3 + \omega^2 3 + 5$.
(d) $\omega^3 + \omega 5 + 2$. **(f)** $\omega^\omega + \omega^4 5$.

5.11.2. **(A)** Use Exercise 5.8.8, which asserts that for any such α and μ, there is an ordinal number δ such that $\mu^\delta \leq \alpha < \mu^{\delta+1}$. Then modify the proof of Theorem 5.11.1 accordingly.

(B) **(a)** $(\omega + 2)3 = \omega 3 + 2 =$
(1) $10^{\omega}3 + 2$, (2) $(\omega + 1)3 + 1$,
(3) $(\omega + 2)3$, (4) $(\omega + 2)3$, i.e., $(\omega^2)^0 \cdot (\omega + 2)3$.
(d) $(\omega + 1)\omega^2 + (\omega + 2)5 = \omega^3 + \omega 5 + 2 =$
(1) $10^{(\omega^3)} + 10^{\omega}5 + 2$, (2) $(\omega + 1)^2\omega + (\omega + 1)5 + 1$,
(3) $(\omega + 2)^2\omega + (\omega + 2)5$, (4) $\omega^2 \cdot \omega + (\omega 5 + 2)$.

5.11.3. Let

$$\alpha = \omega^{\alpha_1}L_1 + \omega^{\alpha_2}L_2 + \cdots + \omega^{\alpha_{K(\alpha)}}L_{K(\alpha)},$$

$$\beta = \omega^{\beta_1}M_1 + \omega^{\beta_2}M_2 + \cdots + \omega^{\beta_{K(\beta)}}M_{K(\beta)},$$

$$\gamma = \omega^{\gamma_1}N_1 + \omega^{\gamma_2}N_2 + \cdots + \omega^{\gamma_{K(\gamma)}}N_{K(\gamma)}$$

be in Cantor normal form, and suppose that $\alpha_1 \le \beta_1 \le \gamma_1$. Consider the possible cases for this inequality. In each case, at least one sum is eliminated. For instance, if $\alpha_1 < \beta_1 = \gamma_1$, then $\alpha + \beta = \beta$ and $\alpha + \gamma = \gamma$; hence, $\alpha + \beta + \gamma = \beta + \gamma = \beta + \alpha + \gamma$.

5.11.4. **(a)** $\omega^7 2 + \omega^4 6 + \omega^2 7 + \omega + 3$.

5.11.5. **(a)** $\omega^{11}6 + \omega^9 4 + \omega^8 11 + \omega^7 4 + \omega^6 21 + \omega^5 3 + \omega^4 19 + \omega^3 5 + \omega^2 12 + \omega + 2$.

5.11.6. **(a)** $\omega^{\omega 2 + 2}3 + \omega^{\omega 2 + 1}2 + \omega^{\omega + 4} + \omega^{\omega + 3}6 + \omega^{\omega + 2}7 + \omega^{\omega + 1}2 + \omega^{\omega}3 + \omega^5 2 + \omega^4 + \omega 6 + 3$.

5.11.9. Both distributive laws hold.

Section 5.12

5.12.4. **(a)** For $N = 1, 2, 3, \ldots$, let $c_N = \{f \in Z(A, B) : \text{range } f \text{ is an } N\text{-element set}\}$.
(b) Enumerate the elements of the sets c_N by means of specific enumerations of the elements of the sets A and B.

5.12.6. **(b)** $\alpha = \omega^{\alpha}$.

Section 5.13

5.13.3. **(b)** No. Every finite subset of X is in Y; yet $X \notin Y$.

5.13.8. (Hausdorff implies (Kuratowski): For every set X, $\langle X, \mathscr{I}_X\rangle$ is a partial ordering; we write "$\langle X, \mathscr{I}\rangle$" instead of "$\langle X, \mathscr{I}_X\rangle$".

Let $\langle A, R\rangle$ be a partially ordered system and let $\langle B, R\rangle$ be an arbitrary linearly ordered subsystem. Let $B^* = \{C : B \subseteq C \subseteq A \text{ and } R \text{ linearly orders } C\}$. By (Hausdorff), the partially ordered system $\langle B^*, \mathscr{I}\rangle$ has a maximal linearly ordered subsystem, $\langle C^*, \mathscr{I}\rangle$. It can be shown that $\cup C^*$ is a subset of A that includes B and that is linearly ordered by R. If $\langle D, R\rangle$ were any linearly ordered subsystem of $\langle A, R\rangle$ such that $\cup C^* \subset D$, then $\langle C^* \cup \{D\}, \mathscr{I}\rangle$ would be a linearly ordered subsystem, $\langle B^*, \mathscr{I}\rangle$. This would contradict the maximality of $\langle C^*, \mathscr{I}\rangle$.

Chapter 6: Order Types and Cardinal Numbers

Section 6.1

6.1.2. $\mathbf{R}_4 = \mathscr{P}(\mathbf{R}_3) = \mathscr{P}(\{0, 1, 2, \{1\}\}) = \{0; \{0\}, \{1\}, \{2\}, \{\{1\}\}; \{0, 1\}, \{0, 2\}, \{0, \{1\}\}, \{1, 2\}, \{1, \{1\}\}, \{2, \{1\}\}; \{0, 1, 2\}, \{0, 1, \{1\}\}, \{0, 2, \{1\}\}, \{1, 2, \{1\}\}; \mathbf{R}_3\} = \mathscr{P}(3) \cup \{\{\{1\}\}, \{0, \{1\}\}, \{1, \{1\}\}, \{2, \{1\}\}, \{0, 1, \{1\}\}, \{0, 2, \{1\}\}, \{1, 2, \{1\}\}, \mathbf{R}_3\}$.

6.1.4. Let $M_0 = 0$; let $M_{N^+} = 2^{M_N}$. For N in ω, $\mathbf{R}_N$ is an M_N-element set.

6.1.7. **(a)** $\{\rho(y) : y \in x\}$ is a set, by the Axiom of Substitution. Furthermore, it is a set of ordinals. Hence, $\cup\{\rho(y) : y \in x\}$ is an ordinal, α. Thus, every element of x has rank $\leq \alpha$. Therefore, $x \subseteq \mathbf{R}_{\alpha^+}$, $x \in \mathbf{R}_{\alpha^{++}}$, and x has rank $\leq \alpha^+$.

(b) Let α be as in part (a). Then $\rho(x) = \alpha^+$ if for some y in x, $\rho(y) = \alpha$; otherwise, $\rho(x) = \alpha$. [$\rho(x)$ is the smallest ordinal $> \rho(y)$ for every y in x.] Note that if $x_1 = \omega$, then $\cup\{\rho(y) : y \in x_1\} = \omega$; $\rho(x_1) = \omega$. If $x_2 = \{\omega\}$, then

$$\cup\{\rho(y) : y \in x_2\} = \omega; \rho(x_2) = \omega^+.$$

6.1.10. **(a)** $\omega + 1$. **(b)** $\omega + 3$. **(c)** 8. **(d)** $\omega + 3$. **(e)** 6. **(f)** ω. **(g)** ω. **(h)** $\omega + 1$.

6.1.11. **(a)** $x \in \mathbf{R}_{\alpha^+}$ implies that if $y \subseteq x$, then $y \subseteq x \subseteq \mathbf{R}_\alpha$. Hence, $y \in \mathbf{R}_{\alpha^+}$. Therefore, $\mathscr{P}(x) \in \mathbf{R}_{\alpha^{++}}$ and $\rho(\mathscr{P}(x)) \leq \alpha^+$. But $\rho(x) = \alpha$ and $x \in \mathscr{P}(x)$ imply $\rho(\mathscr{P}(x)) \geq \alpha^+$—hence $= \alpha^+$.

6.1.12. **(a)** If x_0 lacks rank, then by a previous exercise, some element x_1 of x_0 lacks rank, and some element x_2 of x_1 lacks rank, etc. Let $\mathscr{U}^0(x) = x$ and let $\mathscr{U}^{N^+}(x) = \cup(\mathscr{U}^N(x))$. Let $\mathscr{V} = \cup\{\mathscr{U}^N(x) : N < \omega\}$ and let

$$\mathscr{V}_0 = \{v \in \mathscr{V} : v \text{ lacks rank}\}.$$

Now apply the Axiom of Dependent Choices (Exercise 4.2.8), with reference to the relation R on $\mathscr{V}_0$ defined by $y_1 R y_2$ if $y_2 \in y_1$.

6.1.14. Let $\mathscr{P}^0(A) = A$, and for N in ω, let $\mathscr{P}^{N^+}(A) = \mathscr{P}(\mathscr{P}^N(A))$. Then

$$Z_0 = \cup\{\mathscr{P}^N(\omega) : N < \omega\}$$

is the smallest Zermelo universe. Clearly, Z_0 is a subset of every Zermelo Universe. Note that $\mathbf{R}_\omega \in \mathbf{R}_{\omega 2} \sim Z_0$.

Section 6.2

6.2.2. **(a), (i); (b), (e); (c), (d); (f), (h); (g), (j).**

6.2.3. **(c)** For $N = 0$, this is trivial. Suppose this is true for all $M \leq N$. By the definition of H, q_{N^+} is the Ith element of $\{q_0, q_1, \ldots, q_{N^+}\}$ iff $H(q_{N^+})$ is the Ith element of $\{H(q_0), H(q_1), \ldots, H(q_{N^+})\}$. Suppose that $M \leq N$. Then by the inductive hypothesis, q_M is the Jth element of $\{q_0, q_1, \ldots, q_N\}$ iff $H(q_M)$ is the Jth element of $\{H(q_0), H(q_1), \ldots, H(q_N)\}$. If $J < I$, then both q_M is the Jth element of $\{q_0, q_1, \ldots, q_{N^+}\}$ and $H(q_M)$ is the Jth element of

$$\{H(q_0), H(q_1), \ldots, H(q_{N^+})\}.$$

If $J \geq 1$, then both q_M is the J^+th element of $\{q_0, q_1, \ldots, q_{N^+}\}$ and $H(q_M)$ is the J^+th element of $\{H(q_0), H(q_1), \ldots, H(q_{N^+})\}$.

6.2.4. **(a)** a_5. **(b)** a_8. **(c)** $q_4 <_{\mathcal{Q}} q_7 <_{\mathcal{Q}} q_1$. Then $H(q_4) = a_{12}$ and $H(q_1) = a_1$; there is no $M \leq 12$ such that $a_{12} R a_M R a_1$ and $a_M \notin \{a_1, a_{12}\}$.

6.2.5. All but **(g)**.

6.2.6. Let $A = \mathcal{R}$ and let $A_1 = \{\Gamma \in \mathcal{R} : \Gamma \leq_{\mathcal{R}} \overset{\equiv}{0}\}$.

6.2.9. **(a)**, **(c)**, **(h)**, **(i)**.

6.2.11. If it were, then the collection of ordinals would be a set, by the Axiom(s) of (Subsets and) Substitution.

Section 6.3

6.3.4. **(a)**, **(d)**; **(b)**, **(h)**; **(c)**; **(e)**, **(f)**; **(g)**.

6.3.6. **(a)** It suffices to show that $\mathcal{T}_{A \times \mathcal{Q}, R_{A \times \mathcal{Q}}}$ has properties (6.2.1), (6.2.2), and (6.2.3). First, $\mathcal{Q}$ has neither a first nor a last element. Hence if $\langle a, q \rangle \in A \times \mathcal{Q}$, there is some q, for which $q_1 <_{\mathcal{Q}} q$—hence $\langle a, q_1 \rangle R_{A \times \mathcal{Q}} \langle a, q \rangle$; there is some q_2 for which $q_2 >_{\mathcal{Q}} q$—hence, $\langle a, q \rangle R_{A \times \mathcal{Q}} \langle a, q_2 \rangle$. Thus, (6.2.1) is true. (6.2.2) follows from Theorem 4.3.2. Finally, if $\langle a_3, q_3 \rangle$ and $\langle a_4, q_4 \rangle$ are distinct elements of $A \times \mathcal{Q}$, and if $\langle a_3, q_3 \rangle R_{A \times \mathcal{Q}} \langle a_4, q_4 \rangle$, pick q_5 such that $q_3 <_{\mathcal{Q}} q_5 <_{\mathcal{Q}} q_4$. Then $\langle a_3, q_5 \rangle$ is distinct from $\langle a_3, q_3 \rangle$ and from $\langle a_4, q_4 \rangle$, and $\langle a_3, q_3 \rangle R_{A \times \mathcal{Q}} \langle a_3, q_5 \rangle R_{A \times \mathcal{Q}} \langle a_4, q_4 \rangle$. Thus (6.2.3) is true.

(b) $\eta \cdot_{\mathcal{T}} \eta = \eta$. Also, $\omega \cdot_{\mathcal{T}} \eta \neq \eta$. (Every linear ordered set of type $\omega \cdot_{\mathcal{T}} \eta$ has jumps.)

6.3.9. **(a)** Take as representative systems

$$A = \langle \{\Gamma : \Gamma \text{ is irrational and } \overset{\equiv}{0} <_{\mathcal{R}} \Gamma <_{\mathcal{R}} \overset{\equiv}{1}\}, \leq_{\mathcal{R}} \rangle$$

and

$$B = \langle \{\Gamma : \Gamma \text{ is irrational and } \overset{\equiv}{1} <_{\mathcal{R}} \Gamma <_{\mathcal{R}} \overset{\equiv}{2}\}, \leq_{\mathcal{R}} \rangle.$$

Show that each is similar to $\langle \{\text{Irrationals}\}, \leq_{\mathcal{R}} \rangle$, as is their "ordered union". Map A onto the negative irrationals and B onto the positive irrationals.

6.3.10. By the Axiom of Substitution, there is a set $X' = \{t_x : x \in X\}$. By the (Disjoint Form of the) Axiom of Choice together with the Axiom of Substitution, there is a set, $\{\langle y_x, R_{y_x} \rangle : x \in X\}$, where for each x in X, $\mathcal{T}_{y_x, R_{y_x}} = t_x$. Use the Axiom of Substitution to obtain $Y = \{\langle y_x \times \{u_x\}, R_{y_x}{}^{\#} \rangle : x \in X\}$, where for each x in X, $\langle z_1, u_x \rangle R_{y_x}{}^{\#} \langle z_2, u_2 \rangle$ iff $z_1 R_{y_x} z_2, z_1, z_2 \in y_x$.

Section 6.4

6.4.7. **(a)** ω, ω^+. **(b)** Yes. $\cup A_1 = \omega^+$.

6.4.11. Consider the cardinal numbers of these sets, and apply the Schröder–Bernstein Theorem.

6.4.12. Suppose that the collection of cardinal numbers is a set, A. Let $\kappa = |\mathcal{P}(\cup \cup A)|$; then $\kappa \in A$. Surely, $\kappa \neq \varnothing$; let $B \in \kappa$. Then $B \in \cup A$; hence, $B \subseteq \cup\cup A$. Therefore,

$\kappa = |B| \preceq |\cup\cup A| \prec |\mathscr{P}(\cup\cup A)| = \kappa$. (Contradiction.) Another proof will be considered in the Exercises of Section 6.7.

Section 6.5

6.5.1. By definition, if c and c' are cardinal numbers, then there are sets A_1 and A_2 such that $c = |A_1|$ and $c' = |A_2|$. Then $A \approx A_1 \times \{1\} = A$, $A_2 \approx A_2 \times \{2\} = A'$, and $A \cap A' = \varnothing$.

6.5.5. Let $A = \{\text{irrationals}\}$ and $A' = \{\Gamma \in \mathscr{R} : \Gamma = \bar{\gamma}$ for some γ in $\mathscr{2}\}$. Then $|A| = |\mathscr{R}|$, $|A'| = |\omega|$, $A \cap A' = \varnothing$ and $A \cup A' = \mathscr{R}$.

6.5.6. $\mathscr{R} \approx A = \{\text{negative reals}\} \approx B = \{\text{nonnegative reals}\}$. Let f be a one-one correspondence between $\mathscr{R}$ and B. Clearly, $\mathscr{P}(\mathscr{R})$ is the pairwise disjoint union, $X_1 \cup X_2 \cup X_3$, of the sets $X_1 = \{\{\Gamma\} : \Gamma \in A\}$, $X_2 = \{\{\Gamma\} : \Gamma \in B\}$, and $X_3 = \{X \in \mathscr{P}(\mathscr{R}) : |Y| \neq |1|\}$. Moreover, $\mathscr{P}(\mathscr{R}) \approx X_2 \cup X_3$ because for every subset $Y \subseteq \mathscr{R}$, we can let

$$F(Y) = \begin{cases} \{f(\Gamma)\}, & \text{if } Y = \{\Gamma\}, \Gamma \in \mathscr{R}; \\ Y, & \text{otherwise.} \end{cases}$$

Let $A = X_1$ and let $A' = X_2 \cup X_3$. Then $|\mathscr{R}| + |\mathscr{P}(\mathscr{R})| = |X_1| + |X_2 \cup X_3| = |X_1 \cup X_2 \cup X_3| = |\mathscr{P}(\mathscr{R})|$.

6.5.7. Let $|A| = c$ and let $|B| = c'$; we must show that there is a subset B' of B for which $B' \approx A$.

Let b_1 be any set not in B. Then $c' + |1| = |B \cup \{b_1\}|$. Moreover, $c \prec c' + |1|$; thus, there is a proper subset $B_1 \subset B \cup \{b_1\}$ such that $A \approx B_1$. If $b_1 \notin B_1$, we take $B' = B_1$. Otherwise, $B_1 \sim \{b_1\} \subset B$; let $b_0 \in B \sim B_1$. Clearly,

$$A \approx B_1 = (B_1 \sim \{b_1\}) \cup \{b_1\} \approx (B_1 \sim \{b_1\}) \cup \{b_0\},$$

and $(B_1 \sim \{b_1\}) \cup \{b_0\} \subseteq B$.

6.5.9. Let $A = B = \mathscr{R}$. Then $A \times B \approx \mathscr{C}$, the set of complex numbers.

6.5.11. $\mathscr{R} \times \mathscr{R} \approx \mathscr{C} \approx \mathscr{R}$; hence, $\mathscr{P}(\mathscr{R} \times \mathscr{R}) \approx \mathscr{P}(\mathscr{R})$. Let

$$F : \mathscr{P}(\mathscr{R}) \times \mathscr{P}(\mathscr{R}) \to \mathscr{P}(\mathscr{R} \times \mathscr{R})$$

be defined by $F(\langle A, B\rangle) = A \times B$, $A \subseteq \mathscr{R}$, $B \subseteq \mathscr{R}$. Then F is a one-one mapping of $\mathscr{P}(\mathscr{R}) \times \mathscr{P}(\mathscr{R})$ into $\mathscr{P}(\mathscr{R} \times \mathscr{R})$. Thus,

$$|\mathscr{P}(\mathscr{R})| \cdot |\mathscr{P}(\mathscr{R})| = |\mathscr{P}(\mathscr{R}) \times \mathscr{P}(\mathscr{R})| \preceq |\mathscr{P}(\mathscr{R} \times \mathscr{R})| = |\mathscr{P}(\mathscr{R})|.$$

But

$$\mathscr{P}(\mathscr{R}) \approx \mathscr{P}(\mathscr{R}) \times \{\varnothing\} \subseteq \mathscr{P}(\mathscr{R}) \times \mathscr{P}(\mathscr{R}).$$

Thus,

$$|\mathscr{P}(\mathscr{R})| \preceq |\mathscr{P}(\mathscr{R})| \cdot |\mathscr{P}(\mathscr{R})|.$$

By the Schröder–Bernstein Theorem, we have equality.

Section 6.6

6.6.2. **(c)** For each f in $\mathscr{F}(A, \mathscr{F}(B, C))$, $H_3(f) = h$, where $h(\langle b, a\rangle) = (f(a))(b)$ for $a \in A$, $b \in B$. Suppose that $f_1 \neq f_2$ and let $h_1 = H_3(f_1)$, $h_2 = H_3(f_2)$. Then for some a_0 in A, $f_1(a_0) \neq f_2(a_0)$. Because $f_1(a_0)$ and $f_2(a_0)$ are each functions from B into C, there must be some b_0 in B at which $(f_1(a_0))(b_0) \neq (f_2(a_0))(b_0)$. It follows that $h_1(\langle b_0, a_0\rangle) = (f_1(a_0))(b_0) \neq (f_2(a_0))(b_0) = h_2(\langle b_0, a_0\rangle)$ and that consequently, $h_1 \neq h_2$. Therefore, H_3 is one-one. Now let g be an arbitrary element of $\mathscr{F}(B \times A, C)$; then for each pair $\langle b, a\rangle$ in $B \times A$ there is a unique c in C such that $g(\langle b, a\rangle) = c$. Thus, for each fixed a_0 in A, each b in B determines a unique c in C. In other words, for each fixed a_0 in A, the set of corresponding ordered pairs $\langle b, c\rangle$, for which $\langle\langle b, a_0\rangle, c\rangle \in g$, is a function, g_{a_0}, from B into C. Let $f: A \to \mathscr{F}(B, C)$ be such that for each a_0 in A, $f(a_0) = g_{a_0}$; then $H_3(f) = g$. Hence, H_3 is onto $\mathscr{F}(B \times A, C)$.

6.6.5. $|\omega| \prec |2|^{|\omega|} \preceq c^{|\omega|}$.

6.6.7. **(a)**

$$|\mathscr{R}|^{|\mathscr{R}|} = (|N|^{|\omega|})^{|\mathscr{R}|} = |N|^{|\omega|\cdot|\mathscr{R}|} = |N|^{|\mathscr{R}|} \preceq |\omega|^{|\mathscr{R}|}$$
$$\preceq |\mathscr{R}|^{|\mathscr{R}|}.$$

Equalities replaces the inequalities.

(b) $\mathscr{F}(\mathscr{R}, 2) \approx \mathscr{F}(\mathscr{R}, \mathscr{R})$.

6.6.8. $|\omega|$.

Section 6.7

6.7.3. For every countable ordinal α, $|\alpha| \prec |\mathscr{R}|$. Thus, $\omega_1 = \{\text{countable ordinals}\} \subseteq \mathbf{Y}(|\mathscr{R}|)$. Clearly, $\mathbf{Y}(|\mathscr{R}|)$ is an ordinal; thus, $\omega_1 \leq \mathbf{Y}(|\mathscr{R}|)$.

6.7.4. For each N in ω, $|N| \prec |C|$; thus, $\omega \subseteq \mathbf{Y}(|C|) \subseteq \mathbf{X}(|C|)$. If α were an infinite ordinal in $\mathbf{X}(|C|)$, then $|\omega| \preceq |\alpha| \preceq |C|$. Hence, $|\omega| \preceq |C|$, and consequently, C would have a denumerable subset. This contradiction establishes that $\omega = \mathbf{X}(c) = \mathbf{Y}(c)$. Moreover, $\mathbf{Z}(c) = \mathbf{X}(c) \sim \mathbf{Y}(c) = \varnothing$.

6.7.11. $c \not\prec \aleph_\beta$. For if $c \prec \aleph_\beta$, then c must be an aleph. Thus, $c = \aleph_\alpha$ for some ordinal α with $\alpha < \beta$.

6.7.13. Let $\alpha_0 = \omega$, $\alpha_1 = \omega_\omega$, $\alpha_2 = \omega_{(\omega_\omega)}, \ldots, \alpha_{N^+} = \omega_{\alpha_N}, \ldots$. Let $\gamma = \cup \{\alpha_N : N < \omega\}$.

6.7.15. Use the Cantor normal form to show that there is an ordinal β and a natural number N such that

$$\omega^\beta \leq \omega_\alpha < \omega^\beta \cdot N.$$

But

$$|\omega^\beta| = |\omega^\beta \cdot N|.$$

6.7.16. $\omega^{\omega_\omega} \cdot \omega_1 = \omega^{\omega_\omega} \cdot \omega^{\omega_1} = \omega^{\omega_\omega + \omega_1} \cdot 1.$

Section 6.8

6.8.2. $|2|^{\aleph_\beta}$ is not necessarily an aleph; thus, Theorem 6.8.2(1) is not applicable.

6.8.7. Let $|A| = \aleph$, $|C| = c$, and $|K| = \kappa$; let R be a fixed well-ordering on A. There is a one-one mapping F of A onto a subset B of $C \times K$. Let

$$C^* = \{c_1 \in C : \text{for some } k_1 \text{ in } K, \langle c_1, k_1 \rangle \in B\}$$

and let $K^* = \{k_1 \in K : \text{for some } c_1 \text{ in } C, \langle c_1, k_1 \rangle \in B\}$. Thus, A is in one-one correspondence with a subset of $C^* \times K^*$. Let $F_C : C^* \to A$ be such that $F_C(c_1)$ is the R-first element a in A for which $F(a) = \langle c_1, k_1 \rangle$ for some k_1 in K. Then F_C is one-one; hence, C^* is well-ordered, and $c^* = |C^*|$ is an aleph. Similarly, $\kappa^* = |K^*|$ is an aleph. Consequently, $\aleph \preceq c^* \cdot \kappa^* = \max\{c^*, \kappa^*\}$. Therefore, either $\aleph \preceq c^* \preceq c$ or $\aleph \preceq \kappa^* \preceq \kappa$.

6.8.9. Here, we can let $B = A_1 \cup A_2$, and we can define G on B as follows: Because $\aleph_0 \preceq c$, we can let $\{c_0, c_1, \ldots, c_M, \ldots\}$ be a denumerable subset of C. For f in B, range f is finite; let $G(f) = c_{J_f}$, where J_f is the smallest subscript J such that $c_J \notin$ range f. (The definition of the mapping G on A_3 remains the same.)

6.8.10. **(a)** $c \succeq |5|$. **(b)** $c \succeq |10|$.

6.8.11. **(b)** Let $k_1, k_2 \in K$, where $|K| = \kappa$. Consider the ω-sequences f in K for which $f(N) = k_1$ for $N \neq N_0$, $f(N) = k_2$ for $N = N_0$, $N_0 = 0, 1, 2, \ldots$. Clearly, $\aleph_0 \preceq \kappa^{\aleph_0}$.

(c) (1) c is an aleph.

(2) $\aleph_0 \preceq c$.

(3) There is no cardinal κ satisfying $c \prec \kappa \prec |2|^c$.

Section 6.9

6.9.2. In the proof that (8′) implies the Well-ordering Theorem, replace each occurrence of "$c \cdot \aleph_0$" by "$c^{\aleph_0}$", of "$+$" by "$\cdot$", of "(8′)" by "(8″)", and replace each multiplication by $|2|$ by squaring.

6.9.3. **(a)** Let c be an arbitrary infinite cardinal. Then $\aleph(c) \preceq c + \aleph(c)$. Suppose that $\aleph(c) \prec c + \aleph(c)$; then $\aleph(c) + \aleph(c) = \aleph(c) \prec c + \aleph(c)$. By (9′), $\aleph(c) \prec c$, in contradiction of Hartog's Theorem.

Thus, $\aleph(c) = c + \aleph(c)$; hence, $c \preceq \aleph(c)$. Every set of cardinality c—hence every set—is well-orderable.

6.9.4. Assume the Axiom of Choice, and let c and κ be cardinals with $c \prec \kappa$. It suffices to consider the case where c and κ are both alephs. Then $c + \kappa = \kappa$. If κ_1 is a cardinal other than κ, then $\kappa_1 \prec \kappa$ or $\kappa \prec \kappa_1$. If $\kappa_1 \prec \kappa$, then $c + \kappa_1 = \max\{c, \kappa_1\} \prec \kappa$. If $\kappa \prec \kappa_1$, then $c \prec \kappa_1$; hence, $c + \kappa_1 = \kappa_1 \succ \kappa$.

Assume that whenever c and κ are cardinals with $c \prec \kappa$, $\kappa - c$ exists. Then there is a unique cardinal c' such that $c' + c = \kappa$. Let c be any infinite cardinal.

If $\aleph(c) \prec c + \aleph(c)$, then $(c + \aleph(c)) - \aleph(c)$ is unique. Because $c + \aleph(c) = c + (\aleph(c) + \aleph(c)) = (c + \aleph(c)) + \aleph(c)$, it follows that $c = c + \aleph(c)$. But then $\aleph(c) \preceq c$, in contradiction of Hartog's Theorem.

Therefore, $\aleph(c) \not\prec c + \aleph(c)$, and because $\aleph(c) \preceq c + \aleph(c)$, it follows that $\aleph(c) = c + \aleph(c)$. Consequently, $c \prec \aleph(c)$, and c is an aleph.

Section 6.10

6.10.8. **(a)** $c_\gamma = 1$, $\kappa_\gamma = 2$, $\gamma < \omega$.

6.10.11. Suppose that $\aleph_\omega = |2|^{\aleph_0}$. For each N in ω, $\aleph_N \prec \aleph_\omega$. Hence, by the König–Zermelo Inequality, if $\kappa_N = \aleph_\omega$, $N < \omega$, then

$$\aleph_\omega = \sum_{N<\omega} \aleph_N \prec \prod_{N<\omega} \kappa_N = \aleph_\omega{}^{\aleph_0}$$

$$= (|2|^{\aleph_0})^{\aleph_0} = |2|^{\aleph_0 \cdot \aleph_0} = |2|^{\aleph_0} = \aleph_\omega.$$

(Contradiction.)

Section 6.11

6.11.1. $\alpha = \omega M$, where $0 < M \leq N$.

6.11.2. **(b)** An increasing ω^2-sequence for ω^ω is given by $F(\omega K + N) = (\omega^K N^+)^+$, $K, N < \omega$.

6.11.6. Use Theorem 6.11.7 and the fact that $\mathrm{cf}(\beta) \leq \beta$.

6.11.7. **(a)** ω. **(b)** ω.

6.11.8. **(a)** $\beta = \omega_1 2$. **(b)** $\beta = \omega_1 \omega_0$.

6.11.11. **(c)** $\aleph_0 \prec |2|^{\aleph_0}$. The Axiom of Choice implies that $|2|^{\aleph_0}$ is an aleph—hence that $\aleph_1 \preceq |2|^{\aleph_0}$. Therefore,

$$|2|^{\aleph_0} \preceq \aleph_1{}^{\aleph_0} \preceq (|2|^{\aleph_0})^{\aleph_0} = |2|^{\aleph_0 \cdot \aleph_0} = |2|^{\aleph_0}.$$

Thus, $\aleph_1{}^{\aleph_0} = |2|^{\aleph_0}$.

6.11.12. Generalize the proof of the special case of the Tarski Recursion Formula, Theorem 6.11.12.

Section 6.12

6.12.1. Suppose that $|2|^{\aleph_0} = \aleph_1$. Then $\aleph_1{}^{\aleph_0} = (|2|^{\aleph_0})^{\aleph_0} = |2|^{\aleph_0 \cdot \aleph_0} = |2|^{\aleph_0} = \aleph_1$. Suppose that $\aleph_1{}^{\aleph_0} = \aleph_1$. Then $|2|^{\aleph_0} \preceq \aleph_1{}^{\aleph_0} = \aleph_1$. Therefore, $|2|^{\aleph_0}$ is an aleph satisfying $\aleph_0 \prec 2^{\aleph_0} \preceq \aleph_1$. Consequently, $|2|^{\aleph_0} = \aleph_1$.

6.12.3. Use the König–Zermelo Inequality.

6.12.6. **(a)** $\aleph_1$. **(b)** $\aleph_1$. **(c)** $\aleph_{\omega+1}$. **(d)** $\aleph_3$. **(e)** $\aleph_{\omega+1}$.

6.12.7. **(a)** $\aleph_9$. **(b)** $\aleph_{\omega 2+1}$. **(c)** $\aleph_{\omega 2+1}$. **(d)** $\aleph_{\omega_1}$. **(e)** $\aleph_{\omega^2+\omega+1}$. **(f)** $\aleph_{\omega_\omega+1}$.

6.12.8. Assume that there exists a Jensen sequence, $[\Gamma_\alpha]_{\alpha<\omega_1}$. For each subset $a \subseteq \omega$, define $\sigma_a : \omega_1 \to \omega_1$ as follows:

$$\sigma_a(\gamma) = \begin{cases} \omega + \gamma + 1, & \text{if } \gamma \in a; \\ \omega, & \text{otherwise.} \end{cases}$$

Clearly, if a and b are distinct subsets of ω, $\sigma_a \upharpoonright \omega \neq \sigma_b \upharpoonright \omega$; thus if $\alpha \geq \omega$, $\sigma_a \upharpoonright \alpha \neq \sigma_b \upharpoonright \alpha$. Define $h : \mathscr{P}(\omega) \to \omega_1$ by letting $h(a)$ be the smallest ordinal α for which $\sigma_a \upharpoonright \alpha = \Gamma_\alpha$. Then $h(a) \geq \omega$ for all a in $\mathscr{P}(\omega)$; h is a one-one mapping of $\mathscr{P}(\omega)$ into ω_1.

6.12.10. **(a)** $\{|0|\}$. **(b)** $\{|0|, |1|, |2|\}$. **(c) and (d)** $\{|0|, |2|\}$.

List of Symbols‡

‡ We also include in this list letters used (though not necessarily exclusively) with specific meanings. We sometimes omit parentheses in expressing iterations of unary operations and of associative binary operations. We sometimes omit bars and the subscripts $\mathscr{Z}$, $\mathscr{Q}$, and $\mathscr{R}$.

Symbol	Meaning	Page
$F : A \to B$	F is a function from A to B.	55
$F(x)$	the value of (the function) F at x	55
I_A	the identity function on A	55
$A \approx B$; $A \not\approx B$	A is (resp., is not) in one-one correspondence with B.	56
$F \upharpoonright A$	the restriction of F to A	56
		56
		57
		60, 301
		61, 91
		61
		62
		62
$\binom{N}{I}$	the number of I-element subsets of an N-element set	65
$\in_A$; $\notin_A$	$\{\langle x, y\rangle \in A^{②} : x \in y\}$; resp., $\{\langle x, y\rangle \in A^{②} : x \notin y\}$	66, 182
$\leq$; $<$; $>$; $\geq$		70
$M < N \leq R$	$M < N$ and $N \leq R$	70
ω_+	the set of nonzero natural numbers	70
$N \geq 1$	$N \in \omega_+$	70
$\mathscr{I}_X$	the inclusion relation on X	73
$\mathrm{sect}_R(x)$	the R-section determined by x	75
$aFb = c$	$F(\langle a, b\rangle) = c$	78
$A \,\Delta\, B$	the symmetric difference of A and B	78
F_M; G_M		85
$+$; $\cdot$	the addition and multiplication operations on ω	85
MN	$M \cdot N$	86
$M \cdot P + N \cdot P$	$(M \cdot P) + (N \cdot P)$	86
$\sum_{K=0}^{N} M_K$	$M_0 + M_1 + \cdots + M_N$	86
$\prod_{K=0}^{N} M_K$	$M_0 \cdot M_1 \cdot \ldots \cdot M_N$	86
exp	exponentiation of natural numbers	86
M^N	$\exp(M, N)$	86
$\dot{-}$	the (modified) subtraction on ω	87
(M, N)	the greatest common divisor of M and N	91

$\sqrt{N}$	the square root of $\overset{\equiv}{N}$	119
$+_{\mathscr{R}}$	the addition operation on $\mathscr{R}$	121
$\{L^{-}, U^{-}\}$; $-_{\mathscr{R}}\{L, U\}$	the additive inverse of $\{L, U\}$	123, 124
$-_{\mathscr{R}}$	the subtraction operation on $\mathscr{R}$	124
$<_{\mathscr{R}}, \leq_{\mathscr{R}}, >_{\mathscr{R}}, \geq_{\mathscr{R}}$		125
$\cdot_{\mathscr{R}}$	the multiplication operation on $\mathscr{R}$	126
$\{L, U\}^{\alpha}$	exponentiation on $\mathscr{R}$	127
$\{L, U\}^{-1}$	the multiplicative inverse of (the real number) $\{L, U\}$	127
$\{L_1, U_1\}/_{\mathscr{R}}\{L_2, U_2\}$	$\{L_1, I_1\}$ divided by $\{L_2, U_2\}$	127
i	an "imaginary" root of $x^2 +_{\mathscr{R}} \overset{\equiv}{1} = \overset{\equiv}{0}$	131
$\oplus$; $\odot$; $\ominus$; $\oslash$	the addition, multiplication, subtraction, and division operations on complex numbers (intuitive approach)	131
Γ	$\Gamma \quad \oplus \quad \overset{\equiv}{0} \odot i$	134
Δi	$\overset{\equiv}{0} \quad \oplus \quad \Delta \odot i, \Delta \neq \overset{\equiv}{0}$	134
i	$\overset{\equiv}{1} \odot i$	134
$\sqrt{\Gamma}$	the square root of (the real number) Γ	134
$\cos \Phi$; $\sin \Phi$; $\tan \Phi$	cosine Φ; sine Φ; tangent Φ	135
$\mathrm{Conj}\,(\Gamma \oplus \Delta i)$	the complex conjugate of $\Gamma \oplus \Delta i$	135
$+_{\mathscr{C}}, -_{\mathscr{C}}, \cdot_{\mathscr{C}}, /_{\mathscr{C}}$	the addition, subtraction, multiplication, and division operations on $\mathscr{C}$	135
J	the isomorphic embedding of $\mathscr{R}$ into $\mathscr{C}$	137
$\bar{\Gamma}$	$J(\Gamma)$	137
(A.C.)	Assume the Axiom of Choice (among our axioms).	147
$\mathscr{P}^{(N)}(X)$	the set of N-element subsets of X	149
$[N]$	the Axiom of Choice for N-element sets	149
$\bigcup_{J=0}^{N \dot{-} 1} A_J$ or $A_0 \cup A_1 \cup \cdots \cup A_{N \dot{-} 1}$	$\cup \{A_0, A_1, \ldots, A_{N \dot{-} 1}\}$	150
$\bigcap_{J=0}^{N \dot{-} 1} A_J$ or $A_0 \cap A_1 \cap \cdots \cap A_{N \dot{-} 1}$	$\cap \{A_0, A_1, \ldots, A_{N \dot{-} 1}\}$	150

Index

M

N

O

S

T